AF581658

Effect of Diet and Physical Activity on Cancer Prevention and Control

Effect of Diet and Physical Activity on Cancer Prevention and Control

Editors

Wendy Demark-Wahnefried
Christina Dieli-Conwright

MDPI • Basel • Beijing • Wuhan • Barcelona • Belgrade • Manchester • Tokyo • Cluj • Tianjin

Editors
Wendy Demark-Wahnefried
University of Alabama
Birmingham
USA

Christina Dieli-Conwright
Harvard Medical School
USA

Editorial Office
MDPI
St. Alban-Anlage 66
4052 Basel, Switzerland

This is a reprint of articles from the Special Issue published online in the open access journal *Nutrients* (ISSN 2072-6643) (available at: https://www.mdpi.com/journal/nutrients/special_issues/Diet_Activity_Cancer).

For citation purposes, cite each article independently as indicated on the article page online and as indicated below:

LastName, A.A.; LastName, B.B.; LastName, C.C. Article Title. *Journal Name* **Year**, *Volume Number*, Page Range.

ISBN 978-3-0365-5219-4 (Hbk)
ISBN 978-3-0365-5220-0 (PDF)

Contents

Preface to "Effect of Diet and Physical Activity on Cancer Prevention and Control" ix

Jimikaye Courtney, Eric Handley, Sherry Pagoto, Michael Russell and David E. Conroy
Alcohol Use as a Function of Physical Activity and Golfing Motives in a National Sample of United States Golfers
Reprinted from: *Nutrients* **2021**, *13*, 1856, doi:10.3390/nu13061856 1

Susan M. Schembre, Michelle R. Jospe, Erin D. Giles, Dorothy D. Sears, Yue Liao, Karen M. Basen-Engquist and Cynthia A. Thomson
A Low-Glucose Eating Pattern Improves Biomarkers of Postmenopausal Breast Cancer Risk: An Exploratory Secondary Analysis of a Randomized Feasibility Trial
Reprinted from: *Nutrients* **2021**, *13*, 4508, doi:10.3390/nu13124508 15

Rebekah L. Wilson, Dong-Woo Kang, Cami N. Christopher, Tracy E. Crane and Christina M. Dieli-Conwright
Fasting and Exercise in Oncology: Potential Synergism of Combined Interventions
Reprinted from: *Nutrients* **2021**, *13*, 3421, doi:10.3390/nu13103421 27

Andrew D. Frugé, Kristen S. Smith, Aaron J. Riviere, Rachel Tenpenny-Chigas, Wendy Demark-Wahnefried, Anna E. Arthur, William M. Murrah, William J. van der Pol, Shanese L. Jasper, Casey D. Morrow, Robert D. Arnold and Kimberly Braxton-Lloyd
A Dietary Intervention High in Green Leafy Vegetables Reduces Oxidative DNA Damage in Adults at Increased Risk of Colorectal Cancer: Biological Outcomes of the Randomized Controlled Meat and Three Greens (M3G) Feasibility Trial
Reprinted from: *Nutrients* **2021**, *13*, 1220, doi:10.3390/nu13041220 51

Rebecca A. G. Christensen and Amy A. Kirkham
Time-Restricted Eating: A Novel and Simple Dietary Intervention for Primary and Secondary Prevention of Breast Cancer and Cardiovascular Disease
Reprinted from: *Nutrients* **2021**, *13*, 3476, doi:10.3390/nu13103476 65

Jordin Lane, Nashira I. Brown, Shanquela Williams, Eric P. Plaisance and Kevin R. Fontaine
Ketogenic Diet for Cancer: Critical Assessment and Research Recommendations
Reprinted from: *Nutrients* **2021**, *13*, 3562, doi:10.3390/nu13103562 81

Christian A. Maino Vieytes, Alison M. Mondul, Sylvia L. Crowder, Katie R. Zarins, Caitlyn G. Edwards, Erin C. Davis, Gregory T. Wolf, Laura S. Rozek, Anna E. Arthur and on behalf of the University of Michigan Head and Neck SPORE Program
Pretreatment Adherence to a Priori-Defined Dietary Patterns Is Associated with Decreased Nutrition Impact Symptom Burden in Head and Neck Cancer Survivors
Reprinted from: *Nutrients* **2021**, *13*, 3149, doi:10.3390/nu13093149 97

Cindy K. Blair, Prajakta Adsul, Dolores D. Guest, Andrew L. Sussman, Linda S. Cook, Elizabeth M. Harding, Joseph Rodman, Dorothy Duff, Ellen Burgess, Karen Quezada, Ursa Brown-Glaberman, Towela V. King, Erika Baca, Zoneddy Dayao, Vernon Shane Pankratz, Sally Davis and Wendy Demark-Wahnefried
Southwest Harvest for Health: An Adapted Mentored Vegetable Gardening Intervention for Cancer Survivors
Reprinted from: *Nutrients* **2021**, *13*, 2319, doi:10.3390/nu13072319 117

Cindy L. Carmack, Nathan H. Parker, Wendy Demark-Wahnefried, Laura Shely, George Baum, Ying Yuan, Sharon H. Giordano, Miguel Rodriguez-Bigas, Curtis Pettaway and Karen Basen-Engquist
Healthy Moves to Improve Lifestyle Behaviors of Cancer Survivors and Their Spouses: Feasibility and Preliminary Results of Intervention Efficacy
Reprinted from: *Nutrients* **2021**, *13*, 4460, doi:10.3390/nu13124460 . . . . **131**

Nicholas J. D'Alonzo, Lin Qiu, Dorothy D. Sears, Vernon Chinchilli, Justin C. Brown, David B. Sarwer, Kathryn H. Schmitz and Kathleen M. Sturgeon
WISER Survivor Trial: Combined Effect of Exercise and Weight Loss Interventions on Insulin and Insulin Resistance in Breast Cancer Survivors
Reprinted from: *Nutrients* **2021**, *13*, 3108, doi:10.3390/nu13093108 . . . . **147**

Maura Harrigan, Courtney McGowan, Annette Hood, Leah M. Ferrucci, ThaiHien Nguyen, Brenda Cartmel, Fang-Yong Li, Melinda L. Irwin and Tara Sanft
Dietary Supplement Use and Interactions with Tamoxifen and Aromatase Inhibitors in Breast Cancer Survivors Enrolled in Lifestyle Interventions
Reprinted from: *Nutrients* **2021**, *13*, 3730, doi:10.3390/nu13113730 . . . . **159**

Alexa Lisevick, Brenda Cartmel, Maura Harrigan, Fangyong Li, Tara Sanft, Miklos Fogarasi, Melinda L. Irwin and Leah M. Ferrucci
Effect of the Lifestyle, Exercise, and Nutrition (LEAN) Study on Long-Term Weight Loss Maintenance in Women with Breast Cancer
Reprinted from: *Nutrients* **2021**, *13*, 3265, doi:10.3390/nu13093265 . . . . **175**

Christine N. May, Annabell Suh Ho, Qiuchen Yang, Meaghan McCallum, Neil M. Iyengar, Amy Comander, Ellen Siobhan Mitchell and Andreas Michaelides
Comparing Outcomes of a Digital Commercial Weight Loss Program in Adult Cancer Survivors and Matched Controls with Overweight or Obesity: Retrospective Analysis
Reprinted from: *Nutrients* **2021**, *13*, 2908, doi:10.3390/nu13092908 . . . . **187**

Dorothy W. Pekmezi, Tracy E. Crane, Robert A. Oster, Laura Q. Rogers, Teri Hoenemeyer, David Farrell, William W. Cole, Kathleen Wolin, Hoda Badr and Wendy Demark-Wahnefried
Rationale and Methods for a Randomized Controlled Trial of a Dyadic, Web-Based, Weight Loss Intervention among Cancer Survivors and Partners: The DUET Study
Reprinted from: *Nutrients* **2021**, *13*, 3472, doi:10.3390/nu13103472 . . . . **197**

Dorothy W. Pekmezi, Tracy E. Crane, Robert A. Oster, Laura Q. Rogers, Teri Hoenemeyer, David Farrell, William W. Cole, Kathleen Wolin, Hoda Badr and Wendy Demark-Wahnefried
Correction: Pekmezi et al. Rationale and Methods for a Randomized Controlled Trial of a Dyadic, Web-Based, Weight Loss Intervention among Cancer Survivors and Partners: The DUET Study. *Nutrients* 2021, *13*, 3472
Reprinted from: *Nutrients* **2022**, *14*, 3141, doi:10.3390/nu14153141 . . . . **213**

Marina M. Reeves, Caroline O. Terranova, Elisabeth A.H. Winkler, Nicole McCarthy, Ingrid J. Hickman, Robert S. Ware, Sheleigh P. Lawler, Elizabeth G. Eakin and Wendy Demark-Wahnefried
Effect of a Remotely Delivered Weight Loss Intervention in Early-Stage Breast Cancer: Randomized Controlled Trial
Reprinted from: *Nutrients* **2021**, *13*, 4091, doi:10.3390/nu13114091 . . . . **215**

Mohd Razif Shahril, Nor Syamimi Zakarai, Geeta Appannah, Ali Nurnazahiah, Hamid Jan Jan Mohamed, Aryati Ahmad, Pei Lin Lua and Michael Fenech
'Energy-Dense, High-SFA and Low-Fiber' Dietary Pattern Lowered Adiponectin but Not Leptin Concentration of Breast Cancer Survivors
Reprinted from: *Nutrients* **2021**, *13*, 3339, doi:10.3390/nu13103339 . . . . **233**

Sylvia L. Crowder, Zonggui Li, Kalika P. Sarma and Anna E. Arthur
Chronic Nutrition Impact Symptoms Are Associated with Decreased Functional Status, Quality of Life, and Diet Quality in a Pilot Study of Long-Term Post-Radiation Head and Neck Cancer Survivors
Reprinted from: *Nutrients* **2021**, *13*, 2886, doi:10.3390/nu13082886 . . . . **245**

Ijeamaka C. Anyene, Isaac J. Ergas, Marilyn L. Kwan, Janise M. Roh, Christine B. Ambrosone, Lawrence H. Kushi and Elizabeth M. Cespedes Feliciano
Plant-Based Dietary Patterns and Breast Cancer Recurrence and Survival in the Pathways Study
Reprinted from: *Nutrients* **2021**, *13*, 3374, doi:10.3390/nu13103374 . . . . **257**

Preface to "Effect of Diet and Physical Activity on Cancer Prevention and Control"

Close to 20 million people around the world are diagnosed with cancer on an annual basis. Thus, there is no doubt that the burden of cancer is significant, and it is a leading cause of morbidity and mortality worldwide. Diet, physical activity, and weight management are increasingly recognized as key factors that can influence cancer prevention, progression, and survival. Moreover, these lifestyle factors also play an important role in mitigating symptoms and comorbidities that result from cancer and its treatment. As nutrition and exercise scientists, we are delighted to be joined by others who devote their careers to the discovery of novel dietary and physical activity-related factors, as well as interventions aimed at cancer prevention and control. It is our pleasure to serve as editors for this book, and we are grateful to all the researchers who have submitted their scientific findings to be included in this compendium. We also are grateful to the managing editors who helped to assemble this important work, Ms. Lindsey Guo and Toki He. As you will find, the order of articles follows the cancer continuum. The book begins with the role of diet and exercise in the primary prevention of cancer in both normal and high-risk individuals and then focuses on preventing neoplastic progression in those who are newly diagnosed with the disease. Later chapters center on dietary and physical activity as key factors in cancer survivorship, finally concluding with works attributing dietary and physical activity factors on cancer survival. We hope that you will enjoy and value these offerings.

Wendy Demark-Wahnefried and Christina Dieli-Conwright

Editors

MDPI

Article

Alcohol Use as a Function of Physical Activity and Golfing Motives in a National Sample of United States Golfers

Jimikaye Courtney [1,*], Eric Handley [1], Sherry Pagoto [2], Michael Russell [1] and David E. Conroy [1]

[1] College of Health and Human Development, Pennsylvania State University, University Park, PA 16802, USA; esh12@psu.edu (E.H.); mar60@psu.edu (M.R.); conroy@psu.edu (D.E.C.)

[2] Institute for Collaboration in Health, Interventions, and Policy, University of Connecticut, Storrs, CT 06269, USA; sherry.pagoto@uconn.edu

* Correspondence: jimikaye@psu.edu; Tel.: +1-814-865-7935

Abstract: Alcohol and physical inactivity are risk factors for a variety of cancer types. However, alcohol use often co-occurs with physical activity (PA), which could mitigate the cancer-prevention benefits of PA. Alcohol is integrated into the culture of one of the most popular physical activities for adults in the United States (U.S.), golf. This study examined how alcohol use was associated with total PA, golf-specific PA, and motives for golfing in a national sample of golfers in the U.S. Adult golfers (n = 338; 51% male, 81% White, 46 ± 14.4 years) self-reported alcohol use, golfing behavior and motives, and PA. Most (84%) golfers consumed alcohol, averaging 7.91 servings/week. Golf participation, including days/week, holes/week, and practice hours/week, was not associated with alcohol use. Golfers with stronger social motives were 60% more likely to consume alcohol. Weekly walking (incident risk ratio (IRR) = 7.30), moderate-to-vigorous PA (MVPA; IRR = 5.04), and total PA (IRR = 4.14) were associated with more alcohol servings/week. Golfers' alcohol use may be higher than the general adult population in the U.S. and contributes 775 extra kilocalories/week, a surplus that may offset PA-related energy expenditure and cancer-protective effects. Alcohol use interventions targeting golfers may facilitate weight loss and reduce cancer risk, especially for golfers motivated by social status.

Keywords: alcohol drinking; sports; golf; motivation; cancer prevention; social hierarchy

Citation: Courtney, J.; Handley, E.; Pagoto, S.; Russell, M.; Conroy, D.E. Alcohol Use as a Function of Physical Activity and Golfing Motives in a National Sample of United States Golfers. *Nutrients* **2021**, *13*, 1856. https://doi.org/10.3390/nu13061856

Academic Editor: David C. Nieman

Received: 4 May 2021
Accepted: 27 May 2021
Published: 29 May 2021

Publisher's Note: MDPI stays neutral with regard to jurisdictional claims in published maps and institutional affiliations.

1. Introduction

Alcohol is a Group 1 carcinogen that directly increases the risk of a multitude of cancers, including increasing the risk of breast and colon cancer by 40–60% [1,2]. Alcohol also represents a calorically dense (7 kcals/g), discretionary source of energy that may promote passive over-consumption of calories [3]. It accounts for up to 10% of total energy intake in adults, and heavier alcohol use increases weight gain and obesity risk [3–7]. Overweight and obesity account for 40% of all cancer diagnoses due to the negative effects of obesity on cancer onset, growth, survival, and metastasis [8,9]. The effects of alcohol consumption on obesity indirectly increase cancer risk [8,9]. Given that over half of U.S. adults consume alcohol [10], alcohol consumption may be a potential behavioral target for addressing energy imbalance, obesity, and cancer risk.

While alcohol use is associated with increased risk for obesity and many cancers, physical activity (PA) is associated with lower body mass index and reduced cancer risk [11,12]. Interestingly, despite these opposing effects, research indicates that PA and alcohol use often co-occur, such that physically active adults are more likely to consume alcohol [13–15]. This antagonistic clustering of PA with alcohol use is unusual given that health behaviors typically co-occur synergistically just as health risk behaviors tend to co-occur [16–19]. This antagonistic co-occurrence of PA and alcohol use means that PA may mitigate the obesity and cancer-related risks of alcohol use or that alcohol use may undermine the protective value of PA against obesity and cancer.

The co-occurrence of PA and alcohol use may be more pronounced in certain types of PA, such as golf, that have incorporated alcohol into the sport's culture and where the proverbial "19th hole" involves a trip to the bar. Golf is the second most popular sport among U.S. adults, with more than 24 million Americans participating annually (more than the population of every state except California and Texas) [20]. Golf could provide a unique context for examining the antagonistic clustering of PA and alcohol use, but little is known about alcohol use in golfers. This study aimed to characterize alcohol use in golfers and investigate associations between golfers' alcohol consumption and PA, both in general and specific to golf participation, and whether golfers' alcohol consumption was associated with the reasons people participate in golf, i.e., their golfing motives, such as golfing for fun, health, competition, or social reasons [21–24].

Cross-sectional research in youth, college students, and the general population consistently supports a positive association between moderate alcohol use and PA [13–15]. This association persists across age groups, levels of PA, and levels of drinking. Longitudinal studies examining within-person associations between alcohol use and PA have mixed findings, with findings indicating a positive [25,26], negative [27], or no association [28–30]. Most previous research has focused on adolescents or college-aged students [13,15,28,30] (for an exception see Conroy et al. [25]). Additionally, previous studies have focused on total PA volume or intensity-specific duration [13,15]. Only a handful of studies have tested for differences in alcohol use as a function of sport type, and none has examined associations between alcohol use and the magnitude of participation in a specific type of PA. Given that some types of PA are conducive to alcohol use, examining alcohol use within people who regularly engage in a specific type of PA would shed more light on the PA–alcohol use association.

Golf is potentially compatible with concomitant alcohol consumption given its relatively slow pace and light intensity, and few sports have integrated alcohol as seamlessly into their experience as golf, where golfers can consume alcohol during and after a round. Golf engages a large proportion of the American population, with 24 million Americans golfing at over 14,000 golfing facilities throughout the U.S. [20]. Many golfers cite the social aspect of golfing as a primary reason for their interest and continued engagement in golfing [21–24]. Golf typically involves small groups, golfers walk or ride in a cart together throughout the course, and many courses include clubhouses that afford opportunities for socializing before or after a round [21–24]. Some clubs sell alcohol from both the clubhouse and beverage carts that drive around the course. This combination of the social nature of golf and the ready availability of alcohol at courses makes golf an interesting type of PA to study when examining the co-occurrence of PA and alcohol use in adults.

Golfers' motives for engaging in the sport are broad and include competition with themselves and others, exercise and health benefits, social benefits, relaxation/stress relief, and enjoyment [21–24]. Likewise, common drinking motives include social, enhancement, coping, and conformity motives [14]. Understanding why individuals engage in golf may deepen researchers' understanding of associations between PA and alcohol use, particularly if both behaviors share underlying motivational origins [14,27]. The social nature of golf could also influence alcohol use [13]. We propose that motives for participating in physical activities, such as golf, may be linked with alcohol use. Social motives for golf in particular may be associated with greater alcohol use during or after a round of golf. Social behavior can be oriented either toward getting along (i.e., affiliative motives) or getting ahead (i.e., status or dominance motives) [31]. We contend that the strength of these social motives may be associated with alcohol use.

The purpose of this study was to characterize golfers' alcohol use and identify behavioral and motivational characteristics of golfers associated with higher alcohol use. The three aims involved testing how golfers' alcohol use is associated with (1) participation in golf-specific activity (e.g., practice frequency), (2) golfing motives, and (3) overall PA volume and intensity-specific durations. We hypothesized that golfers would consume more alcohol if they (1) engaged in more golfing-specific activities, (2) reported higher levels of social motives for golfing, and (3) engaged in more PA overall.

2. Materials and Methods

2.1. Participants and Procedures

Participants were adult golfers recruited from a Qualtrics survey panel. We purposively sampled across the adult lifespan (using a rectangular age distribution of individuals 18 to 70 years old), by gender (50% male), and by race/ethnicity (75% non-Hispanic White, 25% Hispanic or minority). Individuals who golfed at least 18 holes per month in June–August 2019 were eligible to participate. Participants completed the Qualtrics survey online from 27 May–3 June 2020. This study was approved by the Institutional Review Board of Pennsylvania State University (Protocol #: STUDY00014980). Participants provided informed consent to participate in the study.

2.2. Measures

2.2.1. Demographics

Demographic characteristics were assessed using self-reports of age in years, sex (male or female), ethnicity (Non-Hispanic or Hispanic), race (American Indian/Alaska Native, Asian, Native Hawaiian/Pacific Islander, Black, White, or two or more races), marital status (married/cohabitate, widowed, divorced, separated, never married), employment status (never employed, not employed but looking, employed part-time, employed full-time, or retired), and education (no schooling, preschool to Grade 12, high school graduate or equivalent, some college/no degree, Associate's degree, Bachelor's degree, Master's degree, professional degree (Medical Doctor [MD], Doctor of Dental Surgery [DDS], etc.), or doctorate degree). Race was collapsed into a dichotomous variable of White or "Other" race. Marital status was collapsed into a dichotomous variable of married/cohabitate or not married. Employment status was collapsed into three categories: employed full-time, employed part-time, or unemployed/retired. Education status was collapsed into three categories: some college/Associate's degree, Bachelor's degree, or post-graduate degree.

2.2.2. Alcohol

Participants reported the total number of servings consumed over the past week by type of alcohol (beer, wine, liquor, non-caffeinated mixed drink, caffeinated mixed drink with regular soda, caffeinated mixed drink with diet soda, and caffeinated mixed drink with an energy drink). Response options ranged from 0 to 15+ servings in increments of 1. Total alcohol servings/week was calculated as the sum of all drink types. Participants were categorized as exceeding the threshold for moderate alcohol consumption if they consumed more than 7 (women) or 14 (men) alcohol servings/week [32].

2.2.3. Golf Participation

Participation in golfing-related activities was assessed separately for spring (March/April/May), summer (June/July/August), and fall (October/November/December). For each season, participants indicated the average number of days/week they played golf (0 to 7 days in increments of one), the average number of holes/week played (0 to 90+ holes in increments of 9), and the average number of hours/week spent practicing golfing at a driving range or putting green (0 to 10+ hours in increments of one). Principal components analysis with oblique rotation revealed three underlying factors for participation in golfing-related activities: (1) days/week, (2) holes/week, and (3) practice hours/week. Based on those results, scores were averaged across seasons to estimate the daily frequency of golfing (α = .83), typical holes/week (α = .89), and the duration of practice hours/week (α = .88).

2.2.4. Golf Motives

Participants rated the importance of 16 different golf motives on a scale ranging from 0 (not at all important) to 5 (extremely important). Sample motives included "playing better than others", "socializing with my playing partners", "preventing injury", "enjoying a leisurely time on the course", and "challenging myself". Principal components analysis with oblique rotation reduced responses to four correlated factors: social status (5 items, α = .76), health (4 items, α = .69), enjoyment (4 items, α = .71), and skill (3 items, α = .66).

2.2.5. Physical Activity

Past-week PA was measured using the International Physical Activity Questionnaire Short Form (IPAQ-SF) [33]. The IPAQ-SF is a widely used self-reported measure of PA demonstrating acceptable reliability and validity across different populations [34,35]. Participants reported the frequency and average daily duration of past-week vigorous-intensity PA, moderate-intensity PA, and walking. Weekly durations at each intensity level (=frequency × daily duration) were screened for outliers greater than three standard deviations below or above the sample mean. Outliers were all above the sample mean and were winsorized to three standard deviations above the mean for each weekly duration variable (vigorous: n = 6, moderate: n = 7, walking: n = 10). Duration of weekly moderate-to-vigorous PA (MVPA) was calculated as the sum of the winsorized values for weekly vigorous and moderate PA. Participants were classified as meeting PA guidelines if they participated in at least 150 min of moderate PA per week, 75 min of vigorous PA per week, or any combination thereof [36]. In accordance with the IPAQ scoring protocol, total weekly PA volume was calculated as the sum of vigorous, moderate, and walking durations weighted for energy expenditure at each intensity [37]. Due to their non-normal, positively skewed distribution, Box-Cox transformations [38] were applied to normalize the distributions of weekly walking duration (λ = 0.26), weekly MVPA duration (λ = 0.34), and total weekly PA volume (λ = 0.30).

2.3. *Data Analyses*

2.3.1. Quality Assurance

Prior to hypothesis testing, data were screened to identify participants whose data should be excluded from the final analytic data set. In total, 366 individuals completed the Qualtrics survey, and 16 participants were removed during primary screening for providing responses that caused concern regarding the validity of their responses. Participants were removed during primary screening due to providing nonsense responses for write-in variables (e.g., "Kzkzmx". "Yshysusy"; n = 12), selecting the maximum value for weekly servings for all types of alcohol beverages (resulting in an implausible total of 105 drinks per week; n = 1), implausible responses to health history questions (n = 2), or all of the above (n = 1). Primary screening resulted in a sample of 350 participants. Next, as a part of secondary screening, participants who reported playing no golf across all three seasons (i.e., 0 days, 0 holes, 0 practice hours; n = 3) were listwise deleted for ineligibility. Secondary screening also examined the primary outcome variable, total alcohol servings/week, for outliers greater than three standard deviations above the mean. Outliers for alcohol servings/week were treated as missing. Due to the fact that alcohol servings/week was the primary outcome variable, any participant with missing data for alcohol servings/week (n = 9) was listwise deleted from the data set.

2.3.2. Hypothesis Testing

Regression models were used to examine the associations between demographic characteristics, golf participation, golf participation plus golf motives, weekly walking duration (hours/week), weekly MVPA duration (hours/week), total PA volume (MET hours/week), and alcohol servings/week. In the first step of the analysis, we estimated a series of intercept-only models of alcohol servings/week to determine the best fitting model. The models tested included Gaussian, Poisson, negative binomial, zero-inflated Poisson (ZIP), and zero-inflated negative binomial (ZINB) [39,40]. Akaike information criterion (AIC) and Bayesian information criterion (BIC) values, the log-likelihood ratio test (LRT, for nested models), and the Vuong test (for non-nested models) were used to determine the model of best fit [40–42].

In the second step of the analysis we extended the model by adding age (sample mean centered), sex, race, and ethnicity to the logit and count models as demographic predictors of alcohol use, due to variability in alcohol use by age, sex, race, and ethnicity. In the third step of the analysis, the demographics model was expanded by adding five sets of predictors of interest: golf participation (days/week, holes/week, practice hours/week), golf participation plus motives, weekly walking duration, weekly MVPA duration, and total weekly PA volume. Golfing variables and PA were added to the logit and count models. All models were fit in R version 4.0.0 [43].

3. Results

The final analytic sample comprised 338 participants (51.2% male, 90% non-Hispanic, 81% White) with a mean age of 46 years (standard deviation [SD] = 14.4). Table 1 summarizes the demographic characteristics of the sample. As shown in Figure 1, the sample was recruited from across the United States.

Table 1. Participant demographics.

Demographics	Participants (n = 338)
Age in years (Mean ± SD)	46 ± 14.4
Sex (n (%))	
Male	173 (51.2)
Female	165 (48.8)
Ethnicity (n (%))	
Non-Hispanic	304 (89.9)
Hispanic	34 (10.1)
Race (n (%))	
White	275 (81.4)
Other Race [1]	63 (18.6)
Marital Status (n (%))	
Married/Cohabitated	224 (66.3)
Not Married [2]	114 (33.7)
Employment Status (n (%))	
Employed Full-Time	205 (60.6)
Employed Part-Time	35 (10.4)
Unemployed/Retired [3]	98 (29.0)
Education (n (%))	
Some College or less [4]	131 (38.8)
Bachelor's Degree	129 (38.2)
Post-Graduate Degree [5]	78 (23.1)

Notes: SD = standard deviation. [1] Other race includes American Indian/Alaska Native (n = 4, 1.2%), Asian (n = 22, 6.5%), Native Hawaiian/Pacific Islander (n = 1, 0.3%), Black (n = 30, 8.9%), and two or more races (n = 6, 1.8%). [2] Not married includes widowed (n = 6, 1.8%), divorced (n = 31, 9.2%), separated (n = 4, 1.2%), and never married (n = 73, 21.6%). [3] Unemployed/Retired includes "never employed" (n = 3, 0.9%), "not employed but looking" (n = 25, 7.4%), and "retired" (n = 72, 21.3%). [4] Some College or less includes "Less than a high school education" (n = 1, 0.3%), "High school education or GED" (n = 30, 8.9%), "some college/no degree" (n = 64, 18.9%), and "Associate's degree" (n = 36, 10.7%). [5] Post-Graduate Degree includes "Master's degree" (n = 62, 18.3%), "Professional degree" (e.g., MD, DDS, etc.) (n = 9, 2.7%), and "Doctorate degree" (n = 7, 2.1%).

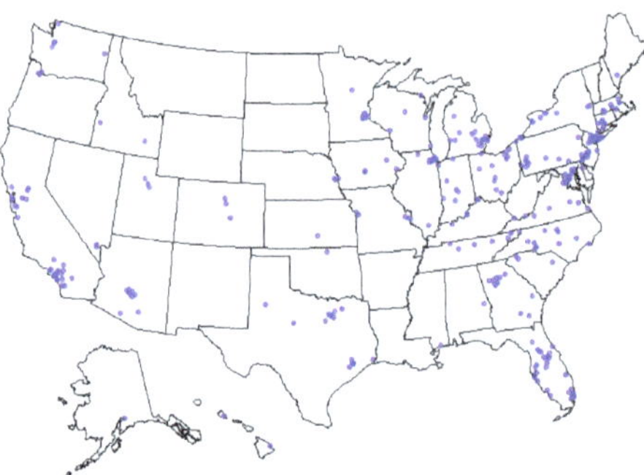

Figure 1. Geographic distribution of sample throughout the U.S.

Table 2 summarizes participants' alcohol servings/week and weekly PA. Most participants (86%) met PA guidelines, and a quarter of all participants exceeded the threshold for moderate alcohol consumption [32].

Table 2. Physical activity and alcohol use behaviors [1].

	Range (Min–Max)	Mean ± SD	Median (1QR, 3QR)
Physical Activity Intensity—Hours per Week			
Walking	0–35.5	7.1 ± 8.0	4.6 (2.0, 8.8)
Moderate Physical Activity	0–37.5	6.4 ± 8.0	4.0 (1.5, 8.0)
Vigorous Physical Activity	0–36.2	6.2 ± 7.1	4.0 (1.5, 8.0)
Moderate to Vigorous Physical Activity [2]	0–64.1	12.8 ± 13.6	8.7 (4.0, 15.6)
Total Physical Activity [3]	0–104.3	19.7 ± 18.1	14.6 (7.3, 24.0)
Physical Activity Volume per Week [4]			
Walking	0–117.2	23.5 ± 26.4	15.1 (6.6, 28.9)
Moderate Physical Activity	0–150.2	25.6 ± 32.2	16.0 (6.0, 32.0)
Vigorous Physical Activity	0–290.0	49.4 ± 56.8	32.0 (12.0, 64.0)
Moderate to Vigorous Physical Activity [5]	0–440.1	75.0 ± 77.2	52.5 (24.0, 96.5)
Total Physical Activity Volume [6]	0–537.1	98.5 ± 92.4	73.2 (35.7, 125.1)
Alcohol Servings per Week			
Beer	0–15	2.2 ± 3.6	1.0 (0.0, 3.0)
Wine	0–15	1.5 ± 2.3	1.0 (0.0, 2.0)
Liquor	0–14	0.8 ± 1.6	0.0 (0.0, 1.0)
Mixed Drink—All [7]	0–26	2.3 ± 4.2	0.0 (0.0, 3.0)
Total Servings per Week [8]	0–31	6.8 ± 7.3	4.0 (1.0, 10.0)

Notes: Min = minimum; Max = maximum; SD = standard deviation; 1QR = first quartile; 3QR = third quartile. [1] n = 338. [2] Sum of winsorized values for vigorous and moderate hours per week. [3] Sum of winsorized values for vigorous, moderate, and walking PA hours per week. [4] Physical activity volume is based on weighted energy expenditure at each intensity level in MET hours per week [37]. [5] Sum of winsorized values for vigorous and moderate MET hours per week. [6] Sum of winsorized values for vigorous, moderate, and walking PA MET hours per week. [7] Sum total of all four types of mixed drinks: non-caffeinated, caffeinated with regular soda, caffeinated with diet soda, and caffeinated with an energy drink. [8] Sum total of all types of alcoholic beverages.

Table 3 summarizes golf participation and motives. Participants' golf participation was similar across the three seasons (days/week: r = 0.59–0.64, $p < .01$; holes/week: r = 0.67–0.79, $p < .01$; practice hours/week: r = 0.70–0.73, $p < .01$).

Fit statistics for the intercept-only model indicated that the ZINB model was the model of best fit. AIC and BIC values were lowest for the ZINB model (AIC = 2004.007, BIC = 2015.476) compared to the Gaussian (AIC = 2307.579, BIC = 2315.225), Poisson (AIC = 3540.811, BIC = 3544.634), negative binomial (AIC = 2014.753, BIC = 2022.399), and ZIP (AIC = 2803.65, BIC = 2811.296). The LRT test confirmed the log-likelihood values were significantly better for the ZINB model than the ZIP model (X^2 = 801.64, $p < .001$). The Vuong test confirmed the ZINB model fit better than the negative binomial model

($z = -1.94$, $p = .03$). Figure 2 shows that the ZINB distribution accounted best for the zero-inflation and overdispersion of alcohol servings/week. Therefore, all subsequent analyses treated the outcome variable following a ZINB modeling approach. Post hoc comparisons of model fit confirmed that, when including predictor variables, the ZINB provided the best fit for all models.

Table 3. Golf participation and motives [1].

	Overall	Spring	Summer	Fall
Days per Week (Mean ± SD)	1.8 ± 1.2	1.9 ± 1.4	2.0 ± 1.3	1.6 ± 1.4
Holes per Week (*n* (%))				
9 holes/week	-	58 (17.2)	53 (15.7)	57 (16.9)
18 holes/week	-	161 (47.6)	157 (46.4)	133 (39.3)
27 holes/week	-	25 (7.4)	33 (9.8)	26 (7.7)
36 holes/week	-	49 (14.5)	51 (15.1)	40 (11.8)
>36 holes/week	-	23 (6.8)	31 (9.2)	19 (5.6)
Not applicable	-	22 (6.5)	13 (3.8)	63 (18.6)
Holes per Week (Mean ± SD) [2]	21.1 ± 13.6	21.5 ± 14.6	23.3 ± 15.1	18.4 ± 15.4
Practice Hours/Week (Mean ± SD)	1.8 ± 1.7	1.9 ± 1.9	1.8 ± 1.8	1.6 ± 1.8
Years Golfed (*n* (%))				
Less than one year	12 (3.6)			
1–5 Years	74 (21.9)			
6–10 Years	74 (21.9)			
11–15 Years	31 (9.2)			
More than 15 Years	147 (43.5)			
Golf Motives [3]				
Social Status (Mean ± SD)	2.4 ± 1.0			
Health (Mean ± SD)	3.6 ± 0.9			
Enjoyment (Mean ± SD)	3.9 ± 0.8			
Skills (Mean ± SD)	3.7 ± 0.9			

Notes: SD = standard deviation. [1] $n = 338$. [2] Calculated by recoding categorical values for holes per week to numeric values (i.e., 0 = 9 holes/week → 9 holes/week), and then averaging the numeric values. [3] Principal components analysis of 16 questions assessing golf motives revealed four underlying factors. Participants' average scores on each factor were calculated by taking their average for all of the items for a given factor.

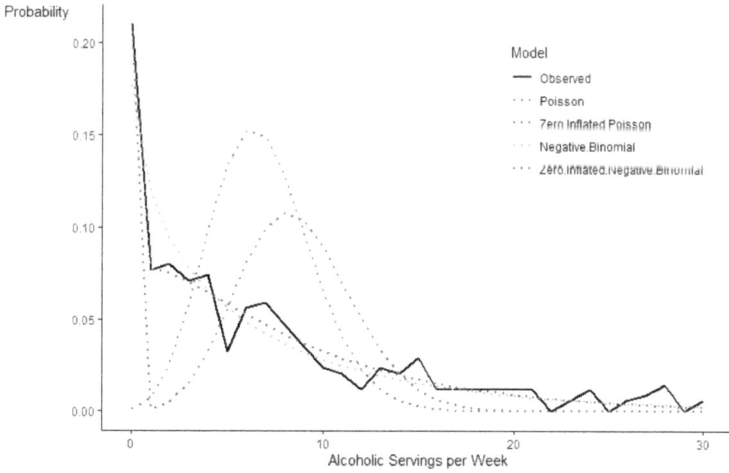

Figure 2. Models for alcohol servings per week.

In the intercept-only model, participants were estimated to consume 7.91 alcohol servings/week (95% CI: 6.95, 8.91). The baseline odds of participants abstaining from alcohol during the prior week was 0.16 (95% CI: 0.08, 0.24). Table 4 presents the estimated model parameters for the demographics, golf participation, and golf participation plus golf motives. The reference group for each model was participants who were the mean age (46 years), female, non-Hispanic, and White. None of the demographic variables was associated with alcohol servings/week or the odds of abstaining from alcohol during the prior week.

3.1. Golf Participation and Motivation

As shown in Table 4, when the three golf participation variables were simultaneously entered into the ZINB model, those variables were not associated with alcohol servings/week or the odds of abstaining from alcohol during the prior week. When golf motives were added to that model, each one unit increase in social status motives increased the odds of a participant consuming alcohol by 60% (95% CI: 0.00, 1.90). None of the other motives for golfing was associated with alcohol servings/week or the odds of abstaining from alcohol during the prior week.

Table 4. Zero-inflated negative binomial models with demographics, golf participation, and golf participation plus motives predicting alcohol servings per week [1].

	Demographics Model	Golf Participation [2]	Golf Participation + Motives [2]
Logit Model	λ (SE)	λ (SE)	λ (SE)
Intercept	−1.70 (0.41) **	−0.97 (0.58)	−2.77 (1.99)
Age [3]	0.03 (0.02)	0.03 (0.02)	0.02 (0.02)
Male	−0.03 (0.44)	−0.09 (0.47)	0.51 (0.61)
Hispanic	−0.44 (1.09)	−0.03 (0.98)	0.22 (0.96)
Other Race	−0.93 (1.01)	−0.89 (1.05)	−14.4 (1027.22)
Golf Days per Week	-	−1.67 (1.70)	−0.79 (0.61)
Golf Holes per Week	-	0.07 (0.08)	0.04 (0.03)
Golf Practice Hours per Week	-	0.09 (0.20)	0.17 (0.22)
Social Status	-	-	−0.52 (0.26) *
Health	-	-	0.81 (0.55)
Enjoyment	-	-	0.17 (0.38)
Skills	-	-	−0.50 (0.42)
Count Model	β (SE)	β (SE)	β (SE)
Intercept	1.88 (0.10) **	1.52 (0.15) **	1.55 (0.31) **
Age [3]	−0.01 (0.01)	−0.01 (0.01)	−0.001 (0.01)
Male	0.20 (0.12)	0.24 (0.12) *	0.20 (0.13)
Hispanic	0.31 (0.19)	0.27 (0.19)	0.28 (0.19)
Other Race	0.16 (0.15)	0.15 (0.15)	0.11 (0.15)
Golf Days per Week	-	0.01 (0.07)	0.02 (0.07)
Golf Holes per Week	-	0.01 (0.01)	0.01 (0.01)
Golf Practice Hours per Week	-	0.07 (0.05)	0.07 (0.05)
Social Status	-	-	0.04 (0.07)
Health	-	-	−0.05 (0.10)
Enjoyment	-	-	−0.07 (0.09)
Skills	-	-	0.09 (0.09)
Log (theta)	0.30 (0.15) *	0.32 (0.18)	0.30 (0.14) *

Notes: SE = standard error; ** $p < .01$; * $p < .05$. [1] $n = 338$. [2] Age, sex, ethnicity, and race were included in all models. The reference group is mean age, female, non-Hispanic, and White. [3] Age was mean-centered so that a one unit increase in age corresponds with a one-year increase in age above the sample mean (46 years of age).

3.2. Physical Activity Duration and Volume

Table 5 presents the estimated model parameters for the weekly walking duration (hours/week), weekly MVPA duration (hours/week), and total weekly PA volume (MET hours/week). Each one unit increase in walking duration (Box-Cox transformed) was associated with 7.30 times more alcohol servings/week (95% CI: 1.79, 32.24). Each one-unit increase in MVPA duration (Box-Cox transformed) was associated with 5.04 times more alcohol servings/week (95% CI: 2.38, 12.32). Each one-unit increase in total PA volume (Box-Cox transformed) was associated with 4.14 times more alcohol servings/week (95% CI: 2.17, 7.97). None of the PA variables was associated with odds of abstaining from alcohol during the prior week.

Table 5. Zero-inflated negative binomial models with physical activity intensity and total physical activity volume predicting alcohol servings per week [1].

	Walking Hours per Week [2]	MVPA Hours per Week [2]	Total PA Volume per Week [2]
Logit Model	λ (SE)	λ (SE)	λ (SE)
Intercept	−2.43 (0.74) **	−1.14 (0.57) *	−1.45 (1.03)
Age [3]	0.03 (0.02)	0.02 (0.02)	0.03 (0.02)
Male	−0.01 (0.45)	−0.03 (0.42)	−0.04 (0.42)
Hispanic	−0.06 (1.02)	−0.53 (1.05)	−0.41 (1.18)
Other Race	−9.50 (85.14)	−0.97 (1.00)	−1.64 (2.80)
Walking Hours per Week (Lambda) [4]	2.99 (2.52)	-	-
MVPA Hours per Week (Lambda) [5]	-	−1.32 (1.31)	-
Total PA Volume per Week (Lambda) [6]	-	-	−0.29 (1.37)
Count Model	β (SE)	β (SE)	β (SE)
Intercept	1.43 (0.17) **	1.26 (0.17) **	0.95 (0.24) **
Age [3]	−0.01 (0.01)	−0.01 (0.01)	−0.01 (0.01)
Male	0.23 (0.12) *	0.20 (0.11)	0.20 (0.11)
Hispanic	0.31 (0.19)	0.26 (0.18)	0.26 (0.19)
Other Race	0.07 (0.15)	0.10 (0.15)	0.06 (0.16)
Walking Hours per Week (Lambda) [4]	1.99 (0.63) **	-	-
MVPA Hours per Week (Lambda) [5]	-	1.62 (0.36) **	-
Total PA Volume per Week (Lambda) [6]	-	-	1.42 (0.33) **
Log (theta)	0.29 (0.13) *	0.44 (0.14) **	0.42 (0.17) *

Notes: SE = standard error; ** $p < .01$; * $p < .05$. [1] $n = 338$. [2] Age, sex, ethnicity, and race were included in all models. The reference group is mean age, female, non-Hispanic, and White. [3] Age was mean-centered so that a one unit increase in age corresponds with a one-year increase in age above the sample mean (46 years of age). [4] Walking hours per week was entered into model using the Box-Cox transformed value. [5] MVPA (moderate-to-vigorous) hours per week was entered into model using the Box-Cox transformed value. [6] Total PA (physical activity) volume per week is based on weighted energy expenditure at each intensity level in MET hours per week [37] and was entered into model using the Box-Cox transformed value.

3.3. Additional Analyses

Exploratory analyses examined potential moderation of PA associations with alcohol servings per week by demographic variables. In the golf participation model, the negative association between golf days/week and the odds of abstaining from alcohol during the prior week was significantly greater with increasing age ($\lambda = 0.10$, standard error [SE] = 0.05, $p = 0.03$) (Table S1). In the walking duration model, the effect of walking on increasing servings was significantly smaller in males than females ($\beta = -2.61$, SE = 1.25, $p = 0.04$) (Table S2). Likewise, total PA volume was positively associated with servings/week, but this association was significantly smaller in males than females ($\beta = -1.22$, SE = 0.57, $p = 0.03$) (Table S3). Additionally, the positive association between total PA volume and alcohol servings/week was significantly smaller with increasing age ($\beta = -0.04$, SE = 0.02, $p < 0.05$) (Table S3). Post hoc comparisons of model fit confirmed that, when including predictor variables, the ZINB provided the model of best fit for all models.

4. Discussion

The objective of this study was to characterize golfers' alcohol use and identify behavioral and motivational characteristics of golfers associated with increased alcohol use. Results provided three new insights into the antagonistic co-occurrence of these behaviors.

First, our findings indicate that alcohol use is common among golfers, with over 80% reporting at least one standard serving of alcohol in the past week. By way of comparison, data from the 2007–2016 National Health and Nutrition Examine Survey (NHANES) and the 2018 National Survey on Drug Use and Health indicated that 66% or 55% of adults in the U.S. used alcohol in the past week or year, respectively [10,44]. Although we did not compare our sample with a matched sample of non-golfers, these results suggest that golfers may be more likely to consume alcohol than the general population. Additionally, our sample of golfers were estimated to consume 7.91 alcohol servings/week; whereas, the 2007–2016 NHANES data indicated that adult drinkers consume a median of 1.8 alcohol servings/week. Similarly, Conroy et al. found that adults across the lifespan reported less than 5 alcohol servings per week [25]. As such, it appears that golfers may be more likely to both consume alcohol and to consume a larger volume of alcohol than the general population, making them a potential target for harm reduction interventions [45]. The level of alcohol consumption among golfers, combined with the prolonged sun exposure associated with this activity [46,47], makes golfers an important target for cancer prevention interventions.

This larger volume of alcohol consumption among golfers is also problematic because alcohol is a calorically dense, discretionary source of energy [3]. Assuming that golfers consumed a standard drink with 14 g of alcohol per serving, our sample consumed an estimated 775 kcal/week of discretionary energy from alcohol alone (7 kcal/g alcohol × 14 g alcohol/serving × 7.91 servings = 755 kcal). Previous research suggests that accumulating small energy imbalances over time may increase the risk for obesity, specifically identifying 100 kcal/day as the energy gap in the U.S. that should be targeted for primary obesity prevention [48]. That the golfers in our sample exceeded this 100 kcal/day from alcohol alone is concerning given the potential for alcohol to increase obesity risk in adults after accounting for other risks factors, including PA participation [3–7]. Furthermore, 31% of female golfers and 20% of male golfers were classified as consuming heavy amounts of alcohol (data not shown), which increases weight gain and obesity risk [3–7]. The association of PA with increasing alcohol consumption was also more pronounced in females than males. These findings suggest that PA-related energy expenditure may not offset alcohol-related energy intake even among individuals who regularly engage in a sport, such as golf, that appears to facilitate alcohol consumption. As such, alcohol use in golfers, and particularly female golfers, may undermine the protective value of PA against obesity and cancer risk [8,9].

A second contribution of this study was the finding that social status motives for golf are associated with alcohol use. The social context has been proposed as a potential explanation for associations between alcohol use and PA [13]. Golf is inherently social, and many golfers participate in golf because of its social nature [21–23]. We assessed four underlying motives for golfing, including social status, health, enjoyment, and skills. Of these four motives, golfing for social status motives increased the odds of consuming alcohol by 60%. The other three motives were not associated with alcohol use. Golfing for social status represents a more extrinsic form of motivation, and previous work has shown that more extrinsic forms of motivation for alcohol use are more strongly associated with drinking than autonomous motivation [14]. It is also possible that individuals who golf to attain social status may share similar motives for their alcohol use, corresponding with both the proposal that PA and alcohol use may derive from common underlying motives [14], and the finding that individuals who report higher levels of social motives for drinking report drinking more frequently [30]. Finally, alcohol consumption could be a coping mechanism for managing the stress created by seeking greater social status [14,49]. Further research examining golfing and alcohol use motives could help identify the mechanisms underlying the association between golfing for social status and alcohol use.

A third contribution of this research involved the examination of participation in a specific type of PA with alcohol use. Other studies have examined associations between alcohol use and sport participation in general [50,51], but, to our knowledge, these are the first findings to characterize alcohol use among participants in a specific type of PA. The findings provide preliminary support for the assertion that the type of PA an individual engages in may be associated with their alcohol use [13,14]. In the case of golf, alcohol use may be facilitated by the culture of the sport. Along with the ready availability of alcohol at clubhouses and beverage carts, several alcoholic beverages are named after professional golfers, and professional golfers celebrate winning The Open Championship by drinking their favorite alcoholic beverage out of the Claret Jug [52]. Similar to golf, some cycling and running events encourage concurrent PA and alcohol use (e.g., beer mile, hash runs). That alcohol consumption is a fixture of golf culture is concerning given the link between alcohol use and cancer risk [1,2,53–55].

The sampling and analytic approaches were noteworthy strengths of this study. This sample was recruited to represent the U.S. population with respect to sex, race, and ethnicity. In contrast, prior work has relied on more homogeneous samples, which were typically limited to adolescents or young adults. The modeling strategy we applied accounted for the distributional features of count-based alcohol use data. This modeling approach, which previous researchers recommended to use when studying MVPA data [56] (which has distributional characteristics similar to alcohol use data), allowed us to address the zero-inflated and over-dispersed nature of alcohol use data. This analytic approach strengthens conclusions by distinguishing the processes that separate teetotalers from drinkers and those that separate light from heavy drinkers. Similar approaches have been applied with intensive longitudinal data but have rarely been applied in cross-sectional analyses of physical activity and alcohol use [25,28–30].

Notwithstanding those strengths, individual minority racial groups were not recruited in large enough samples to test whether race moderated associations. The survey also assessed alcohol use over the prior week, and it is unclear how representative that week was of typical consumption (though error variance due to over-/under-reporting alcohol use was assumed to be random). Given that data were collected during the COVID-19 pandemic, participants' alcohol use may have been higher than normal [57,58], and their PA and golf participation may have been lower than normal [59–62]. A third limitation was that data were collected via self-report at a single occasion, introducing the potential for recall and self-report biases. Future research should consider incorporating device-based measures of PA, alcohol use, or both. Incorporating intensive longitudinal measures of PA and alcohol use in golfers would also help to uncover the temporality of the associations between PA, golf participation, and alcohol use in golfers, as well as the long-term effects of PA and alcohol use on energy balance, obesity, and cancer risk. It is possible that these findings were due to unmeasured third variables associated with golfing and alcohol use, or they could be an artifact of aggregating behaviors across time. Capturing real-time data would help eliminate these possibilities and identify the temporal sequence of the association between alcohol use and golfing.

5. Conclusions

In sum, drawing on a national sample of golfers who were purposively sampled to represent the adult population in sex, race, and ethnicity, this study revealed that alcohol use is frequent among golfers and, consistent with prior research, was greater among those engaged in more overall PA. Although alcohol use was not associated with the intensity of golf engagement, golfers motivated by social status were more likely to consume any alcohol and to consume more alcohol than others. Alcohol use among golfers may represent a public health concern due to its potential to increase discretionary energy intake, promote energy imbalance and weight gain, and mitigate some of the obesity and cancer protective effects of PA. Future research should explore the consequences of alcohol-related energy intake on golfers' body composition and cancer risk to determine whether golf culture warrants a targeted intervention to decrease obesity and cancer risk.

Supplementary Materials: The following are available online at https://www.mdpi.com/article/10.3390/nu13061856/s1, Table S1: Zero-inflated negative binomial models with golf participation predicting alcohol servings per week and moderation by age, Table S2: Zero-inflated negative binomial models with walking hours per week predicting alcohol servings per week and moderation by age, Table S3: Zero-inflated negative binomial models with total physical activity volume per week predicting alcohol servings per week and moderation by age.

Author Contributions: Conceptualization, J.C., E.H., S.P. and D.E.C.; Methodology, J.C. and D.E.C.; Formal Analysis, J.C.; Investigation, D.E.C.; Resources, D.E.C.; Data Curation, J.C.; Writing—Original Draft Preparation, J.C. and D.E.C.; Writing—Review and Editing, E.H., S.P. and M.R.; Visualization, J.C.; Project Administration, D.E.C.; Funding Acquisition, D.E.C. All authors have read and agreed to the published version of the manuscript.

Funding: JC's time was supported by the Prevention and Methodology Training Program (T32 DA017629) with funding from the National Institute on Drug Abuse. The content is solely the responsibility of the authors and does not necessarily represent the official views of the National Institute on Drug Abuse or the National Institutes of Health.

Institutional Review Board Statement: The study was conducted according to the guidelines of the Declaration of Helsinki and approved by the Institutional Review Board of Pennsylvania State University (Protocol code: STUDY00014980; Date of approval: 5/8/2020).

Informed Consent Statement: Informed consent was obtained from all subjects involved in the study.

Data Availability Statement: The datasets analyzed and/or generated during the current study are available from the corresponding author on reasonable request.

Acknowledgments: We would like to acknowledge Deborah Brunke-Reese for her help setting up the Qualtrics survey and her revisions to this manuscript.

Conflicts of Interest: The authors declare no conflict of interest.

References

1. Bagnardi, V.; Rota, M.; Botteri, E.; Tramacere, I.; Islami, F.; Fedirko, V.; Scotti, L.; Jenab, M.; Turati, F.; Pasquali, E.; et al. Alcohol consumption and site-specific cancer risk: A comprehensive dose-response meta-analysis. *Br. J. Cancer* **2015**, *112*, 580–593. [CrossRef] [PubMed]
2. NIAAA. Alcohol Facts and Statistics. Available online: https://www.niaaa.nih.gov/publications/brochures-and-fact-sheets/alcohol-facts-and-statistics (accessed on 29 January 2021).
3. Yeomans, M.R. Alcohol, appetite and energy balance: Is alcohol intake a risk factor for obesity? *Physiol. Behav.* **2010**, *100*, 82–89. [CrossRef]
4. Inan-Eroglu, E.; Powell, L.; Hamer, M.; O'Donovan, G.; Duncan, M.J.; Stamatakis, E. Is there a link between different types of alcoholic drinks and obesity? An Analysis of 280,183 UK Biobank participants. *Int. J. Environ. Res. Public Health* **2020**, *17*, 5178. [CrossRef]
5. Sayon-Orea, C.; Martinez-Gonzalez, M.A.; Bes-Rastrollo, M. Alcohol consumption and body weight: A systematic review. *Nutr. Rev.* **2011**, *69*, 419–431. [CrossRef]
6. Shelton, N.J.; Knott, C.S. Association between alcohol calorie intake and overweight and obesity in English adults. *Am. J. Public Health* **2014**, *104*, 629–631. [CrossRef]
7. Traversy, G.; Chaput, J.-P. Alcohol Consumption and Obesity: An Update. *Curr. Obes. Rep.* **2015**, *4*, 122–130. [CrossRef]
8. CDC. Cancers Associated with Overweight and Obesity Make up 40 Percent of Cancers Diagnosed in the United States. Available online: https://www.cdc.gov/media/releases/2017/p1003-vs-cancer-obesity.html (accessed on 17 February 2021).
9. Steele, C.B.; Thomas, C.C.; Henley, S.J.; Massetti, G.M.; Galuska, D.A.; Agurs-Collins, T.; Puckett, M.; Richardson, L.C. Vital Signs: Trends in incidence of cancers associated with overweight and obesity—United States, 2005–2014. *MMWR Morb. Mortal. Wkly. Rep.* **2017**, *66*, 1052–1058. [CrossRef] [PubMed]
10. SAMHSA. *Results from the 2018 National Survey on Drug Use and Health: Detailed Tables*; Center for Behavioral Health Statistics and Quality, Substance Abuse and Mental Health Services Administration: Rockville, MD, USA, 2019.
11. Lee, I.-M.; Shiroma, E.J.; Lobelo, F.; Puska, P.; Blair, S.N.; Katzmarzyk, P.T. Lancet Physical Activity Series Working Group Effect of physical inactivity on major non-communicable diseases worldwide: An analysis of burden of disease and life expectancy. *Lancet* **2012**, *380*, 219–229. [CrossRef]
12. McTiernan, A.; Friedenreich, C.M.; Katzmarzyk, P.T.; Powell, K.E.; Macko, R.; Buchner, D.; Pescatello, L.S.; Bloodgood, B.; Tennant, B.; Vaux-Bjerke, A.; et al. Physical activity in cancer prevention and survival: A systematic review. *Med. Sci. Sports Exerc.* **2019**, *51*, 1252–1261. [CrossRef]

13. Piazza-Gardner, A.K.; Barry, A.E. Examining physical activity levels and alcohol consumption: Are people who drink more active? *Am. J. Health Promot.* **2012**, *26*, e95–e104. [CrossRef] [PubMed]
14. Leasure, J.L.; Neighbors, C.; Henderson, C.E.; Young, C.M. Exercise and alcohol consumption: What we know, what we need to know, and why it is important. *Front. Psychiatry* **2015**, *6*, 156. [CrossRef]
15. West, A.B.; Bittel, K.M.; Russell, M.A.; Evans, M.B.; Mama, S.K.; Conroy, D.E. A systematic review of physical activity, sedentary behavior, and substance use in adolescents and emerging adults. *Transl. Behav. Med.* **2020**, in press. [CrossRef]
16. Pronk, N.P.; Anderson, L.H.; Crain, A.L.; Martinson, B.C.; O'Connor, P.J.; Sherwood, N.E.; Whitebird, R.R. Meeting recommendations for multiple healthy lifestyle factors. *Am. J. Prev. Med.* **2004**, *27*, 25–33. [CrossRef]
17. Spring, B.; Moller, A.C.; Coons, M.J. Multiple health behaviours: Overview and implications. *J. Public Health* **2012**, *34*, i3–i10. [CrossRef] [PubMed]
18. Meader, N.; King, K.; Moe-Byrne, T.; Wright, K.; Graham, H.; Petticrew, M.; Power, C.; White, M.; Sowden, A.J. A systematic review on the clustering and co-occurrence of multiple risk behaviours. *BMC Public Health* **2016**, *16*, 657. [CrossRef]
19. Spring, B.; King, A.C.; Pagoto, S.L.; Van Horn, L.; Fisher, J.D. Fostering multiple healthy lifestyle behaviors for primary prevention of cancer. *Am. Psychol.* **2015**, *70*, 75–90. [CrossRef] [PubMed]
20. National Golf Foundation. Available online: https://www.ngf.org/ (accessed on 30 July 2020).
21. Siegenthaler, K.L.; O'Dell, I. Older golfers: Serious leisure and successful aging. *World Leis. J.* **2003**, *45*, 45–52. [CrossRef]
22. Wood, L.; Danylchuk, K. Playing our way: Contributions of social groups to women's continued participation in golf. *Leis. Sci.* **2011**, *33*, 366–381. [CrossRef]
23. Stenner, B.J.; Mosewich, A.D.; Buckley, J.D. An exploratory investigation into the reasons why older people play golf. *Qual. Res. Sport Exerc. Health* **2016**, *8*, 257–272. [CrossRef]
24. SportAUS. Sport AUS—Motivations & Reasons for Drop out. Available online: https://app.powerbi.com/view?r=eyJrIjoiZmEwNDRmODgtMDg1Zi00Yjc1LWJkMzQtOWM0NDk3NWY1NzZmIiwidCI6IjhkMmUwZjRjLTU1ZjItNGNiMS04ZWU3LWRhNWRkM2ZmMzYwMCJ9 (accessed on 29 April 2021).
25. Conroy, D.E.; Ram, N.; Pincus, A.L.; Coffman, D.L.; Lorek, A.E.; Rebar, A.L.; Roche, M.J. Daily physical activity and alcohol use across the adult lifespan. *Health Psychol.* **2015**, *34*, 653–660. [CrossRef]
26. Graupensperger, S.; Wilson, O.; Bopp, M.; Blair Evans, M. Longitudinal association between alcohol use and physical activity in US college students: Evidence for directionality. *J. Am. Coll. Health* **2018**, *68*, 155–162. [CrossRef] [PubMed]
27. Abrantes, A.M.; Scalco, M.D.; O'Donnell, S.; Minami, H.; Read, J.P. Drinking and exercise behaviors among college students: Between and within-person associations. *J. Behav. Med.* **2017**, *40*, 964–977. [CrossRef]
28. Cho, D.; Armeli, S.; Weinstock, J.; Tennen, H. Daily- and person-level associations between physical activity and alcohol use among college students. *Emerg. Adulthood.* **2018**, *8*, 428–434. [CrossRef]
29. Gilchrist, J.D.; Conroy, D.E.; Sabiston, C.M. Associations between alcohol consumption and physical activity in breast cancer survivors. *J. Behav. Med.* **2020**, *43*, 166–173. [CrossRef] [PubMed]
30. Henderson, C.E.; Manning, J.M.; Davis, C.M.; Conroy, D.E.; Van Horn, M.L.; Henry, K.; Long, T.; Ryan, L.; Boland, J.; Yenne, E.; et al. Daily physical activity and alcohol use among young adults. *J. Behav. Med.* **2020**, *43*, 365–376. [CrossRef]
31. Hogan, R.; Jones, W.H.; Cheek, J.M. Socioanalytic theory: An alternative to armadillo psychology. In *The Self and Social Life*; McGraw-Hill: New York, NY, USA, 1985; pp. 175–198.
32. U.S. Department of Health and Human Services and U.S. Department of Agriculture. *2015–2020 Dietary Guidelines for Americans*, 8th ed.; 2015. Available online: http://health.gov/dietaryguidelines/2015/guidelines/ (accessed on 29 December 2020).
33. Bauman, A.; Bull, F.; Chey, T.; Craig, C.L.; Ainsworth, B.E.; Sallis, J.F.; Bowles, H.R.; Hagstromer, M.; Sjostrom, M.; Pratt, M.; et al. The International Prevalence Study on Physical Activity: Results from 20 countries. *Int. J. Behav. Nutr. Phys. Act.* **2009**, *6*, 21. [CrossRef]
34. Brown, W.J.; Trost, S.G.; Bauman, A.; Mummery, K.; Owen, N. Test-retest reliability of four physical activity measures used in population surveys. *J. Sci. Med. Sport* **2004**, *7*, 205–215. [CrossRef]
35. Ekelund, U.; Sepp, H.; Brage, S.; Becker, W.; Jakes, R.; Hennings, M.; Wareham, N.J. Criterion-related validity of the last 7-day, short form of the International Physical Activity Questionnaire in Swedish adults. *Public Health Nutr.* **2006**, *9*, 258–265. [CrossRef]
36. U.S. Department of Health and Human Services. *Physical Activity Guidelines for Americans*, 2nd ed.; 2018. Available online: https://health.gov/paguidelines/second-edtion/pdf/Physical_Activity_Guidelines_2nd_edition.pdf (accessed on 29 May 2021).
37. Craig, C.L.; Marshall, A.L.; Sjostrom, M.; Bauman, A.E.; Booth, M.L.; Ainsworth, B.E.; Pratt, M.; Ekelund, U.; Yngve, A.; Sallis, J.F.; et al. International Physical Activity Questionnaire: 12-Country reliability and validity. *Med. Sci. Sports Exerc.* **2003**, *35*, 1381–1395. [CrossRef]
38. Osborne, J.W. Improving your data transformations: Applying the Box-Cox transformation. *Pract. Assess. Res. Eval.* **2010**, *15*, 12.
39. Everitt, B.; Hothorn, T. *A Handbook of Statistical Analyses Using R*, 2nd ed.; CRC Press: Boca Raton, FL, USA, 2010; ISBN 978-1-4200-7933-3.
40. Hilbe, J. *Negative Binomial Regression*; Cambridge University Press: Cambridge, NY, USA, 2011; ISBN 978-0-521-19815-8.
41. Vuong, Q.H. Likelihood ratio tests for model selection and non-nested hypotheses. *Econometrica* **1989**, *57*, 307. [CrossRef]
42. Tüzen, M.F.; Erbaş, S. A comparison of count data models with an application to daily cigarette consumption of young persons. *Commun. Stat. Theory Methods* **2018**, *47*, 5825–5844. [CrossRef]
43. R: The R Project for Statistical Computing. Available online: https://www.r-project.org/ (accessed on 30 July 2020).

44. Centers for Disease Control and Prevention (CDC); National Center for Health Statistics (NCHS). *National Health and Nutrition Examination Survey Data*; Department of Health and Human Serivces, Center for Disease Control and Prevention: Hyattsville, MD, USA, 2007–2016. Available online: wwwn.cdc.gov/nchs/nhanes/ContinuousNhanes/ (accessed on 29 May 2021).
45. Marlatt, G.A.; Witkiewitz, K. Harm reduction approaches to alcohol use. *Addict. Behav.* **2002**, *27*, 867–886. [CrossRef]
46. del Boz, J.; Fernández-Morano, T.; Padilla-España, L.; Aguilar-Bernier, M.; Rivas-Ruiz, F.; de Troya-Martín, M. Skin cancer prevention and detection campaign at golf courses on Spain's Costa del Sol. *Actas Dermosifiliogr.* **2015**, *106*, 51–60. [CrossRef] [PubMed]
47. Dennis, L.K.; Vanbeek, M.J.; Beane Freeman, L.E.; Smith, B.J.; Dawson, D.V.; Coughlin, J.A. Sunburns and risk of cutaneous melanoma: Does age matter? A comprehensive meta-analysis. *Ann. Epidemiol.* **2008**, *18*, 614–627. [CrossRef] [PubMed]
48. Hill, J.O.; Peters, J.C.; Wyatt, H.R. Using the energy gap to address obesity: A commentary. *J. Am. Diet. Assoc.* **2009**, *109*, 1848–1853. [CrossRef]
49. Bechtoldt, M.N.; Schneider, V.K. Predicting stress from the ability to eavesdrop on feelings: Emotional intelligence and testosterone jointly predict cortisol reactivity. *Emotion* **2016**, *16*, 815–825. [CrossRef]
50. Kunz, J.L. Drink and be active? The associations between drinking and participation in sports. *Addict. Res.* **1997**, *5*, 439–450. [CrossRef]
51. Poortinga, W. Associations of physical activity with smoking and alcohol consumption: A sport or occupation effect? *Prev. Med.* **2007**, *45*, 66–70. [CrossRef]
52. BP Staff. The Best Drinks Named After Golfers and Golfers Who Need One. Available online: https://www.bunkersparadise.com/49876/the-best-drinks-named-after-golfers-and-golfers-who-need-one/ (accessed on 7 August 2020).
53. Moore, S.C.; Lee, I.-M.; Weiderpass, E.; Campbell, P.T.; Sampson, J.N.; Kitahara, C.M.; Keadle, S.K.; Arem, H.; Berrington de Gonzalez, A.; Hartge, P.; et al. Association of leisure-time physical activity with risk of 26 types of cancer in 1.44 million adults. *JAMA Intern. Med.* **2016**, *176*, 816. [CrossRef]
54. Schneider, K.L.; Pagoto, S.; Panza, E.; Keeney, J.; Goldberg, D. Skin cancer risk profiles of physically active adults. *Health Behav. Policy Rev.* **2014**, *1*, 324–334. [CrossRef]
55. Snyder, A.; Valdebran, M.; Terrero, D.; Amber, K.T.; Kelly, K.M. Solar ultraviolet exposure in individuals who perform outdoor sport activities. *Sports Med. Open* **2020**, *6*, 42. [CrossRef] [PubMed]
56. Baldwin, S.A.; Fellingham, G.W.; Baldwin, A.S. Statistical models for multilevel skewed physical activity data in health research and behavioral medicine. *Health Psychol.* **2016**, *35*, 552–562. [CrossRef] [PubMed]
57. Pollard, M.S.; Tucker, J.S.; Green, H.D. Changes in adult alcohol use and consequences during the COVID-19 pandemic in the US. *JAMA Netw. Open* **2020**, *3*, e2022942. [CrossRef]
58. Capasso, A.; Jones, A.M.; Ali, S.H.; Foreman, J.; Tozan, Y.; DiClemente, R.J. Increased alcohol use during the COVID-19 pandemic: The effect of mental health and age in a cross-sectional sample of social media users in the U.S. *Prev. Med.* **2021**, *145*, 106422. [CrossRef]
59. Dunton, G.F.; Wang, S.D.; Do, B.; Courtney, J. Early effects of the COVID-19 pandemic on physical activity locations and behaviors in adults living in the United States. *Prev. Med. Rep.* **2020**, *20*, 101241. [CrossRef] [PubMed]
60. Maher, J.P.; Hevel, D.J.; Reifsteck, E.J.; Drollette, E.S. Physical activity is positively associated with college students' positive affect regardless of stressful life events during the COVID-19 pandemic. *Psychol. Sport Exerc.* **2021**, *52*, 101826. [CrossRef] [PubMed]
61. Meyer, J.; McDowell, C.; Lansing, J.; Brower, C.; Smith, L.; Tully, M.; Herring, M. Changes in physical activity and sedentary behavior in response to COVID-19 and their associations with mental health in 3052 US adults. *Int. J. Environ. Res. Public Health* **2020**, *17*, 6469. [CrossRef]
62. Yang, Y.; Koenigstorfer, J. Determinants of physical activity maintenance during the Covid-19 pandemic: A focus on fitness apps. *Transl. Behav. Med.* **2020**, *10*, 835–842. [CrossRef] [PubMed]

MDPI

Article

A Low-Glucose Eating Pattern Improves Biomarkers of Postmenopausal Breast Cancer Risk: An Exploratory Secondary Analysis of a Randomized Feasibility Trial

Susan M. Schembre [1,*], Michelle R. Jospe [1], Erin D. Giles [2], Dorothy D. Sears [3], Yue Liao [4], Karen M. Basen-Engquist [5] and Cynthia A. Thomson [6]

1. Department of Family and Community Medicine, College of Medicine-Tucson, University of Arizona, Tucson, AZ 85721, USA; mjospe@email.arizona.edu
2. Department of Nutrition, Texas A & M University, College Station, TX 77843, USA; egiles@tamu.edu
3. College of Health Solutions, Arizona State University, Tempe, AZ 85287, USA; Dorothy.Sears@asu.edu
4. Department of Kinesiology, College of Nursing and Health Innovation, University of Texas at Arlington, Arlington, TX 76019, USA; Yue.Liao@uta.edu
5. Department of Behavioral Science, University of Texas MD Anderson Cancer Center, Houston, TX 77030, USA; kbasenen@mdanderson.org
6. Department of Health Promotion Sciences, Mel and Enid Zuckerman College of Public Health, University of Arizona, Tucson, AZ 85721, USA; cthomson@arizona.edu

* Correspondence: sschembre@email.arizona.edu

Citation: Schembre, S.M.; Jospe, M.R.; Giles, E.D.; Sears, D.D.; Liao, Y.; Basen-Engquist, K.M.; Thomson, C.A. A Low-Glucose Eating Pattern Improves Biomarkers of Postmenopausal Breast Cancer Risk: An Exploratory Secondary Analysis of a Randomized Feasibility Trial. *Nutrients* **2021**, *13*, 4508. https://doi.org/10.3390/nu13124508

Academic Editors: María-José Sánchez, Cherubini Valentino and Luis A. Moreno

Received: 1 November 2021
Accepted: 11 December 2021
Published: 16 December 2021

Abstract: Postmenopausal breast cancer is the most common obesity-related cancer death among women in the U.S. Insulin resistance, which worsens in the setting of obesity, is associated with higher breast cancer incidence and mortality. Maladaptive eating patterns driving insulin resistance represent a key modifiable risk factor for breast cancer. Emerging evidence suggests that time-restricted feeding paradigms (TRF) improve cancer-related metabolic risk factors; however, more flexible approaches could be more feasible and effective. In this exploratory, secondary analysis, we identified participants following a low-glucose eating pattern (LGEP), defined as consuming energy when glucose levels are at or below average fasting levels, as an alternative to TRF. Results show that following an LGEP regimen for at least 40% of reported eating events improves insulin resistance (HOMA-IR) and other cancer-related serum biomarkers. The magnitude of serum biomarkers changes observed here has previously been shown to favorably modulate benign breast tissue in women with overweight and obesity who are at risk for postmenopausal breast cancer. By comparison, the observed effects of LGEP were similar to results from previously published TRF studies in similar populations. These preliminary findings support further testing of LGEP as an alternative to TRF and a postmenopausal breast cancer prevention strategy. However, results should be interpreted with caution, given the exploratory nature of analyses

Keywords: eating physiology; food intake regulation; blood glucose; metabolism; weight management; obesity; adherence

1. Introduction

High obesity rates among women in the United States and worldwide are leading to a continued rise in obesity-related cancers, most notably postmenopausal breast cancer [1], which is the leading cause of obesity-related cancer deaths among women in the U.S. [2]. Research shows that excessive weight gain and obesity are significant risk factors for postmenopausal breast cancer among women with and without increased genetic risk [3–9]. Postmenopausal breast cancer and obesity are linked through insulin resistance—a key modifiable risk factor. By losing weight, women with obesity improve their metabolic- and cancer-related risk biomarkers, including insulin resistance and insulin-signaling adipokines, and circulating pro-inflammatory cytokines that promote tumorigenesis [10,11].

In a seminal Phase II feasibility study of a 6-month intensive lifestyle intervention conducted in postmenopausal overweight and obese women at increased risk for breast cancer, Fabian et al. demonstrated that weight losses of at least 10% effectively reduced serum biomarkers, including insulin resistance (HOMA-IR), at a magnitude that favorably modulated benign breast tissue biomarkers [12,13]. While intensive lifestyle interventions that promote chronic energy restriction, such as that implemented in Fabian's study, are effective at improving outcomes related to cancer risk in both women at high risk and breast cancer survivors [11,14,15], they are resource-intensive, and similar interventions reported poor long-term adherence. Thus, post-intervention weight regains often hinder long-term treatment effectiveness [16–18]. This and other research suggest that alternative weight loss and cancer prevention approaches with clinically meaningful outcomes are essential.

Intermittent fasting paradigms have become increasingly popular among researchers and health-conscious individuals. These eating paradigms aim to align meal-timing with circadian rhythms. Restricting the consumption of energy intake to a daily timespan of 4–10 h (e.g., time-restricted feeding, TRF) enhances synchronization between the central circadian clock (synchronized by light) and peripheral circadian clocks (entrained by nutrient intake) [19,20]. Desynchronization of the central and peripheral circadian clocks was shown to negatively impact insulin sensitivity [21] and beta-cell function [22–24]. Compared to chronic energy restriction, human and animal models have shown that TRF reduces metabolic disease risk by improving metabolic homeostasis [25]. Despite published support for TRF to improve metabolic outcomes, meta-analyses of research conducted in women and men (mean age range: 21–77 years) with and without metabolic abnormalities over a median of 6 to 8 weeks concluded that TRF has only modest effects on weight (-1.7 to -0.1 kg) and metabolism [26,27], which could limit its utility as a cancer prevention strategy. Moreover, research and healthcare communities acknowledge that TRF and other fasting paradigms might be inappropriate, unacceptable, or result in lower adherence over time among some individuals [28–30]. As such, it is reasonable to explore alternative eating paradigms that are effective and may be more broadly adopted.

Eating when pre-prandial (pre-meal) glucose levels are low ("low-glucose eating pattern") is an evidence-based strategy to improve maladaptive eating patterns. Research shows that eating without physiological hunger is a modifiable health risk behavior associated with excessive weight gain and increased metabolic risk [31,32]. Consistent with this research, we have shown that individuals with obesity are over-sensitive to changes in glucose levels [32] and that low-glucose eating patterns (defined by personalized thresholds) can be taught as an effective self-regulation strategy that promotes weight control [33,34]. Glucose-guided eating (GGE; historically referred to as hunger training) is a timed eating intervention that teaches people to differentiate between physiological hunger and the hedonic desire to eat [35]. In an intervention setting, individuals taught to eat by the GGE paradigm self-monitor their glucose levels using continuous glucose monitors (CGM) or commercially available glucometers and are instructed to eat when two conditions are met: (a) the desire to eat arises and (b) their glucose levels are at or below their personalized threshold. Typically, this training regimen is implemented for 3–4 weeks while people practicing GGE learn to associate symptoms of hunger with their personalized glucose threshold. The GGE paradigm does not rely on glucose as a valid proxy for hunger for GGE. Rather, it is that, to promote metabolic homeostasis, energy intake should not occur when circulating glucose is the primary source of fuel [32].

The modification of glucose eating patterns by GGE is feasible [33,36] and has resulted in clinically significant, average weight loss of 7.4% in 5 months and improvements in eating behavior (including reductions in hedonic eating) and cancer-related risk biomarkers [34,36–39]. GGE has resulted in improvements in whole-body insulin sensitivity by 31% (Matsuda index, 7.1 ± 4.1 to 9.4 ± 5.2) in non-diabetic, lean adults (BMI = 23 ± 4 kg/m^2) [38]. Insulin resistance is the most important modifiable risk factor for postmenopausal breast cancer and is caused by obesity and maladaptive eating patterns.

Insulin resistance has downstream effects on insulin signaling (e.g., IGF-1), adipokines (including adiponectin), and circulating pro-inflammatory cytokines that promote tumorigenesis [40,41]. GGE has shown a beneficial effect on insulin sensitivity is greater than that noted in the study by Fabian et al. [12], suggesting that GGE could be more effective at reducing insulin resistance than weight loss alone. Similar to TRF, GGE has an advantage over intensive lifestyle weight loss programs in that it does not promote chronic energy restriction and it requires minimal human resources. This affords GGE the possibility of wide dissemination. There is a great potential benefit of the GGE intervention in postmenopausal breast cancer prevention, and this needs examination.

The key aspect of the GGE intervention is eating when glucose is low, defined as under one's personalized glucose threshold. The goal of the current study is to explore the impact of low- vs. high-glucose eating patterns on changes in body weight and the selected serum biomarkers of breast cancer risk after 16 weeks and compare these results with those reported in recent TRF studies in similar populations of older women and with the intensive lifestyle intervention conducted by Fabian et al. in postmenopausal overweight and obese women at increased risk for breast cancer. The findings of the current study are intended to support further testing of GGE to promote a low-glucose eating pattern as a strategy to reduce breast cancer risk in postmenopausal women. Therefore, this exploratory, secondary analysis aims to examine the potential effect of a low-glucose eating pattern on postmenopausal breast cancer risk.

2. Materials and Methods

Project Take Charge [42] was a 16-week, 2-arm randomized controlled trial in 50 women at risk of postmenopausal breast cancer. Take Charge aimed to assess the feasibility of adding GGE to a highly disseminated, comprehensive weight-loss intervention, the Diabetes Prevention Program (DPP) [43]. As a standalone intervention, the DPP results in weight losses typically observed in traditional weight-loss interventions of 4–7% [44]. In Project Take Charge, it was hypothesized that, if feasible, the addition of GGE to the DPP versus the DPP alone could synergistically improve weight loss and effects on biomarkers of cancer risk similar to earlier work [12]. Forty-six women completed the Take Charge trial (86%), which found that GGE was feasible, but the planned analyses (group × time ANCOVA adjusting for baseline measures) did not result in a synergistic effect when added to the DPP on changes in body weight or the cancer-related serum biomarkers assessed in the parent study, including those reported in the current study [42]. As such, data from women in the DPP-only and DPP + GGE groups were merged. Interestingly, in post-hoc analyses described in the current study, we found that women assigned to both the DPP-only and the DPP + GGE interventions changed their eating patterns in a manner consistent with GGE.

As part of a randomized feasibility study, GGE was added to a 16-week version of the DPP intervention that targeted women at risk of postmenopausal breast cancer (defined, in part, as Gail model lifetime risk > 20% or 5-year risk > 1.66%) [45]. Participants ($N = 50$) were predominantly White, non-Hispanic older women who were well-educated; lived in the Houston, Texas, metropolitan area; and had a BMI > 27 kg/m^2. This study was approved by the Institutional Review Board and registered at clinicaltrials.gov (NCT03546972). Women provided informed consent prior to initiating the study.

The Project Take Charge protocols were fully described elsewhere [42]. Briefly, as part of Project Take Charge, anthropometric measures (weight and height) and metabolic and cancer risk biomarkers (total cholesterol, HDL, LDL, VLDL, triglycerides, HbA1c, fasting glucose, fasting insulin, insulin resistance by HOMA-IR, CRP, adiponectin, IGF-1) were collected at baseline (week 0) and post-intervention (week 16). Weight (light clothing) and height (without shoes) were measured in duplicate using calibrated equipment to within 0.2 kg and 0.3 cm by trained study staff at baseline, 8 weeks, and 16 weeks. Metabolic and breast cancer risk biomarkers were assessed at baseline and 16 weeks. Fasting blood draws were conducted and processed for analysis according to standardized laboratory protocols at The University of Texas MD Anderson Cancer Center and nearby Labcorp location. Insulin resistance was assessed as HOMA-IR using fasting glucose and insulin levels by the following equation: (Fasting Glucose (mg/dL) X Fasting Insulin (mU/L)/405 [46].

The women enrolled in Project Take Charge additionally provided blinded CGM data using Dexcom G5 (Dexcom, Inc., San Diego, CA, USA) at week 0 (baseline), week 8, and week 16 (post-intervention) for up to 10 days at a time. From the collected CGM data, the mean amplitude of glycemic excursions (MAGE) was calculated using EasyGV [47] as a measure of glycemic variability. The women were trained to record their dietary intake and mealtimes using the combination of a familiar and commercially available diet tracker (MyFitnessPal) and self-captured food photographs shared via email. Diet tracking apps, including MyFitnessPal, were found to be a valid means of assessing energy and nutrient intakes [48,49]. Time-stamped dietary intake was concurrently collected with blinded CGM data for up to 7 days at all three time points. Reported mealtimes were confirmed by the study dietitian using the time-stamped food photos that were matched to MyFitnessPal records. Dietary intake (energy and macronutrient composition) was estimated by transferring the digital diet records into the University of Minnesota Nutrition Data System for Research (NDSR) software. The dietary data transfer was conducted by the study dietitian trained to use NDSR and audited for quality control by the study PI. Dietary records with mealtimes were then merged with the CGM data within 5 min of the time-stamped meals. Discrete eating events were defined as energy intake from foods or beverages of greater than 25 kcals and occurring more than 15 min apart. Women were included in this exploratory analysis if they provided at least 3 valid days of blinded CGM data and time-stamped dietary intake at week 16. A valid day was defined as having at least 2 time-stamped eating events with corresponding CGM data. This resulted in an analytical subgroup of $N = 19$ women.

Women were categorized into eating patterns based on week 16 dietary and CGM data. Those who consumed at least 40% of their recorded meals when their pre-prandial glucose levels were below their personalized threshold will be referred to as following a "low-glucose eating pattern (LGEP)"; whereas those who ate less than 40% of their meals when pre-prandial glucose was below their threshold will be referred to as following a "high-glucose eating pattern (HGEP)". The threshold of 40% eating events was chosen to define the groups post hoc to maximize between-group differences in reductions of HOMA-IR at 16-weeks. LGEP and HGEP were quantified at all three time points using blinded CGM was calculated as the percentage of reported eating occasions where a participant's glucose was equal to or less than their computed, personalized threshold (reflected by the average of two, fasted 5 am glucose levels were collected using blinded CGM during the initial week run-in period).

Outcomes between the original intervention groups were similar. Specifically, the intervention groups (DPP-only vs. DPP + GGE) had comparable changes from baseline to 16 weeks in weight (−5.0 kg vs. −4.9 kg) and HOMA-IR (−0.3 vs. −0.4). Therefore, for this analysis, the data from the DPP-only and DPP + GGE groups were combined. The 16-week changes in weight and serum biomarkers were compared between women in the LGEP and HGEP groups using SPSS version 28. The means, standard deviations, medians, and ranges are reported here and compared to findings from Fabian et al. [12] and published TRF studies in comparable study populations [50,51]. The *p*-values or other tests of significance were not reported due to the secondary and exploratory nature of this analysis. TRF studies were identified using MedLine. Only those articles with predominantly older women at or near an age consistent with the onset of menopause (>45 years) were considered.

3. Results

Table 1 shows that this analytical sample of women ($N = 19$) were predominantly older, White, non-Hispanic, college-educated women, with a BMI in the obese range at baseline. Out of the 19 women with CGM and dietary data, eight (42%) were identified as following a low-glucose eating pattern at post-intervention (week 16) and categorized in the LGEP group. Women in the LGEP were comparable demographically to those in the HGEP group with modestly greater BMI (Table 1).

Table 1. Baseline characteristics of participants according to glucose eating pattern.

	Low-Glucose Eating Pattern	High-Glucose Eating Pattern
N	*N* = 8	*N* = 11
DPP + GGE group, *n* (%)	5 (63%)	6 (55%)
White, non-Hispanic, *n* (%)	8 (100%)	11 (100%)
Married, *n* (%)	7 (88%)	11 (100%)
College educated, *n* (%)	8 (100%)	10 (90%)
Age (years)	59.4 ± 7.0	61.9 ± 4.9
Body mass index (kg/m^2)	32.6 ± 6.2	36.0 ± 7.0

Values are mean ± standard deviation unless otherwise indicated. GGE = Glucose-Guided Eating; DPP = Diabetes Prevention Program.

At baseline (prior to starting the intervention), nearly 70% of reported eating events occurred when glucose was above fasting levels ("HGEP"). By week 8 in the LGEP group, the majority of reported eating events occurred when glucose levels were below the personalized thresholds (approximately 60%). This change in the LGEP group was maintained at week 16. In the HGEP, fewer reported eating events occurred when glucose levels were below personalized thresholds from baseline to mid-intervention to post-intervention (Figure 1).

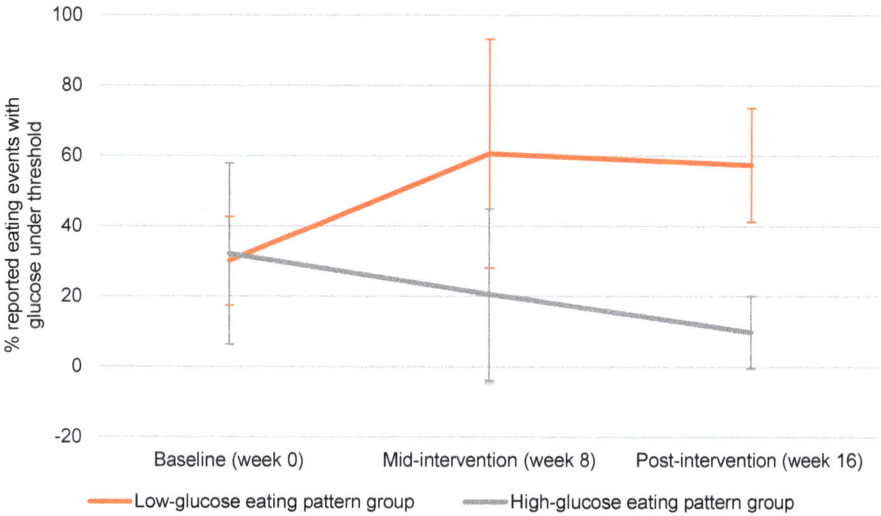

Figure 1. Glucose eating patterns over 16 weeks. Error bars represent standard deviation.

Women in the LGEP group experienced notable improvements in adiponectin, HOMA-IR, fasting insulin, and glycemic variability (calculated as the mean amplitude of glycemic excursions or MAGE) (Figure 2). These changes were evident without substantial differences in energy intake (−323 kcal vs. −445 kcal) or weight loss (−7.4 vs. −5.8 kg) for LGEP vs. HGEP at post-intervention week 16 (Supplementary Table S1). Additionally, LGEP women showed marked reductions in CGM mean glucose levels, as exemplified in Figure 3.

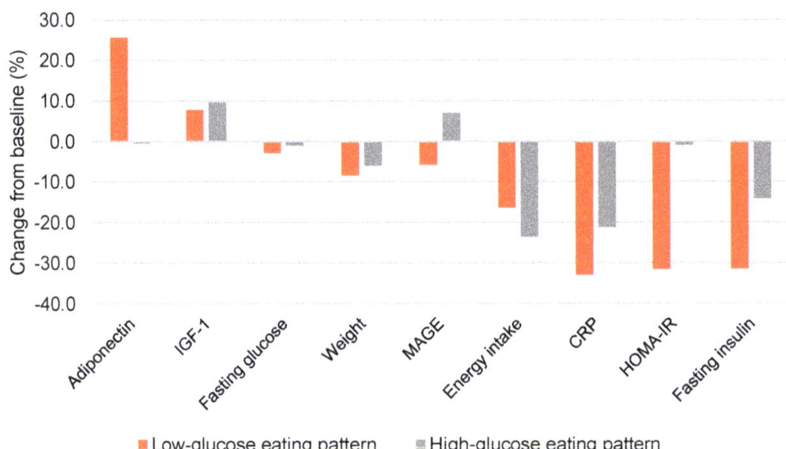

Figure 2. Effect of a low- vs. high-glucose eating pattern on weight, energy intake, and metabolic outcomes at 16 weeks. Bars represented the mean value. IGF-1 = Insulin-like growth factor 1, MAGE = mean amplitude of glycemic excursions, CRP = c-reactive protein, HOMA-IR = homeostasis model assessment-estimated insulin resistance.

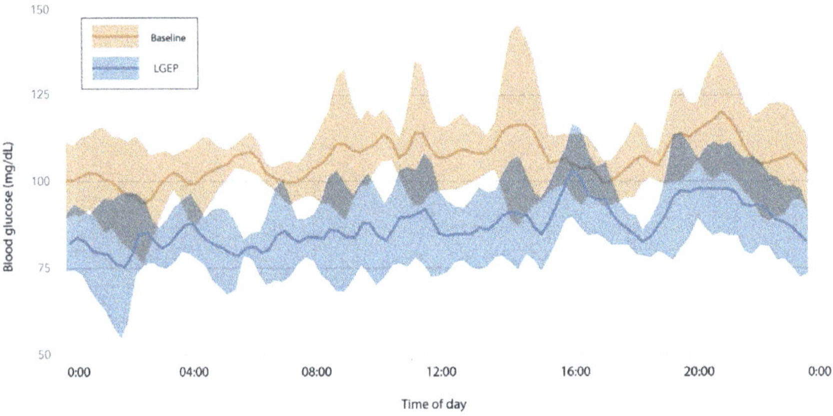

Figure 3. Changes in glucose levels between baseline and 16 weeks for one participant. Summary of 5 days of CGM data at baseline (orange) and 8 days of CGM data at week 8 (blue) after 16 weeks of following a low-glucose eating pattern for a woman with a normal range glycated hemoglobin level (HbA1c = 5.1%) at baseline. Solid lines represent the mean glucose; the shaded areas are standard deviations. This participant followed the low-glucose eating pattern for 18% of eating events at baseline, 91% at week 8, and 76% at week 16. Her baseline fasting glucose of 104 mg/dL and was 84 mg/dL at week 16, a 19% reduction. Her low-glucose eating was present for more eating events than other participants and thus is not representative of all participants, but shows the large change in 2-h glucose levels observed in an individual without prediabetes. low-glucose eating pattern (LGEP).

When results of this study are compared to those from published TRF interventions in similar populations, it suggests that LGEP induces nominally larger average weight loss and reductions in HOMA-IR and fasting insulin that is not explained by changes in energy intake (Table 2). The comparison of LGEP to the intensive lifestyle intervention led by Fabian et al. [12] highlights the potential of an LGEP to induce changes in fasting insulin, HOMA-IR, CRP, and adiponectin, that might similarly translate into favorable modulation of benign breast tissue.

Table 2. Comparison of LGEP to previously published research in similar populations.

	LGEP (4 Months)	TRF 4HR (2 Months)	TRF 6HR (2 Months)	TRF 8HR (3 Months)	ILI, >10% Weight Loss (6 Months)
N	Current study *N* = 7	Cienfuegos, 2020 [50] *N* = 16	Cienfuegos, 2020 [50] *N* = 19	Gabel, 2018 [51] *N* = 23	Fabian, 2013 [12] *N* = 24
Participants	Postmenopausal women at risk for BrCa without DM	90% women	90% women	87% women	Postmenopausal women at risk for BrCa without DM
BMI inclusion (kg/m^2)	>27	>30	>30	>30	>25
Age (years), mean ± SD	59 ± 7	49 ± 2	46 ± 3	50 ± 2	57 ± 5
Body weight (kg)	−7.4 (−8%)	−3.0 (−3%)	−3.0 (−3%)	−3.0 (−3%)	−12.8, (−16%)
Fasting glucose (mg/dL)	−3.3, (−3%)	−5.0 (−6%)	−2.3 (−2%)	+3 (+4%)	−3.0, (−3.0%)
Fasting insulin (µIU/mL)	−6.6, (−32%)	−2.3 (−19%)	−1.9 (12%)	−2.6 (−31%)	−3.7, (−57%)
Insulin resistance (HOMA-IR)	−0.7, (−32%)	−0.8 (−29%)	−0.5 (−12%)	−0.6 (−38%)	−0.5, (−56%)
IGF-1 (nM)	+7.8, (+8%)	NR	NR	NR	+0.6, (+6%)
Adiponectin	+1.8 (+26%)	NR	NR	NR	+3.5, (+31%)
TNF-α (pg/mL)	NR	−2.4 (−29%)	−0.4 (−3%)	NR	−0.2, (−4%)
CRP (µg/mL)	−0.5 (−33%)	NR	NR	NR	−1.0, (−39%)
Energy intake (kcals)	−323 (−16%)	−528 (−30%)	−566 (−29%)	−341 (−20%)	−387, (−21%)
Macronutrient composition as percentage of energy intake (fat, carbohydrates, protein)	36%, 47%, 18%	36%, 46%, 18%	40%, 40%, 20%	37%, 46%, 17%	20%, 60%, 21%

Values are mean (%) unless otherwise indicated. LGEP = Low-glucose eating pattern; TRF = time-restricted eating; ILI = intensive lifestyle intervention; BrCa = breast cancer; DM = diabetes mellitus; BMI = body mass index; HOMA-IR = homeostasis model assessment-estimated insulin resistance; IGF-1 = Insulin-like growth factor 1; TNF-α = Tumor necrosis factor; CRP = c-reactive protein, NR = not reported.

4. Discussion

This study supports the potential efficacy of a low-glucose eating pattern (LGEP) to improve metabolic and cancer risk biomarkers, including insulin resistance, in older women. Importantly, these data support a viable alternative to TRF for improving health outcomes. Furthermore, the positive metabolic effects of an LGEP might be achieved without eating all meals under the personalized glucose threshold, further supporting the flexibility of LGEP and the robust effects of LGEP in relation to metabolic health. Specifically, following LGEP at ≥40% of eating events is associated with significant improvements in weight and serum markers of cancer risk over time. These findings are similar to previously reported findings of the GGE intervention, where modest protocol adherence was associated with clinically relevant weight loss and improvements in eating behavior [38]. However, this is the first analysis to examine the association between LGEP and serum biomarkers of breast cancer risk. Importantly, the magnitude of observed improvements in HOMA-IR in response to LGEP was comparable to those previously shown to impact postmenopausal breast cancer risk at the tissue level. We feel these preliminary findings support further testing of LGEP as a breast cancer prevention strategy.

Comparing our results to those from TRF studies suggests that LGEP could be as effective or more effective at reducing the risk of postmenopausal breast cancer. We hypothesize that GGE could effectively teach women to follow LGEP to achieve these outcomes. The results shown here suggest it is worthwhile to conduct a clinical trial aimed at comparing the effects of these interventions on biomarkers of postmenopausal breast cancer. Key features of such a trial should include adherence for a range of population groups and durability of effects after the intervention has ceased. A previous pilot study showed that GGE is acceptable from a patient perspective and outlined adherence barriers and enablers [39]. Further examination and direct comparison of participants' barriers and challenges to adherence and unwanted side effects in response to GGE and TRF will be needed to confirm GGE as an acceptable alternative to TRF. Comparison of our results to those of Fabian et al. [12] suggests that the magnitude of changes in weight and cancer-related biomarkers produced by LGEP, consistent with GGE, particularly changes in fasting insulin, HOMA-IR, and adiponectin, could have meaningful changes in benign breast tissue indicative of reduced postmenopausal breast cancer risk.

Of note, this and the related Project Take Charge studies to exemplify the benefit of using biological feedback (here glucose levels) to motivate and support effective behavior change (here maladaptive eating patterns). While systematic reviews demonstrated the utility of glucose monitoring in obesity research [52], limited research has been conducted to examine the mechanisms of action by which biological feedback motivates positive health behavior change [53]. One possibility is that GGE may act through the Health Belief Model; wherein, people experience a change in perceived risk by associating their dietary intake to health risk outcomes. Future research will be needed to understand and leverage the use of biological feedback as a cancer prevention strategy better [54].

Strengths of this study include objective quantification of LGEP through passive glucose monitoring and the range of biomarkers tested. However, this analysis is limited by our small, homogeneous sample, which limits the generalizability of our findings. Our findings are most appropriate for hypothesis driving rather than hypothesis testing, and results should be interpreted with caution given the exploratory nature of analyses. Furthermore, the TRF studies were of shorter duration than the current study, which could have implications on comparing the magnitudes of observed effects. It is also unclear why women in Project Take Charge, who were randomized to the DPP-only intervention, changed to their LGEP without additionally receiving the GGE intervention. Future DPP intervention research could test the robustness of these findings. While other clinical trials have tested GGE as a standalone intervention [34,36], following an LGEP, which is promoted by GGE, has not been objectively examined as it was here. As such, it will be important to test the effect of GGE as a standalone intervention (vs. the DPP + GGE) on LGEP and metabolic- and cancer-related biomarkers to ensure the robustness of these preliminary findings in larger and more diverse samples. Furthermore, while 40% of eating events is an achievable change in eating patterns that were sufficient to drive improvements in metabolic outcomes in this group, further research is needed to confirm the adherence level needed for favorable outcomes.

5. Conclusions

This exploratory analysis of the impact of LGEP on weight and metabolic markers offers direction for the next steps in testing GGE as an intervention to prevent postmenopausal breast cancer. The adherence goal of 40% offers a feasible target for future GGE interventions and potential for health benefits, most critically a reduction in risk of postmenopausal breast cancer.

Supplementary Materials: The following is available online at https://www.mdpi.com/article/10.3390/nu13124508/s1, Table S1: Difference in outcomes from baseline to 16 weeks in low- and high-glucose eating patterns.

Author Contributions: Conceptualization, S.M.S.; methodology, S.M.S.; formal analysis, S.M.S.; writing—original draft preparation, S.M.S., M.R.J., E.D.G., D.D.S., Y.L. and C.A.T.; writing—review and editing, S.M.S., M.R.J., E.D.G., D.D.S., Y.L., K.M.B.-E. and C.A.T.; visualization, M.R.J.; supervision, S.M.S. and K.M.B.-E.; project administration, S.M.S.; funding acquisition, S.M.S. All authors have read and agreed to the published version of the manuscript.

Funding: This research was funded by R21CA215415 (PI: Schembre) and Transdisciplinary Research on Energetics and Cancer (TREC) Training Workshop R25CA203650 (PI: Melinda Irwin). Additional support was provided by R00CA169430 (PI: Giles) and the Cancer Prevention Research Institute of Texas (RP180801; PI: Giles).

Institutional Review Board Statement: The study was conducted according to the guidelines of the Declaration of Helsinki and approved by the Institutional Review Board (or Ethics Committee) of The University of Texas MD Anderson Cancer Center Institutional Review Board (2017-0507 and 17 December 2017). It is registered on ClinicalTrials.gov (NCT03546972).

Informed Consent Statement: Informed consent was obtained from all subjects involved in the study.

Data Availability Statement: The data presented in this study are available on request from the corresponding author.

Conflicts of Interest: The authors declare no conflict of interest.

References

1. World Cancer Research Fund; American Institute for Cancer Research. *Food, Nutrition, Physical Activity, and the Prevention of Cancer: A Global Perspective*; The American Institute for Cancer Research: Washington, DC, USA, 2007.
2. American Cancer Society. *Breast Cancer Facts & Figures 2019–2020*; American Cancer Society: Atlanta, GA, USA, 2019.
3. Harvie, M.; Howell, A.; Vierkant, R.A.; Kumar, N.; Cerhan, J.R.; Kelemen, L.E.; Folsom, A.R.; Sellers, T.A. Association of gain and loss of weight before and after menopause with risk of postmenopausal breast cancer in the Iowa women's health study. *Cancer Epidemiol. Biomark. Prev.* **2005**, *14*, 656–661. [CrossRef]
4. Lahmann, P.H.; Schulz, M.; Hoffmann, K.; Boeing, H.; Tjonneland, A.; Olsen, A.; Overvad, K.; Key, T.J.; Allen, N.E.; Khaw, K.T.; et al. Long-term weight change and breast cancer risk: The European prospective investigation into cancer and nutrition (EPIC). *Br. J. Cancer* **2005**, *93*, 582–589. [CrossRef]
5. Eliassen, A.H.; Colditz, G.A.; Rosner, B.; Willett, W.C.; Hankinson, S.E. Adult weight change and risk of postmenopausal breast cancer. *JAMA* **2006**, *296*, 193–201. [CrossRef]
6. Ahn, J.; Schatzkin, A.; Lacey, J.V., Jr.; Albanes, D.; Ballard-Barbash, R.; Adams, K.F.; Kipnis, V.; Mouw, T.; Hollenbeck, A.R.; Leitzmann, M.F. Adiposity, adult weight change, and postmenopausal breast cancer risk. *Arch. Intern. Med.* **2007**, *167*, 2091–2102. [CrossRef]
7. Manders, P.; Pijpe, A.; Hooning, M.J.; Kluijt, I.; Vasen, H.F.; Hoogerbrugge, N.; van Asperen, C.J.; Meijers-Heijboer, H.; Ausems, M.G.; van Os, T.A.; et al. Body weight and risk of breast cancer in BRCA1/2 mutation carriers. *Breast Cancer Res. Treat.* **2011**, *126*, 193–202. [CrossRef] [PubMed]
8. Huang, Z.; Hankinson, S.E.; Colditz, G.A.; Stampfer, M.J.; Hunter, D.J.; Manson, J.E.; Hennekens, C.H.; Rosner, B.; Speizer, F.E.; Willett, W.C. Dual effects of weight and weight gain on breast cancer risk. *JAMA* **1997**, *278*, 1407–1411. [CrossRef] [PubMed]
9. Ballard-Barbash, R.; Hunsberger, S.; Alciati, M.H.; Blair, S.N.; Goodwin, P.J.; McTiernan, A.; Wing, R.; Schatzkin, A. Physical activity, weight control, and breast cancer risk and survival: Clinical trial rationale and design considerations. *J. Natl. Cancer Inst.* **2009**, *101*, 630–643. [CrossRef] [PubMed]
10. Byers, T.; Sedjo, R.L. Does intentional weight loss reduce cancer risk? *Diabetes Obes. Metab.* **2011**, *13*, 1063–1072. [CrossRef]
11. Reeves, M.M.; Terranova, C.O.; Eakin, E.G.; Demark-Wahnefried, W. Weight loss intervention trials in women with breast cancer: A systematic review. *Obes. Rev.* **2014**, *15*, 749–768. [CrossRef]
12. Fabian, C.J.; Kimler, B.F.; Donnelly, J.E.; Sullivan, D.K.; Klemp, J.R.; Petroff, B.K.; Phillips, T.A.; Metheny, T.; Aversman, S.; Yeh, H.W.; et al. Favorable modulation of benign breast tissue and serum risk biomarkers is associated with >10% weight loss in postmenopausal women. *Breast Cancer Res. Treat.* **2013**, *142*, 119–132. [CrossRef]
13. Fabian, C.J.; Klemp, J.R.; Marchello, N.J.; Vidoni, E.D.; Sullivan, D.K.; Nydegger, J.L.; Phillips, T.A.; Kreutzjans, A.L.; Hendry, B.; Befort, C.A.; et al. Rapid Escalation of High-Volume Exercise during Caloric Restriction; Change in Visceral Adipose Tissue and Adipocytokines in Obese Sedentary Breast Cancer Survivors. *Cancers* **2021**, *13*, 4871. [CrossRef]
14. Playdon, M.; Thomas, G.; Sanft, T.; Harrigan, M.; Ligibel, J.; Irwin, M. Weight Loss Intervention for Breast Cancer Survivors: A Systematic Review. *Curr. Breast Cancer Rep.* **2013**, *5*, 222–246. [CrossRef]
15. Birks, S.; Peeters, A.; Backholer, K.; O'Brien, P.; Brown, W. A systematic review of the impact of weight loss on cancer incidence and mortality. *Obes. Rev.* **2012**, *13*, 868–891. [CrossRef]
16. Burgess, E.; Hassmen, P.; Pumpa, K.L. Determinants of adherence to lifestyle intervention in adults with obesity: A systematic review. *Clin. Obes.* **2017**, *7*, 123–135. [CrossRef] [PubMed]
17. Leung, A.W.Y.; Chan, R.S.M.; Sea, M.M.M.; Woo, J. An Overview of Factors Associated with Adherence to Lifestyle Modification Programs for Weight Management in Adults. *Int. J. Environ. Res. Public Health* **2017**, *14*, 922. [CrossRef]
18. Diabetes Prevention Program Research, G.; Knowler, W.C.; Fowler, S.E.; Hamman, R.F.; Christophi, C.A.; Hoffman, H.J.; Brenneman, A.T.; Brown-Friday, J.O.; Goldberg, R.; Venditti, E.; et al. 10-year follow-up of diabetes incidence and weight loss in the Diabetes Prevention Program Outcomes Study. *Lancet* **2009**, *374*, 1677–1686. [CrossRef]
19. Chaix, A.; Manoogian, E.N.C.; Melkani, G.C.; Panda, S. Time-Restricted Eating to Prevent and Manage Chronic Metabolic Diseases. *Ann. Rev. Nutr.* **2019**, *39*, 291–315. [CrossRef]
20. Manoogian, E.N.C.; Panda, S. Circadian rhythms, time-restricted feeding, and healthy aging. *Ageing Res. Rev.* **2017**, *39*, 59–67. [CrossRef] [PubMed]
21. Morris, C.J.; Yang, J.N.; Garcia, J.I.; Myers, S.; Bozzi, I.; Wang, W.; Buxton, O.M.; Shea, S.A.; Scheer, F.A. Endogenous circadian system and circadian misalignment impact glucose tolerance via separate mechanisms in humans. *Proc. Natl. Acad. Sci. USA* **2015**, *112*, E2225–E2234. [CrossRef]
22. Turek, F.W.; Joshu, C.; Kohsaka, A.; Lin, E.; Ivanova, G.; McDearmon, E.; Laposky, A.; Losee-Olson, S.; Easton, A.; Jensen, D.R.; et al. Obesity and metabolic syndrome in circadian Clock mutant mice. *Science* **2005**, *308*, 1043–1045. [CrossRef] [PubMed]
23. Buxton, O.M.; Cain, S.W.; O'Connor, S.P.; Porter, J.H.; Duffy, J.F.; Wang, W.; Czeisler, C.A.; Shea, S.A. Adverse metabolic consequences in humans of prolonged sleep restriction combined with circadian disruption. *Sci. Transl. Med.* **2012**, *4*, 129ra43. [CrossRef] [PubMed]
24. Gale, J.E.; Cox, H.I.; Qian, J.; Block, G.D.; Colwell, C.S.; Matveyenko, A.V. Disruption of circadian rhythms accelerates development of diabetes through pancreatic beta-cell loss and dysfunction. *J. Biol. Rhythm.* **2011**, *26*, 423–433. [CrossRef] [PubMed]
25. Melkani, G.C.; Panda, S. Time-restricted feeding for prevention and treatment of cardiometabolic disorders. *J. Physiol.* **2017**, *595*, 3691–3700. [CrossRef] [PubMed]

26. Moon, S.; Kang, J.; Kim, S.H.; Chung, H.S.; Kim, Y.J.; Yu, J.M.; Cho, S.T.; Oh, C.M.; Kim, T. Beneficial Effects of Time-Restricted Eating on Metabolic Diseases: A Systemic Review and Meta-Analysis. *Nutrients* **2020**, *12*, 1267. [CrossRef] [PubMed]
27. Pellegrini, M.; Cioffi, I.; Evangelista, A.; Ponzo, V.; Goitre, I.; Ciccone, G.; Ghigo, E.; Bo, S. Effects of time-restricted feeding on body weight and metabolism. A systematic review and meta-analysis. *Rev. Endocr. Metab. Disord.* **2020**, *21*, 17–33. [CrossRef] [PubMed]
28. Wilhelmi de Toledo, F.; Buchinger, A.; Burggrabe, H.; Holz, G.; Kuhn, C.; Lischka, E.; Lischka, N.; Lutzner, H.; May, W.; Ritzmann-Widderich, M.; et al. Fasting therapy—An expert panel update of the 2002 consensus guidelines. *Forsch. Komplementmed.* **2013**, *20*, 434–443. [CrossRef]
29. Parr, E.B.; Devlin, B.L.; Radford, B.E.; Hawley, J.A. A Delayed Morning and Earlier Evening Time-Restricted Feeding Protocol for Improving Glycemic Control and Dietary Adherence in Men with Overweight/Obesity: A Randomized Controlled Trial. *Nutrients* **2020**, *12*, 505. [CrossRef]
30. Trepanowski, J.F.; Kroeger, C.M.; Barnosky, A.; Klempel, M.C.; Bhutani, S.; Hoddy, K.K.; Gabel, K.; Freels, S.; Rigdon, J.; Rood, J.; et al. Effect of Alternate-Day Fasting on Weight Loss, Weight Maintenance, and Cardioprotection Among Metabolically Healthy Obese Adults A Randomized Clinical Trial. *JAMA Intern. Med.* **2017**, *177*, 930–938. [CrossRef]
31. French, S.A.; Epstein, L.H.; Jeffery, R.W.; Blundell, J.E.; Wardle, J. Eating behavior dimensions. Associations with energy intake and body weight. A review. *Appetite* **2012**, *59*, 541–549. [CrossRef]
32. Schembre, S.M.; Liao, Y.; Huh, J.; Keller, S. Using pre-prandial blood glucose to assess eating in the absence of hunger in free-living individuals. *Eat. Behav.* **2020**, *38*, 101411. [CrossRef]
33. Jospe, M.R.; Brown, R.C.; Roy, M.; Taylor, R.W. Adherence to hunger training using blood glucose monitoring: A feasibility study. *Nutr. Metab.* **2015**, *12*, 22. [CrossRef]
34. Ciampolini, M.; Lovell-Smith, D.; Sifone, M. Sustained self-regulation of energy intake. Loss of weight in overweight subjects. Maintenance of weight in normal-weight subjects. *Nutr. Metab.* **2010**, *7*, 4. [CrossRef] [PubMed]
35. Ciampolini, M.; Bianchi, R. Training to estimate blood glucose and to form associations with initial hunger. *Nutr. Metab.* **2006**, *3*, 42. [CrossRef]
36. Jospe, M.R.; de Bruin, W.E.; Haszard, J.J.; Mann, J.I.; Brunton, M.; Taylor, R.W. Teaching people to eat according to appetite—Does the method of glucose measurement matter? *Appetite* **2020**, *151*, 104691. [CrossRef]
37. Ciampolini, M.; Lovell-Smith, D.; Bianchi, R.; de Pont, B.; Sifone, M.; van Weeren, M.; de Hahn, W.; Borselli, L.; Pietrobelli, A. Sustained self-regulation of energy intake: Initial hunger improves insulin sensitivity. *J. Nutr. Metab.* **2010**, *2010*, 286952. [CrossRef] [PubMed]
38. Jospe, M.R.; Taylor, R.W.; Athens, J.; Roy, M.; Brown, R.C. Adherence to Hunger Training over 6 Months and the Effect on Weight and Eating Behaviour: Secondary Analysis of a Randomised Controlled Trial. *Nutrients* **2017**, *9*, 1260. [CrossRef]
39. De Bruin, W.E.; Ward, A.L.; Taylor, R.W.; Jospe, M.R. 'Am I really hungry?' A qualitative exploration of patients' experience, adherence and behaviour change during hunger training: A pilot study. *BMJ Open* **2019**, *9*, e032248. [CrossRef] [PubMed]
40. Ertunc, M.E.; Hotamisligil, G.S. Lipid signaling and lipotoxicity in metaflammation: Indications for metabolic disease pathogenesis and treatment. *J. Lipid Res.* **2016**, *57*, 2099–2114. [CrossRef] [PubMed]
41. Rose, D.P.; Gracheck, P.J.; Vona-Davis, L. The Interactions of Obesity, Inflammation and Insulin Resistance in Breast Cancer. *Cancers* **2015**, *7*, 2147–2168. [CrossRef] [PubMed]
42. Schembre, S.M.; Jospe, M.R.; Bedrick, E.J.; Liang, L.; Brewster, A.M.; Levy, E.; Dirba, D.D.; Campbell, M.; Taylor, R.W.; Basen-Engquist, K.M. Hunger Training as a self-regulation strategy in a comprehensive weight loss program for breast cancer prevention: A randomized feasibility study. *Cancer Prev. Res.* **2021**. under review. [CrossRef]
43. Wing, R.R.; Hamman, R.F.; Bray, G.A.; Delahanty, L.; Edelstein, S.L.; Hill, J.O.; Horton, E.S.; Hoskin, M.A.; Kriska, A.; Lachin, J.; et al. Achieving weight and activity goals among diabetes prevention program lifestyle participants. *Obes. Res.* **2004**, *12*, 1426–1434. [CrossRef]
44. Stephens, S.K.; Cobiac, L.J.; Veerman, J.L. Improving diet and physical activity to reduce population prevalence of overweight and obesity: An overview of current evidence. *Prev. Med.* **2014**, *62*, 167–178. [CrossRef]
45. Gail, M.H.; Brinton, L.A.; Byar, D.P.; Corle, D.K.; Green, S.B.; Schairer, C.; Mulvihill, J.J. Projecting individualized probabilities of developing breast cancer for white females who are being examined annually. *J. Natl. Cancer Inst.* **1989**, *81*, 1879–1886. [CrossRef] [PubMed]
46. Matthews, D.R.; Hosker, J.P.; Rudenski, A.S.; Naylor, B.A.; Treacher, D.F.; Turner, R.C. Homeostasis model assessment: Insulin resistance and beta-cell function from fasting plasma glucose and insulin concentrations in man. *Diabetologia* **1985**, *28*, 412–419. [CrossRef] [PubMed]
47. Hill, N.R.; Oliver, N.S.; Choudhary, P.; Levy, J.C.; Hindmarsh, P.; Matthews, D.R. Normal Reference Range for Mean Tissue Glucose and Glycemic Variability Derived from Continuous Glucose Monitoring for Subjects Without Diabetes in Different Ethnic Groups. *Diabetes Technol. Ther.* **2011**, *13*, 921–928. [CrossRef]
48. Pendergast, F.J.; Ridgers, N.D.; Worsley, A.; McNaughton, S.A. Evaluation of a smartphone food diary application using objectively measured energy expenditure. *Int. J. Behav. Nutr. Phys. Act.* **2017**, *14*, 30. [CrossRef] [PubMed]
49. Teixeira, V.; Voci, S.M.; Mendes-Netto, R.S.; da Silva, D.G. The relative validity of a food record using the smartphone application MyFitnessPal. *Nutr. Diet.* **2018**, *75*, 219–225. [CrossRef]

50. Cienfuegos, S.; Gabel, K.; Kalam, F.; Ezpeleta, M.; Wiseman, E.; Pavlou, V.; Lin, S.; Oliveira, M.L.; Varady, K.A. Effects of 4- and 6-h Time-Restricted Feeding on Weight and Cardiometabolic Health: A Randomized Controlled Trial in Adults with Obesity. *Cell Metab.* **2020**, *32*, 366–378.e3. [CrossRef]
51. Gabel, K.; Hoddy, K.K.; Haggerty, N.; Song, J.; Kroeger, C.M.; Trepanowski, J.F.; Panda, S.; Varady, K.A. Effects of 8-h time restricted feeding on body weight and metabolic disease risk factors in obese adults: A pilot study. *Nutr. Healthy Aging* **2018**, *4*, 345–353. [CrossRef]
52. Hegedus, E.; Salvy, S.-J.; Wee, C.P.; Naguib, M.; Raymond, J.K.; Fox, D.S.; Vidmar, A.P. Use of continuous glucose monitoring in obesity research: A scoping review. *Obes. Res. Clin. Pract.* **2021**, *15*, 431–438. [CrossRef] [PubMed]
53. Connell, L.E.; Carey, R.N.; de Bruin, M.; Rothman, A.J.; Johnston, M.; Kelly, M.P.; Michie, S. Links Between Behavior Change Techniques and Mechanisms of Action: An Expert Consensus Study. *Ann. Behav. Med.* **2019**, *53*, 708–720. [CrossRef] [PubMed]
54. Liao, Y.; Basen-Engquist, K.M.; Urbauer, D.L.; Bevers, T.B.; Hawk, E.; Schembre, S.M. Using Continuous Glucose Monitoring to Motivate Physical Activity in Overweight and Obese Adults: A Pilot Study. *Cancer Epidemiol. Biomark. Prev.* **2020**, *29*, 761–768. [CrossRef] [PubMed]

MDPI

Review

Fasting and Exercise in Oncology: Potential Synergism of Combined Interventions

Rebekah L. Wilson [1,2], Dong-Woo Kang [1,2], Cami N. Christopher [1,3], Tracy E. Crane [4] and Christina M. Dieli-Conwright [1,2,*]

[1] Division of Population Sciences, Department of Medical Oncology, Dana-Farber Cancer Institute, Boston, MA 02215, USA; Rebekahl_Wilson@dfci.harvard.edu (R.L.W.); Dong-Woo_Kang@dfci.harvard.edu (D.-W.K.); Cameron_Christopher@dfci.harvard.edu (C.N.C.)
[2] Department of Medicine, Harvard Medical School, Boston, MA 02215, USA
[3] Department of Epidemiology, School of Public Health, Boston University, Boston, MA 02215, USA
[4] Division of Medical Oncology, Miller School of Medicine, Sylvester Comprehensive Cancer Center, University of Miami, Miami, FL 33146, USA; tecrane@med.miami.edu
* Correspondence: ChristinaM_Dieli-Conwright@dfci.harvard.edu; Tel.: +1-617-582-8321

Abstract: Nutrition and exercise interventions are strongly recommended for most cancer patients; however, much debate exists about the best prescription. Combining fasting with exercise is relatively untouched within the oncology setting. Separately, fasting has demonstrated reductions in chemotherapy-related side effects and improved treatment tolerability and effectiveness. Emerging evidence suggests fasting may have a protective effect on healthy cells allowing chemotherapy to exclusively target cancer cells. Exercise is commonly recommended and attenuates treatment- and cancer-related adverse changes to body composition, quality of life, and physical function. Given their independent benefits, in combination, fasting and exercise may induce synergistic effects and further improve cancer-related outcomes. In this narrative review, we provide a critical appraisal of the current evidence of fasting and exercise as independent interventions in the cancer population and discuss the potential benefits and mechanisms of combined fasting and exercise on cardiometabolic, body composition, patient-reported outcomes, and cancer-related outcomes. Our findings suggest that within the non-cancer population combined fasting and exercise is a viable strategy to improve health-related outcomes, however, its safety and efficacy in the oncology setting remain unknown. Therefore, we also provide a discussion on potential safety issues and considerations for future research in the growing cancer population.

Keywords: fasting; nutrition; exercise; cancer

Citation: Wilson, R.L.; Kang, D.-W.; Christopher, C.N.; Crane, T.E.; Dieli-Conwright, C.M. Fasting and Exercise in Oncology: Potential Synergism of Combined Interventions. *Nutrients* **2021**, *13*, 3421. https://doi.org/10.3390/nu13103421

Academic Editor: Ricardo Ribeiro

Received: 8 September 2021
Accepted: 23 September 2021
Published: 28 September 2021

Publisher's Note: MDPI stays neutral with regard to jurisdictional claims in published maps and institutional affiliations.

1. Introduction

The majority of cancer patients at diagnosis, during treatment, and while in remission will experience cancer- and treatment-related physiological and psychological side effects including, but not limited to, undesirable alterations in body composition, increase in cardiometabolic biomarkers, and reductions in quality of life [1]. In many cases, the occurrence of these physiological and psychological outcomes may be more detrimental than the cancer itself and can potentially lead to a poorer prognosis, development of other comorbidities, and pre-mature mortality [2–4]. Despite these risks, the benefits of oncology treatments frequently outweigh the risk of side effects [5,6]. While cancer- and treatment-related side effects are likely to occur for majority of patients, the risks of developing severe side effects are often related to patient characteristics such as obesity, functional status, nutritional intake, presence of comorbidities, and genetic predispositions [7–12]. With the exception of genetics, these characteristics are largely modifiable via energy balance interventions, which can provide patients with the opportunity to prevent or improve cancer- and treatment-related side effects, resulting in improved overall quality of life.

Energy balance interventions include the manipulation of exercise and dietary habits to alter a person's energy expenditure and intake, respectively. The adoption of such strategies, albeit tailored to the specific outcome desired (e.g., calorie deficit diet and high volume exercise for weight loss), is growing across various cancer populations and has been shown to improve cancer- and treatment-induced adverse changes in body composition, functional status, inflammatory environments, and prevent obesity-related comorbidities such as cardiovascular disease, type 2 diabetes, and metabolic syndrome [13,14]. Despite many studies highlighting the effectiveness of energy balance interventions for cancer patients, there is still much debate about the most appropriate prescriptions. Combined fasting and exercise is one such prescription that is growing in interest within the oncology field based on their respective independent benefits.

Fasting is the purposeful avoidance of food, and in some cases drink, for a specific amount of time. Fasting is an ancient practice and has been used for medical purposes since the fifth century B.C. with Hippocrates recommending abstinence from food or drink for patients presenting with specific symptoms of illness [15]. Today, fasting is practiced for a variety of reasons including religious, health, and ritualistic purposes (Table 1). Within oncology, fasting has been evaluated as a potential intervention to alleviate treatment-related toxicities and symptoms, and to potentially impact body composition and cardiometabolic outcomes in people with cancer [16,17]. Despite a limited number of trials with the majority including small sample sizes, studies have found fasting to be a safe intervention while receiving treatment for cancer [18–21]. Long-term fasting (e.g., several weeks/months of reduced energy intake with fasting periods >72 h), may not be feasible in oncology care due to its potential to increase risk of undesirable weight loss in cancer patients [22]. However, shorter-term fasting (e.g., interventions completed over several weeks/months utilizing fasting periods of 12 to 72 h with ad libitum feeding during fed hours) may be feasible for cancer patients. It is important to note that fasting-mimicking diets have been studied as a mechanism to improve risk factors associated with cancer-related outcomes [23,24], however, this type of fasting differs from intermittent fasting in that they do not have specified windows of eating, but rather promote a low-calorie diet with a specific meal plan, as such this type of diet is not discussed in the current review. Exercise has also been deemed a safe and feasible intervention within various oncology settings with the majority of current evidence in breast, prostate, lung, and colorectal patients [14]. While not yet a consistent strategy utilized within oncology standard of care, there is strong evidence indicating exercise improves health-related quality of life, various psychosocial outcomes such as anxiety and depression, cancer-related fatigue, and cardiorespiratory fitness and muscular strength among various cancer patients on treatment and in remission [25,26]. Furthermore, although preliminary, several studies have reported that exercise may play a role in improving treatment tolerance and efficacy (e.g., relative dose intensity and treatment delivery, tumor size, and long-term survival) [27–31].

Table 1. Definitions of different types and concepts of fasting and exercise.

Type/Concept	Definition
Fasting-related	
Intermittent energy restriction	Restricting energy intake to approximately 60–75% below energy requirements for short periods, followed by periods with normal energy intake. One example is the 5:2 diet, consisting of approximately 5 days of eucaloric (a diet that provides the number of calories to maintain your body weight) feeding and approximately 2 days of a very-low-calorie diet per week.
Long-term fasting	Temporarily fasting, typically for a period >72 h.
Short-term fasting	Temporarily fasting, typically for a period between 12 and 72 h. An example of this type of fasting is alternate day fasting.
Time-restricted feeding	Reducing food intake to a set number of hours each day (e.g., eating in a <10 h daily period). One method of time restricted feeding is Prolonged overnight fasting whereby time-restricted feeding occurs overnight.

Table 1. *Cont.*

Type/Concept	Definition
	(Alternate definition) the practice of consuming ad libitum energy within a restricted window of time and fasting thereafter (upwards of 12–16 h).
Religious fasting	Intermittent fasting exists in some religious practices. These include the Black Fast of Christianity most often practiced during Lent, Varta (Hinduism), Ramadan (Islam), Yom Kippur and other fasts (Judaism), Fast Sunday (Latter-day Saints), Jain (Buddhist) fasting. Religious fasting practices may only require abstinence from certain foods or last for a short period of time and cause negligible effects.
Fasting-mimicking diet	Maintaining a fasting-like state by periodically consuming a very-low-calorie, low-protein diet (not necessarily fasting)
Exercise-related	
Exercise	Planned and structured, and repetitive bodily movement in order to improve or maintain physical health outcomes [32].
Physical activity	Any bodily movement produced by skeletal muscles that results in energy expenditure [32].
Physical inactivity	Not performing sufficient amounts of moderate- and vigorous-intensity activity (MVPA), i.e., not meeting specified physical activity guidelines [33].
Sedentary behavior	Any waking behavior characterized by an energy expenditure $\leq$1.5 metabolic equivalent tasks (METs) while in a sitting, reclining or lying posture [33].

As a combined entity, fasting and exercise have been shown to induce synergistic effects on improved metabolic outcomes such as body composition, cholesterol, and insulin sensitivity in non-cancer populations [34]; although much of the research examining the interaction of fasting and exercise has been carried out in sports performance and acute settings [35]. However, there is a growing interest for how combined fasting and exercise interventions may be optimized for health and therapeutic benefits in oncology settings, based on the current evidence in the non-cancer population and their independent benefits established within the cancer population [14,35–37]. In this review, we summarize the current evidence on the independent effects of fasting and exercise in cancer settings and discuss the potential impacts and mechanisms of combined fasting and exercise interventions on cardiometabolic, body composition, patient-reported outcomes, and cancer-related outcomes. We also discuss the potential safety issues of combined exercise and fasting in cancer patients and suggest considerations for future research in this setting.

2. What Metabolic Changes Occur during Fasting and Exercise?

The combination of fasting and exercise can drastically change how our bodies utilize and synthesize fuel sources. It is important to understand the physiological changes that occur during this state before identifying the potential beneficial outcomes that a combined fasting and exercise strategy may induce among cancer patients.

The act of consuming food provides our bodies with nutrients that are broken down and utilized as fuel in order to survive. However, when food is not supplied, the body relies on processes of biosynthesis as well as stored glycogen, fats, and proteins as metabolic fuel substrates [38]. Therefore, the metabolic substrates, and their catabolic or anabolic pathways, differ in a fed state from a fasted state and is likely one of the key mechanisms responsible for many of the changes observed when undertaking fasting compared to a fed intervention [39,40]. In a fed state, the body predominantly utilizes glucose from the recently consumed meal as the primary source of fuel via glycolysis, oxidative phosphorylation, and carbohydrate oxidation [38,41,42]. In a fasted state, glucose and glycogen stores are depleted, so fats become the primary source of fuel via lipolysis and fat oxidation [38,42,43]. Gluconeogenesis and ketogenesis are also increased in a fasted state to ensure the homeostasis of organs that only use glucose or ketones as fuel [38].

The preferential fuel source of the body is further altered when exercise is included and is dictated by exercise completed in a fed or fasted state, the type of meal consumed before exercise, and the intensity and duration of exercise [44,45]. For example, when exercising in a fed state, glucose is the predominant source of fuel; however, when compared to consuming a meal with a high glycemic index (GI), a low GI meal is associated with lower rates of glycolysis [44]. Fuel for exercise in a fasted state comes from increased fat oxidation, particularly the breakdown of intramyocellular triacylglycerol (IMTG) [43]. Regardless of a fed or fasted state, the metabolic substrate utilization is similar at intensities >70% of VO_{2max} or durations of continuous exercise >2 h [45]. These acute differences in metabolic substrate utilization highlight that timing of a fasting period and exercise bout, when undertaking a combined fasting and exercise intervention, may be of critical importance as it is unclear how timing will impact the long-term benefits of a combined intervention [36].

3. Effect of Fasting, Exercise, and Combined Fasting and Exercise

The potential mechanisms and impacts of combined fasting and exercise in cancer patients are illustrated in Figure 1. Briefly, exercise during a fasted state in cancer patients can maximize glucose regulation, lipid oxidization, and systemic inflammation to improve each of the suggested outcomes including cardiometabolic markers, body composition, patient-reported outcomes, and cancer-related outcomes. We discuss below the independent mechanisms and effects of fasting and exercise and then the effects of potential benefits of combined fasting and exercise by each outcome.

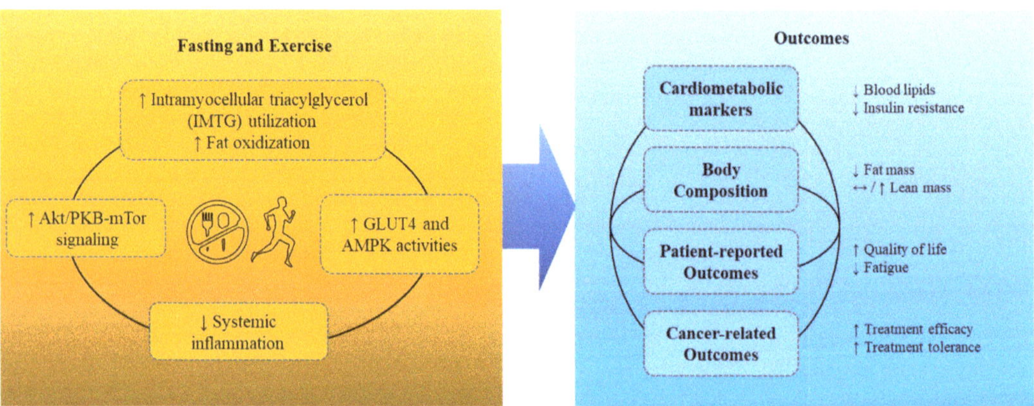

Figure 1. Potential mechanisms and impacts of combined fasting and exercise in cancer patients. Up arrow refers to an increase in the component, down arrow refers to a decrease, and the side ways arros refer to no change.

3.1. Cardiometabolic Biomarkers

With advances in diagnosis programs, treatments, and general awareness of cancers and their signs and symptoms, the 5-year survival rates of many cancer types have improved over recent decades [46]. However, systemic cancer treatments such as chemotherapy and hormonal therapy, may induce cardiometabolic dysregulation (e.g., insulin resistance, dyslipidemia), which can also lead to the development of other comorbidities, poorer quality of life, and sometimes pre-mature mortality [4]. Therefore, it is important to identify interventions that can improve cardiometabolic biomarkers such as insulin-related markers and lipid profiles.

3.1.1. Fasting

As previously discussed, when in the fasted state the body transitions to utilizing glycogen stores for energy in an effort to maintain glucose homeostasis. Insulin-like growth factor-1 (IGF-1) is a primary mediator of growth hormone and has significant metabolic effects. Obesity has been attributed to 15–20% of cancer-related deaths where obese individuals often present with higher levels of IGF-1 which has been identified as a potential mechanism associating obesity with increased cancer risk and disease progression [47–49]. Several studies utilizing short-term fasting have demonstrated a reduction in IGF-1 with some also observing a reduction in insulin [50]. In women with breast cancer, data analysis from the Women's Healthy Eating and Living Study showed that each 2 h increase in overnight fasting was associated with a significant reduction in hemoglobin A1C (HbA1c) ($\beta = -0.37$; 95% CI, -0.72 to -0.01) [51]. A growing body of literature suggests short-term fasting may play a role across the cancer continuum to improve outcomes, such as treatment toxicity and efficacy, through normalization of metabolic markers, with several clinical trials of short-term fasting underway in cancer patients [21]. Importantly, obesity and high levels of insulin and IGF-1, in addition to a diagnosis of diabetes mellitus are associated with worse survival in cancer [52–55]. While the current evidence is mostly focused on insulin and glucose pathways, the effects of fasting on lipid profiles are underexplored in cancer settings, which warrants further investigations.

3.1.2. Exercise

Cardiometabolic changes are widely studied within the oncology field as many systemic cancer treatments result in side effects that alter metabolic homeostasis (e.g., androgen deprivation therapy causing insulin resistance) [56,57]. Such metabolic dysregulations can have a devastating effect and lead to an increased risk of cancer progression, most commonly documented for breast, colorectal, prostate, and endometrial cancers, and development of other comorbidities, in direct and indirect mechanisms including obesity [58,59]. Aerobic exercise increases the rate of glucose uptake into the contracting skeletal muscles, up to 2–3 fold, primarily through the regulation of GLUT4 glucose transporters, which improves insulin sensitivity and lowers circulating insulin levels [60]. Resistance training that induces muscle hypertrophy and qualitative adaptation can also improve insulin resistance by enhancing the expression of glucose transporters and mitochondrial oxidative capacity [61]. More evidence, therefore, supports the synergistic effects of aerobic and resistance training on insulin sensitivity and glucose metabolism [62]. For blood lipid levels, oxidation of triglycerides and fatty acids progressively increases during aerobic exercise to generate energy sources, especially at a lower intensity (~45% of VO_{2max}), however, this process heavily depends on the individuals' fitness and the rates of other substrate oxidation at different exercise intensities [63]. Resistance exercise may elicit lower energy expenditure during a single bout of exercise than aerobic exercise [64], however, the training effects of resistance exercise, including maintenance or improvement of muscle mass and density, higher resting metabolic rate, and increased fat metabolism, also contribute to the reduction in circulating lipid levels (e.g., reduced total cholesterol, low-density lipoprotein (LDL), and triglycerides, and increased high-density lipoprotein (HDL)), which further establish the additive benefits of combined aerobic and resistance exercise [65,66].

3.1.3. Combined Fasting and Exercise

There is no current evidence describing the effect combined fasting and exercise interventions have on lipid levels in the cancer population. Therefore, we draw on non-cancer examples to propose the potential benefits a combined intervention may have on cardiometabolic outcomes of cancer patients. Exercising in the fasted state, compared to a fed state, appears to be more effective at manipulating lipid levels given the increase in lipoprotein lipase activity in the non-cancer population [67]. Bhutani et al. [68] assessed the effect of a 12-week combined fasting and exercise intervention on lipid levels of non-cancer obese participants utilizing aerobic exercise and alternate day fasting where 25%

of their daily energy intake was consumed on the "fast" days between 12–2 p.m. and ate ad libitum on the "fed" days. The authors reported no change in total cholesterol or triglyceride concentrations for any group; however, the combined fasting and exercise group significantly reduced LDL ($-12 \pm 5\%$) and increased HDL ($18 \pm 9\%$) concentrations, compared to no change in the fasting-only, exercise-only, and control groups. The combined fasting and exercise group also demonstrated favorable changes in the proportion of small LDL and HDL particles, further emphasizing the cardio-protective effects of this type of intervention, compared to either fasting or exercise alone. In contrast, Cho et al. [69] did not find any changes in total cholesterol, LDL, and HDL between groups, although they did employ a shorter intervention of 8 weeks utilizing combined aerobic and resistance exercise but had a similar alternate day fasting regime as Bhutani et al. [68]. However, Cho et al. [69] did report a significant difference in triglyceride concentrations where combined fasting and exercise-induced a decrease, compared to an increase for the control group. Both studies significantly decreased fat mass, so the change in lipid levels cannot be attributed to the intervention itself, but is potentially dependent on fat mass change. This has been suggested to be the case among cancer patients where men with prostate cancer on androgen deprivation therapy (ADT) undertaking an exercise-only study demonstrated an improvement in triglycerides, but was dependent on a loss in fat mass [70]. Further research is required into the best prescription to manipulate lipid levels in cancer patients, particularly as it relates to the timing of a meal in a fasting and exercise intervention, and if a loss in fat mass is required to alter the lipid profile.

Evidence suggests that an increase in total fat mass, with a particular impact of high concentrations of IMTG, is a significant contributor to the development of insulin resistance in both cancer and non-cancer populations and could be a key target area when trying to improve metabolic outcomes [36,71–73]. The timing of an exercise bout in relation to the fed-period may be of high importance when trying to optimize the impact of a combined fasting and exercise intervention on fat mass and IMTG to improve insulin resistance and glycemic control given the changes in substrate utilization experienced in the varying fasted, fed, and exercising states [67]. Combined fasting and exercise can significantly reduce total fat mass, as described later in the body composition section [68,69]. Similarly, an acute bout of exercise in a fasted state, compared to a fed state, can induce a ~60% depletion in IMTG in type I muscle fibers, as it is a readily available source of fats for fuel [43,74]. Consequently, these superior changes in fat mass and IMTG may contribute to desired improvements in insulin resistance. For example, within the non-cancer population, Cho et al. [69] reported no between-group differences in HOMA-IR, a measure of insulin resistance, across four assessed groups (combined fasting and exercise, fasting-only, exercise-only, and control), although there was a trend for a between-group difference at baseline ($p = 0.063$) where those in the combined intervention group had a mean score of 2.23 ± 0.72, and the control group 0.97 ± 0.45. However, over the 8-week intervention, the combined fasting and exercise group had a non-significant mean decline of -1.12 ± 0.68, whereas the control group had a significant 1.10 ± 0.97 increase. While there are no standardized cut-off points for insulin resistance as defined by HOMA-IR, a score <1.5 has consistently been identified as insulin sensitive with superior metabolic outcomes [75–78]. Therefore, the observed change, while not statistically significant, may be of clinical relevance and needs to be further evaluated, particularly within the context of fat mass loss given the combined fasting and exercise group lost significantly more fat mass than the control (-3.2 ± 0.5 versus -0.3 ± 0.8 kg), although IMTG was not examined. Furthermore, it is unclear if the exercise bout in the combined fasting and exercise group was completed on fasting days only, feeding days only, or a combination. Understanding the timing of fasting and exercise as well as meal consumption prior to exercise may provide further insights into why we see participants who respond and do not respond to lifestyle interventions and is an important factor to consider during a combined fasting and exercise intervention.

Another potential mechanism for combined fasting and exercise to improve insulin resistance and glycemic control is through the increase in GLUT4 protein and AMPK activity. Van Proeyen et al. [79] recruited young healthy male participants who all undertook a 6-week hypercaloric high-fat diet and were randomized to either high-fat diet control, exercise in a fasted state, or exercise in a carbohydrate-fed state where both exercise groups completed 4 aerobic exercise sessions per week at ~70–80% heart rate maximum. The authors reported that exercise in a fasted state alleviated the negative effects of a high-fat diet on glucose tolerance and insulin sensitivity, and was attributed to the increase in GLUT4 protein and AMPK activity in the fasted exercise group, compared to no change in the carbohydrate-fed exercise, and high-fat diet control groups. However, it is unclear if similar long-term results would be observed if the carbohydrate-fed exercise group were to also undertake a fasting component, yet, still eat prior to exercise. Nevertheless, this study highlights that in the absence of healthy dietary advice and an iso- or hypocaloric state, undertaking exercise after an overnight fast has a positive effect on insulin sensitivity and glycemic control. This type of intervention is worth further exploring for cancer patients who may be eating poorly and do not want to change their dietary intake given they are already undergoing a substantial number of changes, as such a fasting component with no dietary advice may result in increased compliance to the intervention. However, the benefit of improving dietary intake should not be completely dismissed.

3.2. *Body Composition*

An increase in fat mass and decline in lean mass loss often occurs in cancer patients as a result of hospitalization and extended bedrest, increased stress-related eating, decreased physical activity, or as a result of treatment [80,81]. Within the cancer population, the quantity and distribution of fat mass and lean mass are influential in the effectiveness of treatment, development and severity of cancer- and treatment-related side effects, and progression of cancer [80,82–84]. Changes in fat mass and lean mass can also play critical roles in the development of cardiometabolic outcomes, previously described, and contribute to the development of comorbidities and further reducing the health and well-being of a cancer patient [2–4]. Therefore, implementing lifestyle-based intervention strategies, such as fasting and exercise, to improve body composition, or prevent cancer- or treatment-induced worsening of body composition, is critical to a cancer patient's care. Here, we discuss the impact fasting, exercise, and combined fasting and exercise interventions have on fat mass and lean mass among cancer patients and the potential mechanisms involved.

3.2.1. Fasting

Body composition changes as they relate to fasting appear to be associated with the length of time fasting occurs. In a pilot crossover study among cancer patients undergoing chemotherapy, comparing cycles of short-term fasting to normocaloric diet, a significant loss in mean fat mass (measured by bioelectrical impedance) (-0.63 ± 0.23; 95% CI -1.09–(-0.17); $p = 0.008$) was observed which lead to a significant weight loss during moderate short-term fasting (-0.84 ± 0.26; 95% CI -1.35–(-0.33); $p = 0.002$) [50]. Aside from fat mass, body composition remained stable. Mean body weight and mean fat mass were 71.4 ± 12.3 kg and 23.0 ± 8.8 kg at the beginning and 69.8 ± 11.6 kg and 21.4 ± 8.4 kg at the end of the intervention, respectively [50]. Other studies corroborate these findings for lack of weight change using short-term fasting [16]. In these studies, after the fasting days, an increase in weight was commonly resulting in achievement of baseline weight [16,50,85]. These early studies suggest interventions that include short-term fasting carry a low risk of negatively impacting body composition and therefore likely the better option for cancer patients, particularly those at risk of weight loss leading to a poor prognosis (e.g., lung cancer).

3.2.2. Exercise

The changes in body composition followed by exercise training have been widely investigated in cancer settings. Particularly, most exercise oncology research with body composition outcomes has focused on lean mass in prostate cancer patients receiving

ADT [86], where patients often experience significant declines in lean mass [87,88]. A recent meta-analysis of 21 clinical trials in prostate cancer patients reported that exercise, primarily resistance exercise, significantly reduced fat mass by 0.6 kg and increased lean mass by 0.5 kg after a mean intervention period of 20 weeks [89]. Resistance exercise is effective in improving, or at least maintaining, lean mass by counteracting impaired anabolic signal pathways and inhibiting the cellular atrophy mechanism during ADT [90]. Fat mass has been more commonly investigated in patients with breast or colon cancer given the strong links between adipocytokines/obesity-related markers and these cancers [91]. Although several studies have demonstrated the significant loss of fat mass after exercise [92–96], this is commonly a result of the control group continuing to increase fat mass as opposed to the exercise intervention inducing a significant fat mass decline. Additionally, fat mass is more commonly a primary outcome of interest in combined physical activity (e.g., meeting physical activity guidelines) and dietary studies [97,98]. Overall, there is generally a lack of evidence on body composition outcomes other than ADT settings, and the findings are not consistent and heavily depend on the modes and intensities of exercise [86,99].

3.2.3. Combined Fasting and Exercise

Given the strong connection between obesity, or excess fat mass, and cancer development and progression, weight loss, or more importantly fat mass loss, is often a key consideration as part of a cancer patient's care [100]. By combining fasting and exercise, there is some evidence that such an intervention will have a synergistic effect on fat mass loss due to increased fat oxidation and energy expenditure over intake [68,69]. Within the non-cancer obese population, two randomized control studies have been conducted where body composition changes were compared between four groups: combined fasting and exercise, fasting-only, exercise-only, and a control group [68,69]. Cho et al. [69] reported fat mass to significantly decrease in both the combined intervention (-3.2 ± 0.5 kg) and fasting-only (-3.2 ± 0.6 kg) groups compared to control (-0.3 ± 0.8 kg). The exercise group also significantly decreased fat mass (-1.7 ± 0.5 kg) compared to baseline, but not the control group. While Bhutani et al. [68] also reported both the combined intervention and fasting-only groups to reduce fat mass, the combined intervention group had a superior loss (-5.0 ± 1.0 kg versus 2.0 ± 1.0 kg). While the studies utilized similar alternate day fasting regimens, they differed in exercise modes (combined aerobic and resistance versus aerobic-only) and lengths of intervention (8 versus 12 weeks), which likely contributed to the variation in fat mass results. These studies indicate the potential feasibility of a combined fasting and exercise intervention to improve fat mass in obese individuals, but how this translates to the obese cancer population is unknown. It must be highlighted that combined fasting and exercise can induce fat mass loss, independent of total weight loss, by prescribing an energy balance or surplus during feeding hours [79,101]. This is important for the cancer population as weight loss can sometimes be a red flag for poor prognosis (e.g., cachexia), or is not recommended during treatment such as radiation therapy as it could result in day-to-day movement of organs, therefore, decreasing radiation accuracy if image-guided radiation is not used [102,103]. Further research is required into the best prescription of combined fasting and exercise interventions for cancer patients and its effect on fat mass changes both dependent and independent of total weight loss.

The pathways involved in lean mass changes (e.g., Akt/PKB-mTor signaling) are down-regulated in fasting-only interventions leading to an increase in muscle protein breakdown; however, with the addition of exercise, Akt/PKB-mTor signaling is reactivated, leading to lean mass maintenance, although this has only been shown in murine models [34]. The ability to significantly increase lean mass while undertaking a combined fasting and exercise intervention is unclear. Resistance exercise and protein supplementation are known strategies to induce lean mass hypertrophy [104,105]. However, while in an energy deficit state, anabolic suppression (i.e., a blunted training response) during resistance training has been previously demonstrated even in the presence of protein supplementation and adequate daily protein intake of 1.2 $g \cdot kg^{-1}$ body weight and may explain the lack of lean mass hypertrophy reported in combined fasting and exercise interventions [106,107]. In the same way, fasting and exercise interventions may be able to reduce fat mass independent of weight loss by manipulating energy intake during fed hours, the same concept may apply to achieving an increase in lean mass. Tinsley et al. [108] examined non-cancer resistance-trained females and compared three groups over 8 weeks: combined fasting and resistance exercise with a calcium β-hydroxy β-methylbutyrate supplement, combined fasting and resistance exercise with a placebo, where both groups undertook time-restricted fasting regimens, and a non-fasting control diet with a placebo. All groups also received daily protein supplementation. The study demonstrated a significant increase in lean mass (1.0–1.4 kg) over the 8-week period for all groups compared to baseline with no between-group differences. All groups significantly increased their total energy intake during the intervention, which may have contributed to the significant increase in lean mass observed in contrast to other combined fasting and exercise studies that had an energy balance or deficit [40,68,69]. Moreover, this study was conducted in young, trained female participants and its applicability to the male, un-trained, and cancer populations is limited. Further research is required to examine the best prescription of exercise mode, nutrient supplementation, and total energy intake to induce a significant increase in lean mass while undergoing a combined fasting and exercise intervention.

3.3. *Patient-Reported Outcomes*

Pain, fatigue, anxiety, depression, and sleep disturbances are among the most commonly identified detrimental patient-reported outcomes of cancer and cancer-related treatment [109–111]. Here, we discuss the impact of fasting, exercise, and combined fasting and exercise-based interventions have on these outcomes in cancer patients and the potential mechanisms involved.

3.3.1. Fasting

Commonly experienced symptoms as a result of cancer and its treatment including fatigue, gastrointestinal disturbances, and pain have all been preliminarily examined as potential patient-reported outcomes that may be improved as a result of short-term fasting. A case series of 10 patients with various types of cancer demonstrated that fasting in combination with chemotherapy is feasible and eluded to the potential for fasting to reduce fatigue, weakness, and gastrointestinal side effects [20]. In a pilot study among breast and ovarian cancer patients undergoing chemotherapy, women randomized to either undergo short-term fasting in the first half of their chemotherapy cycle followed by their usual diet or vice versa with short-term fasting followed in the second half of the chemotherapy cycles. For both groups in the fasted state, quality of life and fatigue scores both improved [16]. A pilot study by Zorn et al. [50] found a modified short-term fast during chemotherapy reduced stomatitis, headaches, weakness, and overall total toxicities score. Although there have been a limited number of studies, initial findings of the impact of fasting on patient-reported outcomes in cancer are promising.

3.3.2. Exercise

There is strong evidence on the benefits of exercise on numerous patient-reported outcomes during and after cancer treatment [25], such as health-related quality of life [112,113], cancer-related fatigue [114,115], and anxiety and depression [116]. The mechanisms of the positive impacts of exercise on psychosocial distress may include direct psychological interpositions, such as providing a constructive distraction and reducing time on rumination, directing energy positively, and improving the feeling of control over cancer [117], as well as biological pathways, such as releasing β-endorphins and circulating levels of neurotrophic factors (BDNF) [118]. For cancer-related fatigue, engaging in exercise, although counterintuitive, plays a significant role in reducing acute and chronic fatigue, which is superior compared to paratheatrical agents or psychological interventions [115]. Potential mechanisms include reducing elevated pro-inflammatory cytokines, normalizing circadian rhythm dysregulation, and improving impaired muscle oxidative capacity [119].

3.3.3. Combined Fasting and Exercise

Independently, both fasting and exercise are associated with improved patient-reported outcomes (e.g., quality of life, fatigue, depression), in both cancer and non-cancer populations [16,120]. However, the effect of a combined fasting and exercise intervention on patient-reported outcomes is not well described. The study by Albrecht et al. [37], described further in the cancer-related outcomes section, is the only study that describes the effect of combined fasting and exercise on patient-reported outcomes among cancer patients. The ovarian cancer patient that was examined in this case study reported an improvement in feelings of anxiety, perceived stress, and emotional functioning. While this case study highlights the potential for a combined fasting and exercise intervention to improve patient-reported outcomes, it cannot be dismissed that an improvement was observed due to the feeling of hope that may have come from entering a study with the intention of improving disease outcomes.

3.4. *Cancer-Related Outcomes*

Lifestyles that contain an increased amount of physical activity and the consumption of a healthy diet are well-established modifiable factors that decrease a person's risk of cancer development [121]. Given this relationship, research has increased in examining the role of exercise and nutrition after a cancer diagnosis in the progression of cancer and the effectiveness of cancer-related treatment. Termed the Warburg Effect, cancer cells rely on aerobic glycolysis deriving most of their energy from glucose converted to lactate for energy followed by lactate fermentation, even when oxygen is available [122]. As such, a shift in energy metabolism from glycolytic metabolism to oxidative phosphorylation, which occurs in a fasted state, may be a means by which cancer growth rate is altered [123]. Therefore, combined fasting and exercise has the potential to provide this needed change in metabolism to combat cancer [21,124]. Here, we discuss the impact fasting, exercise, and combined fasting and exercise-based interventions have on cancer progression and recurrence, and treatment tolerance and effectiveness and the potential mechanisms involved.

3.4.1. Fasting

Broadly in humans, studies of long-term calorie restriction, including or excluding long-term fasting periods, have demonstrated a reduction in metabolic and hormonal factors associated with cancer risk [125–127]. However, long-term fasting (e.g., >72 h) is not practical in the oncology care space as it may lead to unacceptable weight loss in cancer patients [22]. Short-term fasting (e.g., 12–72 h) may be feasible for cancer patients. In mice, shorter periods of fasting have been shown to slow cancer growth as effectively as long-term fasting without compromising body weight [128–130] with the effects of the short-term fasting improving differential stress response between healthy somatic cells and cancer cells [19,128,129,131,132]. The mechanism by which this is occurring is through a protective response in healthy cells wherein nutrient deprivation (fasting) shuts down pathways promoting growth in order to provide energy in maintenance and repair pathways that contribute to resistance to chemotherapy, a phenomenon knowing as 'differential stress resistance' [16,133,134]. Alternatively, due to mutations in oncogenes, tumor cells are unable to activate this protective response because of uncontrolled activation of growth pathways. In order for tumors to maintain their high rate of growth, an abundance of nutrients are required and thus short-term fasting leads to increased sensitivity of tumor cells to chemotherapy [128–130]. This increased sensitivity is hypothesized to be a promising strategy to enhance the efficacy and tolerability of chemotherapy [19]. For example, in another study examining the feasibility of dose escalation fasting (24, 48, and 72 h) over the course of a chemotherapy cycle, patients who fasted for $\geq$48 h had a trend towards reduced neutropenia compared to patients who only fasted for 24 h periods with the 48 h fasting group also reducing leukocyte damage [18]. In a pilot study of short term fasting in HER-2 negative breast cancer patients, those who were randomized to the short term fasting intervention, compared to unfasted women, experienced reduced hematological toxicities 7 days post-chemotherapy administration (p = 0.007, 95% CI 0.106–0.638 and p = 0.007, 95% CI 38.7–104 for erythrocyte and thrombocyte counts, respectively) [19]. Patients undertaking short-term fasting, compared to non-fasted patients, have also been shown to have fewer postponements of chemotherapy [50]. Finally, in a secondary analysis of women with breast cancer participating in the Women's Healthy Eating and Living Study women who fasted <13 h/night had a 36% increased risk of recurrence (HR, 1.36; 95% CI 1.05–1.76) compared to those who fasted ≥13 h per night [51].

As use of immunotherapy increases in oncology, fasting demonstrates some promise in preclinical studies as a potential modality to bolster antitumor immunity. Prolonged overnight fasting was found to reduce IGF-1 levels and protein kinase A activity in a variety of cell populations in mice leading to signal transduction changes in long-term hematopoietic stem cells [135]. Further, multiple cycles of fasting lessened immunosuppression and chemotherapy-induced mortality. In both in vivo and in vitro studies in mice with colorectal cancer, alternate day fasting for two weeks inhibited tumor growth without causing a reduction in body weight, suppressed M2 polarization of tumor-associated macrophages inhibiting tumor growth through decreased levels of adenosine, and increased autophagy of tumor cells [135,136]. Further research is required into the differing effects that such types of fasting may have on cancer-related outcomes.

3.4.2. Exercise

A body of preclinical evidence has demonstrated the direct impacts of exercise in suppressing tumor progression and metastasis [137,138], yet, evidence within the clinical setting is lacking. The underlying mechanisms are still unclear, however, several plausible mechanisms include the acute increase in the concentrations of immune cells (e.g., natural killer cells, monocytes, and neutrophils), the muscle-to-cancer crosstalk through muscle contraction-derived cytokines (e.g., interleukin-6 and SPARC), and the downregulation of tumorigenesis pathway through catecholamine (e.g., epinephrine). These mechanisms also interdependently suppress tumor growth by enhancing mobilization and redistribution of cytotoxic immune cells into the tumor cells [139,140]. Another mechanism that has been identified in which exercise can improve cancer-related outcomes is through increased tumor vascular permeability and angiogenesis. Recent preclinical studies showed that repeated bouts of aerobic exercise enhanced treatment efficacy and thereby suppressed clinical tumor progression by improving tumor vascular permeability and angiogenesis, which caused oxygen delivery and drug penetration into tumor cells [141,142]. This mechanism is plausible as hypoxic status is one of the key characteristics of tumor microenvironment (TME), which increases treatment resistance to the tumors and can be reversed by the improvements of vascular functions during aerobic exercise. Lastly, emerging evidence has demonstrated that maintaining or improving lean mass during chemotherapy may improve chemotherapy tolerance and completion in cancer patients [27,143]. Systemic cancer drugs are primarily distributed and metabolized (i.e., pharmacokinetics) by blood flow and perfusion in lean tissues, however, treatment dosages are typically determined by estimated total body surface area (BSA) without considering body composition [144,145]. Cancer patients with identical BSA may present substantial differences in body composition, which is associated with chemotherapy toxicity and efficacy [146]. Therefore, resistance training to improve body composition (i.e., increased lean mass and decreased fat mass) as well as potentially muscle quality (i.e., reduced IMTG content) [88] poses a great potential to enhance treatment outcomes. Nevertheless, only preliminary evidence exists and very little is known about how exercise may mediate the response to cancer therapy in patients, where further preclinical and clinical exercise research is warranted.

3.4.3. Combined Fasting and Exercise

Despite the independent impacts of fasting and exercise, the effect of a combined fasting and exercise intervention on cancer-related outcomes is largely unknown. To our knowledge, only one study has been conducted in the cancer population that utilized a combined fasting and exercise intervention [37]. This proof of concept case study, which examined a woman with recurrent stage III ovarian cancer in a watch and reevaluate phase of treatment, evaluated the effect of the intervention on ovarian tumor growth as well as health-related quality of life and psychological symptoms. The intervention involved an 18 h fast, low-fat meal, flaxseed oil and caffeine supplements, and 90 min of treadmill walking repeated daily across a 3-day period where the patient was housed in a research facility and completed once a month for 3 months. This intervention was selected to slow cancer progression based on the proposed mechanisms where it would create the best environment to induce the optimal free fatty acid (FFA) level of 1 to 2 nM maintained over an extended period of time, and that unsaturated fats (flaxseed oil) has a cytotoxic effect having been demonstrated in preclinical studies [147,148]. On the days where emesis did not occur, four out of seven of the study days, FFA concentrations reached this desired level for $\geq$4 h. However, CA125, a marker used to monitor ovarian cancer progression, continued to increase over the course of the study period, although a computed tomography scan indicated no sign of cancer progression. Given the study design, conclusions about the effect a combined fasting and exercise intervention has on tumor outcomes is limited. The role FFAs play in cancer prognosis is complex and the mechanisms are not fully understood [149]. Further research is required into the previously identified independent mechanisms of fasting and exercise, and how combining these interventions may have a superior, synergistic effect, in altering TME and treatment tolerance and efficacy.

4. Safety with Intervention Implementation

Though fasting and exercise have independently been shown to have low adverse events and are generally safe in cancer populations, intervention safety should be addressed for future research and implementation. When various periods of fasting were utilized prior to and up to 24 h post-chemotherapy, patients commonly reported negative symptoms including headaches, nausea, dizziness, and fatigue, though these were not severe enough to be considered an adverse event [18]. Additionally, when intermittent fasting is not managed, it can cause malnutrition, eating disorders, and severe damage to organs [150]. It is unclear how exercise in combination with fasting may escalate these negative outcomes among cancer patients, particularly during treatment. Furthermore, the combination of exercise and fasting may be detrimental in maintaining body composition for patients who already have a low BMI or cachexia. The risks of being underweight, compounded with the possible combined impact of fasting and exercise on weight loss and fat mass loss, may further impair treatment efficacy and result in a poorer prognosis [102,103]. Given these safety concerns, it is crucial that future studies are thoroughly designed to mitigate these risks and to promote the prospective desired health benefits of fasting and exercise among cancer patients.

5. Future Research and Key Considerations

The sparsity of research with multimodal fasting and exercise interventions among cancer survivors lends to a plethora of future investigations to improve research in this area.

5.1. Timing of Intervention Delivery

While studies have shown that exercise after an overnight fast has beneficial effects on insulin sensitivity and glycemic control, it is not clear how the timing of treatment may interact [36,79]. Long-term impacts of the timing of this relationship of fasting and exercise are not well established with respect to treatment. It is also not clear at what point in the cancer diagnosis trajectory that this combined intervention may be most beneficial. Perhaps an opportune time to intervene includes an emphasis on the pre-surgical window.

For example, there may be a benefit to a staggered approach of intermittent fasting prior to surgery followed by exercise or there may be synergy between the two modalities that would prove advantageous to improving the TME between diagnosis and surgery.

5.2. Alternative Intervention Modalities

Consideration of the individual components of a lifestyle intervention and how they are prescribed to best support cancer patients and health outcomes is a key element when considering how to prescribe fasting and exercise. Within the recent Physical Activity Guidelines for Americans, a primary recommendation is to break up a prolonged period of sedentary activities by sitting less and moving more [151]. An intervention that focuses on reducing sedentary behaviors may be easier to implement and be more appealing than a strictly supervised exercise prescription. Likewise, the implementation of fasting, where the patient has to limit their food intake for a certain period of time, as opposed to changing the type of food they consume, may be more appealing and easier to adhere to.

5.3. Treatment and Diagnosis Considerations

Intervention effects may vary by type of cancer diagnosis and cancer-related treatments. Additional scientific exploration requires investigating the effect by diagnosis given the variability in symptom management (i.e., cancers of the gastrointestinal system). Furthermore, variability in intervention benefits may alter based on pre-existing chronic conditions whereby more vulnerable cancer patients with comorbidities such as diabetes or cardiovascular disease may experience a greater benefit. Cancer-related treatment history is of further consideration as said treatments may negatively alter lifestyle behaviors and increase risk of comorbid conditions providing an opportunity to intervene with a combined fasting and exercise approach.

5.4. Cultural Relevancy/Religious Considerations

The mechanisms of fasting in culture are not novel. As mentioned in Table 1, the practice of fasting exists in a variety of religions and cultures. While a few studies have utilized combined fasting and exercise interventions during Ramadan, the impacts of the combined intervention are not clear [40]. To our knowledge, there is a lack of investigations focusing on other types of religious fasting in combination with exercise in our literature search. Therefore, future research needs to consider cultural fasting practices when designing lifestyle intervention studies.

5.5. Age Considerations

The impact of fasting and exercise interventions among cancer patients across the lifespan with particular focus on adolescent and young adults, and older cancer survivors warrants investigations [40,79]. The combined impact of exercise and fasting may be particularly impactful for these more vulnerable, understudied populations at high risk for poor cancer outcomes, premature aging, and exacerbated comorbidities [40].

5.6. Ongoing Trials

Few ongoing trials are underway examining the impacts of combined exercise and fasting among various populations with only one study focusing on cancer patients (Table 2). The intervention designs were heterogeneous, varying in the number of days per week fasting is incorporated, duration of fasting (i.e., number of fasted hours per day), and modalities of exercise (i.e., aerobic, resistance, or both) [152–158]. The target populations were diverse, with the majority of the studies targeting a combined young adult and older adult population. Of the studies we identified, most focused on overweight and obese populations with and without comorbidities (e.g., diabetes). Contrary to the design of the studies we have previously reviewed, only one of the identified ongoing trials targeted healthy, young adults, [157] indicating the importance and expansion of this area of research in clinical populations.

Table 2. Ongoing clinical trials with combined fasting and exercise interventions (http://ClinicalTrials.gov, accessed on 1 September 2021).

Identifier	Study Design	Population	Experimental Groups	Intervention Characteristics	Outcomes of Interest
			Cancer Populations		
NCT04708860 [152]	Single-arm trial	Women ages 18 and older with Metastatic Breast Cancer	Combined POF and exercise	12-week trial POF: Restriction of caloric food/drink after 8 pm, wait 13 h after last meal before eating, fasting 6 days/week Exercise: Moderate-intensity aerobic and strength training, 2 times/week of 30–45 min strength classes, 120 min aerobic activity per week	• Primary: Rate of enrollment, rate of adherence to intervention • Secondary: Change in metabolic markers, quality of life, and patient-reported outcomes
			Other Populations		
NCT04004403 [153]	Randomized Clinical Trail	Obese prediabetic adults ages 18–64 with NAFLD	(1) ADF (2) Exercise (ad libitum fed) (3) ADF + exercise (4) Control (ad libitum fed, no exercise)	24-week trial ADF: Fast day: 25% energy intake (~500 kcal), Feed day: ad libitum fed Exercise: Aerobic exercise training, 5 sessions/week	• Primary: Change in hepatic steatosis, body weight • Secondary: Change in hepatokine profile, hepatic insulin sensitivity, insulin resistance, HbA1c, and other metabolic disease risk factors
NCT04131647 [154]	Randomized Clinical Trail	Overweight and obese older adult (ages 50–70) veterans	(1) Weight Maintenance Only (2) Weight maintenance + IF	24-week program Weight Maintenance Program: Nutrition advice, walking, and resistance training, 12-week program IF + Exercise: One day of IF per week, consisting of 2 small meals/day; the combined program continues for 24-week program, following completion of the 12-week weight maintenance program; walking and resistance exercises	• Primary: Change in body weight • Secondary: Change in gait speed, body fat, lipoprotein lipase

Table 2. *Cont.*

Identifier	Study Design	Population	Experimental Groups	Intervention Characteristics	Outcomes of Interest
			Other Populations		
NCT04585581 [155]	Randomized Clinical Trail	Overweight women getting pregnant in next 6 months	(1) TRE + HIIT Exercise (2) Standard Care	Duration of pregnancy (min. 28 weeks) TRE + Exercise: Minimum of 14 h/day HIIT exercise: 2–3 days/week	• Primary: Plasma glucose concentration at gestational week 28 • Secondary: Maternal and offspring cardiometabolic health measures
NCT04768725 [156]	Randomized Clinical Trail	Obese, postmenopausal women ages 45–59, sedentary lifestyle	(1) IF (2) IF + Physical-Cognitive Exergaming Program (3) Control	12-week trial IF: Self-selected diet with 25–75% of estimated baseline energy requirements for 2 days/week (fast day) along with ad libitum for 5 days/week (feed day) Exercise: Moderate-Vigorous intensity [60–70% of heart rate maximum for aerobic and 60–70% of 1 repetition maximum, 8–12 repetitions/set, 3 sets of each exercise for resistance exercise, 60 min/session, 3 sessions/week (36 total sessions)]	• Primary: Change in cognitive (memory, logic, brain-derived neurotropic factor) outcomes • Secondary: Change in additional cognitive and functional measures, biomarkers, physical function
NCT04834687 [157]	Randomized Clinical Trail	Healthy 17–24-year-old young adults	(1) Exercise-Only (2) TRE (3) Combined TRE + Exercise (4) Control	Trial length not reported Exercise: Aerobic rope-skipping, 3 days/week, 90 min/session TRE: 14 h fast, 10 h eating window, high-fiber diet	• Primary: Change in body weight • Secondary: Change in cardiovascular metabolic markers, executive function, intestinal flora

Table 2. *Cont.*

Identifier	Study Design	Population	Experimental Groups	Intervention Characteristics	Outcomes of Interest
			Other Populations		
NCT04978376 [158]	Non-Randomized Clinical Trial	Overweight, older adults 50–70 years old with pre-diabetes	(1) TRE-only (2) TRE + Endurance Exercise (3) TRE with Resistance Exercise (4) Control	10-week trial TRE is restricted eating with ad libitum eating between 12:00–20:00 Endurance Exercise: 3–5 days/week of supervised exercise Resistance Exercise: 3–5 days/week of supervised exercise	• Primary: Change in body weight • Secondary: Change in body composition, insulin, glucose, and HbA1c

Abbreviations: POF = Prolonged Overnight Fasting; NAFLD = Non-Alcoholic Fatty Liver Disease; ADF = Alternate-Day Fasting; IF = Intermittent Fasting; TRE = Time-Restricted Eating; HIIT = High-Intensity Interval Training.

The identified ongoing studies are investigating outcomes of interest that will be crucial for understanding the physiological effects and implementing combined fasting and exercise interventions. The majority of the ongoing trials examine the impact of a combined intervention on change in cardiometabolic biomarkers, including insulin sensitivity, insulin resistance, HbA1c, hepatic function, glucose concentrations, lipoprotein lipase, and lipid profiles [152–158]. Other notable outcomes of interest are changes in body composition, quality of life, cognitive/memory-related measures, physical function, and other patient-reported outcomes. The impact from these ongoing trials will benefit the collective understanding of the effect of combined fasting and exercise across the lifespan in vulnerable cancer populations, and will be important to inform the effectiveness, safety, and feasibility of these interventions in future trials.

6. Conclusions

Independently, fasting and exercise are well-tolerated among cancer patients, and while they both induce independent benefits, when combined, their additive or synergistic effects on cardiometabolic, body composition, patient-reported, and cancer-related outcomes are unknown within the cancer population. Many cancer patients experience cancer- and treatment-related side effects, many of which have been demonstrated to be managed, improved, or prevented with energy balance interventions. We are proposing combined fasting and exercise as a potentially viable strategy that may benefit cancer patients and improve cardiometabolic, body composition, patient-reported, and cancer-related outcomes, but much research is required in this area before it is deemed safe and feasible within this population.

Author Contributions: R.L.W., D.-W.K. and C.N.C. were responsible for conducting the initial literature search and screening potentially eligible studies for discussion. R.L.W., D.-W.K., C.N.C., T.E.C. and C.M.D.-C. jointly contributed to the conception and design of the work, and were responsible for designing the narrative review protocol, interpreting results, and writing the manuscript. All authors have read and agreed to the published version of the manuscript.

Funding: This research received no external funding.

Institutional Review Board Statement: Not applicable.

Informed Consent Statement: Not applicable.

Data Availability Statement: Not applicable.

Conflicts of Interest: The authors declare no conflict of interest.

References

1. Buchinger, O., Sr. 40 Years of fasting therapy. *Hippokrates* **1959**, *30*, 246–248.
2. Cespedes Feliciano, E.M.; Chen, W.Y.; Bradshaw, P.T.; Prado, C.M.; Alexeeff, S.; Albers, K.B.; Castillo, A.L.; Caan, B.J. Adipose tissue distribution and cardiovascular disease risk among breast cancer survivors. *J. Clin. Oncol.* **2019**, *37*, 2528–2536. [CrossRef]
3. Keating, N.L.; O'Malley, A.J.; Freedland, S.J.; Smith, M.R. Diabetes and cardiovascular disease during androgen deprivation therapy: Observational study of veterans with prostate cancer. *J. Natl. Cancer Inst.* **2010**, *102*, 39–46. [CrossRef]
4. Felicetti, F.; Fortunati, N.; Brignardello, E. Cancer survivors: An expanding population with an increased cardiometabolic risk. *Diabetes Res. Clin. Pract.* **2018**, *143*, 432–442. [CrossRef] [PubMed]
5. Gupta, D.; Lee Chuy, K.; Yang, J.C.; Bates, M.; Lombardo, M.; Steingart, R.M. Cardiovascular and metabolic effects of androgen-deprivation therapy for prostate cancer. *J. Oncol. Pract.* **2018**, *14*, 580–587. [CrossRef]
6. Corremans, R.; Adao, R.; De Keulenaer, G.W.; Leite-Moreira, A.F.; Bras-Silva, C. Update on pathophysiology and preventive strategies of anthracycline-induced cardiotoxicity. *Clin. Exp. Pharmacol. Physiol.* **2019**, *46*, 204–215. [CrossRef] [PubMed]
7. Cardinale, D.; Iacopo, F.; Cipolla, C.M. Cardiotoxicity of anthracyclines. *Front. Cardiovasc. Med.* **2020**, *7*, 26. [CrossRef]
8. Wilson, R.L.; Shannon, T.; Calton, E.; Galvao, D.A.; Taaffe, D.R.; Hart, N.H.; Lyons-Wall, P.; Newton, R.U. Efficacy of a weight loss program prior to robot assisted radical prostatectomy in overweight and obese men with prostate cancer. *Surg. Oncol.* **2020**, *35*, 182–188. [CrossRef]
9. Newton, R.U.; Jeffery, E.; Galvao, D.A.; Peddle-McIntyre, C.J.; Spry, N.; Joseph, D.; Denham, J.W.; Taaffe, D.R. Body composition, fatigue and exercise in patients with prostate cancer undergoing androgen-deprivation therapy. *BJU Int.* **2018**, *122*, 986–993. [CrossRef]

10. Baker, A.M.; Smith, K.C.; Coa, K.I.; Helzlsouer, K.J.; Caulfield, L.E.; Peairs, K.S.; Shockney, L.D.; Klassen, A.C. Clinical care providers' perspectives on body size and weight management among long-term cancer survivors. *Integr. Cancer Ther.* **2015**, *14*, 240–248. [CrossRef]
11. Thomas, R.J.; Holm, M.; Williams, M.; Bowman, E.; Bellamy, P.; Andreyev, J.; Maher, J. Lifestyle factors correlate with the risk of late pelvic symptoms after prostatic radiotherapy. *Clin. Oncol.* **2013**, *25*, 246–251. [CrossRef]
12. Ryan, A.M.; Power, D.G.; Daly, L.; Cushen, S.J.; Ní Bhuachalla, Ē.; Prado, C.M. Cancer-associated malnutrition, cachexia and sarcopenia: The skeleton in the hospital closet 40 years later. *Proc. Nutr. Soc.* **2016**, *75*, 199–211. [CrossRef]
13. Burden, S.; Jones, D.J.; Sremanakova, J.; Sowerbutts, A.M.; Lal, S.; Pilling, M.; Todd, C. Dietary interventions for adult cancer survivors. *Cochrane Database Syst. Rev.* **2019**. [CrossRef] [PubMed]
14. Christensen, J.F.; Simonsen, C.; Hojman, P. Exercise training in cancer control and treatment. *Compr. Physiol.* **2018**, *9*, 165–205. [CrossRef] [PubMed]
15. Cathcart, P.; Craddock, C.; Stebbing, J. Fasting: Starving cancer. *Lancet Oncol.* **2017**, *18*, 431. [CrossRef]
16. Bauersfeld, S.P.; Kessler, C.S.; Wischnewsky, M.; Jaensch, A.; Steckhan, N.; Stange, R.; Kunz, B.; Bruckner, B.; Sehouli, J.; Michalsen, A. The effects of short-term fasting on quality of life and tolerance to chemotherapy in patients with breast and ovarian cancer: A randomized cross-over pilot study. *BMC Cancer* **2018**, *18*, 476. [CrossRef]
17. de Groot, S.; Lugtenberg, R.T.; Cohen, D.; Welters, M.J.P.; Ehsan, I.; Vreeswijk, M.P.G.; Smit, V.T.H.B.M.; de Graaf, H.; Heijns, J.B.; Portielje, J.E.A.; et al. Fasting mimicking diet as an adjunct to neoadjuvant chemotherapy for breast cancer in the multicentre randomized phase 2 DIRECT trial. *Nat. Commun.* **2020**, *11*, 3083. [CrossRef]
18. Dorff, T.B.; Groshen, S.; Garcia, A.; Shah, M.; Tsao-Wei, D.; Pham, H.; Cheng, C.W.; Brandhorst, S.; Cohen, P.; Wei, M.; et al. Safety and feasibility of fasting in combination with platinum-based chemotherapy. *BMC Cancer* **2016**, *16*, 360. [CrossRef]
19. de Groot, S.; Vreeswijk, M.P.; Welters, M.J.; Gravesteijn, G.; Boei, J.J.; Jochems, A.; Houtsma, D.; Putter, H.; van der Hoeven, J.J.; Nortier, J.W. The effects of short-term fasting on tolerance to (neo) adjuvant chemotherapy in HER2-negative breast cancer patients: A randomized pilot study. *BMC Cancer* **2015**, *15*, 652. [CrossRef] [PubMed]
20. Safdie, F.M.; Dorff, T.; Quinn, D.; Fontana, L.; Wei, M.; Lee, C.; Cohen, P.; Longo, V.D. Fasting and cancer treatment in humans: A case series report. *Aging* **2009**, *1*, 988. [CrossRef]
21. De Groot, S.; Pijl, H.; van der Hoeven, J.J.; Kroep, J.R. Effects of short-term fasting on cancer treatment. *J. Exp. Clin. Cancer Res.* **2019**, *38*, 1–14. [CrossRef]
22. Doyle, C.; Kushi, L.H.; Byers, T.; Courneya, K.S.; Demark-Wahnefried, W.; Grant, B.; McTiernan, A.; Rock, C.L.; Thompson, C.; Gansler, T.; et al. Nutrition and physical activity during and after cancer treatment: An American Cancer Society guide for informed choices. *CA Cancer J. Clin.* **2006**, *56*, 323–353. [CrossRef]
23. Wei, M.; Brandhorst, S.; Shelehchi, M.; Mirzaei, H.; Cheng, C.W.; Budniak, J.; Groshen, S.; Mack, W.J.; Guen, E.; Di Biase, S.; et al. Fasting-mimicking diet and markers/risk factors for aging, diabetes, cancer, and cardiovascular disease. *Sci. Transl. Med.* **2017**, *9*. [CrossRef]
24. Di Biase, S.; Lee, C.; Brandhorst, S.; Manes, B.; Buono, R.; Cheng, C.W.; Cacciottolo, M.; Martin-Montalvo, A.; de Cabo, R.; Wei, M.; et al. Fasting-Mimicking Diet Reduces HO-1 to Promote T Cell-Mediated Tumor Cytotoxicity. *Cancer Cell* **2016**, *30*, 136–146. [CrossRef] [PubMed]
25. Campbell, K.L.; Winters-Stone, K.M.; Wiskemann, J.; May, A.M.; Schwartz, A.L.; Courneya, K.S.; Zucker, D.S.; Matthews, C.E.; Ligibel, J.A.; Gerber, L.H.; et al. Exercise Guidelines for Cancer Survivors: Consensus Statement from International Multidisciplinary Roundtable. *Med. Sci. Sports Exerc.* **2019**, *51*, 2375–2390. [CrossRef] [PubMed]
26. Schmitz, K.H. *Exercise Oncology: Prescribing Physical Activity before and after a Cancer Diagnosis*; Springer International Publishing: Berlin/Heidelberg, Germany, 2020.
27. Bland, K.A.; Zadravec, K.; Landry, T.; Weller, S.; Meyers, L.; Campbell, K.L. Impact of exercise on chemotherapy completion rate: A systematic review of the evidence and recommendations for future exercise oncology research. *Crit. Rev. Oncol. Hematol.* **2019**, *136*, 79–85. [CrossRef] [PubMed]
28. Morielli, A.R.; Courneya, K.S. Effects of exercise on cancer treatment completion and efficacy. In *Exercise Oncology: Prescribing Physical Activity before and after a Cancer Diagnosis*; Springer Nature: Cham, Switzerland, 2020; pp. 235–256.
29. Courneya, K.S.; Segal, R.J.; McKenzie, D.C.; Dong, H.; Gelmon, K.; Friedenreich, C.M.; Yasui, Y.; Reid, R.D.; Crawford, J.J.; Mackey, J.R. Effects of exercise during adjuvant chemotherapy on breast cancer outcomes. *Med. Sci. Sports Exerc.* **2014**, *46*, 1744–1751. [CrossRef] [PubMed]
30. Rief, H.; Bruckner, T.; Schlampp, I.; Bostel, T.; Welzel, T.; Debus, J.; Förster, R. Resistance training concomitant to radiotherapy of spinal bone metastases—Survival and prognostic factors of a randomized trial. *Radiat. Oncol.* **2016**, *11*, 1–7. [CrossRef]
31. Wiskemann, J.; Kleindienst, N.; Kuehl, R.; Dreger, P.; Schwerdtfeger, R.; Bohus, M. Effects of physical exercise on survival after allogeneic stem cell transplantation. *Int. J. Cancer* **2015**, *137*, 2749–2756. [CrossRef] [PubMed]
32. Caspersen, C.J.; Powell, K.E.; Christenson, G.M. Physical activity, exercise, and physical fitness: Definitions and distinctions for health-related research. *Public Health Rep.* **1985**, *100*, 126–131.
33. Van der Ploeg, H.P.; Hillsdon, M. Is sedentary behaviour just physical inactivity by another name? *Int. J. Behav. Nutr. Phys. Act.* **2017**, *14*, 142. [CrossRef]
34. Jaspers, R.T.; Zillikens, M.C.; Friesema, E.C.; delli Paoli, G.; Bloch, W.; Uitterlinden, A.G.; Goglia, F.; Lanni, A.; de Lange, P. Exercise, fasting, and mimetics: Toward beneficial combinations? *FASEB J.* **2017**, *31*, 14–28. [CrossRef]

35. Wallis, G.A.; Gonzalez, J.T. Is exercise best served on an empty stomach? *Proc. Nutr. Soc.* **2019**, *78*, 110–117. [CrossRef]
36. Hansen, D.; De Strijcker, D.; Calders, P. Impact of endurance exercise training in the fasted state on muscle biochemistry and metabolism in healthy subjects: Can these effects be of particular clinical benefit to type 2 diabetes mellitus and insulin-resistant patients? *Sports Med.* **2017**, *47*, 415–428. [CrossRef]
37. Albrecht, T.A.; Anderson, J.G.; Jones, R.; Bourguignon, C.; Taylor, A.G. A complementary care study combining flaxseed oil, caffeine, fasting, and exercise in women diagnosed with advanced ovarian cancer: Findings from a case study. *Holist. Nurs. Pract.* **2012**, *26*, 308–316. [CrossRef] [PubMed]
38. Rui, L. Energy metabolism in the liver. *Compr. Physiol.* **2014**, *4*, 177–197. [CrossRef] [PubMed]
39. Stote, K.S.; Baer, D.J.; Spears, K.; Paul, D.R.; Harris, G.K.; Rumpler, W.V.; Strycula, P.; Najjar, S.S.; Ferrucci, L.; Ingram, D.K.; et al. A controlled trial of reduced meal frequency without caloric restriction in healthy, normal-weight, middle-aged adults. *Am. J. Clin. Nutr.* **2007**, *85*, 981–988. [CrossRef] [PubMed]
40. Keenan, S.; Cooke, M.B.; Belski, R. The effects of intermittent fasting combined with resistance training on lean body mass: A systematic review of human studies. *Nutrients* **2020**, *12*, 2349. [CrossRef] [PubMed]
41. Donnelly, R.P.; Finlay, D.K. Glucose, glycolysis and lymphocyte responses. *Mol. Immunol.* **2015**, *68*, 513–519. [CrossRef]
42. Coyle, E.F.; Jeukendrup, A.E.; Wagenmakers, A.J.; Saris, W.H. Fatty acid oxidation is directly regulated by carbohydrate metabolism during exercise. *Am. J. Physiol.* **1997**, *273*, E268–E275. [CrossRef]
43. De Bock, K.; Richter, E.A.; Russell, A.P.; Eijnde, B.O.; Derave, W.; Ramaekers, M.; Koninckx, E.; Leger, B.; Verhaeghe, J.; Hespel, P. Exercise in the fasted state facilitates fibre type-specific intramyocellular lipid breakdown and stimulates glycogen resynthesis in humans. *J. Physiol.* **2005**, *564*, 649–660. [CrossRef] [PubMed]
44. Thomas, D.E.; Brotherhood, J.R.; Brand, J.C. Carbohydrate feeding before exercise: Effect of glycemic index. *Int. J. Sports Med.* **1991**, *12*, 180–186. [CrossRef]
45. Vieira, A.F.; Costa, R.R.; Macedo, R.C.; Coconcelli, L.; Kruel, L.F. Effects of aerobic exercise performed in fasted v. fed state on fat and carbohydrate metabolism in adults: A systematic review and meta-analysis. *Br. J. Nutr.* **2016**, *116*, 1153–1164. [CrossRef] [PubMed]
46. Siegel, R.L.; Miller, K.D.; Jemal, A. Cancer statistics, 2020. *CA Cancer J. Clin.* **2020**, *70*, 7–30. [CrossRef]
47. Weroha, S.J.; Haluska, P. The insulin-like growth factor system in cancer. *Endocrinol. Metab. Clin. N. Am.* **2012**, *41*, 335–350. [CrossRef]
48. Pollak, M.N.; Schernhammer, E.S.; Hankinson, S.E. Insulin-like growth factors and neoplasia. *Nat. Rev. Cancer* **2004**, *4*, 505–518. [CrossRef]
49. Calle, E.E.; Rodriguez, C.; Walker-Thurmond, K.; Thun, M.J. Overweight, obesity, and mortality from cancer in a prospectively studied cohort of U.S. adults. *N. Engl. J. Med.* **2003**, *348*, 1625–1638. [CrossRef]
50. Zorn, S.; Ehret, J.; Schäuble, R.; Rautenberg, B.; Ihorst, G.; Bertz, H.; Urbain, P.; Raynor, A. Impact of modified short-term fasting and its combination with a fasting supportive diet during chemotherapy on the incidence and severity of chemotherapy-induced toxicities in cancer patients-a controlled cross-over pilot study. *BMC Cancer* **2020**, *20*, 578. [CrossRef]
51. Marinac, C.R.; Nelson, S.H.; Breen, C.I.; Hartman, S.J.; Natarajan, L.; Pierce, J.P.; Flatt, S.W.; Sears, D.D.; Patterson, R.E. Prolonged nightly fasting and breast cancer prognosis. *JAMA Oncol.* **2016**, *2*, 1049–1055. [CrossRef] [PubMed]
52. Ferroni, P.; Riondino, S.; Laudisi, A.; Portarena, I.; Formica, V.; Alessandroni, J.; D'Alessandro, R.; Orlandi, A.; Costarelli, L.; Cavaliere, F. Pretreatment insulin levels as a prognostic factor for breast cancer progression. *Oncologist* **2016**, *21*, 1041. [CrossRef] [PubMed]
53. Duggan, C.; Wang, C.Y.; Neuhouser, M.L.; Xiao, L.; Smith, A.W.; Reding, K.W.; Baumgartner, R.N.; Baumgartner, K.B.; Bernstein, L.; Ballard-Barbash, R. Associations of insulin-like growth factor and insulin-like growth factor binding protein-3 with mortality in women with breast cancer. *Int. J. Cancer* **2013**, *132*, 1191–1200. [CrossRef] [PubMed]
54. Meyerhardt, J.A.; Catalano, P.J.; Haller, D.G.; Mayer, R.J.; Macdonald, J.S.; Benson, A.B., 3rd; Fuchs, C.S. Impact of diabetes mellitus on outcomes in patients with colon cancer. *J. Clin. Oncol.* **2003**, *21*, 433–440. [CrossRef]
55. Derr, R.L.; Ye, X.; Islas, M.U.; Desideri, S.; Saudek, C.D.; Grossman, S.A. Association between hyperglycemia and survival in patients with newly diagnosed glioblastoma. *J. Clin. Oncol.* **2009**, *27*, 1082. [CrossRef] [PubMed]
56. Kang, D.W.; Fairey, A.S.; Boule, N.G.; Field, C.J.; Courneya, K.S. Effects of exercise on insulin, IGF axis, adipocytokines, and inflammatory markers in breast cancer survivors: A systematic review and meta-analysis. *Cancer Epidemiol. Biomark. Prev.* **2017**, *26*, 355–365. [CrossRef]
57. Wang, Y.; Jin, B.; Paxton, R.J.; Yang, W.; Wang, X.; Jiao, Y.; Yu, C.; Chen, X. The effects of exercise on insulin, glucose, IGF-axis and CRP in cancer survivors: Meta-analysis and meta-regression of randomised controlled trials. *Eur. J. Cancer Care* **2020**, *29*, e13186. [CrossRef]
58. Arcidiacono, B.; Iiritano, S.; Nocera, A.; Possidente, K.; Nevolo, M.T.; Ventura, V.; Foti, D.; Chiefari, E.; Brunetti, A. Insulin resistance and cancer risk: An overview of the pathogenetic mechanisms. *Exp. Diabetes Res.* **2012**, *2012*, 789174. [CrossRef]
59. Butler, L.M.; Perone, Y.; Dehairs, J.; Lupien, L.E.; de Laat, V.; Talebi, A.; Loda, M.; Kinlaw, W.B.; Swinnen, J.V. Lipids and cancer: Emerging roles in pathogenesis, diagnosis and therapeutic intervention. *Adv. Drug Deliv. Rev.* **2020**, *159*, 245–293. [CrossRef] [PubMed]
60. Goodyear, L.J.; Kahn, B.B. Exercise, glucose transport, and insulin sensitivity. *Annu. Rev. Med.* **1998**, *49*, 235–261. [CrossRef]

61. Pesta, D.H.; Goncalves, R.L.S.; Madiraju, A.K.; Strasser, B.; Sparks, L.M. Resistance training to improve type 2 diabetes: Working toward a prescription for the future. *Nutr. Metab.* **2017**, *14*, 24. [CrossRef] [PubMed]
62. Keshel, T.E.; Coker, R.H. Exercise Training and Insulin Resistance: A current review. *J. Obes. Weight Loss Ther.* **2015**, *5*, S5-003. [CrossRef] [PubMed]
63. Horowitz, J.F.; Klein, S. Lipid metabolism during endurance exercise. *Am. J. Clin. Nutr.* **2000**, *72*, 558S–563S. [CrossRef] [PubMed]
64. Gordon, B.; Chen, S.; Durstine, J.L. The effects of exercise training on the traditional lipid profile and beyond. *Curr. Sports Med. Rep.* **2014**, *13*, 253–259. [CrossRef] [PubMed]
65. Mann, S.; Beedie, C.; Jimenez, A. Differential effects of aerobic exercise, resistance training and combined exercise modalities on cholesterol and the lipid profile: Review, synthesis and recommendations. *Sports Med.* **2014**, *44*, 211–221. [CrossRef]
66. Noland, R.C. Chapter Three—Exercise and Regulation of Lipid Metabolism. In *Progress in Molecular Biology and Translational Science*; Bouchard, C., Ed.; Academic Press: Cambridge, MA, USA, 2015; Volume 135, pp. 39–74.
67. Haxhi, J.; Scotto di Palumbo, A.; Sacchetti, M. Exercising for metabolic control: Is timing important? *Ann. Nutr. Metab.* **2013**, *62*, 14–25. [CrossRef]
68. Bhutani, S.; Klempel, M.C.; Kroeger, C.M.; Trepanowski, J.F.; Varady, K.A. Alternate day fasting and endurance exercise combine to reduce body weight and favorably alter plasma lipids in obese humans. *Obesity* **2013**, *21*, 1370–1379. [CrossRef]
69. Cho, A.R.; Moon, J.Y.; Kim, S.; An, K.Y.; Oh, M.; Jeon, J.Y.; Jung, D.H.; Choi, M.H.; Lee, J.W. Effects of alternate day fasting and exercise on cholesterol metabolism in overweight or obese adults: A pilot randomized controlled trial. *Metabolism* **2019**, *93*, 52–60. [CrossRef] [PubMed]
70. Galvao, D.A.; Taaffe, D.R.; Spry, N.; Joseph, D.; Newton, R.U. Acute versus chronic exposure to androgen suppression for prostate cancer: Impact on the exercise response. *J. Urol.* **2011**, *186*, 1291–1297. [CrossRef]
71. Kelley, D.E.; Goodpaster, B.H. Skeletal muscle triglyceride. An aspect of regional adiposity and insulin resistance. *Diabetes Care* **2001**, *24*, 933–941. [CrossRef]
72. Collins, K.H.; Herzog, W.; MacDonald, G.Z.; Reimer, R.A.; Rios, J.L.; Smith, I.C.; Zernicke, R.F.; Hart, D.A. Obesity, metabolic syndrome, and musculoskeletal disease: Common inflammatory pathways suggest a central role for loss of muscle integrity. *Front. Physiol.* **2018**, *9*, 112. [CrossRef]
73. Cheung, A.S.; Hoermann, R.; Dupuis, P.; Joon, D.L.; Zajac, J.D.; Grossmann, M. Relationships between insulin resistance and frailty with body composition and testosterone in men undergoing androgen deprivation therapy for prostate cancer. *Eur. J. Endocrinol.* **2016**, *175*, 229–237. [CrossRef]
74. van Loon, L.J.; Koopman, R.; Stegen, J.H.; Wagenmakers, A.J.; Keizer, H.A.; Saris, W.H. Intramyocellular lipids form an important substrate source during moderate intensity exercise in endurance-trained males in a fasted state. *J. Physiol.* **2003**, *553*, 611–625. [CrossRef]
75. Qu, H.-Q.; Li, Q.; Rentfro, A.R.; Fisher-Hoch, S.P.; McCormick, J.B. The definition of insulin resistance using HOMA-IR for Americans of Mexican descent using machine learning. *PLoS ONE* **2011**, *6*, e21041. [CrossRef] [PubMed]
76. Tang, Q.; Li, X.; Song, P.; Xu, L. Optimal cut-off values for the homeostasis model assessment of insulin resistance (HOMA-IR) and pre-diabetes screening: Developments in research and prospects for the future. *Drug Discov. Ther.* **2015**, *9*, 380–385. [CrossRef]
77. Yamada, C.; Mitsuhashi, T.; Hiratsuka, N.; Inabe, F.; Araida, N.; Takahashi, E. Optimal reference interval for homeostasis model assessment of insulin resistance in a Japanese population. *J Diabetes Investig.* **2011**, *2*, 373–376. [CrossRef] [PubMed]
78. Lee, C.H.; Shih, A.Z.; Woo, Y.C.; Fong, C.H.; Leung, O.Y.; Janus, E.; Cheung, B.M.; Lam, K.S. Optimal cut-offs of homeostasis model assessment of insulin resistance (HOMA-IR) to identify dysglycemia and type 2 diabetes mellitus: A 15-year prospective study in Chinese. *PLoS ONE* **2016**, *11*, e0163424. [CrossRef]
79. Van Proeyen, K.; Szlufcik, K.; Nielens, H.; Pelgrim, K.; Deldicque, L.; Hesselink, M.; Van Veldhoven, P.P.; Hespel, P. Training in the fasted state improves glucose tolerance during fat-rich diet. *J. Physiol.* **2010**, *588*, 4289–4302. [CrossRef] [PubMed]
80. Cespedes Feliciano, E.M.; Kroenke, C.H.; Caan, B.J. The obesity paradox in cancer: How important is muscle? *Annu. Rev. Nutr.* **2018**, *38*, 357–379. [CrossRef]
81. Galvao, D.A.; Spry, N.A.; Taaffe, D.R.; Newton, R.U.; Stanley, J.; Shannon, T.; Rowling, C.; Prince, R. Changes in muscle, fat and bone mass after 36 weeks of maximal androgen blockade for prostate cancer. *BJU Int.* **2008**, *102*, 44–47. [CrossRef]
82. Lee, K.; Kruper, L.; Dieli-Conwright, C.M.; Mortimer, J.E. The impact of obesity on breast cancer diagnosis and treatment. *Curr. Oncol. Rep.* **2019**, *21*, 41. [CrossRef]
83. Dickerman, B.A.; Torfadottir, J.E.; Valdimarsdottir, U.A.; Giovannucci, E.; Wilson, K.M.; Aspelund, T.; Tryggvadottir, L.; Sigurdardottir, L.G.; Harris, T.B.; Launer, L.J.; et al. Body fat distribution on computed tomography imaging and prostate cancer risk and mortality in the AGES-Reykjavik study. *Cancer* **2019**, *125*, 2877–2885. [CrossRef]
84. De Nardi, P.; Salandini, M.; Chiari, D.; Pecorelli, N.; Cristel, G.; Damascelli, A.; Ronzoni, M.; Massimino, L.; De Cobelli, F.; Braga, M. Changes in body composition during neoadjuvant therapy can affect prognosis in rectal cancer patients: An exploratory study. *Curr. Probl. Cancer* **2020**, *44*, 100510. [CrossRef]
85. Michalsen, A.; Li, C. Fasting therapy for treating and preventing disease—Current state of evidence. *Forsch. Komplement.* **2013**, *20*, 444–453. [CrossRef]
86. Köppel, M.; Mathis, K.; Schmitz, K.H.; Wiskemann, J. Muscle hypertrophy in cancer patients and survivors via strength training. A meta-analysis and meta-regression. *Crit. Rev. Oncol. Hematol.* **2021**, *163*, 103371. [CrossRef]

87. Haseen, F.; Murray, L.J.; Cardwell, C.R.; O'Sullivan, J.M.; Cantwell, M.M. The effect of androgen deprivation therapy on body composition in men with prostate cancer: Systematic review and meta-analysis. *J. Cancer Surviv.* **2010**, *4*, 128–139. [CrossRef]
88. Chang, D.; Joseph, D.J.; Ebert, M.A.; Galvão, D.A.; Taaffe, D.R.; Denham, J.W.; Newton, R.U.; Spry, N.A. Effect of androgen deprivation therapy on muscle attenuation in men with prostate cancer. *J. Med. Imaging Radiat. Oncol.* **2014**, *58*, 223–228. [CrossRef] [PubMed]
89. Lopez, P.; Taaffe, D.R.; Newton, R.U.; Galvão, D.A. Resistance exercise dosage in men with prostate cancer: Systematic review, meta-analysis, and meta-regression. *Med. Sci. Sports Exerc.* **2021**, *53*, 459. [CrossRef] [PubMed]
90. Zdravkovic, A.; Hasenöhrl, T.; Palma, S.; Crevenna, R. Effects of resistance exercise in prostate cancer patients: A systematic review update as of March 2020. *Wien. Klin. Wochenschr.* **2020**, *132*, 452–463. [CrossRef]
91. McTiernan, A. Obesity and cancer: The risks, science, and potential management strategies. *Oncology* **2005**, *19*, 871–881; discussion 881–882, 885–886. [PubMed]
92. Irwin, M.L.; Alvarez-Reeves, M.; Cadmus, L.; Mierzejewski, E.; Mayne, S.T.; Yu, H.; Chung, G.G.; Jones, B.; Knobf, M.T.; DiPietro, L. Exercise improves body fat, lean mass, and bone mass in breast cancer survivors. *Obesity* **2009**, *17*, 1534–1541. [CrossRef]
93. Guinan, E.M.; Connolly, E.M.; Hussey, J. Exercise training in breast cancer survivors: A review of trials examining anthropometric and obesity-related biomarkers of breast cancer risk. *Phys. Ther. Rev.* **2013**, *18*, 79–89. [CrossRef]
94. Thomas, G.A.; Cartmel, B.; Harrigan, M.; Fiellin, M.; Capozza, S.; Zhou, Y.; Ercolano, E.; Gross, C.P.; Hershman, D.; Ligibel, J.; et al. The effect of exercise on body composition and bone mineral density in breast cancer survivors taking aromatase inhibitors. *Obesity* **2017**, *25*, 346–351. [CrossRef] [PubMed]
95. Brown, J.C.; Zemel, B.S.; Troxel, A.B.; Rickels, M.R.; Damjanov, N.; Ky, B.; Rhim, A.D.; Rustgi, A.K.; Courneya, K.S.; Schmitz, K.H. Dose–response effects of aerobic exercise on body composition among colon cancer survivors: A randomised controlled trial. *Br. J. Cancer* **2017**, *117*, 1614–1620. [CrossRef] [PubMed]
96. Dieli-Conwright, C.M.; Mortimer, J.E.; Schroeder, E.T.; Courneya, K.; Demark-Wahnefried, W.; Buchanan, T.A.; Tripathy, D.; Bernstein, L. Effects of aerobic and resistance exercise on metabolic syndrome, sarcopenic obesity, and circulating biomarkers in overweight or obese survivors of breast cancer: A randomized controlled trial. *J. Clin. Oncol.* **2018**, *36*, 875–883. [CrossRef] [PubMed]
97. Reeves, M.M.; Terranova, C.O.; Eakin, E.G.; Demark-Wahnefried, W. Weight loss intervention trials in women with breast cancer: A systematic review. *Obes. Rev.* **2014**, *15*, 749–768. [CrossRef]
98. LeVasseur, N.; Cheng, W.; Mazzarello, S.; Clemons, M.; Vandermeer, L.; Jones, L.; Joy, A.A.; Barbeau, P.; Wolfe, D.; Ahmadzai, N. Optimising weight-loss interventions in cancer patients—A systematic review and network meta-analysis. *PLoS ONE* **2021**, *16*, e0245794. [CrossRef]
99. Lee, J. The effects of resistance training on muscular strength and hypertrophy in elderly cancer patients: A systematic review and meta-analysis. *J. Sport Health Sci.* **2021**. [CrossRef]
100. Deng, T.; Lyon, C.J.; Bergin, S.; Caligiuri, M.A.; Hsueh, W.A. Obesity, inflammation, and cancer. *Annu. Rev. Pathol.* **2016**, *11*, 421–449. [CrossRef]
101. Schoenfeld, B.J.; Aragon, A.A.; Wilborn, C.D.; Krieger, J.W.; Sonmez, G.T. Body composition changes associated with fasted versus non-fasted aerobic exercise. *J. Int. Soc. Sports Nutr.* **2014**, *11*, 54. [CrossRef]
102. Stroup, S.P.; Cullen, J.; Auge, B.K.; L'Esperance, J.O.; Kang, S.K. Effect of obesity on prostate-specific antigen recurrence after radiation therapy for localized prostate cancer as measured by the 2006 Radiation Therapy Oncology Group-American Society for Therapeutic Radiation and Oncology (RTOG-ASTRO) Phoenix consensus definition. *Cancer* **2007**, *110*, 1003–1009. [CrossRef]
103. Fearon, K.; Strasser, F.; Anker, S.D.; Bosaeus, I.; Bruera, E.; Fainsinger, R.L.; Jatoi, A.; Loprinzi, C.; MacDonald, N.; Mantovani, G.; et al. Definition and classification of cancer cachexia: An international consensus. *Lancet Oncol.* **2011**, *12*, 489–495. [CrossRef]
104. Hanson, E.D.; Nelson, A.R.; West, D.W.; Violet, J.A.; O'Keefe, L.; Phillips, S.M.; Hayes, A. Attenuation of resting but not load-mediated protein synthesis in prostate cancer patients on androgen deprivation. *J. Clin. Endocrinol. Metab.* **2017**, *102*, 1076–1083. [PubMed]
105. Cermak, N.M.; Res, P.T.; de Groot, L.C.; Saris, W.H.; van Loon, L.J. Protein supplementation augments the adaptive response of skeletal muscle to resistance-type exercise training: A meta-analysis. *Am. J. Clin. Nutr.* **2012**, *96*, 1454–1464. [CrossRef] [PubMed]
106. Slater, G.J.; Dieter, B.P.; Marsh, D.J.; Helms, E.R.; Shaw, G.; Iraki, J. Is an Energy Surplus Required to Maximize Skeletal Muscle Hypertrophy Associated With Resistance Training. *Front. Nutr.* **2019**, *6*, 131. [CrossRef] [PubMed]
107. Murphy, C.; Koehler, K. Caloric restriction induces anabolic resistance to resistance exercise. *Eur. J. Appl. Physiol.* **2020**, *120*, 1155–1164. [CrossRef] [PubMed]
108. Tinsley, G.M.; Moore, M.L.; Graybeal, A.J.; Paoli, A.; Kim, Y.; Gonzales, J.U.; Harry, J.R.; VanDusseldorp, T.A.; Kennedy, D.N.; Cruz, M.R. Time-restricted feeding plus resistance training in active females: A randomized trial. *Am. J. Clin. Nutr.* **2019**, *110*, 628–640. [CrossRef] [PubMed]
109. Stark, L.; Tofthagen, C.; Visovsky, C.; McMillan, S.C. The symptom experience of patients with cancer. *J. Hosp. Palliat. Nurs.* **2012**, *14*, 61–70. [CrossRef]
110. Weber, D.; O'Brien, K. Cancer and cancer-related fatigue and the interrelationships with depression, stress, and inflammation. *J. Evid. Based Complement. Altern. Med.* **2017**, *22*, 502–512. [CrossRef]
111. Theobald, D.E. Cancer pain, fatigue, distress, and insomnia in cancer patients. *Clin. Cornerstone* **2004**, *6* (Suppl. S1D), S15–S21. [CrossRef]

112. Buffart, L.M.; Kalter, J.; Sweegers, M.G.; Courneya, K.S.; Newton, R.U.; Aaronson, N.K.; Jacobsen, P.B.; May, A.M.; Galvão, D.A.; Chinapaw, M.J.; et al. Effects and moderators of exercise on quality of life and physical function in patients with cancer: An individual patient data meta-analysis of 34 RCTs. *Cancer Treat. Rev.* **2017**, *52*, 91–104. [CrossRef]
113. Mishra, S.I.; Scherer, R.W.; Geigle, P.M.; Berlanstein, D.R.; Topaloglu, O.; Gotay, C.C.; Snyder, C. Exercise interventions on health-related quality of life for cancer survivors. *Cochrane Database Syst. Rev.* **2012**. [CrossRef]
114. Kessels, E.; Husson, O.; Van der Feltz-Cornelis, C.M. The effect of exercise on cancer-related fatigue in cancer survivors: A systematic review and meta-analysis. *Neuropsychiatr. Dis. Treat.* **2018**, *14*, 479. [CrossRef] [PubMed]
115. Mustian, K.M.; Alfano, C.M.; Heckler, C.; Kleckner, A.S.; Kleckner, I.R.; Leach, C.R.; Mohr, D.; Palesh, O.G.; Peppone, L.J.; Piper, B.F.; et al. Comparison of pharmaceutical, psychological, and exercise treatments for cancer-related fatigue: A meta-analysis. *JAMA Oncol.* **2017**, *3*, 961–968. [CrossRef]
116. Brown, J.C.; Huedo-Medina, T.B.; Pescatello, L.S.; Ryan, S.M.; Pescatello, S.M.; Moker, E.; LaCroix, J.M.; Ferrer, R.A.; Johnson, B.T. The efficacy of exercise in reducing depressive symptoms among cancer survivors: A meta-analysis. *PLoS ONE* **2012**, *7*, e30955. [CrossRef] [PubMed]
117. Kang, D.W.; Fairey, A.S.; Boule, N.G.; Field, C.J.; Courneya, K.S. Exercise duRing Active Surveillance for prostatE cancer-the ERASE trial: A study protocol of a phase II randomised controlled trial. *BMJ Open* **2019**, *9*, e026438. [CrossRef]
118. Cartmel, B.; Hughes, M.; Ercolano, E.A.; Gottlieb, L.; Li, F.; Zhou, Y.; Harrigan, M.; Ligibel, J.A.; von Gruenigen, V.E.; Gogoi, R.; et al. Randomized trial of exercise on depressive symptomatology and brain derived neurotrophic factor (BDNF) in ovarian cancer survivors: The Women's Activity and Lifestyle Study in Connecticut (WALC). *Gynecol. Oncol.* **2021**, *161*, 587–594. [CrossRef]
119. i Ferrer, B.-C.S.; van Roekel, E.; Lynch, B.M. The role of physical activity in managing fatigue in cancer survivors. *Curr. Nutr. Rep.* **2018**, *7*, 59–69. [CrossRef]
120. Bourke, L.; Gilbert, S.; Hooper, R.; Steed, L.A.; Joshi, M.; Catto, J.W.; Saxton, J.M.; Rosario, D.J. Lifestyle changes for improving disease-specific quality of life in sedentary men on long-term androgen-deprivation therapy for advanced prostate cancer: A randomised controlled trial. *Eur. Urol.* **2014**, *65*, 865–872. [CrossRef]
121. Kohler, L.N.; Garcia, D.O.; Harris, R.B.; Oren, E.; Roe, D.J.; Jacobs, E.T. Adherence to diet and physical activity cancer prevention guidelines and cancer outcomes: A systematic review. *Cancer Epidemiol. Biomark. Prev.* **2016**, *25*, 1018–1028. [CrossRef]
122. Warburg, O. On the origin of cancer cells. *Science* **1956**, *123*, 309–314. [CrossRef] [PubMed]
123. Romijn, J.; Godfried, M.; Hommes, M.; Endert, E.; Sauerwein, H. Decreased glucose oxidation during short-term starvation. *Metabolism* **1990**, *39*, 525–530. [CrossRef]
124. Pedersen, L.; Christensen, J.F.; Hojman, P. Effects of exercise on tumor physiology and metabolism. *Cancer J.* **2015**, *21*, 111–116. [CrossRef]
125. Fontana, L.; Weiss, E.P.; Villareal, D.T.; Klein, S.; Holloszy, J.O. Long-term effects of calorie or protein restriction on serum IGF-1 and IGFBP-3 concentration in humans. *Aging Cell* **2008**, *7*, 681–687. [CrossRef]
126. Renehan, A.G.; Zwahlen, M.; Minder, C.; T O'Dwyer, S.; Shalet, S.M.; Egger, M. Insulin-like growth factor (IGF)-I, IGF binding protein-3, and cancer risk: Systematic review and meta-regression analysis. *Lancet* **2004**, *363*, 1346–1353. [CrossRef]
127. Walford, R.L.; Mock, D.; Verdery, R.; MacCallum, T. Calorie restriction in biosphere 2: Alterations in physiologic, hematologic, hormonal, and biochemical parameters in humans restricted for a 2-year period. *J. Gerontol. Ser. A Biol. Sci. Med. Sci.* **2002**, *57*, B211–B224. [CrossRef] [PubMed]
128. Raffaghello, L.; Lee, C.; Safdie, F.M.; Wei, M.; Madia, F.; Bianchi, G.; Longo, V.D. Starvation-dependent differential stress resistance protects normal but not cancer cells against high-dose chemotherapy. *Proc. Natl. Acad. Sci. USA* **2008**, *105*, 8215–8220. [CrossRef] [PubMed]
129. Lee, C.; Raffaghello, L.; Brandhorst, S.; Safdie, F.M.; Bianchi, G.; Martin-Montalvo, A.; Pistoia, V.; Wei, M.; Hwang, S.; Merlino, A.; et al. Fasting cycles retard growth of tumors and sensitize a range of cancer cell types to chemotherapy. *Sci. Transl. Med.* **2012**, *4*, 124ra27. [CrossRef] [PubMed]
130. Safdie, F.; Brandhorst, S.; Wei, M.; Wang, W.; Lee, C.; Hwang, S.; Conti, P.S.; Chen, T.C.; Longo, V.D. Fasting enhances the response of glioma to chemo- and radiotherapy. *PLoS ONE* **2012**, *7*, e44603. [CrossRef]
131. Laviano, A.; Rossi Fanelli, F. Toxicity in chemotherapy—When less is more. *N. Engl. J. Med.* **2012**, *366*, 2319–2320. [CrossRef]
132. Longo, V.D.; Mattson, M.P. Fasting: Molecular mechanisms and clinical applications. *Cell Metab.* **2014**, *19*, 181–192. [CrossRef]
133. Fontana, L.; Partridge, L.; Longo, V.D. Extending healthy life span—From yeast to humans. *Science* **2010**, *328*, 321–326. [CrossRef]
134. Bishop, N.A.; Guarente, L. Genetic links between diet and lifespan: Shared mechanisms from yeast to humans. *Nat. Rev. Genet.* **2007**, *8*, 835–844. [CrossRef]
135. Cheng, C.W.; Adams, G.B.; Perin, L.; Wei, M.; Zhou, X.; Lam, B.S.; Da Sacco, S.; Mirisola, M.; Quinn, D.I.; Dorff, T.B.; et al. Prolonged fasting reduces IGF-1/PKA to promote hematopoietic-stem-cell-based regeneration and reverse immunosuppression. *Cell Stem Cell* **2014**, *14*, 810–823. [CrossRef]
136. Sun, P.; Wang, H.; He, Z.; Chen, X.; Wu, Q.; Chen, W.; Sun, Z.; Weng, M.; Zhu, M.; Ma, D.; et al. Fasting inhibits colorectal cancer growth by reducing M2 polarization of tumor-associated macrophages. *Oncotarget* **2017**, *8*, 74649–74660. [CrossRef]
137. Koelwyn, G.J.; Zhuang, X.; Tammela, T.; Schietinger, A.; Jones, L.W. Exercise and immunometabolic regulation in cancer. *Nat. Metab.* **2020**, *2*, 849–857. [CrossRef]
138. Hojman, P.; Gehl, J.; Christensen, J.F.; Pedersen, B.K. Molecular Mechanisms Linking Exercise to Cancer Prevention and Treatment. *Cell Metab.* **2018**, *27*, 10–21. [CrossRef]

139. Pedersen, L.; Idorn, M.; Olofsson, G.H.; Lauenborg, B.; Nookaew, I.; Hansen, R.H.; Johannesen, H.H.; Becker, J.C.; Pedersen, K.S.; Dethlefsen, C.; et al. Voluntary Running Suppresses Tumor Growth through Epinephrine- and IL-6-Dependent NK Cell Mobilization and Redistribution. *Cell Metab.* **2016**, *23*, 554–562. [CrossRef]
140. Idorn, M.; Hojman, P. Exercise-Dependent Regulation of NK Cells in Cancer Protection. *Trends Mol. Med.* **2016**, *22*, 565–577. [CrossRef] [PubMed]
141. Betof, A.S.; Lascola, C.D.; Weitzel, D.; Landon, C.; Scarbrough, P.M.; Devi, G.R.; Palmer, G.; Jones, L.W.; Dewhirst, M.W. Modulation of murine breast tumor vascularity, hypoxia, and chemotherapeutic response by exercise. *JNCI J. Natl. Cancer Inst.* **2015**, *107*. [CrossRef] [PubMed]
142. Morrell, M.B.G.; Alvarez-Florez, C.; Zhang, A.; Kleinerman, E.S.; Savage, H.; Marmonti, E.; Park, M.; Shaw, A.; Schadler, K.L. Vascular modulation through exercise improves chemotherapy efficacy in Ewing sarcoma. *Pediatr. Blood Cancer* **2019**, *66*, e27835. [CrossRef] [PubMed]
143. Yang, L.; Morielli, A.R.; Heer, E.; Kirkham, A.A.; Cheung, W.Y.; Usmani, N.; Friedenreich, C.M.; Courneya, K.S. Effects of Exercise on Cancer Treatment Efficacy: A Systematic Review of Preclinical and Clinical Studies. *Cancer Res.* **2021**. [CrossRef] [PubMed]
144. Pin, F.; Couch, M.E.; Bonetto, A. Preservation of muscle mass as a strategy to reduce the toxic effects of cancer chemotherapy on body composition. *Curr. Opin. Support. Palliat. Care* **2018**, *12*, 420–426. [CrossRef] [PubMed]
145. Cespedes Feliciano, E.M.; Chen, W.Y.; Lee, V.; Albers, K.B.; Prado, C.M.; Alexeeff, S.; Xiao, J.; Shachar, S.S.; Caan, B.J. Body composition, adherence to anthracycline and taxane-based chemotherapy, and survival after nonmetastatic breast cancer. *JAMA Oncol.* **2020**, *6*, 264–270. [CrossRef] [PubMed]
146. Prado, C.M.; Lieffers, J.R.; McCargar, L.J.; Reiman, T.; Sawyer, M.B.; Martin, L.; Baracos, V.E. Prevalence and clinical implications of sarcopenic obesity in patients with solid tumours of the respiratory and gastrointestinal tracts: A population-based study. *Lancet Oncol.* **2008**, *9*, 629–635. [CrossRef]
147. Dupertuis, Y.M.; Meguid, M.M.; Pichard, C. Colon cancer therapy: New perspectives of nutritional manipulations using polyunsaturated fatty acids. *Curr. Opin. Clin. Nutr. Metab. Care* **2007**, *10*, 427–432. [CrossRef]
148. Scheim, D.E. Cytotoxicity of unsaturated fatty acids in fresh human tumor explants: Concentration thresholds and implications for clinical efficacy. *Lipids Health Dis.* **2009**, *8*, 54. [CrossRef]
149. Comba, A.; Lin, Y.H.; Eynard, A.R.; Valentich, M.A.; Fernandez-Zapico, M.E.; Pasqualini, M.E. Basic aspects of tumor cell fatty acid-regulated signaling and transcription factors. *Cancer Metastasis Rev.* **2011**, *30*, 325–342. [CrossRef]
150. Horne, B.D.; Muhlestein, J.B.; Anderson, J.L. Health effects of intermittent fasting: Hormesis or harm? A systematic review. *Am. J. Clin. Nutr.* **2015**, *102*, 464–470. [CrossRef]
151. Piercy, K.L.; Troiano, R.P.; Ballard, R.M.; Carlson, S.A.; Fulton, J.E.; Galuska, D.A.; George, S.M.; Olson, R.D. The Physical Activity Guidelines for Americans. *JAMA* **2018**, *320*, 2020–2028. [CrossRef]
152. Ligibel, J. Pilot Study of the Impact of a Combined Intermittent Fasting and Exercise Intervention on Metabolic Markers in Patients with Advanced, Hormone Receptor Positive Breast Cancer. Available online: https://clinicaltrials.gov/ct2/show/NCT04708860 (accessed on 3 September 2021).
153. Varady, K. Alternate Day Fasting Combined with Exercise for the Treatment of Non-Alcoholic Fatty Liver Disease (NAFLD). Available online: https://clinicaltrials.gov/ct2/show/NCT04004403 (accessed on 3 September 2021).
154. Ryan, A. Promotion of Successful Weight Management in Overweight and Obese Veterans. Available online: https://clinicaltrials.gov/ct2/show/NCT04131647 (accessed on 3 September 2021).
155. Risa, Ø. Before the Beginning: Preconception Lifestyle Interventions to Improve Future Metabolic Health. Available online: https://clinicaltrials.gov/ct2/show/NCT04585581 (accessed on 3 September 2021).
156. Sungkarat, S. A Randomized Controlled Trial Investigating the Effects of Combined Physical-Cognitive Exercise and Dietary Intervention on Cognitive Performance and Changes in Blood Biomarkers of Postmenopausal Obese Women. Available online: https://clinicaltrials.gov/ct2/show/NCT04768725 (accessed on 3 September 2021).
157. Zhu, Y. Effects of Diet and Exercise Interventions on Cardiometabolic Risk Markers, Executive Function, and Intestinal Flora in Undergraduate Students: A Randomized Controlled Trial. Available online: https://clinicaltrials.gov/ct2/show/NCT04834687 (accessed on 3 September 2021).
158. Gabel, K. Time Restricted Eating with Physical Activity for Weight Management. Available online: https://clinicaltrials.gov/ct2/show/NCT04978376 (accessed on 3 September 2021).

MDPI

Article

A Dietary Intervention High in Green Leafy Vegetables Reduces Oxidative DNA Damage in Adults at Increased Risk of Colorectal Cancer: Biological Outcomes of the Randomized Controlled Meat and Three Greens (M3G) Feasibility Trial

Andrew D. Frugé [1,*], Kristen S. Smith [1], Aaron J. Riviere [1], Rachel Tenpenny-Chigas [1], Wendy Demark-Wahnefried [2], Anna E. Arthur [3], William M. Murrah [4], William J. van der Pol [5], Shanese L. Jasper [6], Casey D. Morrow [7], Robert D. Arnold [7] and Kimberly Braxton-Lloyd [8]

1 Department of Nutrition, Dietetics and Hospitality Management, Auburn University, Auburn, AL 36849, USA; kss0034@auburn.edu (K.S.S.); ajr0042@auburn.edu (A.J.R.); raychelynn_11@hotmail.com (R.T.-C.)
2 Department of Nutrition Sciences, University of Alabama at Birmingham, Birmingham, AL 35294, USA; demark@uab.edu
3 Department of Food Science and Human Nutrition, Division of Nutritional Sciences, University of Illinois at Urbana-Champaign, Champaign, IL 61801, USA; aarthur@illinois.edu
4 Department of Educational Foundations, Leadership, and Technology, Auburn University, Auburn, AL 36849, USA; wmm0017@auburn.edu
5 Department of Computational Biology and Bioinformatics, University of Alabama at Birmingham, Birmingham, AL 35294, USA; liamvdp@uab.edu
6 Department of Cell, Developmental and Integrative Biology, University of Alabama at Birmingham, Birmingham, AL 35294, USA; slj0012@auburn.edu
7 Department of Drug Discovery and Development, Auburn University Harrison School of Pharmacy, Auburn, AL 36849, USA; caseym@uab.edu (C.D.M.); rda0007@auburn.edu (R.D.A.)
8 Department of Pharmacy Services, Auburn University Harrison School of Pharmacy, Auburn, AL 36849, USA; lloydkb@auburn.edu
* Correspondence: fruge@auburn.edu; Tel.: +334-844-3271

Citation: Frugé, A.D.; Smith, K.S.; Riviere, A.J.; Tenpenny-Chigas, R.; Demark-Wahnefried, W.; Arthur, A.E.; Murrah, W.M.; van der Pol, W.J.; Jasper, S.L.; Morrow, C.D.; et al. A Dietary Intervention High in Green Leafy Vegetables Reduces Oxidative DNA Damage in Adults at Increased Risk of Colorectal Cancer: Biological Outcomes of the Randomized Controlled Meat and Three Greens (M3G) Feasibility Trial. *Nutrients* **2021**, *13*, 1220. https://doi.org/10.3390/nu13041220

Academic Editor: Francesca Giampieri

Received: 18 February 2021
Accepted: 5 April 2021
Published: 7 April 2021

Abstract: Green leafy vegetables (GLV) may reduce the risk of red meat (RM)-induced colonic DNA damage and colorectal cancer (CRC). We previously reported the primary outcomes (feasibility) of a 12-week randomized controlled crossover trial in adults with habitual high RM and low GLV intake with body mass index (BMI) > 30 kg/m^2 (NCT03582306). Herein, our objective was to report a priori secondary outcomes. Participants were recruited and enrolled in 2018, stratified by gender, and randomized to two arms: immediate intervention group (IG, n = 26) or delayed intervention group (DG, n = 24). During the 4 week intervention period, participants were provided with frozen GLV and counseled to consume 1 cooked cup equivalent daily. Participants consumed their normal diet for the remaining 8 weeks. At each of four study visits, anthropometrics, stool, and blood were taken. Overall, plasma Vitamin K1 (0.50 ± 1.18 ng/mL, $p < 0.001$) increased, while circulating 8OHdG (−8.52 ± 19.05 ng/mL, $p < 0.001$), fecal 8OHdG (−6.78 ± 34.86 ng/mL, $p < 0.001$), and TNFα (−16.95 ± 60.82 pg/mL, $p < 0.001$) decreased during the GLV intervention compared to control periods. Alpha diversity of fecal microbiota and relative abundance of major taxa did not differ systematically across study periods. Further investigation of the effects of increased GLV intake on CRC risk is warranted.

Keywords: chemoprevention; colorectal cancer; diet; green leafy vegetables; red meat; 8-hydroxy-2′deoxyguanosine

1. Introduction

The most recent global estimates of cancer incidence and mortality place colorectal cancer (CRC) as the fourth most prevalent and second deadliest cancer worldwide [1]. The

World Cancer Research Fund International Continuous Update Project (CUP) scientists' 2017 meta-analysis of 111 prospective cohort studies supported the relationships between increased risk of CRC with increased red and processed meat intake, as well as decreased risk of CRC with increased vegetable intake [2]. In 2015, approximately 38.3% of new CRC cases were directly attributed to suboptimal diets in the United States [3]. This high-meat, low-vegetable "Western" dietary pattern is most common in developed and developing countries and is directly associated with CRC risk [4], with a recent meta-analysis of 28 studies indicating a 30% increased risk of CRC for adults consuming this dietary pattern [5].

Excess adiposity has been recognized as another modifiable risk factor for colon cancer for more than two decades [6]. In a 2018 CUP meta-analysis of 47 cohort studies with 7,393,510 participants, increased body weight, body mass index (BMI), waist circumference, and waist-to-hip ratio were all independently associated with increased CRC risk in both men and women [7]. Since dietary behaviors are more easily improved than body composition [8], it is imperative to determine which dietary approaches and public health messages produce the greatest risk reduction. Based on our survey of 990 adults in the United States, a slight majority of respondents indicated they would not be willing to forego red meat (RM) consumption, and only 14.5% of men and 14.9% of women indicated that they did not like green leafy vegetables (GLV) [9]. Therefore, risk reduction may be more feasible via addition of GLV, rather than omission of RM.

A series of preclinical studies indicate that chlorophyll in GLV prevents the cytotoxic and carcinogenic effects of heme in RM [10–13], which is mediated by the microbiota residing in the colon [14,15]. 8-hydroxy-2′-deoxyguanosine (8-OHdG) is a marker of DNA damage associated with increased adenoma risk [16], which we aim to use as a proxy for cytotoxicity observed in preclinical models. High-sensitivity C-reactive protein (hsCRP) [17] is associated with elevated Proteobacteria [18] and has been linked with increased risk of colon cancer [19,20] and mortality [21]. Similarly, interleukin-6 (IL-6) has been implicated in colon cancer prognosis [22], metastasis [23], and mortality [24]. A 23-week plant-based diet (<50 g animal products/day) significantly reduced IL-6 and tumor necrosis factor-α (TNFα) in 89 obese adults [18], which corresponded with a decrease in Proteobacteria and an increase in *Bifidobacterium*. Since TNFα is produced primarily in response to lipopolysaccharide [25], a structural component of Gram-negative bacteria, it may also be a marker for mucosal health. In addition to the ability of chlorophyll to bind heme, it is hypothesized that the high flavonol content of GLV promotes the growth of several short-chain fatty acid-producing bacterial genera, which are associated with cytoprotective effects in the colon [26].

We sought to directly study the preliminary effects of increased GLV consumption in adults with increased BMI consuming a Western dietary pattern. The primary outcomes of this 12-week crossover trial were previously reported, indicating feasibility of accrual and retention, with adherence slightly below target but acceptable [27]. Herein, we report biological outcomes that may be relevant to CRC risk reduction, which include cytokines, gut microbiota, and Vitamin K1 as an objective measure of intervention adherence.

2. Materials and Methods

2.1. Study Design and Aims

Detailed methods have been described previously [27]. The study was approved as protocol #18-180 EP 1806 by the Auburn University Institutional Review Board. It was conducted in accordance with the Declaration of Helsinki and pre-registered on ClinicalTrials.gov (NCT03582306). The aims of this report were determined a-priori and are included in the ClinicalTrials registry. The aim of this paper is to report all biological data relevant to oxidative DNA damage, inflammation, and microbiota.

2.2. Participant Recruitment and Informed Consent

Participants were recruited via email from July to September 2018 in the Auburn-Opelika area in east Alabama. Interested individuals completed an online eligibility survey which included food frequency questionnaire questions to assess habitual RM and GLV consumption and were contacted for follow-up by study staff. Eligibility criteria were (1) current low-GLV consumption (<2 servings/day); (2) current high-RM consumption (>5 servings red meat/week); (3) high BMI (>30 kg/mP^2 P); (4) willing to maintain normal prescription and/or supplement intake; (5) willing to adhere to dietary protocol; (6) ability to store and cook study foods; (7) English speaking and reading ability. Participants were excluded if they had a previous diagnosis of CRC or used oral or IV antibiotics, corticosteroids, immunosuppressive agents, or commercial probiotics within the last four weeks. Written informed consent was obtained prior to any post-screening data collection.

2.3. Randomization and Interventions

Randomization occurred after the completion of baseline assessment. Participants were stratified by gender into blocks of four, with each participant in successive order receiving the gender-specific envelope with the group assignment, which was generated by KSS [19]. Enrollment was conducted by KSS and ADF, and ADF assigned participants to intervention groups. All participants received the intervention in random order: either immediately (first four weeks of the 12-week study) or delayed (last four weeks of the 12-week study). During the intervention period, participants were provided with frozen GLV purchased by study staff directly from local retailers. Participants were given a recipe book and instructed to consume 1 cup cooked GLV daily (including spinach, kale, collards, mustard greens, and turnip greens). Additionally, they were encouraged not to alter any other elements of their diet, including red meat consumption. During the four-week washout and control periods, participants were asked to consume their habitual diet.

2.4. Data Collection

All biological measures were obtained at baseline and repeated every four weeks, with subjective measures obtained at each time point and reported previously [27]. Subjective measures included the Food Acceptability Questionnaire (FAQ) [28–30], Dietary Habits and Colon Cancer Beliefs Survey (DHCCBS) [9], the International Physical Activity Questionnaire (IPAQ) [31]. IPAQ data report frequency and duration of physical activity by intensity level and reports minutes sitting. Thus, we combined physical activity data to report total active minutes and total sitting minutes. Two 24 h dietary recalls were obtained at each timepoint by a dietetics student or Registered Dietitian. Recalls were entered by study staff into the Automated Self-Administered 24-Hour Dietary Assessment tool (ASA24) [32]. Calorie and macronutrient values reported for each time point represent the average of the two recalls obtained.

Height and weight were measured using standard procedures and used to calculate BMI; waist and hip circumferences were measured using a standard tape measure at each time point [33]. Body composition was analyzed using a handheld Body Impedance Analysis (BIA) instrument (Omron HBF-306C, Omron Healthcare, Inc. Lake Forest, IL, USA).

Frozen fecal samples were obtained by study staff at each visit after participants collected them at their home using commode specimen collectors and sterile collection tubes and stored in their home freezers immediately. Patients reported the consistency of their stool using the Bristol Stool Forms Scale (BSFS) [34]. Samples were stored at −80 °C until further processing. Microbial genomic DNA was isolated using standard methods and kits from Zymo Research (Irvine, CA, USA), and the 250 base pair V4 region of the rRNA gene amplified by polymerase chain reaction and sequenced using the Illumina Miseq (San Diego, CA, USA). The informatic analyses were performed using the Quantitative Insight into Microbial Ecology (QIIME) suite, version 1.7 as modified by Kumar et al. (2014) [35,36].

Phlebotomy was performed by a trained phlebotomist; sera and plasma were separated in their respective collection tubes, aliquoted, and frozen at −80 °C until analysis.

Oxidized guanine species (8-hydroxy-2′-deoxyguanosine [8-OHdG], 8-hydroxyguanosine, and 8-hydroxyguanine) were measured in fecal water to determine the genotoxicity of the lumen and concurrent oxidative stress in the plasma via enzyme-linked immunosorbent assay (ELISA) from StressMarq Biosciences (Victoria, Canada). To normalize fecal water concentrations, solid concentration was determined based on BSFS responses. BSFS responses fall into 7 types of stool, which is commonly compressed into 3 categories: 1: hard and lumpy (Types 1–2); 2: normal consistency (Types 3–5); and 3: loose and watery (Types 6–7) [37]. Previous research indicates stool classification type is correlated with water concentration, and estimated water content can be predicted with knowledge of stool type (1: 67%; 2: 72%; 3: 77%) [38]. Therefore, to determine solid concentration, we multiplied stool sample weight by the remaining percentage (out of 100%) and divided the calculated solid weight by total sample weight. This solid concentration value was then used to uniformly dilute samples for fecal water analyses. CRP and IL-6/TNFα were measured via ELISA kits from RayBiotech (Peachtree Corner, GA, USA) and ABCam (Cambridge, UK) respectively.

LC–MS/MS was used to analyze plasma vitamin K1. Phylloquinone (K1) and its deuterated internal standard, Vitamin K1-d7, were purchased from Sigma Aldrich (St. Louis, MO, USA) Stock solutions of the analyte and deuterated internal standard were prepared by dissolving each compound in MeOH: CH3Cl (2:1). Calibration and quality control (QC) samples were made from stock solutions and UV-treated (for vitamin K depletion), pooled plasma. The calibration range consisted of thirteen levels from 0.024–100 ng/mL. At the time of analysis, 300 µL aliquots of plasma were transferred to 1.8 mL polypropylene centrifuge tubes and spiked with 10 µL of the internal standard (30 ng/mL). Ice-cold acetonitrile (900 µL) was added to each tube and vortexed for fifteen seconds. Samples were then centrifuged at 1500 rpm for fifteen minutes at 4 °C, and the supernatant was subsequently transferred to glass vials. The samples were dried under a stream of N2 and reconstituted in 100 µL of MeOH: CH3Cl (2:1). Twelve microliters of the reconstituted solution was injected for analysis. An Agilent Technologies 1290 Infinity UPLC coupled via Agilent Jetstream electrospray ionization (AJ-ESI) to the 6460 Triple Quadrupole Mass Spectrometer (Agilent Technologies, Santa Clara, CA, USA) was used for the plasma analysis of vitamin K1. Samples were injected (2 µL) onto a reversed-phase Zorbax SB-C8 column, 1.8 µm, 2.1 × 50 mm (Agilent Technologies); the column temperature was kept at 40 °C. The mobile phase consisted of [A] 0.1% formic acid in 5mM NH4 formate and [B] 0.1% formic acid in methanol introduced at a flow rate of 0.5 mL/min. Chromatographic separation was achieved using gradient elution. The solvent composition was maintained at 70% [B] for the first 0.5 min, increased and held at 95% [B] from 1.5–10.5 min, and decreased back to 70% from 10.5–11 min. The AJ-ESI ion source was operated in positive ion mode, and the QQQ scan type used for analysis was multiple reaction monitoring (MRM). The transitions used for the quantification of vitamin K1 and internal standard vitamin K1-d7 were 451.3–186.95 and 458.4–194.1, respectively. Nitrogen was used as the drying gas (10 L/min at 350 °C), nebulizer (45 psi), and collision gas; the capillary voltage was set as 4500 V. The LLOQ for the method was determined as 2.9 pg. Intra- and inter-day accuracy and precision were assessed by analyzing six replicates of QC samples at the concentrations of 0.20, 3.13, and 50 ng/mL on three consecutive days. Calculated accuracy and precision within 15% (20% for LLOQ) were considered acceptable.

2.5. Statistical Analysis

The primary outcomes were feasibility, which included accrual, retention, and adherence. Thus, power analysis was based on adherence. Setting alpha = 0.05, beta = 0.20, and n = 44 (assuming 10% attrition), 93% adherence would have resulted in a Cohen's D = 0.80 [39]. Analysis of all biological outcomes were, therefore, exploratory and defined a priori at clinicaltrials.gov (https://clinicaltrials.gov/ct2/show/NCT03582306, accessed on 1 December 2020).

Statistical analyses were conducted in SPSS 24.0 (IBM Corp. Released 2016. IBM SPSS Statistics for Windows, Version 24.0. Armonk, NY, USA: IBM Corp.) with study arm allocation blinded to analysis. Descriptive statistics were obtained for study participants, who were compared using independent sample t-tests for continuous variables and chi-square tests for categorical variables. WMM conducted statistical analyses and was blinded to intervention assignment.

Treatment effects for biomarkers were analyzed with multi-variate analysis of covariance (MANCOVA) models which included fixed factors for treatment condition, study arm, period, and participant. Since participants were block-randomized by gender, this factor was also included in the models. Outcome measures that substantially deviated from normality were log-transformed prior to analysis. Statistical significance threshold was set at $p = 0.05$. Baseline measures at each period for the relevant biomarker measures were included as covariates to increase power and adjust for differences in pretreatment levels. In addition to accounting for the within-subjects design, the MANCOVA modeling also makes possible the simultaneous estimation of treatment effects and carryover effects. The latter was estimated by the coefficients for the study arm. Differences in measures at the beginning of each period across treatment conditions were assessed using t-tests.

Microbiome analysis was conducted as described previously by Frugé et al. [40]. One-way Analysis of Variance (ANOVA) and Kruskal–Wallis tests adjusted with false discovery rates (FDR) were used to compare between-group and group x time differences in microbiota.

3. Results

3.1. Study Participant Characteristics

Fifty participants were recruited and enrolled in the trial lasting from July 2018 to December 2018. The CONSORT diagram can be found in the 2019 report by Frugé et al. [27]. Baseline characteristics of study participants are shown in Table 1. Participants were mostly in their late forties and early fifties and had an average BMI of 36.2 (Class II obesity) and body fat percentage of 38.7. Based on food frequency questionnaire data from the study screener, participants consumed ten servings of red meat and less than one half a serving of green leafy vegetables weekly at baseline. Most participants were female (62%) and non-Hispanic White (80%). Randomization was only stratified by gender, and a larger proportion of African American participants were allocated to the delayed group ($p = 0.035$). Twenty-three out of 50 participants had graduate and/or professional degrees, and eight had an associate degree or lower, with the majority in the latter category being randomized to the delayed intervention. The majority (58%) of participants were married. In the first four weeks of the study, one participant withdrew consent and one participant was lost to follow-up after illness not related to the study—both were in the immediate intervention group. Forty-eight participants completed the study, forty of whom had complete biological sample data for analysis. Over the course of the study, no clinically or statistically significant changes were observed with regard to weight, BMI, and body fat percentage.

At baseline, participants consumed 2083 ± 559 calories, coming from 86 ± 26 g protein, 93 ± 33 g fat, and 226 ± 75 g carbohydrate, with no differences in calories or macronutrients between groups. Additionally, no significant changes in total calories or any macronutrients were observed across time points. Participants reported an average total weekly active time of 928 ± 1494 min, with no significant changes across time points, though total active minutes trended downward over the course of the study (−230 ± 1048, $p = 0.109$). Total sitting time was 429 ± 195 for all participants, with the immediate group reporting more sitting time at baseline (502 ± 198 vs. 356 ± 165 min, $p = 0.011$). No significant differences in sitting time were observed across all other time points.

3.3. Microbial Diversity and Taxa

At baseline, no differences in alpha diversity were observed between groups. Over the course of each four-week period, no changes in alpha diversity beyond those expected due to sampling variability were observed for either group. Figure 1a reports the mean and 95% confidence interval for the observed species at each time point by group.

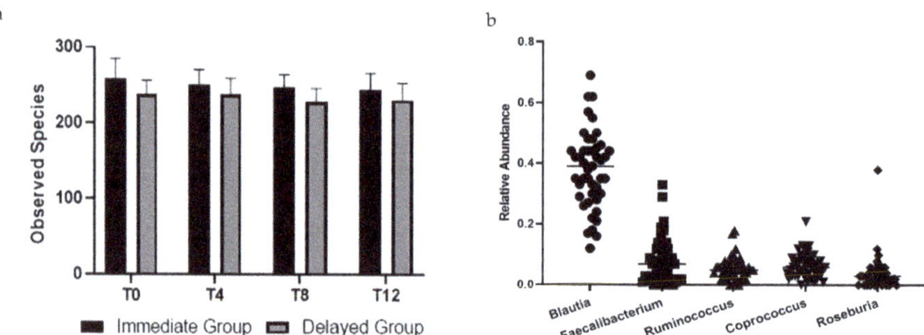

Figure 1. (**a**) Observed species of bacteria in stool samples collected from participants across all times points. There was no effect of time (p = 0.391) or group × time (p = 0.600). Points are means, and bars represent 95% confidence intervals. (**b**) Relative abundance of top five genera in stool samples collected at baseline of participants in a randomized controlled crossover high green leafy vegetable dietary intervention.

There were no differences between groups in bacterial taxa at baseline. Five of the most prevalent genera were within the Firmicutes phylum and are reported in Figure 1b. Relative abundance at the phyla level for all participants were Firmicutes (84.7 ± 11.2%), Actinobacteria (8.9 ± 9.5%), Proteobacteria (3.4 ± 6.6%), Verrucomicrobia (1.4 ± 2.4%), and Bacteroidetes (1.1 ± 1.7%). Seven of the ten most abundant genera were Firmicutes. Baseline relative abundance of the five most abundant genera are reported in Figure 1a. Post-hoc Kruskal–Wallis tests compared operational taxonomic units (OTUs) between groups at each time point and within groups from pre- to post-intervention. Bray-Curtis dissimilarity tests with $p < 0.05$ were only observed between groups at the end of the first four-week period; no OTUs differed between groups after False Discovery Rate correction (Figure 2).

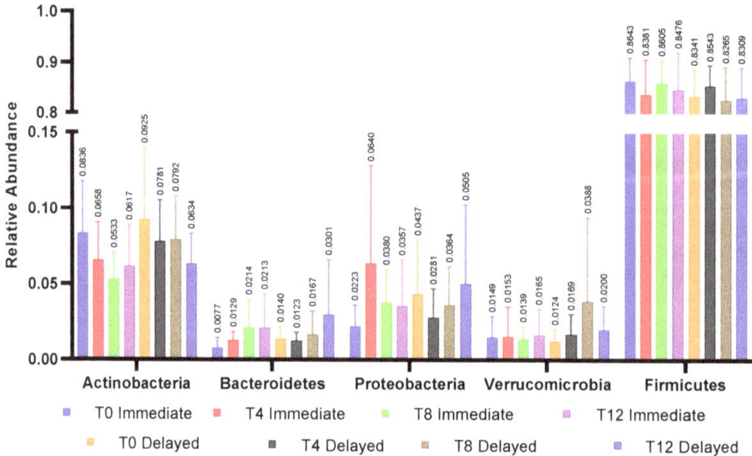

Figure 2. Relative abundance of 5 major phyla within groups across all time points. No significant differences were observed between the immediate and delayed groups within phyla or across timepoints within phyla.

4. Discussion

This is the first trial to assess the effects of a dietary intervention high in GLV in adults with elevated BMI and high habitual RM consumption who are at increased risk for CRC. We report several often-cited biomarkers but capitalize on the novelty of measuring 8OHdG in both plasma and fecal water, the latter of which has not been reported to date. While our primary outcomes report noted subjective increases in GLV and Vitamin K consumption, we additionally measured plasma Vitamin K1 as an objective measure of adherence to the protocol. Finally, we report on the microbial structure of stool samples and the (lack of) changes to the microbiome over the course of the 12-week study.

It is imperative to note that Vitamin K1 analyses indicate that there was a significant drop-in for participants during the control periods (i.e., participants continued or began consumption of GLV off protocol). The raw data suggest that five participants in the immediate group continued high GLV consumption throughout the entire study (anecdotally, reports of decreased indigestion and improved gastrointestinal function led participants to continue their newly formed GLV habit) and ten subjects in the delayed group voluntarily began consuming GLV either prior to or during the first four weeks of the study. A shortcoming in our study implementation was that we did not tell participants to avoid GLV during the recruitment and consent process. Nonetheless, the treatment*arm interaction for Vitamin K1 and 8OHdG is significant because the delayed group was also less adherent during the intervention period and did not experience the significant changes in Vitamin K1 and 8OHdG that were observed in the immediate group. Therefore, the effects of the intervention on biomarkers were likely diminished by non-adherence to the protocol.

Vitamin K1 was above detectable limits in only eight participants at the initial study visit, indicating that our screening method assessing habitual GLV intake was reliable. The significant increase in Vitamin K1 during the intervention for both immediate and delayed groups suggests meaningful adherence to the protocol. A weeklong intervention providing 100 g broccoli daily reported a two-fold increase in plasma Vitamin K1 [41]. In our study, seven of 18 objectively adherent participants in the immediate group had undetectable Vitamin K1 levels after the 4-week washout period. Given the observation that Vitamin K1 levels can be increased in seven days and diminished in 28 days or less, it is clear that the potential benefits of GLV might require sustained consumption. Estimating a minimally therapeutic dose of GLV is warranted, though it is already established that a number of genetic variants affect Vitamin K1 absorption and metabolism [42].

Several DNA damage metabolites including 8OHdG have been measured, most prominently in urine, for decades. Since plasma or urinary levels of these metabolites are heavily influenced by kidney function and excretion, these measures have limited value as a cross-sectional screening tool [43]. Conversely, plasma 8OHdG is a promising biomarker in assessing longitudinal exposure to oxidative stress with relevance for cancer prevention and control [44] given its relevance to mutagen formation and carcinogenesis [45].

Few diets and dietary supplement interventions have assessed changes in 8OHdG in urine and even fewer in plasma. Urinary 8OHdG was decreased ($p = 0.041$) after 4 weeks of agraz (berry juice/nectar, 200 mL) supplementation in 40 women with metabolic syndrome participating in a crossover trial [46]. In a study of 50 men and women with metabolic syndrome randomized to receive 30 g mixed nuts daily for 12 weeks vs. control, urine 8-oxo-7-hydro-2′-deoxyguanosine (8-oxodG) decreased significantly more in the intervention group (−2.42 nmol/mmol creatinine, $p < 0.001$) compared to the control group [47]. Though we did not assess blood pressure or insulin sensitivity to determine metabolic syndrome in our study participants, age and BMI were comparable to participants in both of these studies.

An 8-week vegetable (300 g) and polyunsaturated oil (25 mL) supplemented dietary intervention did not decrease urinary 8-oxodG and 8-oxo-7,8-dihydroguanosine (8-oxoGuo); however, DNA damage assessed through peripheral blood mononuclear cells indicated lower double-stranded DNA breaks in 54 patients with diabetes [48]. The only comparable study assessing plasma 8OHdG was in postmenopausal women (n = 48) receiving 22 g blueberry powder daily vs. placebo, in which investigators observed a decrease in plasma

8OHdG at four ($p = 0.04$) but not eight weeks [49]. In addition to Vitamin K1, GLVs contain glucosinolates, carotenoids, folate, and other DNA-protective compounds [50]. Since all of the above-cited interventions contained phytochemical- and antioxidant-rich foods and supplements, the reduction in oxidative DNA damage observed in our study could also be expected.

In designing this clinical trial, we hypothesized that the dietary fiber in GLV would support the function and integrity of the intestinal epithelium, which would be mediated by bacteria and result in decreased systemic inflammation [51]. In short, lipopolysaccharide from Gram-negative bacteria activate Toll-like receptor 4, increasing intestinal permeability, immune activation, and production of TNFα [25,52]. While CRP is a non-specific cytokine, a recent meta-analysis observed that increased CRP is associated with risk of CRC [19]; elevated IL-6 has been associated with CRC recurrence [22].

These cytokines have been measured in numerous diet and dietary-supplement interventions aimed primarily at cardiometabolic and diabetes-related outcomes. Relevant studies typically observe decreases in one or two, but rarely all three of these cytokines. These interventions include grape and grapeseed extracts, tart cherry juice, and avocado, as well as supplements aimed to modulate microbiota using synbiotics, kefir, and resistant starch [53–62]. Our observed trend in TNFα reduction may be spurious given the relative stability of the gut microbiota during the 12-week study.

Though there were statistically significant changes in some of these biomarkers, minimum clinically important difference (MCID) for prevention of CRC have not been established. Associations between risk, progression, and/or mortality of CRC with these biomarkers have been observed and can inform but not validate MCID at this time. Thus, we propose the following MCIDs: plasma 8OHdG—11 ng/mL [16], serum TNFa—30 pg/mL [63], serum IL6—3 pg/mL [63], serum CRP—1750 pg/mL [64]. Plasma 8OHdG was the only biomarker in our study that had changes close to the proposed MCID, but longitudinal studies are needed to validate each of these.

The relative abundance of Firmicutes in our sample is higher than our lab has previously observed in overweight and obese men and women in the southeastern U.S. [40,65]. Since alpha diversity measured by observed species was also lower, it is unlikely that a potential overgrowth of species after collection caused an increase in relative abundance of Firmicutes. Nonetheless, we hypothesized the exponential increase in dietary fiber and bioactive plant compounds from GLV during the intervention period would increase microbial diversity [66]; however, diversity remained relatively constant for all participants throughout our study.

Three of the top five genera (Figure 2) are known butyrate producers; within these genera are species in Clostridium Clusters IV and XIVa [67]. Increased abundance of these bacteria that convert otherwise unabsorbable carbohydrates (soluble fibers, resistant starches, etc.) into absorbable short-chain fatty acids result in greater amounts of energy harvested from food by the host [68] and may contribute to obesity [69]. More specifically, species within the *Faecalibacterium, Ruminoccocus,* and *Roseburia* genera produce butyrate, which is beneficial to epithelial cells; however, the healthy phenotype results in a marked increase in Bacteroidetes [70], which was not observed in our participants. Recently, abundance of *Blautia* has been inversely associated with visceral fat area in both men and women [71]. Posthoc analysis of our data did not support this finding, possibly because our study population had BMI > 30, leaving no lean comparators.

Limitations

Because participants were recruited primarily from within the faculty and staff of a university in the southeastern United States, adults with a bachelor's degree or higher were overrepresented (48%) compared to the national (31.5%) and state (24.9%) averages [72,73]. Twenty percent of participants were African American, which is lower than the state (26.8%) but higher than the national average (13.4%) [72,73]. Thus, adherence and effects of the intervention may not be generalizable to other populations.

Though the crossover design in this study allowed us to maximize potential effects given our modest sample size, the high rate of drop in both prior to and after intervention periods diluted many of the trends, which did not reach statistical significance in biomarkers. A brief pre-trial run-in control period may have prevented the initial drop-in by some and allowed more time to educate participants on the value of strictly adhering to protocol not only during the intervention but in the control period as well. Since we were the first group to report fecal 8OHdG, our methods may not have been sensitive enough to adjust for the water content of original samples; thus, these results should be interpreted with caution. Nonetheless, though not powered to detect changes in biomarkers, plasma and fecal 8OHdG as well as serum TNFα were significantly decreased by the intervention, which warrants further investigation.

5. Conclusions

This randomized controlled crossover dietary intervention is the first to report potential benefits of increasing green leafy vegetable consumption in adults at increased risk for CRC. Because of the small sample size resulting from powering the study for feasibility, the results are exploratory and should be interpreted with caution. Nonetheless, plasma and fecal 8OHdG, a biomarker of DNA damage, and serum TNFα were decreased by the intervention in conjunction with increased plasma Vitamin K1, the objective measure of dietary adherence. The direction of key effects and the heterogeneity of the effects evidence across individuals suggest that a larger study is warranted. Additionally, it is important to determine whether this can be replicated in a larger, more diverse population and to further explore the relationship between decreased 8OHdG and CRC risk reduction.

Author Contributions: Conceptualization, A.D.F., W.D.-W., A.E.A., and K.B.-L.; Data curation, K.S.S. and R.T.-C.; Formal analysis, W.M.M. and A.D.F.; Funding acquisition, A.D.F. and K.B.-L.; Methodology, A.D.F., K.S.S., W.D.-W., A.E.A., W.J.v.d.P., C.D.M., and R.D.A.; Project administration, A.D.F., K.S.S., and A.J.R.; Writing—original draft, A.D.F.; Writing—review and editing, K.S.S., A.J.R., R.T.-C., W.D.-W., A.E.A., W.M.M., W.J.v.d.P., C.D.M., S.L.J., R.D.A., and K.B.-L. All authors have read and agreed to the published version of the manuscript.

Funding: This study was supported by Auburn University, the Auburn University Research Initiative in Cancer, R25 (CA047888) and UL1TR003096.

Institutional Review Board Statement: The study was conducted according to the guidelines of the Declaration of Helsinki, and approved by the Auburn University Institutional Review Board protocol #18-208 EP 1806, approved 18 June 2018).

Informed Consent Statement: Informed consent was obtained from all subjects involved in the study.

Data Availability Statement: Deidentified data may be obtained from the corresponding author upon written request.

Conflicts of Interest: The authors declare no conflict of interest.

References

1. Bray, F.; Ferlay, J.; Soerjomataram, I.; Siegel, R.L.; Torre, L.A.; Jemal, A. Global cancer statistics 2018: GLOBOCAN estimates of incidence and mortality worldwide for 36 cancers in 185 countries. *CA Cancer J. Clin.* **2018**, *68*, 394–424. [CrossRef]
2. Vieira, A.R.; Abar, L.; Chan, D.S.M.; Vingeliene, S.; Polemiti, E.; Stevens, C.; Greenwood, D.; Norat, T. Foods and beverages and colorectal cancer risk: A systematic review and meta-analysis of cohort studies, an update of the evidence of the WCRF-AICR Continuous Update Project. *Ann. Oncol. Off. J. Eur. Soc. Med. Oncol.* **2017**, *28*, 1788–1802. [CrossRef]
3. Zhang, F.F.; Cudhea, F.; Shan, Z.; Michaud, D.S.; Imamura, F.; Eom, H.; Ruan, M.; Rehm, C.D.; Liu, J.; Du, M.; et al. Preventable Cancer Burden Associated With Poor Diet in the United States. *JNCI Cancer Spectr.* **2019**, *3*. [CrossRef]
4. Alsheridah, N.; Akhtar, S. Diet, obesity and colorectal carcinoma risk: Results from a national cancer registry-based middle-eastern study. *BMC Cancer* **2018**, *18*, 1227. [CrossRef]
5. Garcia-Larsen, V.; Morton, V.; Norat, T.; Moreira, A.; Potts, J.F.; Reeves, T.; Bakolis, I. Dietary patterns derived from principal component analysis (PCA) and risk of colorectal cancer: A systematic review and meta-analysis. *Eur. J. Clin. Nutr.* **2019**, *73*, 366–386. [CrossRef]

6. Giovannucci, E.; Ascherio, A.; Rimm, E.B.; Colditz, G.A.; Stampfer, M.J.; Willett, W.C. Physical Activity, Obesity, and Risk for Colon Cancer and Adenoma in Men. *Ann. Intern. Med.* **1995**, *122*, 327–334. [CrossRef]
7. Abar, L.; Vieira, A.R.; Aune, D.; Sobiecki, J.G.; Vingeliene, S.; Polemiti, E.; Stevens, C.; Greenwood, D.C.; Chan, D.S.M.; Schlesinger, S.; et al. Height and body fatness and colorectal cancer risk: An update of the WCRF-AICR systematic review of published prospective studies. *Eur. J. Nutr.* **2018**, *57*, 1701–1720. [CrossRef]
8. Mozaffarian, D.; Hao, T.; Rimm, E.B.; Willett, W.C.; Hu, F.B. Changes in diet and lifestyle and long-term weight gain in women and men. *N. Engl. J. Med.* **2011**, *364*, 2392–2404. [CrossRef]
9. Smith, K.S.; Raney, S.V.; Greene, M.W.; Frugé, A.D. Development and Validation of the Dietary Habits and Colon Cancer Beliefs Survey (DHCCBS): An Instrument Assessing Health Beliefs Related to Red Meat and Green Leafy Vegetable Consumption. *J. Oncol.* **2019**, *2019*, 2326808. [CrossRef]
10. De Vogel, J.; Jonker-Termont, D.S.; van Lieshout, E.M.; Katan, M.B.; van der Meer, R. Green vegetables, red meat and colon cancer: Chlorophyll prevents the cytotoxic and hyperproliferative effects of haem in rat colon. *Carcinogenesis* **2005**, *26*, 387–393. [CrossRef]
11. De Vogel, J.; Jonker-Termont, D.S.M.L.; Katan, M.B.; van der Meer, R. Natural Chlorophyll but Not Chlorophyllin Prevents Heme-Induced Cytotoxic and Hyperproliferative Effects in Rat Colon. *J. Nutr.* **2005**, *135*, 1995–2000. [CrossRef]
12. De Vogel, J.; van-Eck, W.B.; Sesink, A.L.; Jonker-Termont, D.S.; Kleibeuker, J.; van der Meer, R. Dietary heme injures surface epithelium resulting in hyperproliferation, inhibition of apoptosis and crypt hyperplasia in rat colon. *Carcinogenesis* **2008**, *29*, 398–403. [CrossRef] [PubMed]
13. Jssennagger, N.J.; Derrien, M.; van Doorn, G.M.; Rijnierse, A.; van den Bogert, B.; Muller, M.; Dekker, J.; Kleerebezem, M.; van der Meer, R. Dietary heme alters microbiota and mucosa of mouse colon without functional changes in host-microbe cross-talk. *PLoS ONE* **2012**, *7*, e49868. [CrossRef] [PubMed]
14. Martin, O.C.; Lin, C.; Naud, N.; Tache, S.; Raymond-Letron, I.; Corpet, D.E.; Pierre, F.H. Antibiotic suppression of intestinal microbiota reduces heme-induced lipoperoxidation associated with colon carcinogenesis in rats. *Nutr. Cancer* **2015**, *67*, 119–125. [CrossRef] [PubMed]
15. Ijssennagger, N.; Belzer, C.; Hooiveld, G.J.; Dekker, J.; van Mil, S.W.; Muller, M.; Kleerebezem, M.; van der Meer, R. Gut microbiota facilitates dietary heme-induced epithelial hyperproliferation by opening the mucus barrier in colon. *Proc. Natl. Acad. Sci. USA* **2015**, *112*, 10038–10043. [CrossRef] [PubMed]
16. Sato, T.; Takeda, H.; Otake, S.; Yokozawa, J.; Nishise, S.; Fujishima, S.; Orii, T.; Fukui, T.; Takano, J.; Sasaki, Y.; et al. Increased Plasma Levels of 8-Hydroxydeoxyguanosine Are Associated with Development of Colorectal Tumors. *J. Clin. Biochem. Nutr.* **2010**, *47*, 59–63. [CrossRef]
17. Verdam, F.J.; Fuentes, S.; de Jonge, C.; Zoetendal, E.G.; Erbil, R.; Greve, J.W.; Buurman, W.A.; de Vos, W.M.; Rensen, S.S. Human intestinal microbiota composition is associated with local and systemic inflammation in obesity. *Obesity* **2013**, *21*, E607–E615. [CrossRef] [PubMed]
18. Xiao, S.; Fei, N.; Pang, X.; Shen, J.; Wang, L.; Zhang, B.; Zhang, M.; Zhang, X.; Zhang, C.; Li, M.; et al. A gut microbiota-targeted dietary intervention for amelioration of chronic inflammation underlying metabolic syndrome. *FEMS Microbiol. Ecol.* **2014**, *87*, 357–367. [CrossRef]
19. Zhou, B.; Shu, B.; Yang, J.; Liu, J.; Xi, T.; Xing, Y. C-reactive protein, interleukin-6 and the risk of colorectal cancer: A meta-analysis. *Cancer Causes Control CCC* **2014**, *25*, 1397–1405. [CrossRef] [PubMed]
20. Allin, K.H.; Nordestgaard, B.G. Elevated C-reactive protein in the diagnosis, prognosis, and cause of cancer. *Crit. Rev. Clin. Lab. Sci.* **2011**, *48*, 155–170. [CrossRef] [PubMed]
21. Goyal, A.; Terry, M.B.; Jin, Z.; Siegel, A.B. C-reactive protein and colorectal cancer mortality in U.S. adults. *Cancer Epidemiol. Biomark.* **2014**, *23*, 1609–1618. [CrossRef]
22. Shiga, K.; Hara, M.; Nagasaki, T.; Sato, T.; Takahashi, H.; Sato, M.; Takeyama, H. Preoperative Serum Interleukin-6 Is a Potential Prognostic Factor for Colorectal Cancer, including Stage II Patients. *Gastroenterol. Res. Pract.* **2016**, *2016*, 8. [CrossRef]
23. Tawara, K.; Oxford, J.T.; Jorcyk, C.L. Clinical significance of interleukin (IL)-6 in cancer metastasis to bone: Potential of anti-IL-6 therapies. *Cancer Manag. Res.* **2011**, *3*, 177–189. [CrossRef]
24. Waldner, M.J.; Foersch, S.; Neurath, M.F. Interleukin-6—A Key Regulator of Colorectal Cancer Development. *Int. J. Biol. Sci.* **2012**, *8*, 1248–1253. [CrossRef] [PubMed]
25. Cani, P.D.; Amar, J.; Iglesias, M.A.; Poggi, M.; Knauf, C.; Bastelica, D.; Neyrinck, A.M.; Fava, F.; Tuohy, K.M.; Chabo, C.; et al. Metabolic Endotoxemia Initiates Obesity and Insulin Resistance. *Diabetes* **2007**, *56*, 1761–1772. [CrossRef]
26. Li, Y.; Zhang, T.; Chen, G.Y. Flavonoids and Colorectal Cancer Prevention. *Antioxidants* **2018**, *7*, 187. [CrossRef] [PubMed]
27. Frugé, A.D.; Smith, K.S.; Riviere, A.J.; Demark-Wahnefried, W.; Arthur, A.E.; Murrah, W.M.; Morrow, C.D.; Arnold, R.D.; Braxton-Lloyd, K. Primary Outcomes of a Randomized Controlled Crossover Trial to Explore the Effects of a High Chlorophyll Dietary Intervention to Reduce Colon Cancer Risk in Adults: The Meat and Three Greens (M3G) Feasibility Trial. *Nutrients* **2019**, *11*, 2349. [CrossRef] [PubMed]
28. Barnard, N.D.; Gloede, L.; Cohen, J.; Jenkins, D.J.A.; Turner-McGrievy, G.; Green, A.A.; Ferdowsian, H. A Low-Fat Vegan Diet Elicits Greater Macronutrient Changes, but Is Comparable in Adherence and Acceptability, Compared with a More Conventional Diabetes Diet among Individuals with Type 2 Diabetes. *J. Am. Diet. Assoc.* **2009**, *109*, 263–272. [CrossRef] [PubMed]
29. Moore, W.J.; McGrievy, M.E.; Turner-McGrievy, G.M. Dietary adherence and acceptability of five different diets, including vegan and vegetarian diets, for weight loss: The New DIETs study. *Eat. Behav.* **2015**, *19*, 33–38. [CrossRef] [PubMed]

30. Barnard, N.; Scialli, A.R.; Bertron, P.; Hurlock, D.; Edmunds, K. Acceptability of a therapeutic low-fat, vegan diet in premenopausal women. *J. Nutr. Educ. Behav.* **2000**, *32*, 314–319. [CrossRef]
31. Craig, C.L.; Marshall, A.L.; Sjostrom, M.; Bauman, A.E.; Booth, M.L.; Ainsworth, B.E.; Pratt, M.; Ekelund, U.; Yngve, A.; Sallis, J.F.; et al. International physical activity questionnaire: 12-country reliability and validity. *Med. Sci. Sports Exerc.* **2003**, *35*, 1381–1395. [CrossRef]
32. Subar, A.F.; Kirkpatrick, S.I.; Mittl, B.; Zimmerman, T.P.; Thompson, F.E.; Bingley, C.; Willis, G.; Islam, N.G.; Baranowski, T.; McNutt, S.; et al. The Automated Self-Administered 24-Hour Dietary Recall (ASA24): A Resource for Researchers, Clinicians, and Educators from the National Cancer Institute. *J. Acad. Nutr. Diet.* **2012**, *112*, 1134–1137. [CrossRef]
33. Lohman, T.G.; Roche, A.F.; Martorell, R. *Anthropometric Standardization Reference Manual*; Human Kinetics Books: Champaign, IL, USA, 1988; Volume 177.
34. O'Donnell, L.J.; Virjee, J.; Heaton, K.W. Detection of pseudodiarrhoea by simple clinical assessment of intestinal transit rate. *BMJ* **1990**, *300*, 439–440. [CrossRef] [PubMed]
35. Caporaso, J.G.; Kuczynski, J.; Stombaugh, J.; Bittinger, K.; Bushman, F.D.; Costello, E.K.; Fierer, N.; Pena, A.G.; Goodrich, J.K.; Gordon, J.I.; et al. QIIME allows analysis of high-throughput community sequencing data. *Nat. Meth.* **2010**, *7*, 335–336. [CrossRef]
36. Kumar, R.; Eipers, P.; Little, R.B.; Crowley, M.; Crossman, D.K.; Lefkowitz, E.J.; Morrow, C.D. Getting started with microbiome analysis: Sample acquisition to bioinformatics. *Curr. Protoc. Hum. Genet.* **2014**, *82*, 18.18.11–18.18.29. [CrossRef]
37. Longstreth, G.F.; Thompson, W.G.; Chey, W.D.; Houghton, L.A.; Mearin, F.; Spiller, R.C. Functional Bowel Disorders. *Gastroenterology* **2006**, *130*, 1480–1491. [CrossRef] [PubMed]
38. Blake, M.R.; Raker, J.M.; Whelan, K. Validity and reliability of the Bristol Stool Form Scale in healthy adults and patients with diarrhoea-predominant irritable bowel syndrome. *Aliment. Pharmacol. Ther.* **2016**, *44*, 693–703. [CrossRef] [PubMed]
39. Rosenthal, R.; Rubin, D.B. r equivalent: A simple effect size indicator. *Psychol. Methods* **2003**, *8*, 492–496. [CrossRef]
40. Fruge, A.D.; Ptacek, T.; Tsuruta, Y.; Morrow, C.D.; Azrad, M.; Desmond, R.A.; Hunter, G.R.; Rais-Bahrami, S.; Demark-Wahnefried, W. Dietary Changes Impact the Gut Microbe Composition in Overweight and Obese Men with Prostate Cancer Undergoing Radical Prostatectomy. *J. Acad. Nutr. Diet* **2016**. [CrossRef] [PubMed]
41. O'Sullivan, S.M.; Galvin, K.; Heneghan, C.; Davidson, R.; Clark, I.; Lucey, A.J. Does daily consumption of vitamin K1 from cruciferous vegetables reach the circulation and the knee joint? *Proc. Nutr. Soc.* **2018**, *77*, E68. [CrossRef]
42. Westerman, K.; Kelly, J.M.; Ordovás, J.M.; Booth, S.L.; DeMeo, D.L. Epigenome-wide association study reveals a molecular signature of response to phylloquinone (vitamin K1) supplementation. *Epigenetics* **2020**, 1–12. [CrossRef] [PubMed]
43. Sanchez, M.; Roussel, R.; Hadjadj, S.; Moutairou, A.; Marre, M.; Velho, G.; Mohammedi, K. Plasma concentrations of 8-hydroxy-2′-deoxyguanosine and risk of kidney disease and death in individuals with type 1 diabetes. *Diabetologia* **2018**, *61*, 977–984. [CrossRef] [PubMed]
44. Valavanidis, A.; Vlachogianni, T.; Fiotakis, C. 8-hydroxy-2′-deoxyguanosine (8-OHdG): A critical biomarker of oxidative stress and carcinogenesis. *J. Environ. Sci. Health. Part C* **2009**, *27*, 120–139. [CrossRef]
45. Kasai, H. Analysis of a form of oxidative DNA damage, 8-hydroxy-2′-deoxyguanosine, as a marker of cellular oxidative stress during carcinogenesis. *Mutat. Res.* **1997**, *387*, 147–163. [CrossRef]
46. Espinosa-Moncada, J.; Marín-Echeverri, C.; Galvis-Pérez, Y.; Ciro-Gómez, G.; Aristizábal, J.C.; Blesso, C.N.; Fernandez, M.L.; Barona-Acevedo, J. Evaluation of Agraz Consumption on Adipocytokines, Inflammation, and Oxidative Stress Markers in Women with Metabolic Syndrome. *Nutrients* **2018**, *10*, 1639. [CrossRef]
47. López-Uriarte, P.; Nogués, R.; Saez, G.; Bulló, M.; Romeu, M.; Masana, L.; Tormos, C.; Casas-Agustench, P.; Salas-Salvadó, J. Effect of nut consumption on oxidative stress and the endothelial function in metabolic syndrome. *Clin. Nutr.* **2010**, *29*, 373–380. [CrossRef]
48. Müllner, E.; Brath, H.; Pleifer, S.; Schiermayr, C.; Baierl, A.; Wallner, M.; Fastian, T.; Millner, Y.; Paller, K.; Henriksen, T.; et al. Vegetables and PUFA-rich plant oil reduce DNA strand breaks in individuals with type 2 diabetes. *Mol. Nutr. Food Res.* **2013**, *57*, 328–338. [CrossRef]
49. Johnson, S.A.; Feresin, R.G.; Navaei, N.; Figueroa, A.; Elam, M.L.; Akhavan, N.S.; Hooshmand, S.; Pourafshar, S.; Payton, M.E.; Arjmandi, B.H. Effects of daily blueberry consumption on circulating biomarkers of oxidative stress, inflammation, and antioxidant defense in postmenopausal women with pre- and stage 1-hypertension: A randomized controlled trial. *Food Funct.* **2017**, *8*, 372–380. [CrossRef]
50. Manchali, S.; Chidambara Murthy, K.N.; Patil, B.S. Crucial facts about health benefits of popular cruciferous vegetables. *J. Funct. Foods* **2012**, *4*, 94–106. [CrossRef]
51. Kamada, N.; Núñez, G. Regulation of the immune system by the resident intestinal bacteria. *Gastroenterology* **2014**, *146*, 1477–1488. [CrossRef] [PubMed]
52. Chow, J.C.; Young, D.W.; Golenbock, D.T.; Christ, W.J.; Gusovsky, F. Toll-like receptor-4 mediates lipopolysaccharide-induced signal transduction. *J. Biol. Chem.* **1999**, *274*, 10689–10692. [CrossRef] [PubMed]
53. Rabiei, S.; Hedayati, M.; Rashidkhani, B.; Saadat, N.; Shakerhossini, R. The Effects of Synbiotic Supplementation on Body Mass Index, Metabolic and Inflammatory Biomarkers, and Appetite in Patients with Metabolic Syndrome: A Triple-Blind Randomized Controlled Trial. *J. Diet. Suppl.* **2019**, *16*, 294–306. [CrossRef]
54. Park, E.; Edirisinghe, I.; Burton-Freeman, B. Avocado Fruit on Postprandial Markers of Cardio-Metabolic Risk: A Randomized Controlled Dose Response Trial in Overweight and Obese Men and Women. *Nutrients* **2018**, *10*, 1287. [CrossRef]

55. Mayr, H.L.; Itsiopoulos, C.; Tierney, A.C.; Ruiz-Canela, M.; Hebert, J.R.; Shivappa, N.; Thomas, C.J. Improvement in dietary inflammatory index score after 6-month dietary intervention is associated with reduction in interleukin-6 in patients with coronary heart disease: The AUSMED heart trial. *Nutr. Res.* **2018**, *55*, 108–121. [CrossRef]
56. Martin, K.R.; Burrell, L.; Bopp, J. Authentic tart cherry juice reduces markers of inflammation in overweight and obese subjects: A randomized, crossover pilot study. *Food Funct.* **2018**, *9*, 5290–5300. [CrossRef] [PubMed]
57. López-Chillón, M.T.; Carazo-Díaz, C.; Prieto-Merino, D.; Zafrilla, P.; Moreno, D.A.; Villaño, D. Effects of long-term consumption of broccoli sprouts on inflammatory markers in overweight subjects. *Clin. Nutr.* **2019**, *38*, 745–752. [CrossRef]
58. Kopf, J.C.; Suhr, M.J.; Clarke, J.; Eyun, S.I.; Riethoven, J.M.; Ramer-Tait, A.E.; Rose, D.J. Role of whole grains versus fruits and vegetables in reducing subclinical inflammation and promoting gastrointestinal health in individuals affected by overweight and obesity: A randomized controlled trial. *Nutr. J.* **2018**, *17*, 72. [CrossRef]
59. Hou, Y.Y.; Ojo, O.; Wang, L.L.; Wang, Q.; Jiang, Q.; Shao, X.Y.; Wang, X.H. A Randomized Controlled Trial to Compare the Effect of Peanuts and Almonds on the Cardio-Metabolic and Inflammatory Parameters in Patients with Type 2 Diabetes Mellitus. *Nutrients* **2018**, *10*, 1565. [CrossRef] [PubMed]
60. Estévez-Santiago, R.; Silván, J.M.; Can-Cauich, C.A.; Veses, A.M.; Alvarez-Acero, I.; Martinez-Bartolome, M.A.; San-Román, R.; Cámara, M.; Olmedilla-Alonso, B.; de Pascual-Teresa, S. Lack of a Synergistic Effect on Cardiometabolic and Redox Markers in a Dietary Supplementation with Anthocyanins and Xanthophylls in Postmenopausal Women. *Nutrients* **2019**, *11*, 1533. [CrossRef] [PubMed]
61. Esgalhado, M.; Kemp, J.A.; Azevedo, R.; Paiva, B.R.; Stockler-Pinto, M.B.; Dolenga, C.J.; Borges, N.A.; Nakao, L.S.; Mafra, D. Could resistant starch supplementation improve inflammatory and oxidative stress biomarkers and uremic toxins levels in hemodialysis patients? A pilot randomized controlled trial. *Food Funct.* **2018**, *9*, 6508–6516. [CrossRef]
62. Bardagjy, A.S.; Hu, Q.; Giebler, K.A.; Ford, A.; Steinberg, F.M. Effects of grape consumption on biomarkers of inflammation, endothelial function, and PBMC gene expression in obese subjects. *Arch. Biochem. Biophys.* **2018**, *646*, 145–152. [CrossRef] [PubMed]
63. Chang, P.H.; Pan, Y.P.; Fan, C.W.; Tseng, W.K.; Huang, J.S.; Wu, T.H.; Chou, W.C.; Wang, C.H.; Yeh, K.Y. Pretreatment serum interleukin-1β, interleukin-6, and tumor necrosis factor-α levels predict the progression of colorectal cancer. *Cancer Med.* **2016**, *5*, 426–433. [CrossRef] [PubMed]
64. Erlinger, T.P.; Platz, E.A.; Rifai, N.; Helzlsouer, K.J. C-Reactive Protein and the Risk of Incident Colorectal Cancer. *JAMA* **2004**, *291*, 585–590. [CrossRef] [PubMed]
65. Frugé, A.D.; Van der Pol, W.; Rogers, L.Q.; Morrow, C.D.; Tsuruta, Y.; Demark-Wahnefried, W. Fecal Akkermansia muciniphila Is Associated with Body Composition and Microbiota Diversity in Overweight and Obese Women with Breast Cancer Participating in a Presurgical Weight Loss Trial. *J. Acad. Nutr. Diet* **2020**, *120*, 650–659. [CrossRef] [PubMed]
66. De Angelis, M.; Ferrocino, I.; Calabrese, F.M.; De Filippis, F.; Cavallo, N.; Siragusa, S.; Rampelli, S.; Di Cagno, R.; Rantsiou, K.; Vannini, L.; et al. Diet influences the functions of the human intestinal microbiome. *Sci. Rep.* **2020**, *10*, 4247. [CrossRef] [PubMed]
67. Louis, P.; Flint, H.J. Diversity, metabolism and microbial ecology of butyrate-producing bacteria from the human large intestine. *FEMS Microbiol. Lett.* **2009**, *294*, 1–8. [CrossRef] [PubMed]
68. Davis, C.D. The Gut Microbiome and Its Role in Obesity. *Nutr. Today* **2016**, *51*, 167–174. [CrossRef] [PubMed]
69. Chakraborti, C.K. New-found link between microbiota and obesity. *World J. Gastrointest. Pathophysiol.* **2015**, *6*, 110–119. [CrossRef] [PubMed]
70. Lu, Y.; Fan, C.; Li, P.; Lu, Y.; Chang, X.; Qi, K. Short Chain Fatty Acids Prevent High-fat-diet-induced Obesity in Mice by Regulating G Protein-coupled Receptors and Gut Microbiota. *Sci. Rep.* **2016**, *6*, 37589. [CrossRef] [PubMed]
71. Ozato, N.; Saito, S.; Yamaguchi, T.; Katashima, M.; Tokuda, I.; Sawada, K.; Katsuragi, Y.; Kakuta, M.; Imoto, S.; Ihara, K.; et al. Blautia genus associated with visceral fat accumulation in adults 20–76 years of age. *NPJ Biofilms Microbiomes* **2019**, *5*, 28. [CrossRef]
72. United States Census Bureau. Quickfacts United States. 2020. Available online: https://www.census.gov/quickfacts/fact/table/US/PST045218 (accessed on 1 December 2020).
73. United States Census Bureau. QuickFacts Alabama. 2020. Available online: https://www.census.gov/quickfacts/AL (accessed on 1 December 2020).

MDPI

Review

Time-Restricted Eating: A Novel and Simple Dietary Intervention for Primary and Secondary Prevention of Breast Cancer and Cardiovascular Disease

Rebecca A. G. Christensen [1] and Amy A. Kirkham [2,*]

[1] Dalla Lana School of Public Health, University of Toronto, Toronto, ON M5T 3M7, Canada; r.christensen@mail.utoronto.ca

[2] Faculty of Kinesiology and Physical Education, University of Toronto, Toronto, ON M5S 2C9, Canada

* Correspondence: amy.kirkham@utoronto.ca; Tel.: +1-416-946-4069

Abstract: There is substantial overlap in risk factors for the pathogenesis and progression of breast cancer (BC) and cardiovascular disease (CVD), including obesity, metabolic disturbances, and chronic inflammation. These unifying features remain prevalent after a BC diagnosis and are exacerbated by BC treatment, resulting in elevated CVD risk among survivors. Thus, therapies that target these risk factors or mechanisms are likely to be effective for the prevention or progression of both conditions. In this narrative review, we propose time-restricted eating (TRE) as a simple lifestyle therapy to address many upstream causative factors associated with both BC and CVD. TRE is simple dietary strategy that typically involves the consumption of ad libitum energy intake within 8 h, followed by a 16-h fast. We describe the feasibility and safety of TRE and the available evidence for the impact of TRE on metabolic, cardiovascular, and cancer-specific health benefits. We also highlight potential solutions for overcoming barriers to adoption and adherence and areas requiring future research. In composite, we make the case for the use of TRE as a novel, safe, and feasible intervention for primary and secondary BC prevention, as well as tertiary prevention as it relates to CVD in BC survivors.

Keywords: breast cancer; cardiovascular disease; time-restricted eating; time-restricted feeding; intermittent fasting; metabolic syndrome; fasting

Citation: Christensen, R.A.G.; Kirkham, A.A. Time-Restricted Eating: A Novel and Simple Dietary Intervention for Primary and Secondary Prevention of Breast Cancer and Cardiovascular Disease. *Nutrients* **2021**, *13*, 3476. https://doi.org/10.3390/nu13103476

Academic Editors: Wendy Demark-Wahnefried and Christina Dieli-Conwright

Received: 8 September 2021
Accepted: 28 September 2021
Published: 30 September 2021

Publisher's Note: MDPI stays neutral with regard to jurisdictional claims in published maps and institutional affiliations.

1. Introduction

Breast cancer is the most common malignancy among women worldwide, with 1 in 8 North American women expected to be diagnosed in their lifetime [1]. While there is no single biological target for the primary prevention of breast cancer, diet, adult weight gain, and obesity are estimated to be responsible for up to 50% of cases [2–5]. Metabolic dysfunction, signified by presence of hyperglycemia, dyslipidemia, hypertension, and abdominal obesity, is primary driver in the risk of type 2 diabetes and cardiovascular diseases [6]. However, in the past decade, metabolic dysfunction has also emerged as an underlying determinant of the relationship between obesity and breast cancer risk [7–10]. Other systemic factors associated with overweight/obesity and cardiovascular disease, such as chronic inflammation [11] and oxidative stress [12], are also associated with breast cancer risk [13–17]. Thus, strategies for the primary prevention of breast cancer are also likely to impact the risk of cardiovascular and metabolic diseases.

Fortunately for those who receive a breast cancer diagnosis, the death rate for early stage (non-metastatic) breast cancer has dropped by over 40% in the last 40 years [18]. Concomitant to improved cancer survival, the death rate due to cardiovascular disease has increased and now approaches the rate of cancer death [19]. In fact, women diagnosed with breast cancer are at a 2 to 3-fold elevated risk of cardiovascular-related death relative to the general population perpetually after their diagnosis [19]. While the elevated cardiovascular risk partially arises from the presence of pre-existing shared risk factors, it is also compounded by breast cancer treatments that result in direct toxicity to the heart ('cardiotoxicity'), metabolic dysfunction, as well as lifestyle toxicity (i.e., physical inactivity, poor diet, weight gain [20,21]). The biological and behavioral sequelae resulting from breast cancer treatment can persist long into the survivorship period, as evidenced by the excess risk of cardiovascular death increasing to >5-fold at 10+ years after diagnosis [19]. In the 20 years following a breast cancer diagnosis, the risk of recurrence of breast cancer ranges from 10–38% depending on diagnosis characteristics [22]. Similar to primary prevention of breast cancer, risk factor targets for secondary prevention as it relates specifically to prevention of cancer recurrence, second cancers, and cancer mortality include poor diet [23], physical inactivity [24] and obesity [25,26]. Thus, after a breast cancer diagnosis, lifestyle intervention strategies for secondary prevention will also have overlap with tertiary prevention as it relates to addressing the cardiovascular and metabolic sequelae of treatment and prevention of the related diseases.

The multi-faceted and intertwined risk profile for breast cancer and cardiovascular disease thus creates a shared opportunity to reduce the risk of both conditions by treating their shared underlying biological and behavioral mechanisms. In this context, therapies that target multiple possible biologic and behavioral mechanisms or pathways for these conditions will be the most effective for prevention.

Intermittent fasting is a relatively new dietary intervention that has recently gained substantial public interest. There are multiple different formats including alternative day fasting (i.e., restriction of caloric intake on 2 to 5 days/week alternated with ad libitum consumption), periodic prolonged fasting (i.e., 24 h to one week), and repeated daily fasting. The latter, referred to as Time Restricted Feeding (for animals) or Time-Restricted Eating (TRE) (for humans), involves the consumption of ad libitum energy intake within a set time window, ranging from 4–10 h, but most commonly 8 h. This is followed by a water-only fast for the remaining time in the 24-h period, typically 16 h (i.e., "16:8 TRE"). As 16:8 TRE requires relatively minor lifestyle changes and has simple instructions [27], it may be feasible as a long-term lifestyle intervention. Importantly, TRE has numerous health benefits that are relevant to the primary, secondary, and tertiary prevention of breast cancer. The purpose of this narrative review is to describe all relevant peer reviewed literature on the potential for TRE as a therapy for primary and secondary prevention of breast cancer, secondary prevention as it relates to recurrence or cancer mortality, as well as tertiary prevention as it relates to cardiovascular disease in breast cancer survivors.

2. TRE Health Benefits and Mechanisms

2.1. Body and Fat Mass

Body mass is an important determinant of metabolic health, which is often signified by the absence of elevated cholesterol, triglycerides, blood pressure, blood sugar and waist circumference. Body mass index (BMI) is most often used to classify the level of health risk associated with body mass relative to height. Notably, having even a modest increase in body mass, such that BMI exceeds 25 kg/m^2, as defined by having overweight, is a predictor of postmenopausal breast cancer [28], breast cancer mortality [29] and cardiovascular disease in women [30]. Accumulation of fat mass in particular is the driver of increased risk of breast cancer in postmenopausal women, as this is the main source of estrogen in the body after menopause [31]. Likewise for risk of cardiovascular disease, body fat is a stronger predictor than the more generalized measure of BMI [32]. In addition, weight gain is common after a breast cancer diagnosis, with 50–96% of women reporting weight gain during breast cancer treatment [33]. While weight gain is typically associated with an increase in both fat mass and fat-free mass, following a breast cancer diagnosis, the more common pattern observed is an increase in fat mass, but a decrease in fat-free mass [33,34].

Preclinical data have highlighted the significance of the timing of food intake in weight gain [35,36]. The circadian rhythm is the oscillation of physiological rhythms between activity and rest, feeding and fasting, nutrient utilization and storage drive by the natural day/night cycle [37]. The brain and nearly every peripheral organ have circadian timekeeping mechanisms that impact their function [38]. Late dinner times, late-night snacking, or eating during the night (as is common in night or shift workers) can chronically disrupt circadian rhythms, impacting gene expression and lead to metabolic dysfunction and disease including obesity [37]. Shortening the length of the eating window and skewing intake away from the evening/night helps to re-align food intake with the natural 24-h human cycle of feeding and fasting [39]. TRE thereby resets the body's peripheral clocks, which results in improved oscillations in gene expression and enhanced energy metabolism [37].

A 16:8 ad libitum TRE protocol typically results in 20–30% spontaneous caloric restriction and mild weight loss of 1–4% over 1–12 weeks without the need to count calories [40]. A recent meta-analysis of 12 TRE intervention studies lasting 4–12 weeks with 294 participants reported an overall significant weight reduction of −0.9 kg (95% CI, −1.71 to −0.10) [41]. In a subgroup analysis of five studies including patients with metabolic abnormalities (i.e., overweight/obesity, pre-diabetes, metabolic syndrome), the weight loss was greater (−3.19 kg, 95% CI, −4.62 to −1.77) [41]. Longer duration studies of TRE are needed to determine if the amount of weight lost induced by TRE can meet or exceed the commonly accepted clinically relevant threshold of 5% of baseline weight [42,43]. The same meta-analysis also reported significant reductions in fat mass with TRE (−1.58 kg, 95% CI, −2.64 to −0.51) while preserving fat-free mass (−0.24 kg, 95% CI, −1.15 to 0.67) measured by bioelectrical impedance or dual-energy X-ray absorptiometry [41]. This is an important finding for the long-term viability of TRE as a health intervention, because weight loss interventions typically result in concomitant decreases in both fat and fat-free mass [44].

Preliminary evidence suggests that TRE reduces fat in 'ectopic' regions of the body (e.g., visceral, liver, intermuscular) [38,45–47]. Relative to whole-body fat mass or BMI, these specific locations of fat deposition are much more strongly linked to the primary risk of cardiovascular disease and breast cancer [48,49], and the risk of cardiovascular, breast cancer, and all-cause mortality among breast cancer survivors [50–52]. Given that breast cancer therapies, including chemotherapy, targeted therapy, and hormonal therapy, have been shown to result in rapid and persistent accumulation of visceral, liver, and intermuscular fat [53,54], TRE may be a promising therapy to employ during active treatment to prevent this metabolic toxicity. While there is interest in the use of intermittent fasting during chemotherapy treatment for breast cancer, the strategies employed to-date have involved longer periods of fasting (24–72 h), which are safe but potentially not widely feasible among humans [55]. The approach of a shortened window for eating each day with TRE may be more palatable for patients and may still provide protective effects against treatment toxicity including ectopic fat accumulation. The use of intermittent fasting during active treatment may also be effective for secondary prevention of cancer based on preliminary findings that nutrient deprivation sensitizes cancer cells to the damaging effects of chemotherapy [56,57].

2.2. Oxidative Stress and Inflammation

Oxidative stress and chronic inflammation are unifying features in the pathogenesis and progression in both cancer and cardiovascular disease and their shared risk factor of obesity [58]. Oxidative stress is a disrupted balance between the production of damaging reactive oxygen species and antioxidant defenses [59]. Accumulating evidence implicates the DNA damage and mutations of tumor suppressor genes associated with oxidative stress and reactive oxygen species as critical initial events in carcinogenesisis [60]. One study reported a relationship between higher levels of oxidative stress, as assessed by plasma lipoperoxides, and a two-fold greater risk of breast cancer recurrence (relative risk = 2.10, 95% CI 1.10–4.00) [17]. Elevated oxidative stress is also implicated in various types of cardiovascular disease, primarily through effects on endothelial function and myocardial calcium handling, which contribute to hypertension and/or arrhythmia [61]. Preliminary evidence suggests that TRE may reduce oxidative stress in men with pre-diabetes [62] and healthy men [63], but more evidence is needed in women with chronic disease. Notably, while excess body mass is associated with oxidative stress, TRE-induced weight loss was not a requisite for reduced oxidative stress in prediabetic men [62].

Obesity is said to be a state of chronic inflammation. Chronic inflammation promotes malignant transformation of cells, carcinogenesis, and progression and is a precursor to stroke, as well as mediates all stages of atherosclerosis and other cardiovascular disease events [58,64]. One pilot study of a 4-week 16:8 TRE intervention observed no significant changes in c-reactive protein [65], a marker of systemic inflammation. Conversely, two other studies employing longer durations of daily fasts (14–15 h) or intervention length (8 weeks) have observed significant decreases in the proinflammatory markers interleukin (IL)-6 and IL-1β [66,67]. Further, the reduction in proinflammatory markers was independent of weight loss in one study [66]. More research is needed to better elucidate the dose response effects of TRE on inflammation.

2.3. Metabolic Syndrome

Metabolic syndrome is a constellation of metabolic disturbances including hyperglycemia, dyslipidemia, hypertension, and abdominal obesity, that increase the risk of heart disease, stroke, and type 2 diabetes. Metabolic syndrome also appears to play an important role in breast cancer. A recent meta-analysis showed that women with metabolic syndrome have a 52% increased risk of breast cancer [68]. Among women diagnosed with breast cancer, metabolic syndrome increases the risk of recurrence or distant metastases and breast cancer mortality [69,70].

Glucose metabolism specifically has also recently emerged as a key biological mechanism in breast cancer development [71]. Biologic mechanisms underpinning this relationship include glucose-mediated upregulation of oncogenic pathways in non-malignant breast cells [72] and insulin resistance-related promotion of cellular proliferation and inhibition of apoptosis [73]. A meta-analysis reported that the clinical manifestation of impaired glucose control, type 2 diabetes, increases the risk of breast cancer by 23% [74]. The diagnostic blood marker for the diagnosis of diabetes, hemoglobin A1c, which provides a measure of the average blood glucose concentration over the previous 8–12 weeks, is associated with risk of breast cancer independent of diabetes [75], and the risk of cardiovascular disease in women without diabetes [76].

Two prospective cohort studies illustrate the potential for TRE to improve chronic glucose control as a strategy for primary and tertiary prevention of breast cancer. Among 2212 women with elevated BMI, each 3-h increase in habitual overnight fast time was associated with 19% lower odds of elevated hemoglobin A1c [77]. Among 2413 breast cancer survivors without diabetes, habitual overnight fasting duration was inversely associated with hemoglobin A1c [78]. Few TRE intervention studies have measured hemoglobin A1c, likely because the length of most interventions to-date ($\leq$12 weeks) are not long enough to impact this chronic marker. However, one study with a 10-h eating window in patients with metabolic syndrome reported that hemoglobin A1c significantly decreased among those with elevated baseline levels $\geq$5.7% without a concurrent change in physical activity [79]. Further, a meta-analysis of 10 TRE intervention studies with 238 participants reported a statistically significant but modest reduction in fasting blood glucose (−2.96 mg/dL, 95% CI, −5.60 to −0.33) [41].

The individual components of metabolic syndrome are also linked to both breast cancer and cardiovascular disease. Hypertension is one of the main causal risk factors related to cardiovascular disease [80] with a strong, positive dose–response relationship with the risk of death from ischemic heart disease and stroke [81]. Breast cancer risk is also associated with hypertension, with several meta-analyses reporting a 7–38% higher risk of breast cancer among women with hypertension compared to normotensive women [82,83]. A meta-analysis of six TRE studies with 97 participants found modest but clinically significant decreases in systolic (−3.07 mmHg, 95% CI, −5.76 to −0.37) and diastolic (−1.77 mmHg, 95% CI, −4.51 to 1.07) blood pressure [41]. Importantly TRE may reduce blood pressure independent of weight loss [40].

Dyslipidemia, defined as elevated total or low-density lipoprotein (LDL) cholesterol, or low high-density lipoprotein (HDL) cholesterol, is another important metabolic risk factor. Research regarding the association between blood lipid levels and breast cancer incidence is mixed. Some studies have suggested an inverse relationship between lipid levels and breast cancer risk [84,85] while others have shown a positive association [86,87]. This discrepancy may be explained by inclusion of women taking cholesterol-lowering drugs (statins), as these treatments have been shown to reduce breast cancer incidence, recurrence, and mortality [88]. While there are still some discordant results, research tends to suggest that the level of HDL is inversely associated with breast cancer risk [89,90]. In contrast, the detrimental effect of elevated cholesterol on cardiovascular disease risk is well established [91]. A meta-analysis of 14 TRE studies with 343 participants reported significant reductions in triglycerides (−11.60 mg/dL, 95% CI, −23.30 to −0.27), but highly variable effects on LDL (0.05 mg/dL, 95% CI, −4.77 to 4.87) and HDL (1.01 mg/dL, 95% CI, −1.52 to 3.55). It is possible that favorable changes to LDL would be evident once clinically significant weight loss (>5% from baseline) is attained with longer adherence to TRE [40]. A number of shorter duration (1–8 weeks) TRE studies have reported significantly increased HDL levels, while others have not, potentially related to concomitant changes in metabolism that require further study [40]. While TRE may result in favorable changes to some aspects of the lipid profile, larger sample sizes and studies with longer TRE intervention duration are required to confirm these effects.

The final component of the metabolic syndrome, abdominal obesity, is measured by elevated waist circumference (≥88 cm for women), a simple and practical anthropometric measure. The primary driver of the relationship between abdominal obesity and poor metabolic health is the volume of visceral fat. As discussed earlier, visceral adiposity is strongly linked to the development of cardiovascular disease and breast cancer [48,49] and related mortality [50–52]. Three studies employing 8–10-h eating windows for 12 weeks among individuals with obesity or metabolic syndrome reported a statistically significant decrease or trend in measures of visceral fat via dual-energy X-ray absorptiometry [45,47], and bioelectrical impedance [79]. Given these positive preliminary findings and the importance of this outcome, the effect of TRE on visceral fat merits further study.

2.4. Auxiliary Health Behavior Benefits

Physical activity is an important protective factor for breast cancer incidence [92,93], breast cancer mortality [94] and cardiovascular disease incidence and mortality [95] (including in breast cancer survivors) [96] Therefore, a concomitant reduction in physical activity, as has been known to occur with participation in a moderate or severe calorie restriction diet [97], could attenuate or mute the benefits of calorie restriction on prevention of breast cancer or cardiovascular disease. While evidence is preliminary, TRE does not appear to alter physical activity levels [45,47,98,99]. In fact, two studies have found that a TRE intervention without physical activity may modestly improve physical function in middle-aged and older adults [100,101]. Improved physical function could have downstream effects of increasing habitual physical activity, but this requires longer duration studies in targeted populations with poor physical function.

When TRE or other forms of intermittent fasting or caloric restriction are combined with a purposeful exercise training intervention, additive or synergistic favorable effects on a number of health outcomes relevant to breast cancer and cardiovascular disease have been reported, including cardiorespiratory fitness, body composition, fasting insulin and glucose, and insulin-like growth factor-1 [102–106]. For example, the addition of 16:8 TRE to a structured resistance training intervention in healthy women resulted in a significantly greater reduction in fat mass and percent body fat than resistance training alone and a similar gain in fat-free mass [107]. Therefore, in populations who are physically able, a combined intervention of TRE and exercise training may provide enhanced benefits for the primary, secondary, and tertiary prevention of breast cancer.

Shortening the eating window to follow TRE may incidentally result in dietary behavior changes that are independently linked to breast cancer incidence including reduced caloric intake as already discussed, as well as reductions in alcohol consumption and late-night snacking on sweet [2,3,108–110]. Among 99 healthy individuals or those with obesity, self-reported sleep quality but not duration was improved after following 16:8 TRE for 12 weeks [111].

2.5. Cancer-Specific Biological Effects

Disruption of circadian rhythms can be associated with abnormal cellular division associated with tumorigenesis [112]. Disruptions to the circadian rhythm are linked to breast cancer development through altered expression of circadian genes in the breast tissue in addition to the associated impaired glucose metabolism discussed earlier. Circadian clocks in the breast regulate the expression of numerous genes, and when disrupted can alter breast biology and promote cancer [112]. The potential link between shift work and risk of breast cancer illustrates this relationship. Women who have long-term exposure to rotating night and day work shifts, such as nurses or doctors, may have an increased risk of breast cancer [113]. Re-aligning the circadian clocks will result in improved oscillations in gene expression and enhanced energy metabolism. TRE may help to accomplish this through re-establishing the oscillations in the feeding-fasting cycle, but the effects of TRE have not been studied in shift workers who are also exposed to severe disruptions to oscillations in the sleep-wake cycle.

The regular exposure to a fasting period induced by TRE also has benefits for cellular health. Regular fasting activates cell signaling pathways and integrated adaptive responses between and within organs that increase the expression of antioxidant defenses, DNA repair, protein quality control, mitochondrial biogenesis, autophagy, and reduces inflammation [39]. This adaptive response confers resistance to oxidative and metabolic stress and the removal/repair of damaged molecules [39]. Through these mechanisms, TRE has the potential to modify biological mechanisms in common for a wide range of chronic disorders including, cancer, cardiovascular disease, diabetes, and neurodegenerative disorders [39]. Specific to cancer, there is evidence that repeated fasting can reduce cell proliferation, cancer progression, and metastases [114].

One compelling finding directly relevant to breast cancer is that breast cancer survivors without diabetes who reported habitually fasting overnight for less than <13 h, had a 36% higher risk of breast recurrence (local, regional, or distant recurrence, or new primary) (hazard ratio, 1.36, 95% CI, 1.05 to 1.76) [78]. In this prospective cohort study that followed 2413 breast cancer survivors for 7.3 years, [78] there were trends toward lower hazard of breast cancer-specific mortality (hazard ratio, 1.21, 95% CI, 0.91 to 1.60) and all-cause mortality (hazard ratio, 1.22, 95% CI, 0.95 to1.56) as well [78]. In addition to the total length of the nightly fasting duration, eating after 8 pm appeared to be a potential determinant of risk, as it was associated with increased chronic inflammation and higher BMI among these breast cancer survivors [78].

2.6. Heart Failure

There is preliminary evidence that TRE may be effective in primary and secondary prevention of heart failure. An observational study of 2001 patients undergoing cardiac catheterization without prior myocardial infarction or heart failure reported that prolonged nightly fasting or religious fasting was associated with a 71% reduced incidence of heart failure (adjusted hazard ratio, =0.29, 95% CI, 0.11 to 0.81) and a trend toward reduced incidence of myocardial infarction (adjusted hazard ratio, =0.69, 95% CI, 0.44 to 1.09) [115]. A prospective observational study of 249 individuals with heart failure with reduced ejection fraction during Ramadan reported that stricter adherence to the daily fasting was associated with stabilization of heart failure symptoms [116].

3. TRE Safety

Fasting has been employed for various lengths of times safely for centuries. Fasting for religious reasons is common in two of the most prevalent religions worldwide: Judaism and Islam. Fasting durations employed by individuals following these religious practices is often much more prolonged than for TRE. For example, over the 30 days of Ramadan, individuals fast for anywhere from 10 to 21 h per day depending on their location in the world. In Judaism, personal fasting is undertaken as an act of penance. The most famous fast day of Judaism is Yom Kippur, which consists of a 25-h fast. These religious fasts differ from TRE in that they require total abstinence from food and drink, including water. Water-less fasting has been associated with a state of dehydration [117]. Nonetheless, despite the lack of water, fasting during Ramadan [116,118] and Yom Kippur [119,120] has been found to be relatively safe even for individuals with chronic conditions such as kidney transplant recipients [118], heart failure [116], and diabetes [119,120].

A 2020 systematic review reported that TRE did not cause major adverse events or negatively impact eating disorder symptoms among adults with obesity, metabolic syndrome, or diabetes [40]. Within adults with type 2 diabetes [98,121,122] or pre-diabetes [62], TRE with 15–20 h fasting periods does not cause occurrences of hypoglycemia. In addition, one study reported no impact of TRE on psychological well-being (e.g., depression, anxiety, or stress) [121].

Typically, in the process of losing fat mass through a calorie restricted diet, patients can experience a decrease in fat-free or lean mass that contributes to 20–35% of the total weight lost, depending on baseline weight [123]. However, a meta-analysis of ten TRE intervention studies with 241 participants showed no change in fat-free/lean mass (−0.24 kg, 95% CI, −1.15 to 0.67) without significant heterogeneity among the results of the included studies [41]. However, these findings may differ when TRE is performed for longer duration, especially if it results in clinically significant weight loss. In longer duration TRE interventions or in patient populations with baseline low lean mass or frailty, concurrent prescription of regular physical activity (especially resistance exercise training) and high protein intake (1.25–1.50 times the recommended dietary allowance) are recommended to help to reduce the concomitant loss of lean mass [123].

4. TRE Feasibility

Typically, self-directed dietary regimens involving caloric restriction and/or macronutrient manipulation require patients to self-monitor and adjust their dietary intake, which is highly burdensome for some individuals [124] and can be inaccurate [125]. Self-monitoring and adjusting dietary intake require estimating calorie content or food volume, weighing each individual ingredient, reading food labels, and/or referencing an electronic nutrition database [125]. Individuals commonly underestimate the caloric value of different food items, by as much as 28% on food items over 500 calories in one study [126]. Weighing each ingredient increases accuracy, but requires the purchase of a food scale, and is tedious, and therefore may not be associated with high adherence. Patients following a 'free-living' diet where they purchase their own food and aim to follow a specific prescribed calorie intake or macronutrient ratio will typically need to reference an electronic food database to determine the calorie and macronutrient of different foods. This requires internet access and technological skills that create a barrier in older populations and/or rural areas. An alternative method of reading food labels and self-calculating intake is hampered by reported deficits among adults' understanding of nutrition labels [127]. A dietary program that provides pre-measured meals and snacks offers a high level of convenience and removes the need for self-monitoring but is associated with significant cost.

In contrast, TRE is simple to prescribe and follow. It requires minimal instruction and no specialized training or equipment (e.g., a food scale). This in turn results in minimal administrative time and costs for health care practitioners to prescribe TRE. Contrary to other caloric restrictive diets, which can include changes to the macronutrient content of the diet, TRE diets allow participants to continue enjoying the foods they habitually consume. This makes it easier to implement for patients and does not increase food-related costs. There are also free phone applications available to assist patients to track their eating window (e.g., https://www.zerofasting.com/ and https://www.bodyfast.de/en/), but this is not required to be able to follow TRE. The low cost and simplicity of TRE as an intervention reduce common barriers to the adoption and maintenance of lifestyle therapies.

5. TRE Adherence and Barriers

A recent systematic review demonstrates that adherence to TRE is high, typically 80–90%, including among individuals with obesity, metabolic syndrome, and diabetes for 4–12 weeks [40]. Other studies have reported even higher rates of adherence, including 98% adherence to five weeks of TRE in one small study of 8 men with prediabetes [62]. Furthermore, dropout rates from TRE studies are lower than in other formats of intermittent fasting (~10 vs. 20%) and much lower than caloric restriction (up to 33%) [128]. While it is likely that the lower dropout rate from studies would translate to greater real-world adherence, there is no research evidence to-date to confirm that TRE is associated with long-term adherence. Individuals following TRE have reported feelings of increased energy, well-being, self-awareness, sleep quality, health-related quality of life and enhanced ability to avoid snacking in the evening [99,111]. These positive qualitative experiences may enhance willingness and motivation to maintain adherence to TRE.

The primary barriers to longer-term adherence to TRE include incompatibility with family/social life and work schedules [98]. One potential solution suggested by patients who had followed TRE was allowing a more flexible protocol (e.g., weekends off, customized eating window) [128]. Animal data suggest that time-restricted feeding during the week with ad libitum weekend feeding (even with access to high fat and sugar) is similarly effective to continuous (every day) time-restricted feeding for reducing fat mass, improving insulin resistance and normalizing triglyceride levels [129]. To our knowledge, no studies have been published using a weekday only TRE model in humans. However, the positive results from studies reporting 80–90% TRE adherence suggest that, at minimum, one day off from TRE per week is likely to still provide substantial health benefits. It has been suggested that a priori prescription of a planned hedonic goal deviation (e.g., one 'cheat' day per week) will enhance long-term adherence to the intervention by enhancing motivation to persist, improving emotional experience, and helping with self-regulation [130]. Therefore, a pre-emptive prescription of TRE for only 5–6 days of the week may be an effective strategy that should be further explored.

A wide variety of TRE protocols with different timing and lengths of the eating window and total durations have been shown to offer metabolic health benefits [39,40]. This accumulated evidence can be used to deduce that there is room for flexibility in TRE protocols to address personal preferences. Accounting for preferences through personalized intervention approaches fosters patient autonomy, enjoyment, and adherence [131,132] A personalized TRE protocol has been suggested in the literature as a strategy to enhance long-term adherence [133]. Potential personalization modifications with evidence for health benefits [39] include a) personalizing the eating window time of day as long as it ends ≥3 h prior to bedtime, and if possible, at or prior to 8 pm; b) personalizing the eating window length to 4–10 h (8 being most common); c) performing TRE on 5–7 continuous days per week. Future research is needed to evaluate the effect of combining two or more of these components of personalization to determine the breadth of flexibility possible for the implementation of TRE as a long-term health behavior.

6. Implications and Future Directions

The potential benefits of TRE as a therapy for primary and secondary breast cancer prevention, and tertiary prevention as it relates to cardiovascular disease in breast cancer survivors are three-fold:

1. TRE directly improves many of the biological and behavioral mechanisms underpinning the development of breast cancer and cardiovascular disease. For example, obesity has been found to have a strong causal relationship with primary and secondary prevention of breast cancer and cardiovascular disease and related mortality. Shared features of the pathogenesis and progression of both conditions that may mediate obesity include oxidative stress and chronic inflammation. Metabolic syndrome, a constellation of metabolic disturbances including hyperglycemia, elevated triglycerides, low HDL, hypertension, and abdominal obesity, is well established to increase the risk of cardiovascular disease and has an emerging strong link to breast cancer. While further research is needed to confirm efficacy on all of these specific outcomes, the available evidence suggests that TRE has promising positive effects on inflammation, oxidative stress, and metabolic health.
2. TRE directly addresses some of the safety and feasibility concerns associated with existing dietary interventions. Mainly, while existing weight loss interventions tend to result in loss of lean mass contributing to 20–35% of total weight loss, TRE has been found to result in decreases in fat mass while sparing lean mass. TRE also removes barriers to participating in dietary interventions, by not requiring tedious calorie counting or use of technology. Preliminary evidence and biological plausibility suggest that personalization of a TRE protocol to an individual's preferences or lifestyle may enable long-term adherence while still offering health benefits. There are also no costs associated with this intervention. This may be why adherence rates have been reported to be much higher than other dietary interventions, with one study reporting adherence as high as 98%. To-date, most TRE studies have been 8–12 weeks in duration, but due to its simplicity and potential for high adherence, it could be an effective strategy to ameliorate the well-known issue of long-term adherence to health behaviors, especially with allowance of protocol modifications for personal preferences.
3. TRE is safe. Many studies evaluating the practice of fasting during Ramadan and Yom Kippur suggest that, even without consuming water, it can be safe for individuals with chronic conditions such as diabetes and heart failure. It is also likely safe to perform during chemotherapy treatment for breast cancer, based on evidence that longer periods of fasting have been shown to be safe and tolerable, but this requires further research. In addition, no TRE studies have reported the occurrence of major adverse events nor hypoglycemia even among individuals with diabetes. Instead, individuals following TRE have reported positive feelings of increased energy, well-being, and self-awareness.

This accumulating evidence for the potential of TRE to positively impact the development and progression of breast cancer, cardiovascular, and metabolic diseases merits further research, but leaves a number of remaining questions. First, given the short duration of most TRE studies published to-date, interventions with a longer duration and a longer follow-up period are needed to determine the potential for long-term adherence and sustainability of this dietary intervention. Future research should aim to address barriers identified for TRE, such as incompatibility with social or personal life and work schedules. There are potential solutions to overcome these barriers such as personalizing the eating window length and timing and/or incorporating cheat days, that still need to be empirically tested. Lastly, TRE has not been experimentally tested in patients with cancer or cardiovascular disease. Nonetheless, this review described promising observational evidence in these populations and positive experimental evidence on the effects of TRE on biological and behavioral mechanisms underpinning these conditions. In composite, these data suggest that TRE may be an easy and novel lifestyle interventions for the primary and secondary prevention of breast cancer, as well as tertiary prevention as it relates to cardiovascular disease in breast cancer survivors.

Author Contributions: A.A.K. was responsible for the conceptualization of the research question. R.A.G.C. and A.A.K. were responsible for the writing—original draft and writing—review and editing. All authors have read and agreed to the published version of the manuscript.

Funding: This research received no external funding.

Institutional Review Board Statement: Research ethics review was not undertaken for this study as it exclusively uses tertiary sources and is therefore exempt from ethics review.

Informed Consent Statement: Not applicable.

Data Availability Statement: Not applicable.

Acknowledgments: Not applicable.

Conflicts of Interest: The authors declare no conflict of interest.

References

1. Canadian Cancer Society's Advisory Committee on Cancer Statistics. *Canadian Cancer Statistics 2017*; Canadian Cancer Society's Advisory Committee on Cancer Statistics: Toronto, ON, Canada, 2017.
2. Sprague, B.L.; Trentham-Dietz, A.; Egan, K.M.; Titus-Ernstoff, L.; Hampton, J.M.; Newcomb, P.A. Proportion of Invasive Breast Cancer Attributable to Risk Factors Modifiable after Menopause. *Am. J. Epidemiol.* **2008**, *168*, 404–411. [CrossRef] [PubMed]
3. Wilson, L.F.; Page, A.N.; Dunn, N.A.; Pandeya, N.; Protani, M.M.; Taylor, R.J. Population attributable risk of modifiable risk factors associated with invasive breast cancer in women aged 45–69 years in Queensland, Australia. *Maturitas* **2013**, *76*, 370–376. [CrossRef] [PubMed]
4. Bissell, M.C.; Kerlikowske, K.; Sprague, B.L.; Tice, J.A.; Gard, C.C.; Tossas, K.Y.; Rauscher, G.H.; Trentham-Dietz, A.; Henderson, L.M.; Onega, T.; et al. Breast Cancer Population Attributable Risk Proportions Associated with Body Mass Index and Breast Density by Race/Ethnicity and Menopausal Status. *Cancer Epidemiol. Biomark. Prev.* **2020**, *29*, 2048–2056. [CrossRef] [PubMed]
5. Blot, W.J.; Tarone, R.E. Doll and Peto's Quantitative Estimates of Cancer Risks: Holding Generally True for 35 Years. *J. Natl. Cancer Inst.* **2015**, *107*, djv044. [CrossRef]
6. Bonora, E. The metabolic syndrome and cardiovascular disease. *Ann. Med.* **2006**, *38*, 64–80. [CrossRef]
7. Park, Y.M.; White, A.J.; Nichols, H.B.; O'Brien, K.; Weinberg, C.; Sandler, D.P. The association between metabolic health, obesity phenotype and the risk of breast cancer. *Int. J. Cancer* **2017**, *140*, 2657–2666. [CrossRef]
8. Gunter, M.J.; Xie, X.; Xue, X.; Kabat, G.C.; Rohan, T.E.; Wassertheil-Smoller, S.; Ho, G.Y.; Wylie-Rosett, J.; Greco, T.; Yu, H.; et al. Breast Cancer Risk in Metabolically Healthy but Overweight Postmenopausal Women. *Cancer Res.* **2015**, *75*, 270–274. [CrossRef]
9. Kabat, G.C.; Kim, M.Y.; Lee, J.S.; Ho, G.Y.; Going, S.B.; Beebe-Dimmer, J.; Manson, J.E.; Chlebowski, R.T.; Rohan, T.E. Metabolic Obesity Phenotypes and Risk of Breast Cancer in Postmenopausal Women. *Cancer Epidemiol. Biomark. Prev.* **2017**, *26*, 1730–1735. [CrossRef]
10. Moore, L.L.; Chadid, S.; Singer, M.R.; Kreger, B.E.; Denis, G. Metabolic Health Reduces Risk of Obesity-Related Cancer in Framingham Study Adults. *Cancer Epidemiol. Biomark. Prev.* **2014**, *23*, 2057–2065. [CrossRef]
11. Alderton, G. Obesity and inflammation. *Science* **2020**, *370*, 419. [CrossRef]
12. Marseglia, L.; Manti, S.; D'Angelo, G.; Nicotera, A.G.; Parisi, E.; Di Rosa, G.; Gitto, E.; Arrigo, T. Oxidative Stress in Obesity: A Critical Component in Human Diseases. *Int. J. Mol. Sci.* **2014**, *16*, 378–400. [CrossRef]
13. Pierce, B.; Ballard-Barbash, R.; Bernstein, L.; Baumgartner, R.N.; Neuhouser, M.L.; Wener, M.H.; Baumgartner, K.B.; Gilliland, F.D.; Sorensen, B.E.; McTiernan, A.; et al. Elevated Biomarkers of Inflammation Are Associated With Reduced Survival Among Breast Cancer Patients. *J. Clin. Oncol.* **2009**, *27*, 3437–3444. [CrossRef]
14. O'Hanlon, D.M.; Lynch, J.; Cormican, M.; Given, H.F. The acute phase response in breast carcinoma. *Anticancer. Res.* **2002**, *22*, 1289–1293.
15. Blann, A.D.; Byrne, G.J.; Baildam, A.D. Increased soluble intercellular adhesion molecule-1, breast cancer and the acute phase response. *Blood Coagul. Fibrinolysis* **2002**, *13*, 165–168. [CrossRef]
16. Lee, J.D.; Cai, Q.; Shu, X.O.; Nechuta, S.J. The Role of Biomarkers of Oxidative Stress in Breast Cancer Risk and Prognosis: A Systematic Review of the Epidemiologic Literature. *J. Women's Health* **2017**, *26*, 467–482. [CrossRef]
17. Saintot, M.; Grenier, J.; Simony-Lafontaine, J.; Gerber, M. Oxidant-antioxidant status in relation to survival among breast cancer patients. *Int. J. Cancer* **2001**, *97*, 574–579. [CrossRef]
18. Canadian Cancer Statistics. 2019. Available online: https://cdn.cancer.ca/-/media/files/research/cancer-statistics/2019-statistics/canadian-cancer-statistics-2019-en.pdf (accessed on 23 August 2021).
19. Sturgeon, K.M.; Deng, L.; Bluethmann, S.M.; Zhou, S.; Trifiletti, D.M.; Jiang, C.; Kelly, S.; Zaorsky, N.G. A population-based study of cardiovascular disease mortality risk in US cancer patients. *Eur. Hear. J.* **2019**, *40*, 3889–3897. [CrossRef]
20. Kirkham, A.A.; Beaudry, R.I.; Paterson, D.I.; Mackey, J.R.; Haykowsky, M.J. Curing breast cancer and killing the heart: A novel model to explain elevated cardiovascular disease and mortality risk among women with early stage breast cancer. *Prog. Cardiovasc. Dis.* **2019**, *62*, 116–126. [CrossRef]

21. Dieli-Conwright, C.M.; Wong, L.; Waliany, S.; Bernstein, L.; Salehian, B.; Mortimer, J. An observational study to examine changes in metabolic syndrome components in patients with breast cancer receiving neoadjuvant or adjuvant chemotherapy. *Cancer* **2016**, *122*, 2646–2653. [CrossRef]
22. Pan, H.; Gray, R.; Braybrooke, J.; Davies, C.; Taylor, C.; McGale, P.; Peto, R.; Pritchard, K.I.; Bergh, J.; Dowsett, M.; et al. 20-Year Risks of Breast-Cancer Recurrence after Stopping Endocrine Therapy at 5 Years. *N. Engl. J. Med.* **2017**, *377*, 1836–1846. [CrossRef]
23. Saxe, G.A.; Rock, C.L.; Wicha, M.S.; Schottenfeld, D. Diet and risk for breast cancer recurrence and survival. *Breast Cancer Res. Treat.* **1999**, *53*, 241–253. [CrossRef]
24. Loprinzi, P.D.; Cardinal, B.J.; Winters-Stone, K.; Smit, E.; Loprinzi, C.L. Physical Activity and the Risk of Breast Cancer Recurrence: A Literature Review//Oncology nursing Forum. 2012, p. 39. Available online: https://search.proquest.com/docview/1012271640?accountid=6180%5Cnhttp://dw2zn6fm9z.search.serialssolutions.com/?ctx_ver=Z39.88-2004&ctx_enc=info:ofi/enc:UTF-8&rfr_id=info:sid/ProQ%3Anahs&rft_val_fmt=info:ofi/fmt:kev:mtx:journal&rft.genre=article&rft.jtitl (accessed on 23 August 2021).
25. Ecker, B.L.; Lee, J.Y.; Sterner, C.J.; Solomon, A.C.; Pant, D.K.; Shen, F.; Peraza, J.; Vaught, L.; Mahendra, S.; Belka, G.K.; et al. Impact of obesity on breast cancer recurrence and minimal residual disease. *Breast Cancer Res.* **2019**, *21*, 1–16. [CrossRef]
26. Ligibel, J.A.; Strickler, H.D. Obesity and its impact on breast cancer: Tumor incidence, recurrence, survival, and possible interventions. *Am. Soc. Clin. Oncol. Educ. Book* **2013**, *33*, 52–59. [CrossRef]
27. Morimoto, L.M.; White, E.; Chen, Z.; Chlebowski, R.T.; Hays-Grudo, J.; Kuller, L.; Lopez, A.M.; Manson, J.; Margolis, K.L.; Muti, P.C.; et al. Obesity, body size, and risk of postmenopausal breast cancer: The Women's Health Initiative (United States). *Cancer Causes Control.* **2002**, *13*, 741–751. [CrossRef] [PubMed]
28. Kawai, M.; Minami, Y.; Nishino, Y.; Fukamachi, K.; Ohuchi, N.; Kakugawa, Y. Body mass index and survival after breast cancer diagnosis in Japanese women. *BMC Cancer* **2012**, *12*, 149. [CrossRef] [PubMed]
29. Kannel, W.B.; D'Agostino, R.B.; Cobb, J.L. Effect of weight on cardiovascular disease. *Am. J. Clin. Nutr.* **1996**, *63*, 419S–422S. [CrossRef] [PubMed]
30. Cleary, M.P.; Grossmann, M.E. Obesity and Breast Cancer: The Estrogen Connection. *Endocrinology* **2009**, *150*, 2537–2542. [CrossRef] [PubMed]
31. Zeng, Q.; Dong, S.-Y.; Sun, X.-N.; Xie, J.; Cui, Y. Percent body fat is a better predictor of cardiovascular risk factors than body mass index. *Braz. J. Med Biol. Res.* **2012**, *45*, 591–600. [CrossRef]
32. Vance, V.; Hanning, R.; Mourtzakis, M.; McCargar, L. Weight gain in breast cancer survivors: Prevalence, pattern and health consequences. *Obes. Rev.* **2010**, *12*, 282–294. [CrossRef]
33. Harvie, M.; Campbell, I.; Baildam, A.; Howell, A. Energy Balance in Early Breast Cancer Patients Receiving Adjuvant Chemotherapy. *Breast Cancer Res. Treat.* **2004**, *83*, 201–210. [CrossRef]
34. Arble, D.; Bass, J.; Laposky, A.D.; Vitaterna, M.H.; Turek, F.W. Circadian Timing of Food Intake Contributes to Weight Gain. *Obesity* **2009**, *17*, 2100–2102. [CrossRef]
35. Kohsaka, A.; Laposky, A.D.; Ramsey, K.M.; Estrada, C.; Joshu, C.; Kobayashi, Y.; Turek, F.W.; Bass, J. High-Fat Diet Disrupts Behavioral and Molecular Circadian Rhythms in Mice. *Cell Metab.* **2007**, *6*, 414–421. [CrossRef]
36. Longo, V.D.; Panda, S. Fasting, Circadian Rhythms, and Time-Restricted Feeding in Healthy Lifespan. *Cell Metab.* **2016**, *23*, 1048–1059. [CrossRef]
37. Chaix, A.; Manoogian, E.N.; Melkani, G.C.; Panda, S. Time-Restricted Eating to Prevent and Manage Chronic Metabolic Diseases. *Annu. Rev. Nutr.* **2019**, *39*, 291–315. [CrossRef]
38. de Cabo, R.; Mattson, M.P. Effects of intermittent fasting on health, aging, and disease. *N. Engl. J. Med.* **2019**, *381*, 2541–2551. [CrossRef]
39. Gabel, K.; Varady, K.A. Current research: Effect of time restricted eating on weight and cardiometabolic health. *J. Physiol.* **2020**. [CrossRef]
40. Moon, S.; Kang, J.; Kim, S.H.; Chung, H.S.; Kim, Y.J.; Yu, J.M.; Cho, S.T.; Oh, C.-M.; Kim, T. Beneficial Effects of Time-Restricted Eating on Metabolic Diseases: A Systemic Review and Meta-Analysis. *Nutrients* **2020**, *12*, 1267. [CrossRef]
41. Williamson, D.A.; Bray, G.A.; Ryan, D.H. Is 5% weight loss a satisfactory criterion to define clinically significant weight loss? *Obesity* **2015**, *23*, 2319–2320. [CrossRef]
42. Ryan, D.H.; Yockey, S.R. Weight Loss and Improvement in Comorbidity: Differences at 5%, 10%, 15%, and Over. *Curr. Obes. Rep.* **2017**, *6*, 187–194. [CrossRef]
43. Ryan, A.S.; Pratley, R.E.; Elahi, D.; Goldberg, A.P. Resistive training increases fat-free mass and maintains RMR despite weight loss in postmenopausal women. *J. Appl. Physiol.* **1995**, *79*, 818–823. [CrossRef]
44. Gabel, K.; Hoddy, K.K.; Haggerty, N.; Song, J.; Kroeger, C.M.; Trepanowski, J.F.; Panda, S.; Varady, K.A. Effects of 8-hour time restricted feeding on body weight and metabolic disease risk factors in obese adults: A pilot study. *Nutr. Health Aging* **2018**, *4*, 345–353. [CrossRef]
45. Villanueva, J.E.; Livelo, C.; Trujillo, A.S.; Chandran, S.; Woodworth, B.; Andrade, L.; Le, H.D.; Manor, U.; Panda, S.; Melkani, G.C. Time-restricted feeding restores muscle function in Drosophila models of obesity and circadian-rhythm disruption. *Nat. Commun.* **2019**, *10*, 1–17. [CrossRef]

46. Chow, L.S.; Manoogian, E.N.C.; Alvear, A.; Fleischer, J.; Thor, H.; Dietsche, K.; Wang, Q.; Hodges, J.S.; Esch, N.; Malaeb, S.; et al. Time-Restricted Eating Effects on Body Composition and Metabolic Measures in Humans who are Overweight: A Feasibility Study. *Obesity* **2020**, *28*, 860–869. [CrossRef]
47. Kim, M.S.; Choi, Y.-J.; Lee, Y.H. Visceral fat measured by computed tomography and the risk of breast cancer. *Transl. Cancer Res.* **2019**, *8*, 1939–1949. [CrossRef]
48. Després, J.-P. Body Fat Distribution and Risk of Cardiovascular Disease. *Circulation* **2012**, *126*, 1301–1313. [CrossRef]
49. Caan, B.J.; Feliciano, E.M.C.; Prado, C.M.; Alexeeff, S.; Kroenke, C.H.; Bradshaw, P.; Quesenberry, C.P.; Weltzien, E.K.; Castillo, A.L.; Olobatuyi, T.A.; et al. Association of Muscle and Adiposity Measured by Computed Tomography With Survival in Patients With Nonmetastatic Breast Cancer. *JAMA Oncol.* **2018**, *4*, 798–804. [CrossRef]
50. Feliciano, E.M.C.; Chen, W.Y.; Lee, V.; Albers, K.B.; Prado, C.M.; Alexeeff, S.; Xiao, J.; Shachar, S.S.; Caan, B. Body Composition, Adherence to Anthracycline and Taxane-Based Chemotherapy, and Survival After Nonmetastatic Breast Cancer. *JAMA Oncol.* **2020**, *6*, 264–270. [CrossRef]
51. Feliciano, E.M.C.; Chen, W.Y.; Bradshaw, P.T.; Prado, C.M.; Alexeeff, S.; Albers, K.B.; Castillo, A.L.; Caan, B. Adipose Tissue Distribution and Cardiovascular Disease Risk Among Breast Cancer Survivors. *J. Clin. Oncol.* **2019**, *37*, 2528–2536. [CrossRef]
52. Nguyen, M.; Stewart, R.; Banerji, M.; Gordon, D.; Kral, J. Relationships between tamoxifen use, liver fat and body fat distribution in women with breast cancer. *Int. J. Obes.* **2001**, *25*, 296–298. [CrossRef]
53. Kirkham, A.A.; Pituskin, E.; Thompson, R.B.; Mackey, J.R.; Koshman, S.L.; Jassal, D.; Pitz, M.; Haykowsky, M.J.; Pagano, J.J.; Chow, K.; et al. Cardiac and cardiometabolic phenotyping of trastuzumab-mediated cardiotoxicity: A secondary analysis of the MANTICORE trial. *Eur. Hear. J. Cardiovasc. Pharm.* **2021**. [CrossRef]
54. Lutes, C.; Zelig, R.; Radler, D.R. Safety and Feasibility of Intermittent Fasting During Chemotherapy for Breast Cancer. *Top. Clin. Nutr.* **2020**, *35*, 168–177. [CrossRef]
55. Raffaghello, L.; Lee, C.; Safdie, F.M.; Wei, M.; Madia, F.; Bianchi, G.; Longo, V.D. Starvation-dependent differential stress resistance protects normal but not cancer cells against high-dose chemotherapy. *Proc. Natl. Acad. Sci. USA* **2008**, *105*, 8215–8220. [CrossRef] [PubMed]
56. Lee, C.; Raffaghello, L.; Brandhorst, S.; Safdie, F.M.; Bianchi, G.; Martin-Montalvo, A.; Pistoia, V.; Wei, M.; Hwang, S.; Merlino, A.; et al. Fasting Cycles Retard Growth of Tumors and Sensitize a Range of Cancer Cell Types to Chemotherapy. *Sci. Transl. Med.* **2012**, *4*, 124ra27. [CrossRef] [PubMed]
57. Koene, R.J.; Prizment, A.E.; Blaes, A.; Konety, S.H. Shared Risk Factors in Cardiovascular Disease and Cancer. *Circulation* **2016**, *133*, 1104–1114. [CrossRef]
58. Kang, D.-H. Oxidative Stress, DNA Damage, and Breast Cancer. *Aacn Clin. Issues Adv. Pract. Acute Crit. Care* **2002**, *13*, 540–549. [CrossRef]
59. Davis, J.D. DNA damage and breast cancer. *World J. Clin. Oncol.* **2011**, *2*, 329–338. [CrossRef]
60. Senoner, T.; Dichtl, W. Oxidative Stress in Cardiovascular Diseases: Still a Therapeutic Target? *Nutrients* **2019**, *11*, 2090. [CrossRef]
61. Sutton, E.F.; Beyl, R.; Early, K.S.; Cefalu, W.T.; Ravussin, E.; Peterson, C.M. Early Time-Restricted Feeding Improves Insulin Sensitivity, Blood Pressure, and Oxidative Stress Even without Weight Loss in Men with Prediabetes. *Cell Metab.* **2018**, *27*, 1212–1221.e3. [CrossRef]
62. McAllister, M.J.; Gonzalez, A.E.; Waldman, H.S. Impact of Time Restricted Feeding on Markers of Cardiometabolic Health and Oxidative Stress in Resistance-Trained Firefighters. *J. Strength Cond. Res.* **2020**. [CrossRef]
63. Lindsberg, P.J.; Grau, A.J. Inflammation and Infections as Risk Factors for Ischemic Stroke. *Stroke* **2003**, *34*, 2518–2532. [CrossRef]
64. McAllister, M.J.; Pigg, B.L.; I Renteria, L.; Waldman, H.S. Time-restricted feeding improves markers of cardiometabolic health in physically active college-age men: A 4-week randomized pre-post pilot study. *Nutr. Res.* **2019**, *75*, 32–43. [CrossRef]
65. Moro, T.; Tinsley, G.; Bianco, A.; Marcolin, G.; Pacelli, Q.F.; Battaglia, G.; Palma, A.; Gentil, P.; Neri, M.; Paoli, A. Effects of eight weeks of time-restricted feeding (16/8) on basal metabolism, maximal strength, body composition, inflammation, and cardiovascular risk factors in resistance-trained males. *J. Transl. Med.* **2016**, *14*, 1–10. [CrossRef]
66. Faris, A.-I.E.; Kacimi, S.; Al-Kurd, R.A.; Fararjeh, M.A.; Bustanji, Y.; Mohammad, M.K.; Salem, M.L. Intermittent fasting during Ramadan attenuates proinflammatory cytokines and immune cells in healthy subjects. *Nutr. Res.* **2012**, *32*, 947–955. [CrossRef]
67. Esposito, K.; Chiodini, P.; Capuano, A.; Bellastella, G.; Maiorino, M.I.; Rafaniello, C.; Giugliano, D. Metabolic syndrome and postmenopausal breast cancer. *Menopause* **2013**, *20*, 1301–1309. [CrossRef]
68. Bjørge, T.; Lukanova, A.; Jonsson, H.; Tretli, S.; Ulmer, H.; Manjer, J.; Stocks, T.; Selmer, R.; Nagel, G.; Almquist, M.; et al. Metabolic Syndrome and Breast Cancer in the Me-Can (Metabolic Syndrome and Cancer) Project. *Cancer Epidemiol. Biomark. Prev.* **2010**, *19*, 1737–1745. [CrossRef]
69. Berrino, F.; Villarini, A.; Traina, A.; Bonanni, B.; Panico, S.; Mano, M.P.; Mercandino, A.; Galasso, R.; Barbero, M.; Simeoni, M.; et al. Metabolic syndrome and breast cancer prognosis. *Breast Cancer Res. Treat.* **2014**, *147*, 159–165. [CrossRef]
70. Cho, Y.; Hong, N.; Kim, K.-W.; Cho, S.J.; Lee, M.; Lee, Y.-H.; Lee, Y.-H.; Kang, E.S.; Cha, B.-S.; Lee, B.-W. The Effectiveness of Intermittent Fasting to Reduce Body Mass Index and Glucose Metabolism: A Systematic Review and Meta-Analysis. *J. Clin. Med.* **2019**, *8*, 1645. [CrossRef]
71. Onodera, Y.; Nam, J.-M.; Bissell, M.J. Increased sugar uptake promotes oncogenesis via EPAC/RAP1 and O-GlcNAc pathways. *J. Clin. Investig.* **2013**, *124*, 367–384. [CrossRef]

72. Cohen, D.H.; Leroith, D. Obesity, type 2 diabetes, and cancer: The insulin and IGF connection. *Endocr. Relat. Cancer* **2012**, *19*, F27–F45. [CrossRef]
73. De Bruijn, K.M.J.; Arends, L.R.; Hansen, B.; Leeflang, S.; Ruiter, R.; van Eijck, C.H.J. Systematic review and meta-analysis of the association between diabetes mellitus and incidence and mortality in breast and colorectal cancer. *BJS* **2013**, *100*, 1421–1429. [CrossRef]
74. de Beer, J.C.; Liebenberg, L. Does cancer risk increase with HbA1c, independent of diabetes? *Br. J. Cancer* **2014**, *110*, 2361–2368. [CrossRef]
75. Blake, G.J.; Pradhan, A.D.; Manson, J.E.; Williams, G.R.; Buring, J.; Ridker, P.M.; Glynn, R.J. Hemoglobin A1c Level and Future Cardiovascular Events Among Women. *Arch. Intern. Med.* **2004**, *164*, 757–761. [CrossRef]
76. Marinac, C.R.; Natarajan, L.; Sears, D.D.; Gallo, L.C.; Hartman, S.J.; Arredondo, E.; Patterson, R.E. Prolonged Nightly Fasting and Breast Cancer Risk: Findings from NHANES (2009–2010). *Cancer Epidemiol. Biomark. Prev.* **2015**, *24*, 783–789. [CrossRef]
77. Marinac, C.R.; Nelson, S.H.; Breen, C.I.; Hartman, S.J.; Natarajan, L.; Pierce, J.P.; Flatt, S.W.; Sears, D.D.; Patterson, R.E. Prolonged Nightly Fasting and Breast Cancer Prognosis. *JAMA Oncol.* **2016**, *2*, 1049–1055. [CrossRef]
78. Wilkinson, M.J.; Manoogian, E.N.; Zadourian, A.; Lo, H.; Fakhouri, S.; Shoghi, A.; Wang, X.; Fleischer, J.; Navlakha, S.; Panda, S.; et al. Ten-Hour Time-Restricted Eating Reduces Weight, Blood Pressure, and Atherogenic Lipids in Patients with Metabolic Syndrome. *Cell Metab.* **2019**, *31*, 92–104.e5. [CrossRef] [PubMed]
79. Fuchs, F.D.; Whelton, P.K. High Blood Pressure and Cardiovascular Disease. *Hypertension* **2020**, *75*, 285–292. [CrossRef]
80. Jamrozik, K. Age-specific relevance of usual blood pressure to vascular mortality: A meta-analysis of individual data for one million adults in 61 prospective studies. *Lancet* **2002**, *360*, 1903–1913. [CrossRef]
81. Seretis, A.; Cividini, S.; Markozannes, G.; Tseretopoulou, X.; Lopez, D.S.; Ntzani, E.E.; Tsilidis, K.K. Association between blood pressure and risk of cancer development: A systematic review and meta-analysis of observational studies. *Sci. Rep.* **2019**, *9*, 1–12. [CrossRef]
82. Han, H.; Guo, W.; Shi, W.; Yu, Y.; Zhang, Y.; Ye, X.; He, J. Hypertension and breast cancer risk: A systematic review and meta-analysis. *Sci. Rep.* **2017**, *7*, srep44877. [CrossRef]
83. Carter, P.; Uppal, H.; Chandran, S.; Bainey, K.; Potluri, R. 3106Patients with a diagnosis of hyperlipidaemia have a reduced risk of developing breast cancer and lower mortality rates: A large retrospective longitudinal cohort study from the UK ACALM registry. *Eur. Hear. J.* **2017**, *38*. [CrossRef]
84. His, M.; Zelek, L.; Deschasaux, M.; Pouchieu, C.; Kesse-Guyot, E.; Hercberg, S.; Galan, P.; Latino-Martel, P.; Blacher, J.; Touvier, M. Prospective associations between serum biomarkers of lipid metabolism and overall, breast and prostate cancer risk. *Eur. J. Epidemiol.* **2014**, *29*, 119–132. [CrossRef] [PubMed]
85. Potluri, R.; Lavu, D.; Uppal, H.; Chandran, S. P740Hyperlipidaemia as a risk factor for breast cancer? *Cardiovasc. Res.* **2014**, *103*. [CrossRef]
86. Kitahara, C.M.; De González, A.B.; Freedman, N.D.; Huxley, R.; Mok, Y.; Jee, S.H.; Samet, J.M. Total Cholesterol and Cancer Risk in a Large Prospective Study in Korea. *J. Clin. Oncol.* **2011**, *29*, 1592–1598. [CrossRef] [PubMed]
87. Garcia-Estevez, L.; Moreno-Bueno, G. Updating the role of obesity and cholesterol in breast cancer. *Breast Cancer Res.* **2019**, *21*, 1–8. [CrossRef]
88. Touvier, M.; Fassier, P.; His, M.; Norat, T.; Chan, D.S.M.; Blacher, J.; Hercberg, S.; Galan, P.; Druesne-Pecollo, N.; Latino-Martel, P. Cholesterol and breast cancer risk: A systematic review and meta-analysis of prospective studies. *Br. J. Nutr.* **2015**, *114*, 347–357. [CrossRef]
89. Kucharska-Newton, A.M.; Rosamond, W.D.; Mink, P.J.; Alberg, A.J.; Shahar, E.; Folsom, A.R. HDL-Cholesterol and Incidence of Breast Cancer in the ARIC Cohort Study. *Ann. Epidemiol.* **2008**, *18*, 671–677. [CrossRef]
90. Levy, R.I. Cholesterol and cardiovascular disease: No longer whether, but rather when, in whom, and how? *Circulation* **1985**, *72*, 686–691. [CrossRef]
91. McTiernan, A.; Kooperberg, C.L.; White, E.; Wilcox, S.; Coates, R.J.; Adams-Campbell, L.L.; Woods, N.F.; Ockene, J.K. Recreational Physical Activity and the Risk of Breast Cancer in Postmenopausal Women. *JAMA* **2003**, *290*, 1331–1336. [CrossRef]
92. Chan, D.S.M.; Abar, L.; Cariolou, M.; Nanu, N.; Greenwood, D.C.; Bandera, E.V.; McTiernan, A.; Norat, T. World Cancer Research Fund International: Continuous Update Project—systematic literature review and meta-analysis of observational cohort studies on physical activity, sedentary behavior, adiposity, and weight change and breast cancer risk. *Cancer Causes Control* **2019**, *30*, 1183–1200. [CrossRef]
93. Lahart, I.; Metsios, G.S.; Nevill, A.; Carmichael, A.R. Physical activity, risk of death and recurrence in breast cancer survivors: A systematic review and meta-analysis of epidemiological studies. *Acta Oncol.* **2015**, *54*, 635–654. [CrossRef]
94. Zhang, X.; Cash, R.E.; Bower, J.K.; Focht, B.C.; Paskett, E.D. Physical activity and risk of cardiovascular disease by weight status among U.S adults. *PLoS ONE* **2020**, *15*, e0232893. [CrossRef]
95. Jones, L.W.; Habel, L.; Weltzien, E.; Castillo, A.; Gupta, D.; Kroenke, C.H.; Kwan, M.L.; Jr, C.P.Q.; Scott, J.; Sternfeld, B.; et al. Exercise and Risk of Cardiovascular Events in Women With Nonmetastatic Breast Cancer. *J. Clin. Oncol.* **2016**, *34*, 2743–2749. [CrossRef]
96. Redman, L.M.; Heilbronn, L.K.; Martin, C.K.; De Jonge, L.; Williamson, D.A.; Delany, J.P.; Ravussin, E. for the Pennington CALERIE Team Metabolic and Behavioral Compensations in Response to Caloric Restriction: Implications for the Maintenance of Weight Loss. *PLoS ONE* **2009**, *4*, e4377. [CrossRef]

97. Hutchison, A.T.; Regmi, P.; Manoogian, E.N.; Fleischer, J.; Wittert, G.; Panda, S.; Heilbronn, L.K. Time-Restricted Feeding Improves Glucose Tolerance in Men at Risk for Type 2 Diabetes: A Randomized Crossover Trial. *Obesity* **2019**, *27*, 724–732. [CrossRef]
98. Parr, E.B.; Devlin, B.L.; Radford, B.E.; Hawley, J.A. A Delayed Morning and Earlier Evening Time-Restricted Feeding Protocol for Improving Glycemic Control and Dietary Adherence in Men with Overweight/Obesity: A Randomized Controlled Trial. *Nutrients* **2020**, *12*, 505. [CrossRef]
99. Anton, S.D.; Lee, S.A.; Donahoo, W.T.; McLaren, C.; Manini, T.; Leeuwenburgh, C.; Pahor, M. The Effects of Time Restricted Feeding on Overweight, Older Adults: A Pilot Study. *Nutrients* **2019**, *11*, 1500. [CrossRef]
100. Martens, C.R.; Rossman, M.J.; Mazzo, M.R.; Jankowski, L.R.; Nagy, E.E.; Denman, B.A.; Richey, J.J.; Johnson, S.A.; Ziemba, B.; Wang, Y.; et al. Short-term time-restricted feeding is safe and feasible in non-obese healthy midlife and older adults. *GeroScience* **2020**, *42*, 667–686. [CrossRef]
101. Kirkham, A.A.; Beka, V.; Prado, C.M. The effect of caloric restriction on blood pressure and cardiovascular function: A systematic review and meta-analysis of randomized controlled trials. *Clin. Nutr.* **2020**, *40*, 728–739. [CrossRef]
102. Kitzman, D.W.; Brubaker, P.H.; Morgan, T.M.; Haykowsky, M.J.; Hundley, G.; Kraus, W.E.; Eggebeen, J.; Nicklas, B.J. Effect of Caloric Restriction or Aerobic Exercise Training on Peak Oxygen Consumption and Quality of Life in Obese Older Patients With Heart Failure With Preserved Ejection Fraction. *JAMA* **2016**, *315*, 36–46. [CrossRef]
103. Foster-Schubert, K.E.; Alfano, C.M.; Duggan, C.; Xiao, L.; Campbell, K.L.; Kong, A.; Bain, C.E.; Wang, C.; Blackburn, G.L.; McTiernan, A. Effect of Diet and Exercise, Alone or Combined, on Weight and Body Composition in Overweight-to-Obese Postmenopausal Women. *Obesity* **2012**, *20*, 1628–1638. [CrossRef]
104. Rice, B.; Janssen, I.; Hudson, R.; Ross, R. Effects of aerobic or resistance exercise and/or diet on glucose tolerance and plasma insulin levels in obese men. *Diabetes Care* **1999**, *22*, 684–691. [CrossRef]
105. Nemet, D.; Connolly, P.H.; Pontello-Pescatello, A.M.; Rose-Gottron, C.; Larson, J.K.; Galassetti, P.; Cooper, D.M. Negative energy balance plays a major role in the IGF-I response to exercise training. *J. Appl. Physiol.* **2004**, *96*, 276–282. [CrossRef]
106. Tinsley, G.M.; Moore, M.; Graybeal, A.; Paoli, A.; Kim, Y.; Gonzales, J.U.; Harry, J.R.; A VanDusseldorp, T.; Kennedy, D.N.; Cruz, M.R. Time-restricted feeding plus resistance training in active females: A randomized trial. *Am. J. Clin. Nutr.* **2019**, *110*, 628–640. [CrossRef]
107. Tavani, A.; Giordano, L.; Gallus, S.; Talamini, R.; Franceschi, S.; Giacosa, A.; Montella, M.; La Vecchia, C. Consumption of sweet foods and breast cancer risk in Italy. *Ann. Oncol.* **2005**, *17*, 341–345. [CrossRef]
108. Silvera, S.A.N.; Jain, M.; Howe, G.R.; Miller, A.B.; Rohan, T.E. Energy balance and breast cancer risk: A prospective cohort study. *Breast Cancer Res. Treat.* **2005**, *97*, 97–106. [CrossRef]
109. Bradshaw, P.T.; Sagiv, S.K.; Kabat, G.C.; Satia, J.A.; Britton, J.A.; Teitelbaum, S.L.; Neugut, A.I.; Gammon, M.D. Consumption of sweet foods and breast cancer risk: A case-control study of women on Long Island, New York. *Cancer Causes Control* **2009**, *20*, 1509–1515. [CrossRef]
110. Kesztyüs, D.; Fuchs, M.; Cermak, P.; Kesztyüs, T. Associations of time-restricted eating with health-related quality of life and sleep in adults: A secondary analysis of two pre-post pilot studies. *BMC Nutr.* **2020**, *6*, 1–8. [CrossRef]
111. Blakeman, V.; Williams, J.; Meng, Q.-J.; Streuli, C.H. Circadian clocks and breast cancer. *Breast Cancer Res.* **2016**, *18*, 1–9. [CrossRef]
112. Wegrzyn, L.R.; Tamimi, R.M.; Rosner, B.A.; Brown, S.B.; Stevens, R.G.; Eliassen, A.H.; Laden, F.; Willett, W.C.; Hankinson, S.E.; Schernhammer, E.S. Rotating Night-Shift Work and the Risk of Breast Cancer in the Nurses' Health Studies. *Am. J. Epidemiol.* **2017**, *186*, 532–540. [CrossRef]
113. Lv, M.; Zhu, X.; Wang, H.; Wang, F.; Guan, W. Roles of Caloric Restriction, Ketogenic Diet and Intermittent Fasting during Initiation, Progression and Metastasis of Cancer in Animal Models: A Systematic Review and Meta-Analysis. *PLoS ONE* **2014**, *9*, e115147. [CrossRef] [PubMed]
114. Bartholomew, C.; Muhlestein, J.B.; Anderson, J.L.; May, H.T.; Knowlton, K.; Bair, T.L.; Horne, B.D. Intermittent Fasting Lifestyle and Incidence of Heart Failure and Myocardial Infarction in Cardiac Catheterization Patients. *Circulation* **2019**, *140* (Suppl. 1), A10043. Available online: https://www.ahajournals.org/doi/abs/10.1161/circ.140.suppl_1.10043 (accessed on 23 August 2021).
115. Abazid, R.M.; Khalaf, H.H.; Sakr, H.I.; Altorbak, N.A.; Alenzi, H.S.; Awad, Z.M.; Smettei, O.A.; Elsanan, M.A.; Widyan, A.M.; Azazy, A.S.; et al. Effects of Ramadan fasting on the symptoms of chronic heart failure. *Saudi Med. J.* **2018**, *39*, 395–400. [CrossRef] [PubMed]
116. Vitolins, M.Z.; Rand, C.S.; Rapp, S.R.; Ribisl, P.M.; Sevick, M.A. Measuring Adherence to Behavioral and Medical Interventions. *Control. Clin. Trials* **2000**, *21*, S188–S194. [CrossRef]
117. Burke, L.E.; Warziski, M.; Starrett, T.; Choo, J.; Music, E.; Sereika, S.; Stark, S.; Sevick, M.A. Self-Monitoring Dietary Intake: Current and Future Practices. *J. Ren. Nutr.* **2005**, *15*, 281–290. [CrossRef]
118. Mixon, H.; Davis, M.E. Thinking about food: An analysis of calorie estimation accuracy. *J. Integr. Soc. Sci.* **2020**, *10*, 102–125.
119. Rothman, R.L.; Housam, R.; Weiss, H.; Davis, D.; Gregory, R.; Gebretsadik, T.; Shintani, A.; Elasy, T.A. Patient Understanding of Food Labels: The Role of Literacy and Numeracy. *Am. J. Prev. Med.* **2006**, *31*, 391–398. [CrossRef]
120. Rothschild, J.; Hoddy, K.K.; Jambazian, P.; A Varady, K. Time-restricted feeding and risk of metabolic disease: A review of human and animal studies. *Nutr. Rev.* **2014**, *72*, 308–318. [CrossRef]
121. Leiper, J.B.; Molla, A.M. Effects on health of fluid restriction during fasting in Ramadan. *Eur. J. Clin. Nutr.* **2003**, *57*, S30–S38. [CrossRef]

122. Boobes, Y.; Bernieh, B.; Al Hakim, M.R. Fasting Ramadan in kidney transplant patients is safe. *Saudi J. Kidney Dis. Transplant.* **2009**, *20*, 198. Available online: https://www.sjkdt.org/article.asp?issn=1319-2442;year=2009;volume=20;issue=2;spage=198;epage=200;aulast=boobes (accessed on 19 August 2021).
123. Becker, M.; Karpati, T.; Valinsky, L.; Heymann, A. The impact of the Yom Kippur fast on emergency room visits among people with diabetes. *Diabetes Res. Clin. Pract.* **2013**, *99*, e12–e13. [CrossRef]
124. Metzger, M.; Lederhendler, L.; Corcos, A. Blinded Continuous Glucose Monitoring During Yom Kippur Fasting in Patients with Type 1 Diabetes on Continuous Subcutaneous Insulin Infusion Therapy. *Diabetes Care* **2015**, *38*, e34–e35. [CrossRef]
125. Parr, E.B.; Devlin, B.L.; Lim, K.H.C.; Moresi, L.N.Z.; Geils, C.; Brennan, L.; Hawley, J.A. Time-Restricted Eating as a Nutrition Strategy for Individuals with Type 2 Diabetes: A Feasibility Study. *Nutrients* **2020**, *12*, 3228. [CrossRef]
126. Arnason, T.G.; Bowen, M.W.; Mansell, K.D. Effects of intermittent fasting on health markers in those with type 2 diabetes: A pilot study. *World J. Diabetes* **2017**, *8*, 154–164. [CrossRef]
127. Cava, E.; Yeat, N.C.; Mittendorfer, B. Preserving Healthy Muscle during Weight Loss. *Adv. Nutr.* **2017**, *8*, 511–519. [CrossRef]
128. Antoni, R.; Robertson, T.M.; Robertson, M.D.; Johnston, J.D. A pilot feasibility study exploring the effects of a moderate time-restricted feeding intervention on energy intake, adiposity and metabolic physiology in free-living human subjects. *J. Nutr. Sci.* **2018**, *7*. [CrossRef]
129. Chaix, A.; Zarrinpar, A.; Miu, P.; Panda, S. Time-Restricted Feeding Is a Preventative and Therapeutic Intervention against Diverse Nutritional Challenges. *Cell Metab.* **2014**, *20*, 991–1005. [CrossRef]
130. Vale, R.C.D.; Pieters, R.; Zeelenberg, M. The benefits of behaving badly on occasion: Successful regulation by planned hedonic deviations. *J. Consum. Psychol.* **2016**, *26*, 17–28. [CrossRef]
131. Mills, N.; Donovan, J.L.; Wade, J.; Hamdy, F.C.; Neal, D.E.; Lane, J.A. Exploring treatment preferences facilitated recruitment to randomized controlled trials. *J. Clin. Epidemiol.* **2011**, *64*, 1127–1136. [CrossRef]
132. Allen, J.D.; Stewart, M.D.; Roberts, S.A.; Sigal, E.V. The Value of Addressing Patient Preferences. *Value Health* **2017**, *20*, 283–285. [CrossRef]
133. Regmi, P.; Heilbronn, L.K. Time-Restricted Eating: Benefits, Mechanisms, and Challenges in Translation. *iScience* **2020**, *23*, 101161. [CrossRef]

 MDPI

Review

Ketogenic Diet for Cancer: Critical Assessment and Research Recommendations

Jordin Lane [1], Nashira I. Brown [1], Shanquela Williams [1], Eric P. Plaisance [2] and Kevin R. Fontaine [1,*]

1 Department of Health Behavior, School of Public Health, University of Alabama, Birmingham, AL 35294, USA; jalane@uab.edu (J.L.); Nashirab@uab.edu (N.I.B.); iamshan@uab.edu (S.W.)
2 Department of Human Studies, School of Education, University of Alabama, Birmingham, AL 35294, USA; plaisep@uab.edu
* Correspondence: kfontai1@uab.edu; Tel.: +(205)-975-8397

Abstract: Despite remarkable improvements in screening, diagnosis, and targeted therapies, cancer remains the second leading cause of death in the United States. It is increasingly clear that diet and lifestyle practices play a substantial role in cancer development and progression. As such, various dietary compositions have been proposed for reducing cancer risk and as potential adjuvant therapies. In this article, we critically assess the preclinical and human trials on the effects of the ketogenic diet (KD, i.e., high-fat, moderate-to-low protein, and very-low carbohydrate content) for cancer-related outcomes. The mechanisms underlying the hypothesized effects of KD, most notably the Warburg Effect, suggest that restricting carbohydrate content may impede cancer development and progression via several pathways (e.g., tumor metabolism, gene expression). Overall, although preclinical studies suggest that KD has antitumor effects, prolongs survival, and prevents cancer development, human clinical trials are equivocal. Because of the lack of high-quality clinical trials, the effects of KD on cancer and as an adjunctive therapy are essentially unknown. We propose a set of research recommendations for clinical studies examining the effects of KD on cancer development and progression.

Keywords: ketogenic; cancer; adjuvant therapy

Citation: Lane, J.; Brown, N.I.; Williams, S.; Plaisance, E.P.; Fontaine, K.R. Ketogenic Diet for Cancer: Critical Assessment and Research Recommendations. *Nutrients* **2021**, *13*, 3562. https://doi.org/10.3390/nu13103562

Academic Editors: Christina Dieli-Conwright and Wendy Demark-Wahnefried

Received: 7 September 2021
Accepted: 4 October 2021
Published: 12 October 2021

1. Introduction

Despite continued advances in screening, early diagnosis, and treatment, cancer remains the most dreaded of human maladies [1]. Surpassed only by heart disease as the leading cause of death in the United States, it is estimated that there will be 1,898,160 new cases and 608,570 deaths in 2021. The most common cancer sites are prostate, lung and colorectal for men; and breast, lung, and colorectal for women [2]. Lung cancer is the leading cause of cancer deaths for both sexes and is projected to remain so until 2040 [3], and likely well beyond.

Although tobacco remains the primary contributing factor for cancer development, other environmental factors, such as diet and lifestyle, play an extensive role. In 2015, it was estimated that diet accounts for approximately 30% of the attributable risk for cancer [4,5]. In 2017, the CDC estimated that 40% of all cancers are related to overweight and obesity (55% in women and 24% in men), with at least 13 different types of cancer linked to obesity (the most strongly linked were liver, endometrial, esophageal, and kidney) [6]. Although it is well-established that obesity associates strongly with both cancer incidence and mortality, it is less clear whether adiposity itself is the cause of or marker (byproduct) of underlying metabolic dysregulation that creates the conditions where cancer can develop and thrive [7]. The prevailing view has long been that positive energy balance resulting from excess energy consumption, lower energy expenditure, or both contribute to excess adiposity and subsequent manifestations of chronic disease, including cancer. However, emerging evidence suggests that dietary macronutrient composition may play a more extensive role than excess adiposity, per se in the development of cancer [8–12]. For example, high carbohydrate diets that are highly processed with added sugar have been shown to produce a hormonal milieu and metabolic derangements which promote the development of cancer and other chronic diseases. By extension, one might ask whether regulation of the quantity and/or quality of carbohydrates might mitigate cancer risk and/or cancer-related outcomes in those who develop cancer.

While several diets (e.g., vegan, Mediterranean) and dietary regimens (e.g., caloric restriction, intermittent fasting [13]) have been proposed as strategies for cancer prevention and as adjuvant therapies to standard-of-care cancer treatments, we provide a theoretical framework and preliminary evidence from preclinical and clinical studies on how the ketogenic diet (KD) may provide benefits in the prevention and treatment of cancer. Because there are several recent narrative, systematic, and meta-analytic reviews of KD for cancer [14–18], we focus on critically evaluating the state of the knowledge and provide a set of research recommendations to enhance the rigor and replicability of KD–cancer clinical applications and randomized clinical trials.

2. Ketosis and Spectrum of Ketogenic Diets (KD)

Nutritional ketosis has been defined as "the intentional restriction of dietary carbohydrate intake to accelerate the production of ketones and to induce a metabolic effect that stabilizes blood sugar, minimizes insulin release, and thereby mitigates the downstream anabolic and tumorigenic effects of longstanding insulin resistance [19] (p. 99)." Because maintaining stable blood glucose levels is essential for survival, even in the context of severe carbohydrate restriction, glucose can be synthesized from non-glucose substrates (e.g., certain amino acids) by hepatic gluconeogenesis (GNG). As part of a strategy to reduce the deleterious consequences and potentially lethal effects of unregulated protein depletion, mammals (including humans) evolved an efficient method to store excess energy. In a period of excess energy consumption, triglycerides consumed in the diet and produced from glucose and/or glucose in liver are transported to adipose where they are mobilized during prolonged fasting or starvation. Fatty acids released from triglycerides in adipose tissue are then transported to liver where they enter mitochondria and are partially diverted for ketone production—a primary source of energy in the brain during starvation as free fatty acids are unable to cross the blood–brain barrier and thus provide only a small amount of energy.

The classic KD is characterized by high-fat, moderate-to-low protein, and very-low carbohydrate content [20]. This translates into a dietary composition of about 90% fat, 2% carbohydrate, and 8% protein. As implied above, KD received its name because this diet induces physiologic ketosis which is manifested by increased concentrations of ketone bodies and decreased glucose and insulin concentrations in blood [21]. KD's beneficial effects have been observed in a range of conditions including epilepsy and other neurologic diseases, obesity, type 2 diabetes, polycystic ovary syndrome, and cardiovascular disease (see [22,23] for a recent review).

The classic ketogenic diet consists of a ratio between fats and non-fats (carbohydrates + proteins) of 3:1 or 4:1. The major variations include: (1) Very Low-Calorie Ketogenic Diet is time-limited (~12 weeks) calorically restrictive (600–800 kcal), characterized by a minimum protein content (≥75 g/day), limited carbohydrate content (30–50 g/day), and a fixed amount of fat (20 g/day, mainly from olive oil and omega-3 fatty acids); and (2) the Low Glycemic Index Diet characterized by intake of a higher quantity of carbohydrates (60–80 g/day) from low glycemic index sources (e.g., lentils, chickpeas, bran cereals, carrots). Although not, strictly speaking, a KD, the Low Glycemic Index Diet has been effective in treating some forms of epilepsy and headaches [24] (it is thought that this diet, with its less restrictive carbohydrate intake, is unlikely to have beneficial effects on cancer [25]).

3. KD as a Therapeutic for Cancer: Hypothesized Mechanisms

While it is beyond the scope of this work to provide a comprehensive review of the proposed biological mechanisms by which a KD might confer benefits as a cancer therapy, (see [7,14,26–28] for more detailed expositions), we provide a brief and highly simplified overview, with particular emphasis on the rationale for proposing the potential value of a KD.

Despite their rapid proliferation, cancer cells use no more oxygen than non-cancer cells for oxidative purposes. Instead, they use about 10 times more glucose and produce about 70 times the rate of lactic acid than do normal cells. In other words, even with ample oxygen available, most cancer types derive energy from anaerobic glycolysis [29]. The reason that the vast majority (about 80%) of all cancers shift from oxidative phosphorylation to glycolysis (i.e., the Warburg Effect [30]) is unknown although it is speculated that doing so must confer a survival advantage (perhaps the acidic environment imposed by lactic acid is well tolerated by cancer cells, promoting further growth and spread to other organs [31]). Because the shift to glycolysis is manifested at the onset of tumorigenesis, many consider it one of the hallmarks of cancer [32]. Indeed, the Warburg Effect indirectly contributed to PET imaging, as the scan measures glucose disposal by cells (cancer cells take up far more glucose than surrounding cells, allowing contrasts in imaging).

Other factors, so-called nutrient sensors (e.g., insulin, insulinlike growth factor (IGF-1), mammalian target of rapamycin (mTOR), AMP-activated protein kinase (AMPK)) operate in the Warburg Effect, with their pathways playing important and complimentary roles in cellular proliferation and cancer expression [33–37].

Other potential metabolic pathways proposed as to why KD may confer benefits include the possibility that severely restricting carbohydrate intake alters mitochondrial function, the regulation of gene expression, the production of reactive oxygen species, the amino acid metabolism of cancer cells, angiogenesis and the vascularization of the tumor environment [38,39].

In summary, the primary rationale for proposing a KD as prevention or for treatment of cancer is to deprive cancer cells of their primary energy source, glucose, thereby interrupting the elaborate processes of nutrient sensors and other factors that are activated by the presence of glucose and insulin and appear to play important roles in their development and proliferation.

4. Preclinical Studies of KD for Cancer

Some animal models of cancer suggest that KD might be an efficacious cancer therapy when used alone or as an adjuvant to conventional therapies [14]. Specifically, some studies report that KD delays tumor development, slows growth, and increases survival time (e.g., [40,41]). Another set of studies show that KD may make tumor cells more vulnerable to the combination of chemotherapy and radiation as well as enhance the effects of targeted therapy (i.e., PI3K inhibitors) in tumor models [42]. However, other studies report increased tumor growth in rat models of kidney cancer [43] and mouse models of BRAF V600E-positive melanoma [44].

Li and colleagues [45] recently conducted a meta-analysis of 17 published animal studies to estimate KD's potential antitumor effects. They found that KD, alone or in combination with caloric restriction, significantly reduced both tumor weight (standard mean difference [SMD] −2.45, $p = 0.027$) and volume (SMD = −0.76, $p = 0.012$) as well as prolonging survival time (SMD = 1.76, $p = 0.003$). Additional analyses suggested that KD ratio of 4:1 (i.e., severe carbohydrate restriction) was associated with the greatest increase in survival time (see also, [14,46,47]). Finally, the authors found that KD's efficacy varied as a function of several factors, prompting them to conclude, "In summary, the pre-clinical evidence pointed toward an overall antitumor effect of the KD in animal studies currently available with limited tumor types. The efficacy of KD on tumorigenesis appears to be influenced by several factors, including cancer type or subtype, genetic background, cell line and/or model system, composition of the KD and tumor-associated syndromes.

Therefore, more preclinical studies should be performed to elaborate the antitumor effect of KD in the future [45] (p. 11)."

5. Clinical Studies of KD and Cancer

Despite the promising results of KD from preclinical studies, there have been few human trials to isolate the effects of KD on cancer-related outcomes (most have focused on tolerability and safety [48]). For example, in a 4-week pilot study Fine et al. evaluated the safety and feasibility of a KD in 10 patients with different cancers [27]. Among the patients whose disease remained stable or partially remitted, they found ketone levels (i.e., serum beta-hydroxybutyrate [βHB]) on average, that were threefold higher compared with those with progressive disease. To date, most applications of KD in human cancers has been as an adjunctive therapy in conjunction with standard of care (i.e., chemotherapy, radiotherapy, and/or surgery). Recent evaluations of the literature conducted by Weber and associates (29 trials) [14], Talib et al. (14 trials) [48] and Yang and colleagues (6 trials) [15] Sremanakova and associates [18] (11 trials), Plotti et al. [49] (4 trials), and Romer and associates [16] (45 trials) among patients, virtually all being adults (i.e., 18 years of age and older), with a variety of cancers (e.g., glioblastoma, glioblastoma and gliomatosis cerebri, breast cancer, liver, pancreato-biliary cancer, lung and pancreatic, head and neck, colorectal cancer, and mixed cancer sites reported a wide range of favorable outcomes including progression-free survival, increased survival rate, increased rates of response to conventional treatment (i.e., stable disease after 6-week diet) [49], and enhanced quality of life (please see Table 1 for summary of clinical trials). While safe and well-tolerated by the majority of patients, some report side effects, including nausea, constipation, vomiting, hypoglycemia, and fatigue that may compromise adherence to KD [13,20]. Overall, while is has been found that KD may be beneficial for varying types of cancers as it relates to tumor characteristics, survival and side effects [50], it is important to underscore that, as described below, the trials were of varying methodological quality, which inhibits our ability to draw definitive conclusions on the effects of KD as an adjunctive therapy.

Table 1. Summary of Clinical Studies of KD and Cancer.

Cancer Type(s)	Sample Size	Dietary Intervention	Study Duration	Results/Outcomes	References
Prostate	N = 45 Arm A: N = 27 Arm B: N = 18	Arm A: A low-carbohydrate diet, goal: (≤20 g per day), estimated actual carbohydrate intake: 37 g/day; Arm B: Control group (no dietary intervention)	6 months	-Weight loss -BMI reduction -Waist circumference reduction	[51]
Breast cancer	N = 60 Arm A: N = 30 Arm B: N = 30	Arm A: Medium-chain triglycerides (MCT) based ketogenic diet (6% calories from Carbohydrates [CHO], 19% protein, 20% MCT, 55% fat); Patients received 500 mL of MCT oil from the Nutricia Company every 2 weeks Arm B: Standard Diet (55% CHO, 15% protein, and 30% fat)	3 months	-Weight loss -BMI reduction -Reduction in body fat	[52]
Ovarian/endometrial cancer	N = 45 Arm A: N = 25 Arm B: N = 20	Arm A: Ketogenic diet (70% (≥125 g): 25% (≤100 g): 5% (<20 g) energy per day from fat, protein, and carbohydrates) Arm B: American Cancer Society diet (ACS: high in fiber, low in fat) Individual diet advice from certified dietitians. Weekly emails or phone calls. One face-to-face meeting after baseline assessment	3 months	-Self-reported improvement in energy levels (intervention group) -Fewer cravings for starchy foods and fast-food fats -Reduction in total body	[53,54]
Rectal cancer, head and neck cancer Breast cancer	N = 81 Arm A: N = 20 Arm B: N = 61	Arm A: ketogenic diet with additional consumption of non-glucogenic amino acids Arm B: no dietary intervention	30–40 days	-Decreased fat mass	[55]

Table 1. *Cont.*

Cancer Type(s)	Sample Size	Dietary Intervention	Study Duration	Results/Outcomes	References
Pancreatic cancer Duodenal cancer Common bile duct cancer Ampulla of Vater cancer Cholangio-carcinoma Neuroendocrine tumor	*N* = 19 Arm A: *N* = 10 Arm B: *N* = 9	Arm A: Ketogenic diet (3–6%, 14–27%; 70–80% energy per day from carbohydrates, protein, and fat) served as three meals and three snacks per day Arm B: usual Korean diet (55–65%, 7–20%, 15–30% energy per day from carbohydrates, protein and fat) served as three meals per day	12 days	-Decreased body cell mass higher in General Diet arm	[56]
Glioblastoma multiforme	*N* = 53 Arm A: *N* = 6 Arm B: *N* = 47	Arm A: self-administered KD Arm B: unspecified standard American diet	Duration: 3–12 months	- Two patients with grade 1 constipation, 4 patients with grade 1 fatigue, 1 patient with grade 2 fatigue, 1 patient with deep venous thrombosis during treatment, 1 patient with asymptomatic hypoglycemia, 1 patient with nephrolithiasis no grade 3 and higher toxicities or symptomatic hypoglycemia -Weight loss on non-calorie-restricted KD: 1 to 27 Ibs -Weight loss on calorie-restricted KD: 46 Ibs	[57]
Fearon et al. [44] Ovarian, Lung, Gastric	*N* = 5	Crossover study: Nasogastric tube feeding: normal, balanced regimen on days 1–6 KD containing same total calorie and protein on days 7–13	13 days	-Increase in body weight	[58]
Diverse	Recruited patients *N* = 12 Analyzed patients *N* = 10	KD with targeted CHO intake below 5% of total energy intake, written menus and samples of CHO-restriction products were provided	28 days	-Five patients with grade 2 fatigue, 5 patients with grade 1 constipation, 1 patient with grade 1 leg cramps -Weight loss - Decreased caloric intake -Adherence: 5 of 12 patients completed all 28 days of the diet	[27]
Diverse	Analyzed patients *N* = 78 Arm A: *N* = 7 Arm B: *N* = 6 Arm C: *N* = 65	Arm A: full adoption of a non-specified KD, patients informed about a single company producing KD-related food Arm B: partial adoption of a non-specified KD, patients informed about a single company producing KD related food Arm C: patients who did not adopt a KD	Not specified	1. Reduction in TKTL 1 was associated with adopting a KD; 2. Correlation between improvement in cancer status category and full adoption of a KD ($\chi 2 = 33.26$; df = 4; $p = 0.00001$	[59]
Diverse	Analyzed patients *N* = 6	Self-administered KD (recommended CHO intake < 50 g/day) during the course of RT/RCT; patients received basic information on KD; counseling at least once per week	Patient-dependent from 32 to 73 days	-Decreased fat mass	[60]
Glioblastoma	Assessed for eligibility: *N* = 57 Randomized: *N* = 12 Arm A: *N* = 6 Arm B: *N* = 6 Retention at 12 weeks. *N* = 4 Arm A: *N* = 3 Arm B: *N* = 1	Arm A: MCTKD (75%; 15%; 10% of energy per day from fat, protein and carbohydrates, with 30% of fat from MCT nutritional products) Arm B: MKD (80%; 15%; 5% of energy per day from fat, protein and carbohydrates)	12 weeks	1. Arm A: Three patients retained for 3 months (drop-out = 50%) Arm B: One patient retained for 3 months (drop-out = 83%) 2. GHS at baseline: Arm A: patients who later withdrew: 72.2 ± 20.7; patients who retained: 75 ± 6.8 Arm B: patients who later withdrew: 70 ± 13.8; patients who retained: 80 ± 0 GHS: at week 6: Arm A: patients who withdrew at week 6: 41.7 ± 0; patients who retained: 66.7 ± 0 Arm B: patients who withdrew at week 6: 50 ± 0; patients who retained: 100 ± 0 3. Adverse events during the first 6 weeks: Arm A: diarrhea (*n* = 1, CTCAE grade 1), nausea (*n* = 1, CTCAE grade 1), vomiting (*n* = 1, CTCAE grade 2), dyspepsia (*n* = 1, CTCAE grade 1) Arm B: vomiting (*n* = 1, CTCAE grade 1), dry mouth (*n* = 1 MKD, CTCAE grade 1)	[61]

Table 1. *Cont.*

Cancer Type(s)	Sample Size	Dietary Intervention	Study Duration	Results/Outcomes	References
Glioblastoma	Enrolled: $N = 6$ Completed intervention: $N = 4$	MKD (70%: 3–5% (≤20 g) energy per day from fat and carbohydrates; protein consumption was not restricted	12 weeks	-Constipation in two patients, resolved with dietary modification	[62]
Glioblastoma	Included patients $N = 20$ Evaluable for efficiency $N = 17$	KD with CO intake < 60 g/day, additionally highly fermented yoghurt drinks and two different plant oils were provided to be consumed at will. No calorie restriction, patients were instructed to always eat to satiety	Until progression of the disease	-Three out of 20 patients discontinued the diet after 2–3 weeks without progression, due to reduced QoL - Body weight reduction -Diarrhea, constipation, hunger and/or demand for glucose were present in some patients during the diet	[63]
Diverse	Enrolled: $N = 16$ Completed intervention: $N = 5$	KD with CHO limited to 70 g per day and 20 g per meal Two oil–protein shakes consumed in the morning and in the afternoon	12 weeks	-11/16 Patients discontinued the diet - 3/11 were unable to adhere to the diet, -6/11 discontinued due to progressive disease -2/11 died from progressive disease - reported side effects included increase in appetite loss, constipation, diarrhea and fatigue during the diet - QoL was low at baseline and stayed relatively stable during the intervention; worsening of fatigue, pain, dyspnea and role function but emotional functioning and insomnia improved slightly	[64]
Diverse	Enrolled: $N = 17$ Drop-out before first analysis: $N = 6$ Completed intervention: $N = 4$	Modified Atkins Diet with 20 to 40 g of CHO and restricted consumption of high CHO foods no restrictions for calories, protein or fats	16 weeks	-13/17 patients discontinued the diet before 16 weeks -weight loss -Reported adverse effects included: hyperuricemia ($N = 7$), hyperlipidemia ($N = 2$), pedal edema ($N = 2$), anemia ($N = 2$), halitosis ($N = 2$), pruritus ($N = 2$), hypoglycemia ($N = 2$), hyperkalemia ($N = 2$), hypokalemia ($N = 2$), hypomagnesemia ($N = 2$), flulike symptoms/fatigue ($N = 2$)	[65]
Glioblastoma multiforme	Phase A: $N = 9$ Phase B: $N = 8$ Completed intervention $N = 6$	Phase A: Fluid KD with a 4:1 ratio (4 g fat versus 1 g protein plus carbohydrates, 90% energy from fat) Patients were allowed a snack with the same 4:1 diet ratio once a day Phase B: Solid-food KD (diet ratio 1.5–2.0:1) with MCT; (70% energy from fat with the consistency of an emulsion)	14 weeks	-6/9 patients included in phase A completed the 14 weeks KD - Reported adverse effects included: constipation ($n = 7$), nausea/vomiting ($n = 2$), hypercholesterolemia ($n = 1$), hypoglycemia ($n = 1$), low carnitine ($n = 1$) and diarrhea ($n = 1$). CTCAE grade 2: hallucinations ($n = 1$), allergic reaction ($n = 1$) and wound infection ($n = 1$)	[66]
Glioma	$N = 29$	MAD with a 0.8–1:1 ratio (0.8-1 g fat to 1 g carbohydrate plus protein Duration: 6 weeks	6 weeks	-28/29 patients completed the 6-week diet - Reported adverse events: Grade 2 constipation (n = 1), grade 1 fatigue and nausea were present in the patients -Decreased BMI for all patients	[67]
Lung	Enrolled patients: $N = 7$ Completed intervention: $N = 2$	KD with 90%; 8%; 2% of energy per day from fat, protein and carbohydrates. All meals prepared for the patients	42 days	-Weight loss - Reported adverse events included: constipation, diarrhea, nausea, vomiting and fatigue; hyperuricemia	[68]
Pancreas	$N = 2$	KD with 90%; 8%; 2% of energy per day from fat, protein and carbohydrates. All meals readily prepared for the patients	34 days	-1/2 patients completed the intervention 2. Reported adverse events included: Constipation, diarrhea, nausea and vomiting, 1 patient experienced dehydration -Weight loss	[68]

Table 1. *Cont.*

Cancer Type(s)	Sample Size	Dietary Intervention	Study Duration	Results/Outcomes	References
Desmoid tumor	*N* = 1	TPN consisting of 28 kcal fat/kg body weight/day, 1.5 g protein/kg body weight/day; 40 g glucose/day	Desmoid tumor	-Body weight increased	[69]
Glioma	*N* = 2	ERKD: with a 3:1 ratio of ingested nutrients (3 g fat versus 1 g protein plus carbohydrates) 20% restriction of calories per day	12 months	-Adherence: 1/2 patients completed the intervention -Reported headaches -Initial body weight decrease in both patients and remained stable afterward	[70]
Glioblastoma multiforme	*N* = 1	ERKD delivering 600 kcal per day, consisting of 42 g fat, 32 g protein and 10 g CHO per day	56 days	-Bodyweight decreased in the first 14 days of the diet - Grade 4 hyperuricemia reported, resulted in diet change to calorie restricted non-ketogenic diet	[71]
Rectal	*N* = 1	Paleolithic KD, nutrients consumed in a fat:protein ratio of 2:1 animal fat, red meats and organ meats were encouraged, root vegetables were allowed, all other foods were prohibited	24 months	-Decreased bodyweight -Initial decrease in volume after concomitant radiotherapy -Tumor volume remained stable but four hepatic metastases were detected at the end of the diet	[72]
Diverse	*N* = 12	Single 3 h infusion of glucose-based (GTPN) or a lipid-based TPN (LTPN) containing 4 mg glucose/kg/min or 2 mg lipid/kg/min, respectively	3 h	-No statistically significant stimulation or suppression of FDG uptake	[73]
Recurrent Breast	*N* = 1	Self-administered high doses of oral vitamin D3 (10,000 IU/day), and KD rich in oleic acid. Duration: 3 weeks	3 weeks	-Progesterone receptor status positivity increased -HER2 positivity decreased	[74]
Astrocytoma	*N* = 2	KD with 60%; 20%; 10%, 10% of energy per day from MCT oil, protein, carbohydrates and dietary fat plus additional supplements	8 weeks	-Dose uptake ratio tumor: decreased normal cortex decreased -Adherence: 100% patients were able to complete the dietary intervention	[75]
Esophagus Stomach Colon-rectum	*N* = 27 Arm A: *N* = 9 Arm B: *N* = 9 Arm C: *N* = 9	Arm A: glucose-based TPN (100% of the calorie from dextrose); Arm B: lipid-based TPN (80% of the calorie from fat, 20% from dextrose); Arm C: oral diet All diets were iso-caloric and isonitrogenous. Duration: 2 weeks	2 weeks	No statistically significant changes	[76]
Head and neck	*N* = 12	Unspecified Western diet followed by unspecified KD	Variable, up to 4 days	Decline of mean lactate concentration in the tumor tissue during the KD	[77]
Brain	Included: *N* = 9 intervention: *N* = 5 retrospectively added control *N* = 4	KD based on ready-made formula, with a 4:1 ratio of ingested nutrients (4 g fat versus 1 g protein plus carbohydrates)	variable from 2 to 31 months	-Diet tolerated by 4/5 patients,(strict adherence only in 2 patients) -Four out of 50 MRI spectroscopy scans detected ketone bodies in the brains of the patients following the KD	[78]
Lung	*N* = 44	Mild KD (patients were encouraged to avoid high CHO food) in combination with HBO, hyperthermia and polychemotherapy administered during induced hypoglycemia	24 weeks	-Adverse events reported—during treatment period: grade 5 neutropenia (N = 1), grade 3 neutropenia (N = 3), grade 3 anemia (N = 10), grade 4 thrombocytopenia (N = 3), grade 3 fatigue (N = 5), grade 3 diarrhea (N = 8), grade 3 neuropathy (N = 1), all of which were attributed to chemotherapy	[79]
Pancreas	*N* = 25	Mild KD (patients were encouraged to avoid high CHO food) in combination with HBO, hyperthermia and polychemotherapy administered during induced hypoglycemia	Duration: mean follow-up: 25 months	-Adverse events reported: during treatment period: grade 3/4 neutropenia (N = 9), febrile neutropenia (N = 1), grade 3 anemia (N = 7), grade 4 thrombocytopenia (N = 4), grade 3 diarrhea (N = 2), all of which were attributed to chemotherapy	[80]
Brain	*N* = 8	MAD with20g CHO/day restriction	2-24 months: mean-13 months	-7/8 completed intervention -Decreased body weight -Reduction in seizure frequency per week	[81]

Table 1. *Cont.*

Cancer Type(s)	Sample Size	Dietary Intervention	Study Duration	Results/Outcomes	References
Glioblastoma multiforme	*N* = 1	Energy-restricted KD with a 4:1 ratio of calorie intake (fat versus protein plus carbohydrates) Total calories calculated 25% below BMR	4 months	-No metabolically active tumor detected	[82]
Glioblastoma multiforme	*N* = 1	KD with a 4:1 ratio of calorie intake (fat versus protein plus carbohydrates), delivered as calorie-restricted diet, combined with intermittent fasting, HBOT, other novel therapies and SOC treatment	20 months	-Good surgical outcome and regressive changes in histopathology -Decreased body weight	[83]
Diverse	*N* = 6	Very low CHO diet (not further specified) with a multitude of supplements, including amino acids and Vitamin D^3 combined with SOC therapy	Varied	-Shrinkage of tumor or stable disease was reported during the intervention -Subjective improvement reported in some cases	[84]
Head and neck	*N* = 14	KD with as little CHO as possible (estimated < 50 g per day), combined with insulin administration 3 × per day	Not specified	Visible remission after 2–3 weeks, but rebound effect after 2–3 months on the diet	[85]
Extra-cranial	*N* = 30	KD with as little CHO as possible (estimated < 50 g per day), combined with insulin administration 3 × per day	Not specified	Tumor shrinkage in some cases Improvement in general condition and positive effects on clinical symptoms	[86]
Exra-cranial	*N* = 23	KD with as little CHO as possible (estimated < 50 g per day), combined with insulin administration 3 × per day	Not specified	-Reduced pain severity, fatigue but deteriorated orientation	[87]
Pancreatic cancer Duodenal cancer Common bile duct cancer Ampulla of Vater cancer Neuroendocrine tumor	*N* = 18	LCKD: Energy content: 1500 kcal/d, provided 4% from carbohydrate, 16% from protein and 80% from fat. Ketogenic ratio of 1.75:1 (F: C + P w/w)	4 weeks	-Patients were in a poorer nutrition state after surgery, but this was alleviated at week 4; - LCKD induced ketone body production -Week 4, there were no significant differences in ketone levels	[88]
Glioma	*N* = 13 *newly diagnosed= 6* *recurrent=7*	KD + MCT + Metformin 850	6 weeks (recurrent) 2 weeks (newly diagnosed)	Increase in survival rate. Synergistic interaction between radiation therapy and KD.	[89]
Invasive Rectal	*N* = 359	KD ≥ 40% kcal fat and <100 g/day glycemic load (48)	Not specified	Reduced risk of cancer-specific deaths	[90]
Glioblastoma	*N* = 32	KD 50% kcal fat, 25% kcal CHO, 1.5 g/kg protein (17), CD (15)	3 months	No change in glucose increased ketosis No change in body weight	[91]

6. Limitations of Current Literature

Overall, the clinical trial literature on the use of KD as an adjunctive cancer therapy in humans has several important limitations that severely undermines our ability to make causal inferences concerning the effects of KD on cancer. The common theme of the limitations revolve around heterogeneity. That is, dramatic variations, within and between trials on many characteristics such as cancer type, time since diagnosis, patient characteristics (e.g., age, sex, overall health) KD variations, trial duration, study design, and outcomes assessment makes it impossible to draw conclusions on the effects on KD. In a sense, having so much variation in the published trials is a worse state-of-affairs than simply having an absence of trials because of the challenge in trying to draw conclusions from inconsistent findings, at least partly driven by the vast heterogeneity and varying methodological quality. Indeed, because of the vast heterogeneity of the human clinical trial literature, the validity of the published systematic reviews and meta-analytic reviews is highly questionable. For this reason, although preclinical evidence suggests favorable effects of KD, the human trials, to date, are equivocal regarding potential beneficial effects of KD as an adjunctive therapy, let alone as an intervention to impede cancer growth or improve survival.

7. Conclusions and Future Directions

Preclinical studies in multiple strains of mice and types of cancer provide extensive evidence that the KD decreases tumor growth, prolongs survival, and reverses the process of cancer cachexia [14]. Clinical studies in humans are much more limited and have largely focused on small pilot or case studies and few clinical trials (see Table 2 for a summary of the strength of evidence for pre-clinical and human studies). Because of the promising effects in preclinical rodent models and the limited number of rigorous human clinical trials, it is clear that studies are needed in preclinical models and humans to understand the molecular mechanisms of KD and other low-carbohydrate diets in multiple forms of cancer. The hypothesized benefit of any low carbohydrate or low glycemic index diet is that the removal of processed foods containing sugar, added sugar, and lowering of starch-based carbohydrates reduce the amount of insulin required to clear a meal in the postprandial state. Since humans spend over 2/3 of their time in a postprandial state, it is logical to move forward under the supposition that lowering insulin could serve as a strategy to reduce risk of and progression of cancer. Presumably, decreasing the presentation of glucose by dietary carbohydrate restriction at the cellular and the epigenetic programming resulting from elevated insulin concentrations would be expected to reduce tumorigenesis and progression of cancer. In addition, insulin rapidly activates protein synthesis by activating components of protein translation such as eukaryotic initiation and elongation factors along with increasing the cellular content of ribosomes to augment the capacity for protein synthesis.

Table 2. Overview of Strength of Evidence for Beneficial Effects of the Ketogenic Diet for Cancer and Related Outcomes in Pre-Clinical and Clinical Studies.

	Strength of Evidence			
	Strong	**Moderate**	**Weak**	**Unknown**
Pre-Clinical Studies				
Tumor weight		X		
Antitumor effect/Tumor growth	X			
Progression-free survival				X
Tumor volume		X		
Overall survival time		X		
Cells' responsiveness to therapy		X		
Body composition		X		
Clinical Studies as an Adjunctive Therapy				
Tumor weight				X
Antitumor effect/Tumor growth				X
Progression-free survival				X
Tumor volume				X
Overall survival time				X
Cells' responsiveness to therapy				X
Quality of life			X	
Body composition		X		

Studies are also needed to examine the effects of KD in multiple forms of cancer to determine whether the diet provides synergistic or additive benefits as an adjuvant therapy. Based on the sparse data available, there is reason to predict that KD could serve as an adjuvant to reduce tumor formation and progression. In addition, it will be important to examine tolerability of the KD in different types of cancers and treatments. If certain forms of cancers and/or treatments reduce palatability to the point where compliance is lost, then studies will be severely limited in scope and inference as it relates to the interpretation of findings. Thus, it will be important that future studies clearly define and test different levels of carbohydrate on low carbohydrate diets to improve the likelihood of success and to properly evaluate the effects of these diets on cancer risk and progression. It should be considered that the few studies which have examined the effects of KD on some forms of cancer and cancer treatment have observed an attenuation of skeletal muscle loss. While the mechanisms for this response are not entirely clear, preclinical studies from our group and others suggest that the ketone, beta-hydroxybutyrate (βHB), inhibits histone deacetylases which have been shown to preserve muscle in aging rodents [92]. These findings suggest that KD may reduce cancer cachexia and potentially improve functional capacity and quality of life while undergoing treatment.

Finally, with the commercial availability of exogenous ketone supplements, future studies are also needed to examine whether these supplements decrease cancer risk or progression. Little is known about the long-term effects of exogenous ketones in humans, but ketone esters and salt supplements transiently raise serum ketones, providing utility as a potential adjuvant treatment. Non-published observations from our group demonstrate that ketones consumed at or near the postprandial period reduce circulating levels of ketones and presumably have little effect on circulating insulin concentrations. Therefore, innovative dietary strategies with, perhaps, KD with ketone supplementation may be a favored strategy to increase circulating ketones while reducing insulin concentrations. Studies are needed to determine whether ketone supplements alone are sufficient, and at what dose and timing, to improve cancer and cancer-related outcomes.

8. Research Recommendations for Moving the Field Forward

Despite the metabolic rationale and relatively promising results in animal models, human trials testing KD as an adjunctive cancer therapy have been equivocal, indicating that we have a long way to go before drawing conclusions about the value of this diet. As noted above, the few human trials conducted thus far are fraught with methodological limitations, including, but not limited to, small sample sizes of heterogeneous patients (e.g., different cancer sites, disease durations, age, sex, comorbidities, among others), the absence of randomization and control groups, use of different and poorly described KD protocols, poor assessments of dietary adherence, short durations, and poorly defined and measured outcomes. The lack of high-quality trials, therefore, impedes both our scientific understanding and efforts to begin to translate a KD intervention into clinical practice. Without efforts to resolve these methodological limitations, the potential effects of KD on any cancer-related variables or outcomes will remain unknown. As noted by Romer and associates, "To form a final judgment about the efficiency of a KD in Oncology, a randomized controlled trial with a well-designed control group and sufficient power to also detect evidence for absence of antitumor effects is necessary [16] (p. 33)." Of course, not only would high-quality trials be required to detect potential antitumor effects but also on other important variables such as body composition, circulating insulin and inflammatory marker concentrations, side effects, functional capacity, survival time, and quality of life.

As such, we suggest that the following research recommendations may be useful in moving us toward a greater understanding of the effects of KD on cancer and related outcomes in humans (see also Figure 1).

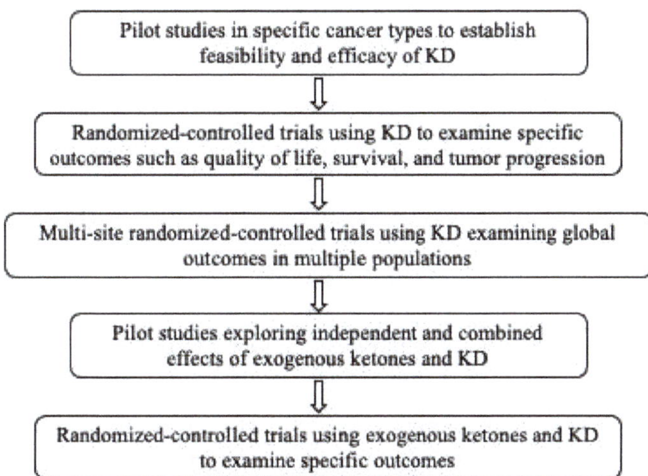

Figure 1. Sequential Research Recommendations for Investigating the Effects of the Ketogenic Diet (KD) on Human Cancers.

- Conduct small, rigorous non-randomized trials with homogeneous patient groups and common cancer sites to assess whether KD produces a "signal" on selected outcomes (particularly those related to response to standard care (e.g., effectiveness, side effects)) that would justify the conduct of larger, randomized-controlled trials.
- In randomized-controlled trials, provide sufficient detail of the KD and control diets (ensuring that they are comparable on vitamins, mineral and other nutrients) so they could be replicated by other investigators.
- Develop a standardized method to monitor and quantify adherence and tolerance to the KD (e.g., [93]).
- Develop a set of standardized assessments and outcome measures that include the full array of relevant variables (e.g., imaging of tumor characteristics, body composition, quality of life, and survival).
- Distinguish trials based on whether they attempt to isolate the unique effects of KD versus those which seek to estimate its effects as an adjunctive therapy.
- Examine the effects of exogenous ketones, alone and in conjunction with a KD, to determine whether they have synergistic or additive effects.
- Because it is unlikely that KD will cure cancer, trials should focus on whether KD reduces cancer progression or recurrence in those who experience remission through standard care.

Although outside of the scope of this paper, future studies should also address qualitative data and patient perceptions, such as quality of life assessments, that can be conducted alongside clinical trials.

Overall, the potential efficacy of KD for human cancers has yet to be determined. The vast heterogeneity of patients studied, in conjunction with the generally poor methodological quality of published trials has clouded our ability to estimate KD's effects on the range of possible cancer-related outcomes. Until there is investment in providing adequate funding to conduct high-quality clinical trials, along with consensus and standardization around "best practices" among investigators, it is hard to see how our understanding of the effects of KD on cancer will advance.

Author Contributions: All authors contributed to the conceptualization and preparations of the manuscript equally. All authors have read and agreed to the published version of the manuscript.

Funding: Fontaine is supported by NIH grants P01CA229997, R01DK118939 and R01DK115483. He also serves on the scientific advisory board to Simply Good Foods USA, Inc., Denver, CO, USA.

Conflicts of Interest: The authors declare no conflict of interest.

References

1. Mukherjee, S.; Ghoshal, S. The Emperor of All Maladies: A Biography of Cancer. *J. Postgrad. Med. Educ. Res.* **2012**, *46*, 112. [CrossRef]
2. Siegel, R.L.; Miller, K.D.; Fuchs, H.E.; Jemal, A. Cancer Statistics, 2021. *CA Cancer J. Clin.* **2021**, *71*, 7–33. [CrossRef]
3. Rahib, L.; Wehner, M.R.; Matrisian, L.M.; Nead, K.T. Estimated Projection of US Cancer Incidence and Death to 2040. *JAMA Netw. Open* **2021**, *4*, e214708. [CrossRef]
4. Doll, R.; Peto, R. The causes of cancer: Quantitative estimates of avoidable risks of cancer in the United States today. *J. Natl. Cancer Inst.* **1981**, *66*, 1192–1308. [CrossRef]
5. Blot, W.J.; Tarone, R.E. Doll and Peto's quantitative estimates of cancer risks: Holding generally true for 35 years. *J. Natl. Cancer Inst.* **2015**, *107*, djv044. [CrossRef]
6. Steele, C.B.; el Huda, N.; Abdullah, I. Vital Signs: Trends in Incidence of Cancers Associated with Overweight and Obesity—United States, 2005–2014. *MMWR Morb. Mortal. Wkly. Rep.* **2017**, *66*, 1052–1058. [CrossRef]
7. Seyfried, T.N.; Shelton, L.M. Cancer as a metabolic disease. *Nutr. Metab.* **2010**, *7*, 7. [CrossRef]
8. Steck, S.E.; Murphy, E.A. Dietary patterns and cancer risk. *Nat. Rev. Cancer* **2019**, *20*, 125–138. [CrossRef] [PubMed]
9. Kant, A.K. Dietary patterns and health outcomes. *J. Am. Diet. Assoc.* **2004**, *104*, 615–635. [CrossRef] [PubMed]
10. Fabiani, R.; Minelli, L.; Bertarelli, G.; Bacci, S. A Western Dietary Pattern Increases Prostate Cancer Risk: A Systematic Review and Meta-Analysis. *Nutrients* **2016**, *8*, 626. [CrossRef] [PubMed]
11. Wang, H.-F.; Yao, A.-L.; Sun, Y.-Y.; Zhang, A.-H. Empirically derived dietary patterns and ovarian cancer risk: A meta-analysis. *Eur. J. Cancer Prev.* **2018**, *27*, 493–501. [CrossRef]
12. Xiao, Y.; Xia, J.; Li, L.; Ke, Y.; Cheng, J.; Xie, Y.; Chu, W.; Cheung, P.; Kim, J.H.; Colditz, G.A.; et al. Associations between dietary patterns and the risk of breast cancer: A systematic review and meta-analysis of observational studies. *Breast Cancer Res.* **2019**, *21*, 16. [CrossRef] [PubMed]
13. Gray, A.; Dang, B.N.; Moore, T.B.; Clemens, R.; Pressman, P. A review of nutrition and dietary interventions in oncology. *SAGE Open Med.* **2020**, *8*. [CrossRef] [PubMed]
14. Weber, D.D.; Aminzadeh-Gohari, S.; Tulipan, J. Ketogenic diet in the treatment of cancer—Where do we stand? *Mol Metab.* **2020**, *33*, 102–121. [CrossRef]
15. Yang, Y.-F.; Mattamel, P.; Joseph, T.; Huang, J.; Chen, Q.; Akinwunmi, B.; Zhang, C.; Ming, W.-K. Efficacy of Low-Carbohydrate Ketogenic Diet as an Adjuvant Cancer Therapy: A Systematic Review and Meta-Analysis of Randomized Controlled Trials. *Nutrients* **2021**, *13*, 1388. [CrossRef]
16. Römer, M.; Dörfler, J.; Huebner, J. The use of ketogenic diets in cancer patients: A systematic review. *Clin. Exp. Med.* **2021**, *21*, 1–36.
17. Oliveira, C.L.; Mattingly, S.; Schirrmacher, R.; Sawyer, M.B.; Fine, E.J.; Prado, C.M. A Nutritional Perspective of Ketogenic Diet in Cancer: A Narrative Review. *J. Acad. Nutr. Diet.* **2017**, *118*, 668–688. [CrossRef] [PubMed]
18. Sremanakova, J.; Sowerbutts, A.M.; Burden, S. A systematic review of the use of ketogenic diets in adult patients with cancer. *J. Hum. Nutr. Diet.* **2018**, *31*, 793–802. [CrossRef]
19. Gershuni, V.M.; Yan, S.L.; Medici, V. Nutritional Ketosis for Weight Management and Reversal of Metabolic Syndrome. *Curr. Nutr. Rep.* **2018**, *7*, 97–106. [CrossRef]
20. Allen, B.G.; Bhatia, S.K.; Anderson, C.; Eichenberger-Gilmore, J.M.; Sibenaller, Z.A.; Mapuskar, K.A.; Schoenfeld, J.D.; Buatti, J.; Spitz, D.; Fath, M. Ketogenic diets as an adjuvant cancer therapy: History and potential mechanism. *Redox Biol.* **2014**, *2*, 963–970. [CrossRef]
21. Paoli, A.; Rubini, A.; Volek, J.S.; Grimaldi, K.A. Beyond weight loss: A review of the therapeutic uses of very-low-carbohydrate (ketogenic) diets. *Eur. J. Clin. Nutr.* **2013**, *67*, 789–796. [CrossRef]
22. Lane J, F.K. Ketogenic Diet and Health. *OBM Integr. Complementary Med.* **2021**, *6*. [CrossRef]
23. Messer, R.D.K.; Eric, H. *Ketogenic Diets for the Treatment of Epilepsy. Bioactive Nutraceuticals and Dietary Supplements in Neurological and Brain Disease: Prevention and Therapy*; Academic Press: Cambridge, MA, USA, 2015; pp. 441–448.
24. Di Lorenzo, C.; Ballerini, G.; Barbanti, P.; Bernardini, A.; D'Arrigo, G.; Egeo, G.; Frediani, F.; Garbo, R.; Pierangeli, G.; Prudenzano, M.; et al. Applications of Ketogenic Diets in Patients with Headache: Clinical Recommendations. *Nutrients* **2021**, *13*, 2307. [CrossRef] [PubMed]
25. Martuscello, R.T.; Vedam-Mai, V.; McCarthy, D.J.; Schmoll, M.E.; Jundi, M.A.; Louviere, C.D.; Griffith, B.G.; Skinner, C.L.; Suslov, O.; Deleyrolle, L.P.; et al. A Supplemented High-Fat Low-Carbohydrate Diet for the Treatment of Glioblastoma. *Clin. Cancer Res.* **2015**, *22*, 2482–2495. [CrossRef] [PubMed]
26. Prince, A.; Zhang, Y.; Croniger, C.; Puchowicz, M. Oxidative Metabolism: Glucose Versus Ketones. *Adv. Exp. Med. Biol.* **2013**, *789*, 323–328. [PubMed]

27. Fine, E.J.; Segal-Isaacson, C.; Feinman, R.D.; Herszkopf, S.; Romano, M.C.; Tomuta, N.; Bontempo, A.F.; Negassa, A.; Sparano, J.A. Targeting insulin inhibition as a metabolic therapy in advanced cancer: A pilot safety and feasibility dietary trial in 10 patients. *Nutrition* **2012**, *28*, 1028–1035. [CrossRef]
28. Youm, Y.-H.; Nguyen, K.Y.; Grant, R.; Goldberg, E.L.; Bodogai, M.; Kim, D.; D'Agostino, D.; Planavsky, N.J.; Lupfer, C.; Kanneganti, T.-D.; et al. The ketone metabolite β-hydroxybutyrate blocks NLRP3 inflammasome–mediated inflammatory disease. *Nat. Med.* **2015**, *21*, 263–269. [CrossRef] [PubMed]
29. Otto, A.M. Warburg effect(s)—A biographical sketch of Otto Warburg and his impacts on tumor metabolism. *Cancer Metab.* **2016**, *4*, 1–8. [CrossRef]
30. Warburg, O. On the Origin of Cancer Cells. *Science* **1956**, *123*, 309–314. [CrossRef] [PubMed]
31. Trabold, O.; Wagner, S.; Wicke, C.; Bs, H.S.; Hussain, M.Z.; Rosen, N.; Bs, A.S.; Becker, H.D.; Hunt, T.K. Lactate and oxygen constitute a fundamental regulatory mechanism in wound healing. *Wound Repair Regen.* **2003**, *11*, 504–509. [CrossRef]
32. Hanahan, D.; Weinberg, R.A. Hallmarks of Cancer: The Next Generation. *Cell* **2011**, *144*, 646–674. [CrossRef]
33. Lien, E.C.; Lyssiotis, A.; Cantley, L.C. Metabolic Reprogramming by the PI3K-Akt-mTOR Pathway in Cancer. *Recent Results Cancer Res.* **2016**, *207*, 39–72.
34. Heiden, M.G.V.; Cantley, L.C.; Thompson, C.B. Understanding the Warburg Effect: The Metabolic Requirements of Cell Proliferation. *Science* **2009**, *324*, 1029–1033. [CrossRef]
35. DeNicola, G.; Cantley, L.C. Cancer's Fuel Choice: New Flavors for a Picky Eater. *Mol. Cell* **2015**, *60*, 514–523. [CrossRef]
36. Wu, N.; Zheng, B.; Shaywitz, A.; Dagon, Y.; Tower, C.; Bellinger, G.; Shen, C.-H.; Wen, J.; Asara, J.; McGraw, T.E.; et al. AMPK-Dependent Degradation of TXNIP upon Energy Stress Leads to Enhanced Glucose Uptake via GLUT1. *Mol. Cell* **2013**, *49*, 1167–1175. [CrossRef]
37. Carracedo, A.; Cantley, L.; Pandolfi, P.P. Cancer metabolism: Fatty acid oxidation in the limelight. *Nat. Rev. Cancer* **2013**, *13*, 227–232. [CrossRef]
38. Vidali, S.; Aminzadeh, S.; Lambert, B.; Rutherford, T.; Sperl, W.; Kofler, B.; Feichtinger, R.G. Mitochondria: The ketogenic diet—A metabolism-based therapy. *Int. J. Biochem. Cell Biol.* **2015**, *63*, 55–59. [CrossRef]
39. Branco, A.F.; Ferreira, A.; Simões, R.; Magalhaes-Novais, S.; Zehowski, C.; Cope, E.; Silva, A.M.; Pereira, D.; Sardao, V.; Cunha-Oliveira, T. Ketogenic diets: From cancer to mitochondrial diseases and beyond. *Eur. J. Clin. Investig.* **2016**, *46*, 285–298. [CrossRef]
40. Otto, C.; Kaemmerer, U.; Illert, B.; Muehling, B.; Pfetzer, N.; Wittig, R.; Voelker, H.U.; Thiede, A.; Coy, J.F. Growth of human gastric cancer cells in nude mice is delayed by a ketogenic diet supplemented with omega-3 fatty acids and medium-chain triglycerides. *BMC Cancer* **2008**, *8*, 122. [CrossRef]
41. Lussier, D.M.; Woolf, E.C.; Johnson, J.L.; Brooks, K.S.; Blattman, J.N.; Scheck, A.C. Enhanced immunity in a mouse model of malignant glioma is mediated by a therapeutic ketogenic diet. *BMC Cancer* **2016**, *16*, 310. [CrossRef]
42. Hopkins, B.D.; Pauli, C.; Du, X.; Wang, D.G.; Li, X.; Wu, D.; Amadiume, S.C.; Goncalves, M.; Hodakoski, C.; Lundquist, M.R.; et al. Suppression of insulin feedback enhances the efficacy of PI3K inhibitors. *Nature* **2018**, *560*, 499–503. [CrossRef]
43. Liśkiewicz, A.; Kasprowska, D.; Wojakowska, A.; Polański, K.; Lewin–Kowalik, J.; Kotulska-Jozwiak, K.; Jędrzejowska–Szypułka, H. Long-term High Fat Ketogenic Diet Promotes Renal Tumor Growth in a Rat Model of Tuberous Sclerosis. *Sci. Rep.* **2016**, *6*, 21807. [CrossRef]
44. Xia, S.; Lin, R.; Jin, L.; Zhao, L.; Kang, H.-B.; Pan, Y.; Liu, S.; Qian, G.; Qian, Z.; Konstantakou, E.; et al. Prevention of Dietary-Fat-Fueled Ketogenesis Attenuates BRAF V600E Tumor Growth. *Cell Metab.* **2017**, *25*, 358–373. [CrossRef]
45. Li, J.; Zhang, H.; Dai, Z. Cancer Treatment With the Ketogenic Diet: A Systematic Review and Meta-analysis of Animal Studies. *Front. Nutr.* **2021**, *8*, 315. [CrossRef]
46. Klement, R.J.; Champ, C.; Otto, C.; Kämmerer, U. Anti-Tumor Effects of Ketogenic Diets in Mice: A Meta-Analysis. *PLoS ONE* **2016**, *11*, e0155050. [CrossRef]
47. Lv, M.; Zhu, X.; Wang, H.; Wang, F.; Guan, W. Roles of Caloric Restriction, Ketogenic Diet and Intermittent Fasting during Initiation, Progression and Metastasis of Cancer in Animal Models: A Systematic Review and Meta-Analysis. *PLoS ONE* **2014**, *9*, e115147. [CrossRef]
48. Talib, W.; Mahmod, A.; Kamal, A.; Rashid, H.; Alashqar, A.; Khater, S.; Jamal, D.; Waly, M. Ketogenic Diet in Cancer Prevention and Therapy: Molecular Targets and Therapeutic Opportunities. *Curr. Issues Mol. Biol.* **2021**, *43*, 42. [CrossRef]
49. Plotti, F.; Terranova, C.; Luvero, D.; Bartolone, M.; Messina, G.; Feole, L.; Cianci, S.; Scaletta, G.; Marchetti, C.; Di Donato, V.; et al. Diet and Chemotherapy: The Effects of Fasting and Ketogenic Diet on Cancer Treatment. *Chemotherapy* **2020**, *65*, 77–84. [CrossRef]
50. Mittelman, S.D. The Role of Diet in Cancer Prevention and Chemotherapy Efficacy. *Annu. Rev. Nutr.* **2020**, *40*, 273–297. [CrossRef]
51. Freedland, S.J.; Allen, J.; Jarman, A.; Oyekunle, T.; Armstrong, A.J.; Moul, J.W.; Sandler, H.M.; Posadas, E.M.; Levin, D.; Wiggins, E.; et al. A Randomized Controlled Trial of a 6-Month Low-Carbohydrate Intervention on Disease Progression in Men with Recurrent Prostate Cancer: Carbohydrate and Prostate Study 2 (CAPS2). *Clin. Cancer Res.* **2020**, *26*, 3035–3043. [CrossRef]
52. Khodabakhshi, A.; Akbari, M.E.; Mirzaei, H.R.; Mehrad-Majd, H.; Kalamian, M.; Davoodi, S.H. Feasibility, Safety, and Beneficial Effects of MCT-Based Ketogenic Diet for Breast Cancer Treatment: A Randomized Controlled Trial Study. *Nutr. Cancer* **2019**, *72*, 627–634. [CrossRef]

53. Cohen, C.W.; Fontaine, K.R.; Arend, R.C.; Soleymani, T.; Gower, B.A. Favorable Effects of a Ketogenic Diet on Physical Function, Perceived Energy, and Food Cravings in Women with Ovarian or Endometrial Cancer: A Randomized, Controlled Trial. *Nutrients* **2018**, *10*, 1187. [CrossRef]
54. Cohen, C.W.; Fontaine, K.R.; Arend, R.C.; Alvarez, R.D.; Iii, C.A.L.; Huh, W.K.; Bevis, K.S.; Kim, K.H.; Straughn, J.M.; A Gower, B. A Ketogenic Diet Reduces Central Obesity and Serum Insulin in Women with Ovarian or Endometrial Cancer. *J. Nutr.* **2018**, *148*, 1253–1260. [CrossRef] [PubMed]
55. Klement, R.J.; Schäfer, G.; Sweeney, R.A. A ketogenic diet exerts beneficial effects on body composition of cancer patients during radiotherapy: An interim analysis of the KETOCOMP study. *J. Tradit. Complement. Med.* **2019**, *10*, 180–187. [CrossRef] [PubMed]
56. Ok, J.H.; Lee, H.; Chung, H.-Y.; Lee, S.H.; Choi, E.J.; Kang, C.M.; Lee, S.M. The Potential Use of a Ketogenic Diet in Pancreatobiliary Cancer Patients After Pancreatectomy. *Anticancer. Res.* **2018**, *38*, 6519–6527. [CrossRef]
57. Champ, C.E.; Palmer, J.; Volek, J.S.; Werner-Wasik, M.; Andrews, D.W.; Evans, J.J.; Glass, J.; Kim, L.; Shi, W. Targeting metabolism with a ketogenic diet during the treatment of glioblastoma multiforme. *J. Neuro-Oncology* **2014**, *117*, 125–131. [CrossRef]
58. Fearon, K.C.; Borland, W.; Preston, T.; Tisdale, M.J.; Shenkin, A.; Calman, K.C. Cancer cachexia: Influence of systemic ketosis on substrate levels and nitrogen metabolism. *Am. J. Clin. Nutr.* **1988**, *47*, 42–48. [CrossRef]
59. Jansen, N.; Walach, H. The development of tumours under a ketogenic diet in association with the novel tumour marker TKTL1: A case series in general practice. *Oncol. Lett.* **2015**, *11*, 584–592. [CrossRef]
60. Klement, R.J.; Sweeney, R.A. Impact of a ketogenic diet intervention during radiotherapy on body composition: I. Initial clinical experience with six prospectively studied patients. *BMC Res. Notes* **2016**, *9*, 143. [CrossRef]
61. Martin-McGill, K.; Cherry, G.; Marson, A.; Smith, C.T.; Jenkinson, M. ACTR-29. KETOGENIC DIETS AS AN ADJUVANT THERAPY IN GLIOBLASTOMA (KEATING): A MIXED METHOD APPROACH TO ASSESSING TRIAL FEASIBILITY. *Neuro-Oncology* **2018**, *20*, vi17. [CrossRef]
62. Martin-McGill, K.J.; Marson, A.; Smith, C.T.; Jenkinson, M.D. The Modified Ketogenic Diet in Adults with Glioblastoma: An Evaluation of Feasibility and Deliverability within the National Health Service. *Nutr. Cancer* **2018**, *70*, 643–649. [CrossRef]
63. Rieger, J.; Bähr, O.; Maurer, G.D.; Hattingen, E.; Franz, K.; Brucker, D.; Walenta, S.; Kämmerer, U.; Coy, J.F.; Weller, M.; et al. ERGO: A pilot study of ketogenic diet in recurrent glioblastoma Erratum in/ijo/45/6/2605. *Int. J. Oncol.* **2014**, *44*, 1843–1852. [CrossRef] [PubMed]
64. Schmidt, M.; Pfetzer, N.; Schwab, M.; Strauss, I.; Kämmerer, U. Effects of a ketogenic diet on the quality of life in 16 patients with advanced cancer: A pilot trial. *Nutr. Metab.* **2011**, *8*, 54. [CrossRef] [PubMed]
65. Tan-Shalaby, J.L.; Carrick, J.; Edinger, K.; Genovese, D.; Liman, A.D.; Passero, V.A.; Shah, R.B. Modified Atkins diet in advanced malignancies—Final results of a safety and feasibility trial within the Veterans Affairs Pittsburgh Healthcare System. *Nutr. Metab.* **2016**, *13*, 1–12. [CrossRef]
66. Van der Louw, E.J. Ketogenic diet treatment as adjuvant to standard treatment of glioblastoma multiforme: A feasibility and safety study. *Therapeutic Adv. Med. Oncol.* **2019**, *11*, 1758835919853958. [CrossRef]
67. Woodhouse, C.; Ward, T.; Gaskill-Shipley, M.; Chaudhary, R. Feasibility of a Modified Atkins Diet in Glioma Patients During Radiation and Its Effect on Radiation Sensitization. *Curr. Oncol.* **2019**, *26*, 433–438. [CrossRef] [PubMed]
68. Zahra, A.; Fath, M.; Opat, E.; Mapuskar, K.A.; Bhatia, S.K.; Rodman, S.; Iii, S.N.R.; Snyders, T.P.; Chenard, C.A.; Eichenberger-Gilmore, J.M.; et al. Consuming a Ketogenic Diet while Receiving Radiation and Chemotherapy for Locally Advanced Lung Cancer and Pancreatic Cancer: The University of Iowa Experience of Two Phase 1 Clinical Trials. *Radiat. Res.* **2017**, *187*, 743–754. [CrossRef] [PubMed]
69. Bozzetti, F.; Cozzaglio, L.; Gavazzi, C.; Brandi, S.; Bonfanti, G.; Lattarulo, M.; Gennari, L. Total nutritional manipulation in humans: Report of acancer patient. *Clin. Nutr.* **1996**, *15*, 207–209. [CrossRef]
70. Schwartz, K.; Chang, H.T.; Nikolai, M.; Pernicone, J.; Rhee, S.; Olson, K.; Kurniali, P.C.; Hord, N.G.; Noel, M. Treatment of glioma patients with ketogenic diets: Report of two cases treated with an IRB-approved energy-restricted ketogenic diet protocol and review of the literature. *Cancer Metab.* **2015**, *3*, 1–10. [CrossRef]
71. Zuccoli, G.; Marcello, N.; Pisanello, A.; Servadei, F.; Vaccaro, S.; Mukherjee, P.; Seyfried, T.N. Metabolic management of glioblastoma multiforme using standard therapy together with a restricted ketogenic diet: Case Report. *Nutr. Metab.* **2010**, *7*, 33. [CrossRef]
72. Tóth, C.; Clemens, Z. Treatment of Rectal Cancer with the Paleolithic Ketogenic Diet: A 24-months Follow-up. *Am. J. Med Case Rep.* **2017**, *5*, 205–216. [CrossRef]
73. Bozzetti, F.; Gavazzi, C.; Mariani, L.; Crippa, F. Glucose-based total parenteral nutrition does not stimulate glucose uptake by humans tumours. *Clin. Nutr.* **2004**, *23*, 417–421. [CrossRef]
74. Branca, J.J.V.; Pacini, S.; Ruggiero, M. Effects of Pre-surgical Vitamin D Supplementation and Ketogenic Diet in a Patient with Recurrent Breast Cancer. *Anticancer. Res.* **2015**, *35*, 5525–5532.
75. Nebeling, L.C.; Miraldi, F.; Shurin, S.B.; Lerner, E. Effects of a ketogenic diet on tumor metabolism and nutritional status in pediatric oncology patients: Two case reports. *J. Am. Coll. Nutr.* **1995**, *14*, 202–208. [CrossRef] [PubMed]
76. Fanelli, F.R.; Franchi, F.; Mulieri, M.; Cangiano, C.; Cascino, A.; Ceci, F.; Muscaritoli, M.; Seminara, P.; Bonomo, L. Effect of energy substrate manipulation on tumour cell proliferation in parenterally fed cancer patients. *Clin. Nutr.* **1991**, *10*, 228–232. [CrossRef]
77. Schroeder, U.; Himpe, B.; Pries, R.; Vonthein, R.; Nitsch, S.; Wollenberg, B. Decline of Lactate in Tumor Tissue After Ketogenic Diet: In Vivo Microdialysis Study in Patients with Head and Neck Cancer. *Nutr. Cancer* **2013**, *65*, 843–849. [CrossRef] [PubMed]

78. Artzi, M.; Liberman, G.; Vaisman, N.; Bokstein, F.; Vitinshtein, F.; Aizenstein, O.; Ben Bashat, D. Changes in cerebral metabolism during ketogenic diet in patients with primary brain tumors: 1H-MRS study. *J. Neuro-Oncology* **2017**, *132*, 267–275. [CrossRef]
79. Iyikesici, M.S. Feasibility study of metabolically supported chemotherapy with weekly carboplatin/paclitaxel combined with ketogenic diet, hyperthermia and hyperbaric oxygen therapy in metastatic non-small cell lung cancer. *Int. J. Hyperth.* **2019**, *36*, 445–454. [CrossRef]
80. Iyikesici, M.S. Long-Term Survival Outcomes of Metabolically Supported Chemotherapy with Gemcitabine-Based or FOLFIRINOX Regimen Combined with Ketogenic Diet, Hyperthermia, and Hyperbaric Oxygen Therapy in Metastatic Pancreatic Cancer. *Complement. Med. Res.* **2019**, *27*, 31–39. [CrossRef]
81. Strowd, R.E.; Cervenka, M.C.; Henry, B.J.; Kossoff, E.H.; Hartman, A.L.; Blakeley, J.O. Glycemic modulation in neuro-oncology: Experience and future directions using a modified Atkins diet for high-grade brain tumors. *Neuro-Oncology Pr.* **2015**, *2*, 127–136. [CrossRef]
82. Moore, K. Using the restricted ketogenic diet for brain cancer management: Comments from neuro-oncologist. In *Cancer as a Metabolic Disease: Management and Prevention of Cancer*, 1st ed.; John Wiley & Sons Inc.: New Jersey, NJ, USA, 2012; pp. 397–400.
83. Elsakka, A.M.A.; Bary, M.A.; Abdelzaher, E.; Elnaggar, M.; Kalamian, M.; Mukherjee, P.; Seyfried, T.N. Management of Glioblastoma Multiforme in a Patient Treated With Ketogenic Metabolic Therapy and Modified Standard of Care: A 24-Month Follow-Up. *Front. Nutr.* **2018**, *5*, 20. [CrossRef]
84. Schwalb, M.; Taubmann, M.; Hines, S.; Reinwald, H.; Ruggiero, M. Clinical Observation of a Novel, Complementary, Immunotherapeutic Approach based on Ketogenic Diet, Chondroitin Sulfate, Vitamin D3, Oleic Acid and a Fermented Milk and Colostrum Product. *Am. J. Immunol.* **2016**, *12*, 91–98. [CrossRef]
85. Brünings, W. Beitraege zum Krebsproblem. Mitteilung: Ueber eine diaetetisch-hormonale Beeinflussung des Krebses. *Münch. Med. Wschr.* **1941**, *88*, 117–123.
86. Brünings, W. Beitraege zum Krebsproblem. Mitteilung: Klinische Anwendungen der diaetetisch-hormonalen Krebsbeeinflussung ("Entzuckerungsmethode"). *Münchener Med. Wochenschr.* **1942**, *89*, 71–76.
87. Schulte, G.; Schütz, H. Insulin in der Krebsbehandlung. *Münch. Med. Wschr.* **1942**, *89*, 648–650.
88. Kang, C.M.; Yun, B.; Kim, M.; Song, M.; Kim, Y.-H.; Lee, S.H.; Lee, H.; Lee, S.M.; Lee, S.-M. Postoperative serum metabolites of patients on a low carbohydrate ketogenic diet after pancreatectomy for pancreatobiliary cancer: A nontargeted metabolomics pilot study. *Sci. Rep.* **2019**, *9*, 1–11. [CrossRef] [PubMed]
89. Porper, K.; Shpatz, Y.; Plotkin, L.; Pechthold, R.G.; Talianski, A.; Hemi, R.; Mardor, Y.; Jan, E.; Genssin, H.; Symon, Z.; et al. A Phase I clinical trial of dose-escalated metabolic therapy combined with concomitant radiation therapy in high-grade glioma. *Neuro-Oncology Adv.* **2021**, *3*, i10. [CrossRef]
90. Kato, I.; Dyson, G.; Snyder, M.; Kim, H.-R.; Severson, R.K. Differential effects of patient-related factors on the outcome of radiation therapy for rectal cancer. *J. Radiat. Oncol.* **2016**, *5*, 279–286. [CrossRef]
91. Santos, J.G.; Da Cruz, W.M.S.; Schï¿½Nthal, A.; Salazar, M.D.; Fontes, C.A.P.; Quirico-Santos, T.; Da Fonseca, C.O. Efficacy of a ketogenic diet with concomitant intranasal perillyl alcohol as a novel strategy for the therapy of recurrent glioblastoma. *Oncol. Lett.* **2017**, *15*, 1263–1270. [CrossRef]
92. Davis, R.A.H.; Deemer, S.E.; Bergeron, J.M.; Little, J.T.; Warren, J.L.; Fisher, G.; Smith, D.L.; Fontaine, K.R.; Dickinson, S.L.; Allison, D.B.; et al. Dietary R, S-1,3-butanediol diacetoacetate reduces body weight and adiposity in obese mice fed a high-fat diet. *FASEB J.* **2019**, *33*, 2409–2421. [CrossRef]
93. Meidenbauer, J.J.; Mukherjee, P.; Seyfried, T.N. The glucose ketone index calculator: A simple tool to monitor therapeutic efficacy for metabolic management of brain cancer. *Nutr. Metab.* **2015**, *12*, 12. [CrossRef] [PubMed]

MDPI

Article

Pretreatment Adherence to a Priori-Defined Dietary Patterns Is Associated with Decreased Nutrition Impact Symptom Burden in Head and Neck Cancer Survivors

Christian A. Maino Vieytes [1], Alison M. Mondul [2], Sylvia L. Crowder [3,4], Katie R. Zarins [5], Caitlyn G. Edwards [1,6], Erin C. Davis [1,7], Gregory T. Wolf [8], Laura S. Rozek [5,8], Anna E. Arthur [1,4,9,*] and on behalf of the University of Michigan Head and Neck SPORE Program †

1 Division of Nutritional Sciences, University of Illinois at Urbana-Champaign, Urbana, IL 61801, USA; cam17@illinois.edu (C.A.M.V.); cke5143@psu.edu (C.G.E.); erin_davis1@urmc.rochester.edu (E.C.D.)
2 Department of Epidemiology, University of Michigan, Ann Arbor, MI 48109, USA; amondul@umich.edu
3 Department of Health Outcomes and Behavior, Moffitt Cancer Center, Tampa, FL 33612, USA; sylvia.crowder@moffitt.org
4 Department of Food Science and Human Nutrition, University of Illinois at Urbana-Champaign, Urbana, IL 61801, USA
5 Department of Environmental Health Sciences, University of Michigan, Ann Arbor, MI 48109, USA; kmrents@umich.edu (K.R.Z.); rozekl@umich.edu (L.S.R.)
6 Department of Nutritional Sciences, The Pennsylvania State University, University Park, PA 16802, USA
7 Department of Pediatrics, Division of Allergy and Immunology, School of Medicine and Dentistry, University of Rochester, Rochester, NY 14642, USA
8 Department of Otolaryngology, University of Michigan, Ann Arbor, MI 48109, USA; Gregwolf@med.umich.edu
9 Department of Dietetics and Nutrition, University of Kansas Medical Center, Kansas City, KS 66160, USA
* Correspondence: aarthur4@kumc.edu; Tel.: +1-913-588-5355
† Membership of the University of Michigan Head and Neck SPORE Program is provided in the Acknowledgments.

Citation: Maino Vieytes, C.A.; Mondul, A.M.; Crowder, S.L.; Zarins, K.R.; Edwards, C.G.; Davis, E.C.; Wolf, G.T.; Rozek, L.S.; Arthur, A.E.; on behalf of the University of Michigan Head and Neck SPORE Program. Pretreatment Adherence to a Priori-Defined Dietary Patterns Is Associated with Decreased Nutrition Impact Symptom Burden in Head and Neck Cancer Survivors. *Nutrients* **2021**, *13*, 3149. https://doi.org/10.3390/nu13093149

Academic Editor: Christina Dieli-Conwright

Received: 12 August 2021
Accepted: 7 September 2021
Published: 9 September 2021

Publisher's Note: MDPI stays neutral with regard to jurisdictional claims in published maps and institutional affiliations.

Abstract: Dietary intake is understood to contribute to nutrition impact symptoms (NIS) in patients with head and neck squamous cell carcinoma (HNSCC). The purpose of this study was to evaluate the performance of four a priori-defined diet quality indices on the presence of NIS 1 year following diagnosis using data on 323 participants from the University of Michigan Head and Neck Specialized Program of Research Excellence (UM-SPORE). Pretreatment dietary intake was measured before treatment initiation using a food frequency questionnaire. NIS were measured along seven subdomains. Multivariable binary logistic regression models were constructed to evaluate relationships between pretreatment scores on a priori-defined diet quality indices (AHEI-2010, aMED, DASH, and a low-carbohydrate score) and the presence of individual symptoms in addition to a composite "symptom summary score" 1 year postdiagnosis. There were several significant associations between different indices and individual NIS. For the symptom summary score, there were significant inverse associations observed for aMED ($OR_{Q5\text{-}Q1}$: 0.36, 95% CI: 0.14–0.88, p_{trend} = 0.04) and DASH ($OR_{Q5\text{-}Q1}$: 0.38, 95% CI: 0.15–0.91, p_{trend} = 0.02) and the presence of NIS 1-year postdiagnosis. Higher adherence to the aMED and DASH diet quality indices before treatment may reduce NIS burden at 1-year postdiagnosis.

Keywords: survivorship; cancer; nutritional epidemiology; nutrition impact symptoms

1. Introduction

Head and neck squamous cell carcinoma (HNSCC) accounts for roughly 4% of all new cancer diagnoses in the United States [1]. HNSCC is commonly diagnosed in the oral cavity, oropharynx, hypopharynx, and larynx and is associated with lifetime exposure to tobacco and alcohol consumption and infection with particular strains of human papillomavirus (HPV) implicated, generally, in tumors affecting the oropharynx [2,3]. In addition to symptomatology arising from tumor morphology and location, side effects that impact food and oral intake due to cancer treatment are also highly prevalent in this population. Nutrition impact symptoms (NIS), as they are termed, include but are not limited to dysgeusia, ageusia, xerostomia, pain, dysphagia, dental problems, mucositis, and trismus [4]. An estimated 90% of HNSCC patients develop acute NIS due to their cancer treatment [5]. Consequently, this symptom burden may perpetuate significant physical, emotional, and psychological issues, hampering the overall quality of life (QOL) and QOL around eating [6,7]. Nonetheless, evidence on chronic NIS following treatment remains scant. The report by Ganzer et al. found that NIS, including dysphagia, xerostomia, and altered taste, persisted after three years postchemotherapy in a mixed-methods study of 10 long-term HNSCC survivors [8]. Chronic NIS may pose significant nutritional consequences, including reduced nutrient intake and impaired nutritional status, which is noteworthy, given that this population is disproportionately affected by high rates of cancer cachexia [9,10].

A posited etiologic mechanism for NIS involves inflammation secondary to cancer treatment and, in particular, treatment with targeted radiation [11]. While previous studies have provided that consuming a balanced and healthful diet rich in fruits and vegetables before treatment may abate inflammation and chronic NIS, research has yet to identify a generalizable dietary pattern that confers protection by mitigating inflammation after diagnosis and throughout cancer treatment [9]. Using a multivariate approach for characterizing a posteriori dietary patterns, our research team reported associations between consuming a primarily "prudent" diet, rich in fruits and vegetables, at pretreatment and reduced symptom burden at 1-year postdiagnosis [9]. Nevertheless, the purpose of the present analysis was to assess the capacity of a priori-defined diet quality indices to predict symptom burden 1-year postdiagnosis, which, for many participants, comes after the initiation and completion of treatment protocols.

A priori diet quality indices are generally used to measure adherence to a set of dietary recommendations or guidelines, in contrast to a posteriori methods implemented to characterize eating behaviors from sample dietary data [12]. An advantage that a priori indices have over their a posteriori counterparts is their generalizability and their facility for policy adaptation. Moreover, index scores calculated on different study samples are directly comparable, whereas a posteriori patterns are not, since they characterize a given sample. In this analysis, we chose to examine the performance of four a priori diet quality indices reported previously in the scientific literature. The results of this analysis could be harnessed to tailor dietary recommendations for HNSCC patients to abrogate the incidence of NIS and would form, to our knowledge, the first study of a priori diet quality indices and their relationship to NIS in HNSCC following diagnosis and treatment. The study hypotheses were that higher adherence to each of the a priori diet quality indices examined corresponds to lower self-reported chronic NIS 1-year postdiagnosis.

2. Materials and Methods

2.1. Study Population

This was a secondary analysis of clinical and dietary data gathered on participants in the University of Michigan Head and Neck Specialized Program of Research Excellence (UM-SPORE) cohort. UM-SPORE is a prospective, longitudinal cohort study of newly diagnosed patients with HNSCC who presented with primary malignancies in the oral cavity, oropharynx, hypopharynx, or larynx and entered the study before the initiation of any treatment for primary HNSCC. Recruitment was conducted through the UM Hospital System and took place between November 2008 through October 2014, whereby newly diagnosed HNSCC cases were screened and solicited consent for inclusion into the study. Informed consent was obtained from all subjects involved in the study. Exclusion criteria for the study are detailed as: (i) age less than 18 years; (ii) being pregnant; (iii) being a non-English speaker; (iv) having a previously diagnosed mental disorder; (v) previous or concomitant diagnosis of a tumor in the non-upper aerodigestive tract; and (vi) previous diagnosis with another form of primary HNSCC within the last five years. Upon entry, participants completed a baseline (pretreatment) food frequency questionnaire (FFQ) and survey questionnaires ascertaining lifestyle and epidemiologic characteristics. Survey measures included history of other identified comorbid conditions, smoking status, drinking status, sleep, physical activity, and depression. All baseline (pretreatment) data collection was conducted before the initiation of any treatment protocol for HNSCC, and participants were subsequently followed longitudinally. Annual reviews of electronic medical records were used to extract clinical factors, including cancer stage, site, and treatment protocol data.

There were 380 participants with baseline and 1-year NIS data, which also had complete baseline/pretreatment FFQ data. Further exclusions included those with missing body mass index (BMI) data (n = 8) and those missing data on any other covariates used in the study (n = 3). Subjects reporting caloric intakes of >5000 kcal/d or <500 kcal/d (n = 5) were excluded on the premise that these levels of intake are likely implausible, making these observations unreliable, which may bias the final results [13]. Furthermore, participants with tumors at sites other than the larynx, oropharynx, hypopharynx, or oral cavity (n = 31), missing full pages of their pretreatment FFQ (n = 9), and having greater than 70 blank responses on their FFQ (n = 1) were excluded from the analysis [14]. The final analytic sample comprised 323 participants. All study procedures were executed in compliance with standards approved by the University of Michigan Institutional Review Board (IRB approval number, for which consent was granted for obtaining and analyzing the data, is HUM00042189) and complied with the Helsinki Declaration of 1975.

2.2. Predictors: Pretreatment a Priori Diet Quality Index Scores

Baseline dietary intake data were collected using the self-administered 2007 Harvard Adult FFQ, a 131-item semiquantitative FFQ formulated to assess the usual intake of select foods, beverages, and supplements and is used to compute a profile of average nutrient intake for a given participant [15,16]. This method affords a practical approach for ranking the participant sample based on relative food and nutrient intake. Participants were asked to complete the questionnaire based on what they believe their usual intakes for select foods and beverages were over the past year. This was prompted through inquiries that accounted for standard portion sizes and frequency (e.g., 2–4 times per week, 1 medium banana). Nutrient intakes were computed by taking proportional weights corresponding to the frequency of intake selected for a given food item, multiplying by the nutrient value for the portion/serving size established on the questionnaire, and then summing across all foods [15]. Nutrient composition values were estimated using the Harvard nutrient database.

Four frequently cited a priori-defined diet quality indices were chosen for the analysis. These included the Dietary Approaches to Stop Hypertension (DASH), the Alternate Mediterranean Diet Index (aMED), the Alternative Healthy Eating Index-2010 (AHEI-2010), and a low-carbohydrate diet index. The choice to use these particular indices arose from their widespread use in the nutritional epidemiology literature and, specifically, within the context of chronic disease risk and management [17–19]. Nutrient- and item-specific intake levels were estimated from the administered Harvard FFQ data and used to calculate diet quality index scores.

The DASH diet has previously been described and is extensively documented as a treatment protocol in hypertension. This dietary pattern emphasizes fruits, vegetables, whole grains, low-fat dairy, nuts, and legumes while limiting intakes of red meat, sweets, and sugar-sweetened beverages [20]. Concerning nutrient intakes, the DASH diet is characterized by reduced intakes of salt (sodium chloride), saturated and total fat, and increased intake of foods with high mineral (primarily potassium and magnesium) and micronutrient value. Calculation of the DASH diet scores was adapted to this cohort using the framework described by Fung et al. [21]. The operationalization of this dietary protocol ranks participants according to their average intake in 8 select food group components: fruits, vegetables, nuts/legumes, low-fat dairy products, whole grains, sodium, red and processed meats, and sugar-sweetened beverages. Scores for each of the first five listed components were taken as the quintile ranking for a participant for that food group. Component scores for the latter three components were assigned antagonistically. Individuals scoring within the highest quintile of intake were given a score of "1", whereas those residents to the lowest quintile of intake were given a score of "5". Summing scores across all components allowed us to arrive at the final composite score, which had a maximum value of 40.

The aMED diet quality index is based on the operationalization provided by Fung et al. [18]. The traditional Mediterranean diet pattern has been characterized by high intakes of fruits, vegetables, breads, cereals, legumes, high-quality fats (primarily olive oil) [22]. Moderate to low intakes of red meat, fish, low-fat dairy, and alcohol (primarily wine) also make up intrinsic components. This dietary pattern is further stipulated by its limiting of foods with a processed origin. Calculation of the aMED score considers intake levels of 9 components that were obtained from participant FFQ data: vegetables, legumes, fruits, nuts, whole grains, red or processed meats, fish, alcohol, and the ratio of monounsaturated/saturated (UFA/SFA) fat intake. Component scores were based on a participant's rank relative to the median intake for that component. That is, those with intakes greater than the median were given a score of "1", while those falling below the median were given a score of "0". For the meat component, falling above the median intake resulted in a score of "0", while ranking below the median gave participants a score of "1". Alcohol intakes between 5 and 15 g/d were designated a score of "1" for the alcohol component. The final composite score was computed by summing scores across all of the 9 components, with a maximum score of 9.

The AHEI was developed in 2002 as an alternative to the Healthy Eating Index (HEI), which operationalized the 1995 iteration of the Dietary Guidelines for Americans, and was tailored with the intention of being a more robust indicator of chronic disease risk [19]. This diet quality index was subsequently updated in 2010 and emphasized similar food components to the aforementioned indices with additional foci on trans fat (as a percentage of total energy intake), polyunsaturated fatty acids (as a percentage of total energy intake), and $n-3$ (EPA + DHA) fatty acid intake. Similar to the calculation of aMED, it awards points for the moderate consumption of alcohol. Operationalizing the index relies on mapping intakes for each food category to a scale ranging from 0 to 10. The scoring algorithm has been previously described by Chiuve et al., and the maximum attainable score for any given participant is 110 [19].

Finally, a low-carbohydrate index, standing in as a proxy for a ketogenic diet, was computed as previously described by Halton et al. [23]. Briefly, percentages of energy intake from each of carbohydrate, fat, and protein were calculated for the study subjects, and they were subsequently partitioned and ranked according to quantiles of intake for each category. For the protein and fat categories, scores were allocated congruently with participant rank (i.e., a rank of "1" was commensurate to a score of "1"). For the carbohydrate score, scores were allocated antagonistically (i.e., a rank of 10 resulted in a score of "0"). A theoretical maximum score of 30 was attainable for this index.

2.3. Covariates

Sociodemographic covariates included age (modeled continuously), sex (modeled dichotomously), and education status (coded as less than or equal to high school or some college or more). Behavioral characteristics included in our models consisted of smoking status (modeled categorically as never, former, or current). The clinical variables were BMI (modeled dichotomously as <25—normal or underweight—or ≥ 25—overweight or obese), tumor HPV infection status (modeled as positive, negative, or equivocal/missing test), cancer stage (modeled dichotomously as 0, I, II or III, IV), and tumor site (modeled categorically as larynx, oral cavity, oropharynx, or hypopharynx). All models examining 1-year NIS variables as their outcome were adjusted for their baseline categorical groupings, derived from their corresponding scale values at baseline (groups were dichotomized as outlined below). Lastly, all models adjusted for total energy intake by including total calories (kcal) as a continuous variable. Treatment modality, sex, and drinking status were given a priori consideration for inclusion but were omitted given that they previously were shown to be highly correlated with other covariates among this patient cohort [24]. Nonetheless, subanalyses included examining associations amongst the different treatment levels to account for any varying effects of particular treatment protocols on NIS, as detailed below.

2.4. Outcomes: NIS at 1-Year Postdiagnosis

Six levels of NIS (trismus, xerostomia, dysphagia with liquids, dysphagia with solid foods, difficulty chewing, and taste perception) were assessed and quantified using the UM Head and Neck Quality of Life (QOL) Questionnaire developed and validated for use in this patient population by Terrell et al. [25]. This 37-item survey evaluates the landscape of HNSCC patient QOL by emphasizing four meaningful domains: communication, eating, emotion, and pain. Six items encompassing the eating subdomain were used to measure the aforementioned outcome variables, and available responses were provided on a discrete 5-point scale from "not at all bothered" (given a numerical score of "1") to "extremely bothered" (indicating a score of "5"). An additional item, assessing the burning pain and discomfort that characterizes mucositis, was included and extracted from the pain domain of the questionnaire. Mucositis has been shown to impact dietary intake and thus was included in the analysis for this reason [9,26]. Responses to each of these individual items were dichotomized into categories "not at all bothered" and "slightly to extremely bothered", as was previously done by members of our research team [27]. A composite symptom summary measure using 1-year NIS data was developed by taking the sum of participant responses across these seven items. A maximum value of 35 was attainable on this index, representing the most severe symptom burden. Subsequently, participants were dichotomized into groups based on a median-split, with NIS symptom summary score <12 defined as a low-symptom burden and scores ≥ 12 defining those experiencing a high-symptom burden. This threshold value was chosen and based on a previous operationalization of this scale [27].

2.5. Statistical Analysis

Descriptive analyses were performed, examining frequencies and means across demographic, behavioral, and clinical factors. The mean 1-year NIS summary score was evaluated across relevant epidemiologic characteristics and tabulated. One-way analysis of variance (ANOVA) was used to assess for significant differences in mean 1-year NIS summary scores across levels of the characteristics examined. Tukey's post-hoc mean separation test was implemented to partition significantly different groups within the different characteristics examined. Bivariate relationships were evaluated with Pearson correlation coefficients computed amongst the four a priori diet quality indices chosen for the analysis.

Continuous and discrete scores from the four a priori diet quality indices were categorized by quintiles. Multivariable binary logistic regression models were fit to evaluate the associations between each a priori diet quality index score and the relevant outcomes. In total, there were eight models constructed for each diet quality index score: (i) a separate model for each of the seven symptoms introduced above and (ii) a model examining the dichotomized 1-year NIS summary score rank as the outcome of interest. All primary analytical models adjusted for participant age, smoking status, BMI category, total calories, educational status, HPV status, cancer stage, tumor site, and the corresponding symptom score group at baseline. All analyses used the lowest quintile (Q1) of intake as the referent group. Odds ratios (OR) and their 95% confidence intervals were computed and tabulated. Tests for linear trend were assessed by assigning the median value of a participant's corresponding quintile and modeling that term as a continuous variable. To assess whether single food group or nutrient categories (measured as either servings per day or total mass in grams) would be able to recapitulate results from models using the composite indices, sixteen additional models were fit, using quintiles of intake for different categories of foods and nutrients that, together, make up the components of the calculated indices.

Stratified analyses (for the outcome of 1-year NIS symptom summary score) were conducted and examined the tested associations across strata for baseline BMI, smoking status, cancer stage, education, tumor site, and treatment modality (radiation versus no radiation used). Stratified models used a truncated set of covariates (age, smoking status, stage, total calories, and HPV status) to ensure adequate model fit and preserve statistical power with the smaller subsets. A sensitivity analysis, used to evaluate for the potential of reverse causality explaining our results, was conducted, where models were fitted separately on subjects reporting no symptoms at study entry/pretreatment (n = 72) and those with at least some degree of symptomatology at study entry (n = 251). Furthermore, restricted cubic spline models were fit to visually ascertain the observed relationships, examine linearity, and assess dose-dependence between each dietary predictor and the odds of 1-year NIS summary score $\geq$12. These models used four interior knots, set at the scores corresponding to quintiles of the respective diet quality index. The median of the lowest quintile of intake for each diet quality index was used as the referent value when computing odds ratios from these models. All analyses were conducted at α = 0.05 and performed in RStudio version 1.4.

3. Results

3.1. Sample Characteristics

Table 1 provides descriptive statistics and means for the analytic cohort. The average age of the analyzed sample was 60.4 years. Generally, age was larger in the highest quintile of intake relative to the lowest for each of the four diet quality indices examined. This cohort contained a majority of males (n = 254; 78.6%). There tended to be a higher proportion of females within the highest quintile of the diet quality indices compared to the lowest quintile. Of note, most participants identified as non-Hispanic white (n = 310; 96.9%). BMI was variable across the different indices, and it had no clear relationship with quintiles of the diet quality indices. Regarding the behavioral variables, proportions of current smokers tended to be higher within the lowest quintile of the examined indices and were most pronounced for AHEI-2010 and DASH, whereas former and never smoker proportions were higher in the highest quintiles of the aMED, AHEI-2010, and DASH indices. Drinking status across quintiles of the indices also suggested higher proportions of current consumption within the lowest quintile compared to the highest (this was true for all indices but most pronounced for aMED and DASH). The differences seen in HPV status across quintiles of those indices appeared to follow the patterns in smoking status, to an extent except for the low-carbohydrate index. There were no other appreciable differences in distributions of participant characteristics.

Table 1. Demographic, clinical, and behavioral characteristics of the study participants (n = 323).

Characteristic		AHEI-2010		aMED		DASH		Low-Carbohydrate	
	Survivors # (%)	Q1 (n = 65)	Q5 (n = 64)	Q1 (n = 76)	Q5 (n = 46)	Q1 (n = 65)	Q5 (n = 59)	Q1 (n = 68)	Q5 (n = 57)
Age (y)									
Mean (SD)	60.4 (10.7)	56.6 (10.9)	62.1 (10.1)	58.1 (11.1)	59.8 (8.5)	57.9 (10.2)	62.3 (9.1)	57.2 (9.8)	62.3 (10.3)
Min/Max	29/95	29/78	34/83	29/85	34/83	30/78	43/81	30/78	43/85
Sex									
Male	254 (78.6)	52 (80.0)	46 (71.9)	65 (85.5)	37 (80.4)	57 (87.7)	45 (76.3)	51 (75.0)	40 (70.2)
Female	69 (21.4)	13 (20.0)	18 (28.1)	11 (14.5)	9 (19.6)	8 (12.3)	14 (23.7)	17 (25.0)	17 (29.8)
Education									
High school or less	91 (28.2)	25 (38.5)	11 (17.2)	24 (31.6)	5 (10.9)	31 (47.7)	7 (11.9)	20 (29.4)	16 (28.1)
Some college or more	232 (71.8)	40 (61.5)	53 (82.8)	52 (68.4)	41 (89.1)	34 (52.3)	52 (88.1)	48 (70.6)	41 (71.9)
Race/Ethnicity									
Non-Hispanic white	310 (96.9)	62 (95.4)	62 (96.9)	75 (98.7)	44 (97.8)	60 (93.8)	55 (94.8)	64 (94.1)	55 (98.2)
Other	7 (2.2)	3 (4.6)	1 (1.6)	1 (1.3)	1 (2.2)	4 (6.2)	2 (3.4)	2 (2.9)	1 (1.8)
Unknown	3 (0.9)	0 (0.0)	1 (1.6)	0 (0.0)	0 (0.0)	0 (0.0)	1 (1.7)	2 (2.9)	0 (0.0)
BMI (kg/m^2)									
Underweight and normal weight (<25)	101 (31.3)	21 (32.3)	18 (28.1)	18 (23.7)	16 (34.8)	23 (35.4)	21 (35.6)	21 (30.9)	14 (24.6)
Overweight and obese (≥25)	222 (68.7)	44 (67.7)	46 (71.9)	58 (76.3)	30 (65.2)	42 (64.6)	38 (64.4)	47 (69.1)	43 (75.4)
Site									
Larynx	66 (20.4)	12 (18.5)	10 (15.6)	21 (27.6)	6 (13.0)	19 (29.2)	5 (8.5)	9 (13.2)	11 (19.3)
Oral cavity	96 (29.7)	22 (33.8)	19 (29.7)	23 (30.3)	14 (30.4)	16 (24.6)	13 (22)	20 (29.4)	22 (38.6)
Oropharynx	157 (48.6)	30 (46.2)	35 (54.7)	32 (42.1)	26 (56.5)	28 (43.1)	40 (67.8)	39 (57.4)	23 (40.4)
Hypopharynx	4 (1.2)	1 (1.5)	0 (0.0)	0 (0.0)	0 (0.0)	2 (3.1)	1 (1.7)	0 (0.0)	1 (1.8)
Stage									
0, I, II	104 (32.2)	18 (27.7)	20 (31.2)	26 (34.2)	13 (28.3)	20 (30.8)	14 (23.7)	24 (35.3)	18 (31.6)
III, IV	219 (67.8)	47 (72.3)	44 (68.8)	50 (65.8)	33 (71.7)	45 (69.2)	45 (76.3)	44 (64.7)	39 (68.4)

Table 1. *Cont.*

Characteristic		AHEI-2010		aMED		DASH		Low-Carbohydrate	
	Survivors # (%)	**Q1** (*n* = 65)	**Q5** (*n* = 64)	**Q1** (*n* = 76)	**Q5** (*n* = 46)	**Q1** (*n* = 65)	**Q5** (*n* = 59)	**Q1** (*n* = 68)	**Q5** (*n* = 57)
HPV Status									
HPV-negative	92 (28.5)	20 (30.8)	17 (26.6)	25 (32.9)	10 (21.7)	17 (26.2)	12 (20.3)	17 (25.0)	23 (40.4)
HPV-positive	71 (22.0)	8 (12.3)	15 (23.4)	16 (21.1)	13 (28.3)	11 (16.9)	15 (25.4)	20 (29.4)	7 (12.3)
Unknown	160 (49.5)	37 (56.9)	32 (50.0)	35 (46.1)	23 (50.0)	37 (56.9)	32 (54.2)	31 (45.6)	27 (47.4)
Treatment									
Surgery only	74 (22.9)	14 (21.5)	20 (31.2)	15 (19.7)	9 (19.6)	10 (15.4)	8 (13.6)	18 (26.5)	17 (29.8)
Surgery + adjuvant radiation	50 (15.5)	12 (18.5)	8 (12.5)	14 (18.4)	8 (17.4)	11 (16.9)	6 (10.2)	4 (5.9)	9 (15.8)
Radiation only	27 (8.4)	3 (4.6)	3 (4.7)	10 (13.2)	5 (10.9)	7 (10.8)	5 (8.5)	5 (7.4)	3 (5.3)
Chemotherapy + radiation	155 (48.0)	29 (44.6)	29 (45.3)	33 (43.4)	23 (50.0)	32 (49.2)	35 (59.3)	35 (51.5)	25 (43.9)
Chemotherapy only	7 (2.2)	2 (3.1)	2 (3.1)	3 (3.9)	1 (2.2)	2 (3.1)	2 (3.4)	2 (2.9)	2 (3.5)
Palliative or unknown	10 (3.1)	5 (7.7)	2 (3.1)	1 (1.3)	0 (0.0)	3 (4.6)	3 (5.1)	4 (5.9)	1 (1.8)
Smoking Status									
Current	106 (32.8)	34 (52.3)	11 (17.2)	33 (43.4)	8 (17.4)	37 (56.9)	11 (18.6)	24 (35.3)	16 (28.1)
Former	118 (36.5)	11 (16.9)	26 (40.6)	23 (30.3)	20 (43.5)	16 (24.6)	25 (42.4)	23 (33.8)	21 (36.8)
Never	99 (30.7)	20 (30.8)	27 (42.2)	20 (26.3)	18 (39.1)	12 (18.5)	23 (39)	21 (30.9)	20 (35.1)
Drinking Status									
Current	230 (71.2)	49 (75.4)	49 (76.6)	57 (75.0)	39 (84.8)	50 (76.9)	44 (74.6)	48 (70.6)	37 (64.9)
Former	71 (22.0)	11 (16.9)	13 (20.3)	15 (19.7)	6 (13.0)	14 (21.5)	13 (22)	18 (26.5)	16 (28.1)
Never	22 (6.8)	5 (7.7)	2 (3.1)	4 (5.3)	1 (2.2)	1 (1.5)	2 (3.4)	2 (2.9)	4 (7.0)

Percentages may not add to 100% given rounding.

3.2. NIS Symptom Summary Scores and Potential Confounders

Concerning the 1-year NIS symptom summary score, the primary analytical outcome, individuals with lower educational background (≤high school completed) tended to have a significantly, albeit slightly, higher symptom burden score ($p < 0.01$). A significant association within tumor site was also identified ($p < 0.01$). More pronounced symptomatology was reported in participants with oropharynx vs larynx tumors. Subjects classified with stage III or IV tumors had significantly higher NIS symptom scores than those with tumors staged 0, I, or II ($p < 0.01$). Additionally, several significant differences in symptom scores were noted across treatment classes ($p < 0.01$). All of these described differences are documented in Table 2.

Table 2. Mean 1-year NIS summary score across select demographic, clinical, and behavioral characteristics (n = 323).

Characteristic	n	Mean NIS Summary Score (SD)	p [a]
Age (y)			
≥60	169	13.6 (6.0)	0.09
<60	154	14.7 (6.2)	
Sex			
Male	254	13.9 (5.8)	0.18
Female	69	15 (7.2)	
Education			
High school or less	91	15.8 (6.4)	<0.01 **
Some college or more	232	13.5 (5.9)	
Race			
Non-Hispanic white	310	14.2 (6.2)	0.72
Other	7	13.9 (6.4)	
Unknown	3	13 (4.6)	
BMI (kg/m^2)			
Underweight and normal weight (<25)	101	15 (6.6)	0.08
Overweight and obese (≥25)	222	13.7 (5.9)	
Site			
Larynx	66	12.4 (5.7) †	<0.01 **
Oral cavity	96	13.8 (6.5)	
Oropharynx	157	15.1 (6.0) †	
Hypopharynx	4	13.2 (3.8)	
Stage			
0, I, II	104	11.9 (5.7)	<0.01 **
III, IV	219	15.2 (6.0)	
HPV Status			
HPV negative	92	14.4 (6.5)	0.44
HPV positive	71	14.5 (5.9)	
Unknown	160	13.8 (6.0)	
Treatment			
Surgery only	74	11 (5.2) †‡¥	<0.0001 **
Surgery + adjuvant radiation	50	16.3 (6.7) †ψ	
Radiation only	27	11.2 (4.5) ψ,ξ,ω	
Chemotherapy + radiation	155	15.1 (5.7) ‡ξ	
Chemotherapy only	7	21 (8.0) ¥ω	
Palliative or unknown	10	15.3 (5.4)	

Table 2. *Cont.*

Characteristic	*n*	Mean NIS Summary Score (SD)	*p* [a]
Smoking Status			
Current	106	15.3 (6.5)	0.58
Former	118	13.8 (6.0)	
Never	99	13.3 (5.7)	
Drinking Status			
Current	230	13.8 (5.9)	0.10
Former	71	15.3 (6.1)	
Never	22	13.7 (7.8)	

[a] *p*-value from ANOVA, modeling 1-year NIS summary score by indicated levels of characteristics; ** $p < 0.01$; In groups with >2 groups, means sharing a superscript (†, ‡, ψ, Ꝣ, ω, or, ¥) in common are significantly different from one another.

3.3. Diet Quality Scores

Summary statistics for each of the diet quality indices examined and results of the correlational analysis are found in Table 3 and Table S1, respectively. Median scores for the AHEI-2010, aMED, DASH, and the low-carbohydrate index with the analyzed sample were 58.54 (Min: 22.85, Max: 89.14), 4 (Min: 0, Max: 9), 24 (Min: 10, Max: 37), and 15 (Min: 0, Max: 30), respectively, and were generally commensurate to the estimated sample means. aMED, DASH, and AHEI-2010 scores all shared Pearson correlation coefficients suggestive of moderately strong and positive relationships. aMED and DASH scores were weakly and inversely correlated with the low-carbohydrate score, while the AHEI-2010 shared a weak, positive correlation with the low-carbohydrate index.

Table 3. Summary statistics for the diet quality index scores within this sample.

Index	Mean (SD)	Median	Minimum	Maximum	Theoretical (Max, Min)
AHEI-2010	57.99 (58)	58.54	22.85	89.14	(0, 110)
aMED	4.08 (58)	4	0	9	(0, 9)
DASH	23.98 (58)	24	10	37	(8, 40)
Low-Carbohydrate	14.98 (58)	15	0	30	(0, 30)

AHEI-2010: the Alternative Healthy Eating Index-2010; aMED: the Alternate Mediterranean Diet Index; DASH: the Dietary Approaches to Stop Hypertension.

3.4. NIS Symptom Burden 1-Year Postdiagnosis

We evaluated the associations between consumption along a priori diet quality scores, derived from FFQs, using multivariable binary logistic regression. These results are referenced from Table 4. When examining the associations between baseline aMED diet quality index scores and responses to the seven NIS symptom scales at 1-year postdiagnosis, it was found that higher consumption along this index was strongly and inversely associated with dysphagia of liquids and, to a lesser extent, with dysphagia of solids, difficulty chewing, xerostomia, and mucositis. A strong and significant inverse relationship with the NIS 1-year symptom summary score was also observed ($OR_{Q5\text{-}Q1}$: 0.36, 95% CI: 0.14–0.88, $p_{trend} = 0.04$). Closer adherence to the DASH protocol was significantly inversely associated with all 1-year symptom scales besides trismus, mucositis, and dysphagia solids. However, the parameter estimates in all of these models for each of quintiles 2–5 were suggestive of a protective association that failed to meet the threshold for statistical significance. The strongest inverse relationship was that with xerostomia ($OR_{Q5\text{-}Q1}$: 0.27, 95% CI: 0.08–0.85, $p_{trend} = 0.04$). Higher consumption along the DASH index was also potently and significantly inversely associated with the 1-year NIS summary score ($OR_{Q5\text{-}Q1}$: 0.38, 95% CI: 0.15–0.91, $p_{trend} = 0.02$).

Associations along the AHEI-2010 index were more modest, with the strongest inverse association seen in xerostomia ($OR_{Q5\text{-}Q1}$: 0.42, 95% CI: 0.14–1.21, p_{trend} = 0.04). There was no significant linear trend observed between consumption along AHEI-2010 and the 1-year symptom summary score, though the parameter estimate for the highest quintile of intake suggested a nonsignificant inverse association with a blunted effect estimate relative to the aMED and DASH indices. No significant inverse associations were noted between higher consumption along the low-carbohydrate index and any of the eight outcomes examined. Notably, there was a positive association between higher pretreatment consumption of the low-carbohydrate index and the odds of experiencing dysphagia with liquids at the 1-year mark ($OR_{Q5\text{-}Q1}$: 2.47, 95% CI: 1.06-5.91, p_{trend} = 0.19). Most of the parameter estimates for the highest quintile of intake in the models considering individual symptoms as outcomes and low-carbohydrate diet score as the explanatory variable were greater than 1 except for trismus, taste, and mucositis. Though failing to meet the threshold for statistical significance, the parameter estimates for each of the second, third, and fifth quintiles of the low-carbohydrate index were all suggestive of a positive association with the 1-year NIS symptom summary score. The results of the set of restricted cubic splines analyses, modeling each diet quality index as a continuous variable and subsequently mapping index scores to their respective odds ratios from spline estimates, are visualized in Figure 1.

Table 4. Multivariable [a] ORs and 95% CI for association between quintiles of select diet quality index scores with being slightly to extremely bothered by at 1-year postdiagnosis (adjusted for baseline symptom levels) (n = 323).

Index/Symptom	Q1	Q2	Q3	Q4	Q5	p_{trend}	$p_{Q5\text{-}Q1}$
AHEI-2010	n = 65	n = 65	n = 64	n = 65	n = 64		
Trismus	1.00	0.90 (0.39–2.06)	0.93 (0.41–2.11)	0.51 (0.21–1.17)	0.65 (0.28–1.54)	0.16	0.33
Xerostomia	1.00	1.07 (0.37–3.04)	0.95 (0.32–2.76)	0.50 (0.17–1.38)	0.42 (0.14–1.21)	0.04 *	0.11
Difficulty chewing	1.00	0.93 (0.40–2.18)	0.86 (0.36–2.01)	0.38 (0.16–0.88) *	0.55 (0.24–1.26)	0.03 *	0.16
Dysphagia liquids	1.00	0.59 (0.26–1.32)	0.58 (0.26–1.28)	0.48 (0.21–1.09)	0.47 (0.19–1.09)	0.07	0.08
Dysphagia solids	1.00	0.77 (0.34–1.77)	0.77 (0.34–1.76)	0.44 (0.19–0.99) *	0.65 (0.28–1.50)	0.15	0.32
Taste	1.00	0.76 (0.31–1.87)	1.03 (0.41–2.63)	0.40 (0.16–0.96) *	0.49 (0.20–1.20)	0.04 *	0.12
Mucositis	1.00	1.45 (0.67–3.16)	1.19 (0.55–2.58)	0.64 (0.30–1.39)	0.81 (0.37–1.78)	0.18	0.61
NIS summary score [b]	1.00	0.85 (0.37–1.93)	0.89 (0.39–2.03)	0.44 (0.20–0.99) *	0.65 (0.28–1.49)	0.12	0.31
aMED	n = 76	n = 55	n = 112	n = 34	n = 46		
Trismus	1.00	0.76 (0.33–1.72)	0.69 (0.34–1.39)	0.30 (0.09–0.85) *	0.84 (0.34–2.07)	0.27	0.71
Xerostomia	1.00	0.69 (0.26–1.85)	0.58 (0.24–1.39)	0.28 (0.08–0.95) *	0.30 (0.10–0.92) *	0.01 *	0.04 *
Difficulty chewing	1.00	0.71 (0.31–1.65)	0.58 (0.28–1.18)	0.29 (0.11–0.76) *	0.32 (0.13–0.79) *	<0.01 **	0.01 *
Dysphagia liquids	1.00	0.40 (0.18–0.88) *	0.37 (0.18–0.73) **	0.44 (0.17–1.13)	0.13 (0.04–0.38) **	<0.01 **	<0.01 **
Dysphagia solids	1.00	0.27 (0.11–0.61) **	0.40 (0.19–0.81) *	0.21 (0.07–0.56) **	0.34 (0.13–0.84) *	0.02 *	0.02 *
Taste	1.00	0.44 (0.18–1.07)	0.30 (0.14–0.65) **	0.57 (0.20–1.64)	0.64 (0.24–1.72)	0.60	0.37
Mucositis	1.00	0.69 (0.32–1.47)	0.82 (0.42–1.56)	1.23 (0.50–3.04)	0.39 (0.16–0.91) *	0.18	0.03 *
NIS summary score [b]	1.00	0.36 (0.16–0.81) *	0.35 (0.17–0.72) **	0.33 (0.12–0.86) *	0.36 (0.14–0.88) *	0.04 *	0.03 *

Table 4. *Cont.*

Index/Symptom	Q1	Q2	Q3	Q4	Q5	p_{trend}	$p_{Q5\text{-}Q1}$
DASH	$n = 65$	$n = 83$	$n = 76$	$n = 40$	$n = 59$		
Trismus	1.00	0.62 (0.28–1.36)	0.56 (0.25–1.24)	0.50 (0.18–1.33)	0.81 (0.34–1.93)	0.48	0.63
Xerostomia	1.00	0.34 (0.12–0.92) *	0.66 (0.23–1.89)	0.27 (0.08–0.86) *	0.27 (0.08–0.85) *	0.04 *	0.03 *
Difficulty chewing	1.00	0.48 (0.21–1.07)	0.40 (0.17–0.89) *	0.50 (0.18–1.33)	0.39 (0.16–0.95) *	0.06	0.04 *
Dysphagia liquids	1.00	0.48 (0.22–1.03)	0.51 (0.23–1.10)	0.49 (0.19–1.25)	0.37 (0.15–0.90) *	0.04 *	0.03 *
Dysphagia solids	1.00	0.46 (0.21–1.01)	0.43 (0.19–0.95) *	0.45 (0.17–1.18)	0.54 (0.22–1.28)	0.19	0.17
Taste	1.00	0.39 (0.15–0.93) *	0.29 (0.11–0.71) **	0.16 (0.05–0.44) **	0.50 (0.18–1.37)	0.04 *	0.18
Mucositis	1.00	0.84 (0.41–1.74)	0.54 (0.26–1.14)	0.61 (0.24–1.48)	0.61 (0.27–1.37)	0.12	0.23
NIS summary score [b]	1.00	0.50 (0.22–1.11)	0.31 (0.13–0.68) **	0.35 (0.13–0.92) *	0.38 (0.15–0.91) *	0.02 *	0.03 *
Low-Carbohydrate	$n = 68$	$n = 69$	$n = 72$	$n = 57$	$n = 57$		
Trismus	1.00	1.13 (0.52–2.46)	1.12 (0.51–2.46)	0.60 (0.25–1.41)	0.68 (0.29–1.59)	0.20	0.37
Xerostomia	1.00	2.64 (0.94–7.68)	0.69 (0.26–1.81)	1.57 (0.54–4.69)	1.35 (0.47–3.93)	0.94	0.58
Difficulty chewing	1.00	0.87 (0.40–1.91)	1.17 (0.54–2.55)	0.43 (0.18–0.97) *	1.22 (0.53–2.84)	0.85	0.65
Dysphagia liquids	1.00	2.03 (0.89–4.72)	1.37 (0.60–3.19)	0.92 (0.37–2.26)	2.47 (1.06–5.91) *	0.19	0.04 *
Dysphagia solids	1.00	1.37 (0.63–2.98)	1.21 (0.56–2.65)	0.87 (0.39–1.96)	1.25 (0.55–2.83)	0.90	0.60
Taste	1.00	0.71 (0.30–1.66)	0.58 (0.25–1.33)	0.47 (0.19–1.16)	0.59 (0.24–1.42)	0.15	0.24
Mucositis	1.00	0.75 (0.36–1.57)	0.68 (0.33–1.40)	0.65 (0.30–1.39)	0.87 (0.40–1.89)	0.59	0.72
NIS summary score [b]	1.00	1.66 (0.77–3.65)	1.24 (0.58–2.65)	0.80 (0.36–1.77)	1.25 (0.56–2.80)	0.92	0.59

[a] Adjusted for age, tumor site, BMI, education status, cancer stage, smoking status, HPV status, BMI, total calories, and corresponding baseline symptom score; [b] Outcome modeled was NIS symptom summary score (generated by taking a subject's sum of their individual NIS scores) ≥12 1-year postdiagnosis; * $p < 0.05$; ** $p < 0.01$; All evaluated NIS were measured using a discrete scale from "1" (indicating "not at all bothered") to "5" (indicating "extremely bothered") and dichotomized as "not at all bothered" and "slightly to extremely bothered".

Table 5. Multivariable [a] ORs and 95% CI for association between quintiles of select food groups and nutrient categories with NIS symptom summary score ≥12 1-year postdiagnosis ($n = 323$).

Food Group	Q1	Q2	Q3	Q4	Q5	p_{trend}	$p_{Q5\text{-}Q1}$
Legumes (servings/d)	1.00	0.87 (0.44–1.72)	0.79 (0.34–1.81)	0.83 (0.38–1.83)	0.69 (0.33–1.43)	0.34	0.32
Nuts (servings/d)	1.00	0.68 (0.31–1.46)	0.56 (0.25–1.23)	0.61 (0.28–1.32)	0.33 (0.15–0.72) **	0.01 *	0.01 **
Whole Grains (g/d)	1.00	0.80 (0.36–1.74)	0.73 (0.33–1.58)	0.61 (0.27–1.37)	0.61 (0.25–1.44)	0.24	0.26
Alcohol (g/d)	1.00	0.75 (0.34–1.62)	0.50 (0.23–1.06)	0.90 (0.42–1.92)	1.13 (0.52–2.48)	0.24	0.75
Red and Processed Meats (servings/d)	1.00	1.94 (0.91–4.18)	1.58 (0.75–3.32)	1.87 (0.83–4.27)	1.48 (0.63–3.47)	0.54	0.37
Total Fruit (servings/d)	1.00	0.53 (0.23–1.19)	0.27 (0.12–0.59) **	0.41 (0.18–0.91) *	0.32 (0.14–0.74) **	0.03 *	0.01 **

Table 5. *Cont.*

Food Group	Q1	Q2	Q3	Q4	Q5	p_{trend}	$p_{Q5\text{-}Q1}$
Total Vegetables (servings/d)	1.00	1.65 (0.76–3.67)	1.41 (0.66–3.02)	0.90 (0.42–1.95)	1.32 (0.59–2.98)	0.95	0.50
UFA/SFA Ratio	1.00	1.16 (0.53–2.55)	0.63 (0.29–1.35)	0.86 (0.39–1.87)	0.66 (0.30–1.44)	0.22	0.29
Sugar-Sweetened Beverages (servings/d)	1.00	1.33 (0.60–2.92)	0.95 (0.44–2.04)	0.92 (0.42–2.03)	0.54 (0.23–1.24)	0.05 *	0.15
Total Low-Fat Dairy (servings/d)	1.00	0.81 (0.37–1.78)	0.62 (0.29–1.31)	0.65 (0.29–1.43)	0.59 (0.27–1.28)	0.27	0.18
Total Sodium (mg/d)	1.00	0.79 (0.35–1.73)	0.99 (0.41–2.39)	1.11 (0.41–3.03)	0.65 (0.17–2.38)	0.67	0.51
n-3 Fatty Acids (g/d)	1.00	1.05 (0.48–2.29)	0.93 (0.43–2.01)	0.75 (0.35–1.59)	0.51 (0.23–1.09)	0.04 *	0.08
Trans Fat (% of total kcal)	1.00	0.46 (0.21–0.99) *	0.68 (0.31–1.47)	0.77 (0.35–1.68)	1.10 (0.49–2.48)	0.50	0.82
Total Carbohydrate (g/d)	1.00	1.18 (0.52–2.70)	1.24 (0.49–3.15)	1.45 (0.48–4.41)	0.96 (0.22–4.24)	0.98	0.96
Total Protein (g/d)	1.00	1.09 (0.47–2.50)	0.94 (0.38–2.31)	0.90 (0.32–2.57)	0.47 (0.13–1.74)	0.19	0.26
Total Fat (g/d)	1.00	0.96 (0.42–2.19)	1.21 (0.50–2.95)	1.03 (0.36–2.94)	0.72 (0.18–2.88)	0.70	0.65

[a] Adjusted for age, tumor site, BMI, education status, cancer stage, smoking status, HPV status, total calories, and baseline NIS symptom summary score; * $p < 0.05$; ** $p < 0.01$.

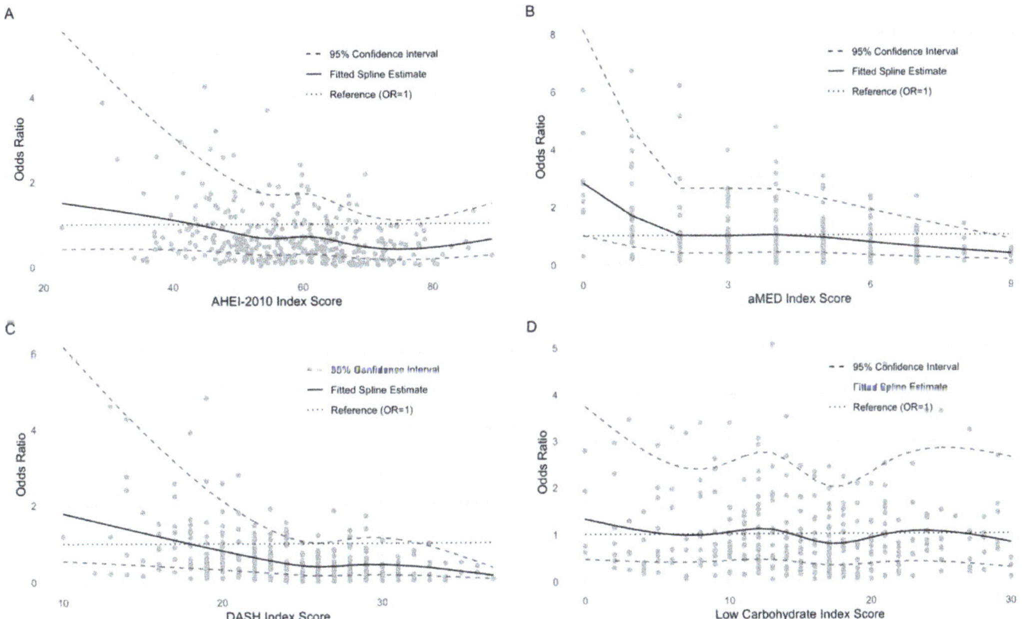

Figure 1. Dose–response relationship between each dietary index, modeled as a continuous variable, and NIS symptom summary score ≥12 1-year postdiagnosis. All multivariable models were adjusted for the same set of covariates in Table 5. Restricted cubic splines models were fit to mappings of each of the observations, according to their (**A**) AHEI-2010, (**B**) aMED, (**C**) DASH, or (**D**) low-carbohydrate index scores, to their respective odds. The odds corresponding to the median intake level for each dietary index were set as referents. Dashed lines indicate 95% confidence bounds. The dotted line indicates OR = 1, which was included for reference.

3.5. Analyses Using Single Nutrient Explanatory Variables

Modeling nutrient categories or food groups in place of the diet quality indices resulted in an abundance of non-significant figures (Table 5). There were a few notable exceptions. Total fruit consumption was strongly and inversely associated with the 1-year NIS summary score. Moreover, total nut consumption saw a very similar magnitude of association compared to fruit intake as the explanatory variable. A weaker inverse association was found for a model considering total n-3 fatty acid intake.

3.6. Subgroup and Sensitivity Analyses

Stratified analyses revealed disparities in several reported associations between diet quality and 1-year NIS symptom summary score among levels of BMI, smoking status, cancer stage, HPV status, education attained, tumor site, and treatment modality. The results of this analysis are found in Table S2. There were few marked distinctions in the magnitude of associations compared to the results from primary analytical models. Those reporting lower educational status had significantly stronger magnitudes of association relative to those in higher education levels for the AHEI-2010, aMED, and DASH indices. Remarkably, the effect sizes in all tumor site levels were commensurate with the estimates from the overall model except for the association of the DASH index with 1-year NIS in the larynx subgroup (note that hypopharynx was omitted from this part of the analysis due to small sample size, $n = 4$). When examining levels of stage, the directionality and magnitude of the associations observed from the primary models were similar to those seen in either of the binary categories of stage, and the same can be said for the levels of HPV status examined. In consideration of participant treatment protocol, when evaluating models on participants who either received some form of radiation or those who did not, effect sizes for all indices were similar, and no substantial differences were appreciated. Finally, when models examining the relationships of baseline dietary indices on NIS 1-year postdiagnosis were fit on a sample subset of individuals entering the study without any symptoms at baseline, it was found that the parameter estimates were generally consistent with those from the primary analytical models including all sample subjects (Table S3). Again, the strongest inverse associations within this subset were found in the higher quantiles of the DASH and aMED pretreatment index indices. In the case of the DASH index, these findings suggest a 76% reduction in the odds of significant NIS symptoms 1-year postdiagnosis for those entering the study without significant symptomatology and adhering closest to the DASH index compared to those also entering the study with no NIS but with the poorest adherence to the DASH protocol at study entry. The results were similar in those presenting with NIS at study entry, and this analysis was performed to evaluate for the possibility of reverse causation, whereby NIS present at pretreatment may have affected dietary adherence to the indices and, thus, distorted the relationship between explanatory variables and the outcomes in this analysis.

4. Discussion

We evaluated associations between four a priori-defined diet quality indices and found that greater adherence to the aMED and DASH dietary protocols during the year before treatment were each associated with diminished risks of experiencing self-reported NIS 1-year postdiagnosis. When analyzing individual NIS separately, it was found that several NIS were inversely associated with closer adherence to these indices. Overall, the aMED and DASH indices exhibited the most robust sets and the most numerous inverse associations when examining individual and overall NIS. The highest adherence category in each of the aMED and DASH indices exhibited a 64% and 62% reduction, respectively, in the odds of experiencing significant NIS burden at 1-year postdiagnosis relative to the lowest quintile of intake. These associations were followed in magnitude and significance by, to an even lesser extent, the AHEI-2010 index. In contrast, adhering to a ketogenic style, low-carbohydrate pattern was not associated with mitigations in self-reported NIS at the 1-year time point either when examining the NIS summary score or when analyzing any individual NIS. It was apparent that there may potentially be a detrimental influence of this low-carbohydrate diet quality index on symptom burden within the HNSCC population. However, these findings would need to be replicated and investigated further in clinical settings before these conclusions are ascertained.

When stratifying results by relevant participant characteristics reported at baseline, we found those significant associations were more pronounced in subjects with higher self-reported attained education status. Nevertheless, it should be acknowledged that participants in the group reporting greater attained education status (some college or more) had higher mean scores on all indices examined (results not shown). It is conceivable to hypothesize that participants in the lower education level were more likely to benefit from closer adherence to these indices, given potentially lower overall adherence. Reported associations for subjects who received radiation as part of their treatment protocol did not appear to diverge appreciably from those who did not receive radiation in their treatment regimen. This observed phenomenon appears to be somewhat inconsistent with reports in the literature of radiation-induced dysphagia and impaired swallowing function [12,28]. Notably, no single food groups, nutrients, or nutrient categories, other than total fruit and nut intakes, demonstrated associations as or more potent as those reported for the aMED or DASH indices. This highlights the ability of a priori diet quality indices to act as multidimensional factors that capture synergisms between food components and consequently significant associations that would otherwise go undetected in single nutrient or food group analyses.

We previously reported the associations between two a posteriori-derived dietary patterns and the same outcomes studied herein [27]. To our knowledge, this is the first study examining the relationships between the four chosen a priori-defined diet quality indices and those outcomes in the head and neck cancer patient population. Whereas empirically derived a posteriori dietary patterns, computed through methods such as principal components analysis or reduced rank regression, are patterns that are specific to and describe the populations under scrutiny, a priori-defined composites are routinely based on sets of predefined or established guidelines, thus underscoring the practicality of these dietary patterns across different populations. Consequently, identifying a priori indices that are particularly applicable and beneficial, in a population-specific manner, facilitates public health messaging and subsequent adoption of those protocols for given populations. The shrewdness of a priori or a posteriori indices has been extolled for its ability to more accurately model the complexity of diet as an epidemiologic or clinical exposure in human studies [29]. There are correlations or interactions amongst nutrients or foods that either blunt or bolster certain associations. Indeed, we observed this phenomenon in our study by modeling each nutrient or food category individually and finding that nutrients or food groups alone did not yield the degree of associations that were, instead, capitulated by modeling each diet quality indices. This analysis identified the aMED index as the most robust indicator of baseline diet quality that predicted reduced symptomatology 1 year following diagnosis. This performance was followed by that of the DASH index with very similar results.

The DASH diet was developed and demonstrated in 1997 as a dietary intervention for curbing hypertension-related sequelae [30]. Since its inception, the diet has been validated in numerous capacities and has drawn noteworthy recognition, being a focal point of national guidelines, particularly intended for those with hypertension [31]. Nonetheless, the tenets of the diet are typically in line with what many consider to be a "healthy" diet not only for hypertension, but for overall well-being and longevity. A Mediterranean diet is among the most frequently cited dietary patterns in the literature on longevity and chronic illness [32]. Though the guidelines for adhering to this regimen remain, in some respects, arcane and ambiguous, the principal emphasis of this pattern lies in the consumption of foods primarily of plant-based origin, olive oil as the chief lipid source, with minimal amounts of animal-derived foods and products [33]. Several methods of quantifying the eating patterns of populations inhabiting the Mediterranean Sea regions have been reported in the literature, and the aMED index represents one iteration. In many respects, DASH and aMED are similar, especially regarding how scores of adherence were tallied in this analysis. Both indices positively reward the consumption of fruits, vegetables, nuts, and legumes and castigate the consumption of red and processed meats. However, there are some notable differences. The DASH index includes additional components for low-fat dairy and sodium consumptions while not including components for fish and alcohol intake and a score for the ratio of fat sources in the diet like the aMED index stresses. In our study, food group subanalyses were not strongly associated with the outcome of NIS burden 1-year postdiagnosis. There was a nonsignificant downward trend in NIS burden across quintiles of low dairy intake. However, other food groups that differ across the indices, such as alcohol intake and sodium, were not predictive of the study's primary analytical outcome. Again, this lack of association may relate to the relative advantage of using dietary patterns over single food groups. However, we posit that the difference in index algorithms is likely at play when the disparities between the performance of each of these indices are appreciated.

Alcohol intake is a known independent risk factor for incident HNSCCs, a family of cancers with strong ties to environmental etiologies, particularly in those cases lacking HPV seropositivity [34,35]. There have been few findings reported in how alcohol consumption impacts NIS for HNSCC cases with continuing alcohol use. Nevertheless, considering the findings that the aMED performed best amongst all indices examined and had an alcohol component appears to be somewhat consistent with those results reported by Potash et al. In their study of 283 HNSCC patients at 1-year postdiagnosis, it was reported that current "social drinkers" had the highest proportion of oral eating function compared to all other groups [36]. However, those labeled as "problem drinkers" were found to have compromised oral function compared to the social-drinking group. We can postulate variability in index performance is due to differences in the way alcohol consumption is rewarded. However, given the equivocal nature of the current evidence to back this conjecture, we propose that more research is warranted for delineating the effects of continued alcohol consumption, postdiagnosis, on HNSCC symptom burden. Further, it should also be noted that subject classification according to AHEI-2010 is based on absolute values of intake, whereas aMED and DASH are based on quantiles of intake within the analyzed sample, which reduces the risk of bias due to misclassification. The use of a method based on quantile classifications, rather than absolute values of intake, is substantiated by the fact that FFQ data, which was employed in this analysis, typically underperforms when the aim is to quantify absolute values of intake accurately but remains a viable method for ranking or distinguishing study subjects based on relative intake [14].

The potential beneficial effects imparted by higher adherence to either aMED or DASH indices are presumably mediated by the high consumption of plant-based foods, providing a food matrix that is ubiquitously filled with anti-inflammatory nutrients and phytochemical components. The consumption of these foods at baseline is plausibly linked to blunted symptomatology 1-year postdiagnosis by way of quenched reactive oxygen species (ROS) that may, otherwise, perpetuate NIS. Mechanistically, the selective and antagonistic effects of phytochemical agents and other dietary bioactive compounds on HNSCC in preclinical study designs have been previously described [37,38]. Moreover, results from a set of randomized control trials investigating the effects of α-tocopherol and β-carotene on radiation-induced toxicities in HNSCC patients suggested a potential therapeutic role of foods rich in those nutrients for mitigating NIS [39,40]. This hypothesis may be further substantiated by the results of analyses considering the low-carbohydrate index as the primary predictor, which was void of any significant association suggestive of a protective effect. Given that this index is composed of nutrient components rather than food groups, it is difficult to ascertain what types of foods contribute to higher adherence scores. Yet, it is valid to assume that the foods highest in fat and protein and lowest in dietary carbohydrates are those of animal origin, which include red and processed meats. Indeed, it was found in bivariate analyses (results not shown here) that total red and processed meat intake was positively correlated with the low-carbohydrate index ($r = 0.28$, $p < 0.001$) and that the index baring the strongest inverse relationship with red and processed meat intake was DASH ($r = -0.38$, $p < 0.001$). Red and processed meats are inherently devoid of phytochemical constituents that produce the anti-inflammatory effects we discuss and may, instead possess proinflammatory potential [40]. Interestingly, we did find a significant positive association between closer adherence to the low-carbohydrate index and increased odds of having dysphagia accompanying liquids, though there was no significant association with total symptom burden 1-year postdiagnosis. These results were similar to those reported between the "western" dietary pattern and the same outcome studied here in our analysis of a posteriori-derived dietary patterns [27].

There are some limitations to our study that are worth stating. Though the longitudinal design is a strength of the study, the use of baseline data to predict outcomes 1-year postdiagnosis may be confounded by diet changes implemented within that period. It is germane to posit that several participants may have adopted "healthier" diet changes in the intervening time window following their diagnosis. This would, potentially, explain why significant associations within the low-carbohydrate index were not ascertained, for instance. Though the 2007 Harvard FFQ, utilized for dietary collection in this study, has been validated for use in the general population, it has yet to be validated for the HNSCC population. Moreover, recall bias and other systematic biases accompanying the FFQ as the principal means of collecting dietary data should not be overlooked. Likewise, although the UM Head and Neck QOL Questionnaire has been validated in this patient population, the use of the NIS symptom summary score has not. Lastly, as is the case with any observational study design, the possibility of residual confounding and reverse causality cannot be ruled out.

5. Conclusions

Dietary patterns adhering to the guidelines forwarded by the aMED or DASH protocols were significantly associated with decreased odds of aggregate NIS symptom burden 1-year postdiagnosis. Other indices, AHEI-2010 and a low-carbohydrate index, showed attenuated, null, or in some cases positive associations. In summary, our findings suggest that promoting consumption of a diet abundant in fruits, vegetables, whole grains, low-fat dairy, legumes, nuts, while minimal in red and processed meat and sodium levels may ameliorate aggregate symptom burden in newly diagnosed HNSCC patients.

Supplementary Materials: The following are available online at https://www.mdpi.com/article/10.3390/nu13093149/s1, Table S1: Pearson Correlation Coefficients (r) matrix with the diet quality indices included in this analysis, Table S2: Stratified ORs and 95% CI for associations between quintiles of a priori-defined diet quality index scores with NIS symptom score $\geq$12 1-year postdiagnosis. Table S3: ORs and 95% CI for associations between quintiles of a priori-defined diet quality index scores with NIS symptom summary score $\geq$ 12 1-year postdiagnosis stratified on the presence of significant pretreatment NIS.

Author Contributions: Conceptualization, C.A.M.V., S.L.C., A.M.M., G.T.W., L.S.R., and A.E.A.; methodology, C.A.M.V., S.L.C., C.G.E., E.C.D., K.R.Z., A.M.M., G.T.W., L.S.R., and A.E.A.; software, C.A.M.V.; formal analysis, C.A.M.V.; investigation, C.A.M.V., S.L.C., C.G.E., E.C.D., A.M.M., and A.E.A.; resources, G.T.W., L.S.R., and A.E.A.; data curation, K.R.Z., A.E.A., and C.A.M.V.; writing—original draft preparation, C.A.M.V.; writing—review and editing, C.A.M.V., S.L.C., C.G.E., E.C.D., A.M.M., L.S.R., G.T.W., and A.E.A.; visualization, C.A.M.V.; supervision, A.M.M. and A.E.A.; project administration, A.E.A.; funding acquisition, G.T.W. All authors have read and agreed to the published version of the manuscript.

Funding: This research was supported by NIH/NCI P50CA097248 and USDA-NIFA Hatch Project 1011487. C.A.M.V. was supported, in part, by a scholarship from the Health Policy Research Scholars Program at the Robert Wood Johnson Foundation. S.L.C. was supported by a T32CA090314 (MPIs: Brandon/Vadaparampil) and an NCI Cancer Prevention and Control Training Grant: 5T32CA09314-17 (MPIs: Susan Vadaparampil and Thomas Brandon).

Institutional Review Board Statement: The study was conducted according to the guidelines of the Declaration of Helsinki and approved by the Institutional Review Board of the University of Michigan (Protocol Code: HUM00042189).

Informed Consent Statement: Informed consent was obtained from all subjects involved in the study.

Data Availability Statement: The data presented in this study are available on request from the corresponding author. The data are not publicly available due to privacy concerns.

Acknowledgments: We thank the patients, clinicians, and principal investigators of the individual projects at the UM Head and Neck SPORE program, who provided access to the longitudinal clinical database and were responsible for the recruitment, treatment, and follow-up of patients included in this report. These investigators included Avraham Eisbruch, Theodore Lawrence, Mark Prince, Jeffrey Terrell, Shaomeng Wang, and Frank Worden.

Conflicts of Interest: The authors declare no conflict of interest. The funders had no role in the design of the study; in the collection, analyses, or interpretation of data; in the writing of the manuscript, or in the decision to publish the results.

References

1. Siegel, R.L.; Miller, K.D.; Jemal, A. Cancer Statistics, 2020. *CA Cancer J. Clin.* **2020**, *70*, 7–30. [CrossRef] [PubMed]
2. Marur, S.; Forastiere, A.A. Head and Neck Squamous Cell Carcinoma: Update on Epidemiology, Diagnosis, and Treatment. *Mayo Clin. Proc.* **2016**, *91*, 386–396. [CrossRef]
3. Crowder, S.L.; Douglas, K.G.; Yanina Pepino, M.; Sarma, K.P.; Arthur, A.E. Nutrition Impact Symptoms and Associated Outcomes in Post-Chemoradiotherapy Head and Neck Cancer Survivors: A Systematic Review. *J. Cancer Surviv.* **2018**, *12*, 479–494. [CrossRef]
4. Rosenthal, D.I.; Mendoza, T.R.; Chambers, M.S.; Burkett, V.S.; Garden, A.S.; Hessell, A.C.; Lewin, J.S.; Ang, K.K.; Kies, M.S.; Gning, I.; et al. The M. D. Anderson Symptom Inventory-Head and Neck Module, a Patient-Reported Outcome Instrument, Accurately Predicts the Severity of Radiation-Induced Mucositis. *Int. J. Radiat. Oncol. Biol. Phys.* **2008**, *72*, 1355–1361. [CrossRef] [PubMed]
5. Gellrich, N.C.; Handschel, J.; Holtmann, H.; Krüskemper, G. Oral Cancer Malnutrition Impacts Weight and Quality of Life. *Nutrients* **2015**, *7*, 2145–2160. [CrossRef] [PubMed]
6. Crowder, S.L.; Najam, N.; Sarma, K.P.; Fiese, B.H.; Arthur, A.E. Quality of Life, Coping Strategies, and Supportive Care Needs in Head and Neck Cancer Survivors: A Qualitative Study. *Support. Care Cancer* **2021**, *29*, 4349–4356. [CrossRef] [PubMed]
7. Crowder, S.L.; Najam, N.; Sarma, K.P.; Fiese, B.H.; Arthur, A.E. Head and Neck Cancer Survivors' Experiences with Chronic Nutrition Impact Symptom Burden after Radiation: A Qualitative Study. *J. Acad. Nutr. Diet.* **2020**, *120*, 1643–1653. [CrossRef]
8. Ganzer, H.; Rothpletz-Puglia, P.; Byham-Gray, L.; Murphy, B.A.; Touger-Decker, R. The Eating Experience in Long-Term Survivors of Head and Neck Cancer: A Mixed-Methods Study. *Support. Care Cancer* **2015**, *23*, 3257–3268. [CrossRef]
9. Crowder, S.L.; Sarma, K.P.; Mondul, A.M.; Chen, Y.T.; Li, Z.; Yanina Pepino, M.; Zarins, K.R.; Wolf, G.T.; Rozek, L.S.; Arthur, A.E. Pretreatment Dietary Patterns Are Associated with the Presence of Nutrition Impact Symptoms 1 Year after Diagnosis in Patients with Head and Neck Cancer. *Cancer Epidemiol. Biomark. Prev.* **2019**, *28*, 1652–1659. [CrossRef]
10. Gorenc, M.; Kozjek, N.R.; Strojan, P. Malnutrition and Cachexia in Patients with Head and Neck Cancer Treated with (Chemo)Radiotherapy. *Rep. Pract. Oncol. Radiother.* **2015**, *20*, 249–258. [CrossRef]
11. Berggren, K.L.; Restrepo Cruz, S.; Hixon, M.D.; Cowan, A.T.; Keysar, S.B.; Craig, S.; James, J.; Barry, M.; Ozbun, M.A.; Jimeno, A.; et al. MAPKAPK2 (MK2) Inhibition Mediates Radiation-Induced Inflammatory Cytokine Production and Tumor Growth in Head and Neck Squamous Cell Carcinoma. *Oncogene* **2019**, *38*, 7329–7341. [CrossRef] [PubMed]
12. Hu, F.B. Dietary Pattern Analysis: A New Direction in Nutritional Epidemiology. *Curr. Opin. Lipidol.* **2002**, *13*, 3–9. [CrossRef]
13. Arthur, A.E.; Goss, A.M.; Demark-Wahnefried, W.; Mondul, A.M.; Fontaine, K.R.; Chen, Y.T.; Carroll, W.R.; Spencer, S.A.; Rogers, L.Q.; Rozek, L.S.; et al. Higher Carbohydrate Intake Is Associated with Increased Risk of All-Cause and Disease-Specific Mortality in Head and Neck Cancer Patients: Results from a Prospective Cohort Study. *Int. J. Cancer* **2018**, *143*, 1105–1113. [CrossRef] [PubMed]
14. Willett, W. *Nutritional Epidemiology*, 3rd ed.; Oxford University Press: New York, NY, USA, 2013.
15. Willett, W.C.; Sampson, L.; Browne, M.L.; Stampfer, M.J.; Rosner, B.; Hennekens, C.H.; Speizer, F.E. THE USE OF A SELF-ADMINISTERED QUESTIONNAIRE TO ASSESS DIET FOUR YEARS IN THE PAST. *Am. J. Epidemiol.* **1988**, *127*, 188–199. [CrossRef] [PubMed]
16. Rimm, E.B.; Giovannucci, E.L.; Stampfer, M.J.; Colditz, G.A.; Litin, L.B.; Willett, W.C. Reproducibility and Validity of an Expanded Self-Administered Semiquantitative Food Frequency Questionnaire among Male Health Professionals. *Am. J. Epidemiol.* **1992**, *135*, 1114–1126. [CrossRef]

17. Sacks, F.M.; Svetkey, L.P.; Vollmer, W.M.; Appel, L.J.; Bray, G.A.; Harsha, D.; Obarzanek, E.; Conlin, P.R.; Miller, E.R.; Simons-Morton, D.G.; et al. Effects on Blood Pressure of Reduced Dietary Sodium and the Dietary Approaches to Stop Hypertension (DASH) Diet. *N. Engl. J. Med.* **2001**, *344*, 3–10. [CrossRef] [PubMed]
18. Fung, T.T.; Chiuve, S.E.; McCullough, M.L.; Rexrode, K.M.; Logroscino, G.; Hu, F.B. Adherence to a DASH-Style Diet and Risk of Coronary Heart Disease and Stroke in Women. *Arch. Intern. Med.* **2008**, *168*, 713–720. [CrossRef]
19. Fung, T.T.; McCullough, M.L.; Newby, P.K.; Manson, J.E.; Meigs, J.B.; Rifai, N.; Willett, W.C.; Hu, F.B. Diet-Quality Scores and Plasma Concentrations of Markers of Inflammation and Endothelial Dysfunction. *Am. J. Clin. Nutr.* **2005**, *82*, 163–173. [CrossRef]
20. Boucher, J.L. Mediterranean Eating Pattern. *Diabetes Spectr. Publ. Am. Diabetes Assoc.* **2017**, *30*, 72–76. [CrossRef]
21. Chiuve, S.E.; Fung, T.T.; Rimm, E.B.; Hu, F.B.; McCullough, M.L.; Wang, M.; Stampfer, M.J.; Willett, W.C. Alternative Dietary Indices Both Strongly Predict Risk of Chronic Disease. *J. Nutr.* **2012**, *142*, 1009–1018. [CrossRef]
22. Halton, T.L.; Willett, W.C.; Liu, S.; Manson, J.E.; Albert, C.M.; Rexrode, K.; Hu, F.B. Low-Carbohydrate-Diet Score and the Risk of Coronary Heart Disease in Women. *N. Engl. J. Med.* **2006**, *355*, 1991–2002. [CrossRef] [PubMed]
23. Maino Vieytes, C.A.; Mondul, A.M.; Li, Z.; Zarins, K.R.; Wolf, G.T.; Rozek, L.S.; Arthur, A.E. Dietary Fiber, Whole Grains, and Head and Neck Cancer Prognosis: Findings from a Prospective Cohort Study. *Nutrients* **2019**, *11*, 2304. [CrossRef]
24. Terrell, J.E.; Nanavati, K.A.; Esclamado, R.M.; Bishop, J.K.; Bradford, C.R.; Wolf, G.T. Head and Neck Cancer—Specific Quality of Life: Instrument Validation. *JAMA Otolaryngol.-Head Neck Surg.* **1997**, *123*, 1125–1132. [CrossRef] [PubMed]
25. Zahn, K.L.; Wong, G.; Bedrick, E.J.; Poston, D.G.; Schroeder, T.M.; Bauman, J.E. Relationship of Protein and Calorie Intake to the Severity of Oral Mucositis in Patients with Head and Neck Cancer Receiving Radiation Therapy. *Head Neck* **2012**, *34*, 655–662. [CrossRef]
26. Bressan, V.; Stevanin, S.; Bianchi, M.; Aleo, G.; Bagnasco, A.; Sasso, L. The Effects of Swallowing Disorders, Dysgeusia, Oral Mucositis and Xerostomia on Nutritional Status, Oral Intake and Weight Loss in Head and Neck Cancer Patients: A Systematic Review. *Cancer Treat. Rev.* **2016**, *45*, 105–119. [CrossRef]
27. Denaro, N.; Merlano, M.C.; Russi, E.G. Dysphagia in Head and Neck Cancer Patients: Pretreatment Evaluation, Predictive Factors, and Assessment during Radio-Chemotherapy, Recommendations. *Clin. Exp. Otorhinolaryngol.* **2013**, *6*, 117–126. [CrossRef]
28. Ozkaya Akagunduz, O.; Eyigor, S.; Kirakli, E.; Tavlayan, E.; Erdogan Cetin, Z.; Kara, G.; Esassolak, M. Radiation-Associated Chronic Dysphagia Assessment by Flexible Endoscopic Evaluation of Swallowing (FEES) in Head and Neck Cancer Patients: Swallowing-Related Structures and Radiation Dose-Volume Effect. *Ann. Otol. Rhinol. Laryngol.* **2019**, *128*, 73–84. [CrossRef] [PubMed]
29. Appel, L.J.; Moore, T.J.; Obarzanek, E.; Vollmer, W.M.; Svetkey, L.P.; Sacks, F.M.; Bray, G.A.; Vogt, T.M.; Cutler, J.A.; Windhauser, M.M.; et al. A Clinical Trial of the Effects of Dietary Patterns on Blood Pressure. *N. Engl. J. Med.* **1997**, *336*, 1117–1124. [CrossRef]
30. Steinberg, D.; Bennett, G.G.; Svetkey, L. The DASH Diet, 20 Years Later. *JAMA* **2017**, *317*, 1529. [CrossRef]
31. Lassale, C.; Batty, G.D.; Baghdadli, A.; Jacka, F.; Sánchez-Villegas, A.; Kivimäki, M.; Akbaraly, T. Healthy Dietary Indices and Risk of Depressive Outcomes: A Systematic Review and Meta-Analysis of Observational Studies. *Mol. Psychiatry* **2019**, *24*, 965–986. [CrossRef]
32. Willett, W.C.; Sacks, F.; Trichopoulou, A.; Drescher, G.; Ferro-Luzzi, A.; Helsing, E.; Trichopoulos, D. Mediterranean Diet Pyramid: A Cultural Model for Healthy Eating. *Am. J. Clin. Nutr.* **1995**, *61*, 1402S–1406S. [CrossRef]
33. Viswanathan, H.; Wilson, J.A. Alcohol–the Neglected Risk Factor in Head and Neck Cancer. *Clin. Otolaryngol. Allied Sci.* **2004**, *29*, 295–300. [CrossRef] [PubMed]
34. Maier, H.; Dietz, A.; Gewelke, U.; Heller, W.D.; Weidauer, H. Tobacco and Alcohol and the Risk of Head and Neck Cancer. *Clin. Investig.* **1992**, *70*, 320–327. [CrossRef] [PubMed]
35. Potash, A.E.; Karnell, L.H.; Christensen, A.J.; Vander Weg, M.W.; Funk, G.F. Continued Alcohol Use in Patients with Head and Neck Cancer. *Head Neck* **2010**, *32*, 905–912. [CrossRef] [PubMed]
36. Shrotriya, S.; Deep, G.; Gu, M.; Kaur, M.; Jain, A.K.; Inturi, S.; Agarwal, R.; Agarwal, C. Generation of Reactive Oxygen Species by Grape Seed Extract Causes Irreparable DNA Damage Leading to G2/M Arrest and Apoptosis Selectively in Head and Neck Squamous Cell Carcinoma Cells. *Carcinogenesis* **2012**, *33*, 848–858. [CrossRef]
37. Eastham, L.L.; Howard, C.M.; Balachandran, P.; Pasco, D.S.; Claudio, P.P. Eating Green: Shining Light on the Use of Dietary Phytochemicals as a Modern Approach in the Prevention and Treatment of Head and Neck Cancers. *Curr. Top. Med. Chem.* **2018**, *18*, 182–191. [CrossRef] [PubMed]
38. Bairati, I.; Meyer, F.; Gélinas, M.; Fortin, A.; Nabid, A.; Brochet, F.; Mercier, J.-P.; Têtu, B.; Harel, F.; Abdous, B.; et al. Randomized Trial of Antioxidant Vitamins to Prevent Acute Adverse Effects of Radiation Therapy in Head and Neck Cancer Patients. *J. Clin. Oncol. Off. J. Am. Soc. Clin. Oncol.* **2005**, *23*, 5805–5813. [CrossRef] [PubMed]
39. Ferreira, P.R.; Fleck, J.F.; Diehl, A.; Barletta, D.; Braga-Filho, A.; Barletta, A.; Ilha, L. Protective Effect of Alpha-Tocopherol in Head and Neck Cancer Radiation-Induced Mucositis: A Double-Blind Randomized Trial. *Head Neck* **2004**, *26*, 313–321. [CrossRef] [PubMed]
40. Arthur, A.E.; Peterson, K.E.; Shen, J.; Djuric, Z.; Taylor, J.M.G.; Hebert, J.R.; Duffy, S.A.; Peterson, L.A.; Bellile, E.L.; Whitfield, J.R.; et al. Diet and Proinflammatory Cytokine Levels in Head and Neck Squamous Cell Carcinoma. *Cancer* **2014**, *120*, 2704–2712. [CrossRef]

MDPI

Article

Southwest Harvest for Health: An Adapted Mentored Vegetable Gardening Intervention for Cancer Survivors

Cindy K. Blair [1,2,*], Prajakta Adsul [1,2], Dolores D. Guest [1,2], Andrew L. Sussman [2,3], Linda S. Cook [1,2], Elizabeth M. Harding [4], Joseph Rodman [2], Dorothy Duff [5], Ellen Burgess [2], Karen Quezada [2], Ursa Brown-Glaberman [1,2], Towela V. King [6], Erika Baca [6], Zoneddy Dayao [1,2], Vernon Shane Pankratz [1,2], Sally Davis [7,8] and Wendy Demark-Wahnefried [9,10]

1 Department of Internal Medicine, University of New Mexico, MSC07-4025, Albuquerque, NM 87131, USA; padsul@salud.unm.edu (P.A.); dguest@salud.unm.edu (D.D.G.); lcook@salud.unm.edu (L.S.C.); ubrown-glaberman@salud.unm.edu (U.B.-G.); zdayao@salud.unm.edu (Z.D.); vpankratz@salud.unm.edu (V.S.P.)
2 University of New Mexico Comprehensive Cancer Center, Albuquerque, NM 87102, USA; asussman@salud.unm.edu (A.L.S.); jrodman@salud.unm.edu (J.R.); emburgess@salud.unm.edu (E.B.); kaquezada@salud.unm.edu (K.Q.)
3 Department of Family and Community Medicine, University of New Mexico, Albuquerque, NM 87131, USA
4 Department of Rehabilitation and Movement Science, University of Vermont, Burlington, VT 05405, USA; elizabeth.harding@med.uvm.edu
5 Albuquerque Area Extension Master Gardener Program, NMSU Cooperative Extension Service, Albuquerque, NM 87107, USA; duffsdales@gmail.com
6 School of Medicine, University of New Mexico, Albuquerque, NM 87131, USA; tvking@salud.unm.edu (T.V.K.); eabaca@salud.unm.edu (E.B.)
7 Department of Pediatrics, University of New Mexico, Albuquerque, NM 87131, USA; sdavis@salud.unm.edu
8 University of New Mexico Prevention Research Center, Albuquerque, NM 87131, USA
9 Department of Nutrition Sciences, University of Alabama at Birmingham, Birmingham, AL 35294, USA; demark@uab.edu
10 O'Neal Comprehensive Cancer Center at the University of Alabama at Birmingham, Birmingham, AL 35294, USA
* Correspondence: CiBlair@salud.unm.edu; Tel.: +1-505-925-7907

Citation: Blair, C.K.; Adsul, P.; Guest, D.D.; Sussman, A.L.; Cook, L.S.; Harding, E.M.; Rodman, J.; Duff, D.; Burgess, E.; Quezada, K.; et al. Southwest Harvest for Health: An Adapted Mentored Vegetable Gardening Intervention for Cancer Survivors. *Nutrients* **2021**, *13*, 2319. https://doi.org/10.3390/nu13072319

Academic Editor: Francesca Giampieri

Received: 6 June 2021
Accepted: 2 July 2021
Published: 6 July 2021

Publisher's Note: MDPI stays neutral with regard to jurisdictional claims in published maps and institutional affiliations.

Abstract: Harvest for Health is a home-based vegetable gardening intervention that pairs cancer survivors with Master Gardeners from the Cooperative Extension System. Initially developed and tested in Alabama, the program was adapted for the different climate, growing conditions, and population in New Mexico. This paper chronicles the feasibility, acceptability, and preliminary efficacy of "Southwest Harvest for Health". During the nine-month single-arm trial, 30 cancer survivor-Master Gardener dyads worked together to establish and maintain three seasonal gardens. Primary outcomes were accrual, retention, and satisfaction. Secondary outcomes were vegetable and fruit (V and F) intake, physical activity, and quality of life. Recruitment was diverse and robust, with 30 survivors of various cancers, aged 50–83, roughly one-third minority, and two-thirds females enrolled in just 60 days. Despite challenges due to the COVID-19 pandemic, retention to the nine-month study was 100%, 93% reported "good-to-excellent" satisfaction, and 87% "would do it again." A median increase of 1.2 servings of V and F/day was documented. The adapted home-based vegetable gardening program was feasible, well-received, and resulted in increased V and F consumption among adult cancer survivors. Future studies are needed to evaluate the effectiveness of this program and to inform strategies to increase the successful implementation and further dissemination of this intervention.

Keywords: cancer survivors; gardening; vegetable; horticultural therapy; quality of life

1. Introduction

In 2019, there were 16.9 million cancer survivors living in the United States (U.S.) [1]. This number is expected to increase to 22.1 million by the year 2030 due to the growth and aging of the population [1,2] and could be substantially higher with further improvements in screening rates, access to care, and effective treatments. Due to the tremendous improvements in early detection and treatment, 56% of cancer survivors are living 10 years or more beyond their diagnosis [1]. However, many cancer survivors are at increased risk for treatment-related comorbidities, including cardiovascular disease, diabetes, and reduced quality of life [3–9]. This has led to preventive health being an important aspect of cancer survivorship [10]. Adherence to a healthy lifestyle has been recommended to improve health outcomes and quality of life and for reduce cancer recurrence and premature mortality [11].

Guidelines for cancer survivorship provide recommendations for adherence to a healthy lifestyle for individuals "living with, through, and beyond cancer" [11,12]. These guidelines encourage cancer survivors to achieve and maintain a healthy lifestyle through weight management, eating a diet low in red meats, sugars, and refined grains, and high in whole grains and vegetables and fruit (V and F), and engaging in regular physical activity. Non-adherence to these recommendations has been associated with increased risk of second malignancies, diabetes, cardiovascular disease, disability, and premature mortality [13–15]. Although several interventions have proven efficacious in improving diet and physical activity for cancer survivors [16–18], very few have been successfully adapted for different populations or settings, which limits their potential to impact population health [18–20].

Harvest for Health is a home-based vegetable gardening intervention that pairs cancer survivors with certified Master Gardeners from the Cooperative Extension System (Extension). Extension is a program within the U.S. Department of Agriculture that operates through the education and outreach arm of land-grant universities nationwide [21]. The Master Gardener Program [22], one of Extension's many education and outreach programs, provides research-based education and training in horticulture to U.S. residents nationwide. Upon completion of their training, the certified Master Gardeners educate and serve their local communities through various projects. The Harvest for Health program has tremendous potential for sustainability, and widespread dissemination since Master Gardener programs exist in all states and territories of the U.S. and typically require 50–100 h of volunteer service annually to maintain certification [21–23]. The intervention, developed and initially tested in Alabama by Demark-Wahnefried and colleagues, has resulted in increased vegetable consumption and leisure-time physical activity and improvements in health-related quality of life (HRQOL) and physical functioning in cancer survivors [24–26].

Widespread adoption and implementation of evidence-based interventions are critical for achieving population-level impact. To achieve widespread implementation, evidence-based interventions need to be adapted to different populations and contexts. We adapted Harvest for Health for the different climate, growing conditions, and population of New Mexico (NM), which was reported previously [27]. We then pilot tested the adapted intervention, Southwest Harvest for Health, and examined the feasibility, acceptability, and preliminary efficacy. Our primary objective is to determine the feasibility and acceptability of the mentored gardening intervention by assessing recruitment, retention, and adherence rates, monitoring adverse events, and evaluating satisfaction with the program in a different population and context. The secondary objective is to explore changes in V and F intake, physical activity, and HRQOL.

2. Methods

A detailed description of the study protocol was published previously [27]. Briefly, this was a single-arm pilot study whereby all participants received the nine-month mentored vegetable gardening intervention. The study was conducted from February 2020 through November 2020. The baseline assessment preceded the COVID-19 pandemic; the six- and nine-month follow-up assessments both occurred during the COVID-19 pandemic. The

University of New Mexico (UNM) Health Sciences Center Institutional Review Board approved this study. Written informed consent was obtained from all participants prior to the baseline assessment.

2.1. Study Participants

The targeted sample size for this pilot study was 30 adult cancer survivors. Oncologists and nurse navigators referred cancer survivors to the study by giving them a study flyer. Additionally, recruitment flyers were distributed via cancer survivor groups, community centers, and other community locations. Interested individuals contacted study staff by email or telephone and were screened for eligibility. Eligible individuals were then scheduled for their baseline assessment.

Eligibility included residence in Bernalillo or Sandoval counties (together comprising most of the Albuquerque-area population). Adults aged 50 years and older with a diagnosis of any type of cancer were eligible. Patients with metastatic cancer were eligible with physician approval. Additional eligibility criteria included: (1) resided in a location that could accommodate a 1.2 m × 2.4 m raised bed garden or four (62.2 cm × 52.1 cm) garden containers, and have access to outdoor running water; (2) able to speak, read, and understand English; and (3) able to participate in the 9-month intervention. Exclusion criteria included: (1) any medical condition that substantially limited activities of daily living (e.g., bending, stooping, walking) that would preclude gardening; (2) eating more than five daily servings of V and F; (3) spending more than 150 min per week in moderate-to-vigorous intensity physical activity; and (4) recent experience (within the past year) with vegetable gardening.

2.2. Harvest for Health Gardening Intervention

A detailed description of the Southwest Harvest for Health intervention has been previously published [27]. Similar to the original Harvest for Health study developed in Alabama [24–26,28], the current pilot study is a community-based, mutually beneficial partnership between UNM and the New Mexico State University Extension Master Gardener Program [29–31]. Harvest for Health pairs each cancer survivor with a certified Master Gardener from Extension [24–26,28]. Together, the participant/Master Gardener dyads work to establish and maintain three seasonal gardens at the participants' homes. Participants receive gardening supplies, plants and seeds, and print materials (study notebook). The study notebook includes articles on safety tips while gardening (e.g., arthritis, protecting hands and feet), instructions for assembling the garden boxes or raised beds, helpful gardening resources from Extension (e.g., "Home Vegetable Gardening in New Mexico" publication), and a planning guide for suggested crops to grow each season. However, most of the gardening knowledge is acquired by working with their Master Gardener mentor. Dyads are asked to communicate every two weeks throughout the intervention, alternating between home visits and telephone or email. The Master Gardener mentor provides information and support related to plants and care of the garden (care of the soil, insect/pest management, watering crops) and helps troubleshooting problems that develop (insects/pests, too little water, too much water or wind, slow growth, etc.).

Due to statewide public health rules implemented during the COVID-19 pandemic, the following changes were made to the study design. The statewide stay-at-home order (March 2020) resulted in issues with scheduling the home deliveries of the larger gardening supplies. Instead, a "drive-through" distribution center was established, and members of the study team loaded the gardening supplies, plants, and seeds into the participants' vehicles. All participants received four gardening containers (62.2 cm × 52.1 cm each; easier to transport than the larger raised bed kits) and a smaller selection of seedlings (limited access to/hours of nurseries and gardening stores). Monthly home visits by the Master Gardeners to their participants' gardens were replaced with an extra telephone call or email, which occurred for the duration of the nine-month study. Participants were encouraged to send photos of their garden to their Master Gardener.

2.3. Primary Outcomes and Measures: Feasibility and Acceptability

The feasibility and acceptability of the home-based mentored vegetable gardening intervention were determined by achieving the following goals: (1) recruitment of 30 adult cancer survivors; (2) retention of ≥80% of the participants; (3) achievement of ≥80% adherence to the intervention; (4) absence of serious adverse events either attributable or possibly attributable to the gardening intervention; and (5) achievement of high acceptability/satisfaction rates with the intervention (≥75%). The cut-points for retention and adherence were selected a priori for comparison with the earlier Harvest for Health studies [25,26].

Retention was calculated as the percentage of participants who completed the post-intervention assessment. Intervention adherence was assessed by the number of completed monthly surveys on garden status, the number of monthly garden photos that were emailed or texted to the study team, and the self-reported frequency of communicating with their Master Gardener mentor (≥2 times per month was specified). Overall satisfaction with the program was assessed via a debriefing survey that was mailed to the study participants. Questions included the following: (1) "How would you rate your experience with the Southwest Harvest for Health study?" (6 response items ranging from excellent to very poor); (2) "Based on your experience, would you do it again?" (5 response items ranging from "yes, most definitely" to "no, not at all"); and (3) "How likely are you to recommend this program to someone else?" (5 response items ranging from "very likely" to "very unlikely"). Additional questions elicited the perceived effect of the intervention on V and F intake, physical activity, and psychosocial well-being, as well as intention to continue gardening on their own.

2.4. Secondary Outcomes and Measures: Health and Lifestyle Outcomes

The intervention health and lifestyle outcomes were assessed at baseline during a home visit. The six- and nine-month follow-up assessments were conducted via telephone and paper or digital surveys to accommodate pandemic restrictions.

Daily V and F consumption was assessed using the "Eating at America's Table Screener" (EATS) either in person (baseline visit; with food props) or via telephone (follow-up visits; with show cards mailed to participants). The EATS screener [32], developed by the National Cancer Institute, comprises 10 questions on frequency (ranging from never to multiple times per day) and quantity (ranging from none to more than two cups) for selected foods. The total number of servings of V and F (fresh, canned, frozen, or 100% juice) were calculated according to the screener scoring recommendations [33]. Questions related to the consumption of white potatoes, fried potatoes, beans and legumes, and mixed vegetable dishes were not included in the computation.

Device-based measures of physical activity and sedentary behavior were measured using an inclinometer/accelerometer. Participants were asked to wear the activPAL3, a small device attached mid-thigh, day and night for seven days at three time points: at the beginning, at six months, and at the end of the study. Participants recorded the following in their sleep diary: the time the device was attached, the time it was removed and reattached (if applicable), and the time they went to bed at night and woke up the next morning. The activPAL monitor provides accurate measures of sedentary time (sitting or lying), standing, and stepping [34–37]. The a priori outcomes of interest were changes in steps per day, time spent stepping at both light-intensity and moderate-intensity cadence, and time spent engaged in sedentary behavior.

Self-reported physical activity was assessed using Godin's Leisure-Time Physical Activity Questionnaire via telephone (after wearing the activPAL3 monitor). The Godin questionnaire assesses the amount of structured exercise (e.g., walking, sports) completed in sessions lasting ten minutes or longer in duration [38,39]. The frequency (times per week) and average duration (minutes) is recorded for types of exercise based on three levels of intensity: mild exercise (minimal effort, no perspiration; example: easy walking), moderate exercise (not exhausting, light perspiration; example: fast walking), and strenuous exercise (heart beats rapidly, sweating; example: jogging or running). Since this questionnaire only assesses structured exercise, it is not directly comparable to the number of steps or time spent stepping measured by a research-grade device. Self-reported Sedentary Behavior was assessed using the Sedentary Behavior Questionnaire (SBQ; paper survey completed after wearing the activPAL3 monitor). The SBQ survey assesses time spent in nine common activities, such as watching television, using a computer, reading, or doing artwork/crafts [40]. Frequency options for each type of activity include none, 15 min or less, 30 min, or 1, 2, 3, 4, or 5 h, or 6 h or more. Time spent in sedentary behavior is assessed for a typical weekday and for a typical weekend day.

HRQOL was measured using PROMIS (Patient-Reported Outcomes Measurement Information System) measures [41]. The 8-item short forms were used to assess domains in mental health (anxiety and depression), physical health (physical function, fatigue, pain, sleep disturbance, and sleep impairment), and social health (satisfaction with social roles and activities, i.e., social functioning). These instruments are valid and reliable for use in diverse clinical samples [42–45]. Surveys were scored using the free Health Measures Scoring Service (https://www.assessmentcenter.net/ac_scoringservice; accessed 3 July 2021). The service provides T-scores, which represent a linear transformation of the raw scores normed to the general population, with a mean of 50 and a standard deviation of 10. For physical function and social functioning, higher scores indicate better functioning; for the remaining domains, higher scores indicate worse functioning.

2.5. Other Outcomes and Measures

The Social Provisions Scale was used to assess participants' perceived level of social support [46]. This survey includes six subscales: reassurance of worth (how other people recognize one's value), social integration (sense of belonging), guidance (information/advice), nurturance (sense of being needed by others), attachment (emotional closeness), and reliable alliance (assurance that other people will provide assistance if needed). Scores on each item range from one (strongly disagree) to four (strongly agree), with subscale scores ranging from four to sixteen. Total perceived social support is the sum of the six subscales (range 24 to 96). Higher scores represent greater support.

2.6. Data Analysis

The primary outcomes of this pilot study were the accrual and retention of the cancer survivors and paired Master Gardeners throughout the nine-month intervention, as well as satisfaction with the program. Secondary outcomes included trends in V and F consumption, physical activity, and HRQOL. The processing of the activPAL data to calculate time spent engaged in physical activities and sedentary activities has been previously described [47]. All activPAL variables were standardized to a 15-h day to limit the effect of within and between-person variability in awake/wear time. Descriptive characteristics of the enrolled cancer survivors are presented as frequencies and percentages or medians with interquartile range (IQR). Similar to most pilot studies, our pilot study was not powered to detect significant nor clinically meaningful changes in measures of V and F intake, physical activity, and HRQOL. However, estimates of the pre-post changes will be useful in planning for a future larger trial. As diet and physical activity are seasonally influenced [18,19], the changes observed since baseline are of primary interest; however, change for both the mid- and post-intervention follow-up are included in the results. Data were summarized as medians and IQRs, and the 6- and 9-month differences were presented as medians and IQRs. The Wilcoxon Signed-Rank Test was used to evaluate the 6- and 9-month change. SAS (version 9.4) was used to perform the statistical analyses.

3. Results

3.1. Feasibility

Of the 40 cancer survivors who were screened, 10 were ineligible, and the remaining 30 were enrolled in the pilot study. The top two reasons for ineligibility included current and successful experience with vegetable gardening and living outside of the study catchment area. Retention in this nine-month intervention was 100%. No adverse events were attributable or possibly attributable to the gardening intervention.

The characteristics of the 30 cancer survivors enrolled in this study are included in Table 1. The median age at study enrollment was 68 years (range 50 to 83 years). Most study participants were female (70%), non-Hispanic white (73%), and slightly over half had graduated from college. Eighty-four percent reported their health as good, very good, or excellent, while the median number of comorbidities reported was 3 (range 0 to 8). The median time since cancer diagnosis was 5 years (range 1 to 17 years). While a variety of cancer types were represented, the most common were breast, prostate, and lung.

Table 1. Baseline characteristics of the cancer survivors participating in Southwest Harvest for Health.

Characteristics	Median (IQR) or Frequency (%)
Age (range 50 to 83)	68 (64, 72)
Sex	
Female	21 (70%)
Male	9 (30%)
Race-ethnicity	
Non-Hispanic White	22 (73%)
Hispanic White	6 (20%)
Other	2 (7%)
Education	
No college degree	13 (43%)
College degree	17 (57%
Cancer type	
Breast	11 (37%)
Prostate	6 (20%)
Lung	4 (13%)
Other [a]	9 (30%)
Treatment received [b]	
Surgery	23 (77%)
Radiation	22 (73%)
Chemotherapy	10 (33%)
Hormone therapy	12 (40%)
Other	2 (7%)
Years since cancer diagnosis (range 1 to 17)	5 (2, 8)
Self-reported general health	
Excellent	2 (7%)
Very good	5 (17%)
Good	18 (60%)
Fair	5 (17%)
Poor	0 (0%)
Number of comorbidities (range 0 to 8)	3 (2, 4)
BMI (kg/m^2)	28.8 (24.4, 32.1)

[a] Colorectal, melanoma, endometrial, lymphoma, ovarian, Merkel cell carcinoma. [b] Percentages do not total 100%, since some participants may have had more than one type of treatment.

3.2. Adherence

Adherence during the intervention was moderately high for completing the monthly gardening activity surveys. Eighty percent of participants completed all six surveys, and the remaining 20% completed five surveys. However, adherence was only modest for sending photos of the garden to the study team (average = 56%; range of 47% to 63%). Most participants (89%) reported communicating with their Master Gardener at least twice a month; on average, 72% reported three or more times per month. The remainder (11%) reported less frequent communication. On average, 40% and 50% of study participants reported working in their garden several times a day and once a day, respectively. The majority (60%) of participants reported working 15–29 min each time they worked in their garden.

3.3. Acceptability

Upon completion of the study, 83% of participants responded "probably yes" or "yes, most definitely" to planning to continue the garden and plant on their own; 13% responded "maybe", and 3% responded "probably no". When asked if they planned to expand their garden, 69% responded "probably yes" or "yes, most definitely", 14% responded "maybe", and 17% responded "probably no". However, among the latter two groups (31%), half of the participants had already expanded their garden during the study, based on the monthly surveys or photos of their garden. Most participants (90%) rated their experience with Southwest Harvest for Health as "very good" or "excellent" (3% "good", 7% "fair") and were "likely" or "very likely" to recommend the program to another cancer survivor (10% "neutral"). Eighty-seven percent responded "yes, most definitely" or "probably yes" that they would "do it again" based on their experience (the remaining 13% were divided between "maybe" and "probably no").

3.4. Secondary Outcomes

Physical activity, V and F intake, and HRQOL scores are reported in Table 2. At study completion, the greatest improvement was observed for the number of servings per day of V and F. Compared to baseline, the median change at post-intervention follow-up was 1.2 additional servings per day. The median change in device-measured physical activity was a decrease of 478 steps per day that corresponded to 1.1 and 4.2 fewer minutes of stepping at a light-intensity and moderate-intensity, respectively. Sedentary behavior increased by 14.8 min per day. On average, there was no appreciable change in physical or mental quality of life; however, there was a modest improvement in social functioning (median: 3.3 points; IQR: -1.4, 10.9).

Table 2. Change in health-related outcomes during the mentored gardening study.

	Pre−Intervention (Pre-COVID−19) Median (IQR)	**Mid-Intervention (during COVID-19) Median (IQR)**	**Post-Intervention (during COVID−19) Median (IQR)**	**Pre-Mid Median Difference (IQR) *p*-Value [a]**	**Pre-Post Median Difference (IQR) *p*-Value [a]**
Lifestyle Behaviors					
V and F (servings per day)	3.8 (2.5, 6.3)	5.5 (3.6, 7.2)	5.3 (3.7, 6.3)	0.9 (−0.3, 2.2) $p = 0.006$	1.2 (−0.4, 2.2) $p = 0.03$
Physical activity [a] Self−Report (Minutes per day)					
Light intensity	11.3 (2.1, 17.1)	4.3 (0, 25.7)	5.3 (0, 25.7)	−1.4 (−10, 4.3) $p = 0.39$	−0.7 (−10, 6.3) $p = 0.66$
Moderate intensity	0 (0, 7.1)	0 (0, 8.6)	0 (0, 12.9)	0 (0, 0) $p = 0.28$	0 (0, 2.9) $p = 0.19$
Device−based Measures					
Steps per day	6781 (5523, 8633)	6403 (4796, 7854)	5831 (4287, 8038)	−792 (−1631, 464) $p = 0.02$	−478 (−1832, 312) $p = 0.05$
Minutes per day:					
Standing	256.4 (201.5, 286.0)	248.1 (181.5, 324.1)	250.3 (180.6, 314.8)	0.5 (−26.8, 29.0) $p = 0.82$	5.8 (−44.3, 49.1) $p = 0.49$
Light intensity	39.6 (30.7, 51.2)	38.3 (29.2, 48.8)	37.3 (26.8, 47.6)	−0.1 (−5.8, 3.0) $p–0.46$	−1.1 (−10.6, 3.4) $p = 0.20$
Moderate intensity	49.5 (38.5, 70.6)	49.3 (36.9, 64.4)	45.8 (30.4, 64.3)	−6.1 (−15.1, 4.0) $p = 0.01$	−4.2 (−13.8, 3.7) $p = 0.06$
Sedentary behavior Self-Report (Minutes per day)	517.8 (379.2, 623.4)	469.2 (355.8, 591.6)	507.0 (379.2, 618.0)	−54.0 (−114.0, 6.0) $p = 0.06$	12.0 (−114.0, 90) $p = 0.87$

Table 2. *Cont.*

	Pre−Intervention (Pre-COVID−19) Median (IQR)	Mid-Intervention (during COVID-19) Median (IQR)	Post-Intervention (during COVID−19) Median (IQR)	Pre-Mid Median Difference (IQR) *p*-Value [a]	Pre-Post Median Difference (IQR) *p*-Value [a]
Device-based Measure (Minutes per day)	440.5 (361.4, 505.8)	457.2 (366.9, 529.0)	457.8 (405.2, 510.3)	27.7 (−40.8, 51.1) $p = 0.50$	14.8 (−35.8, 54.6) $p = 0.63$
HRQOL [b] Physical					
Physical function	47.0 (42.9, 53.1)	44.8 (33.9, 52.4)	45.6 (39.3, 52.8)	−0.1 (−4.9, 2.0) $p = 0.50$	0.0 (−7.0, 4.3) $p = 0.39$
Fatigue	50.5 (45.6, 58.2)	50.7 (46.9, 59.2)	49.8 (43.0, 57.5)	1.1 (−2.1, 3.7) $p = 0.41$	−0.9 (−5.7, 1.5) $p = 0.28$
Pain	55.4 (40.7, 58.3)	54.4 (50.3, 57.5)	50.6 (40.7, 58.7)	0.0 (−2.4, 0.3) $p = 0.57$	0.0 (−7.1, 0.5) $p = 0.09$
Sleep disturbance	50.4 (48.7, 53.2)	51.3 (50.2, 53.5)	52.0 (49.3, 52.9)	−0.7 (−2.3, 3.5) $p = 0.76$	−0.4 (−1.7, 2.4) $p = 0.92$
Sleep impairment	49.1 (40.2, 56.6)	50.8 (39.9, 55.5)	48.8 (40.6, 52.9)	0.7 (−2.1, 4.9) $p = 0.46$	−0.2 (−6.0, 5.1) $p = 0.71$
Mental					
Anxiety	49.3 (37.1, 53.5)	50.9 (37.1, 56.5)	47.8 (38.3, 56.3)	(−1.2, 3.0) $p = 0.41$	0.0 (−3.6, 3.0) $p = 0.85$
Depression	47.2 (38.2, 54.2)	44.5 (38.2, 53.4)	49.9 (38.2, 54.2)	0.0 (−1.5, 3.6) $p = 0.50$	0.0 (−1.7, 4.8) $p = 0.32$
Social Satisfaction with social roles and activities [c]	50.5 (45.3, 58.0)	51.7 (42.7, 58.0)	51.3 (46.6, 65.4)	−1.6 (−5.6, 6.0) $p = 0.77$	3.3 (−1.4, 10.9) $p = 0.14$

[a] *p*-values for the change scores are from the Wilcoxon Signed-Rank Test. [b] There was no vigorous-intensity stepping cadence according to the activPAL monitor. [c] Higher scores indicate better functioning for physical function and social functioning (i.e., satisfaction with social roles and activities); however, for the remaining domains, higher scores indicate worse functioning.

3.5. Other Outcomes

There were no appreciable changes in any of the social support subscales for both the 6- and 9-month follow-up assessments (data not shown). For five of the six subscales at both time points, the median change was zero; for nurturance, the median change was one point (both time points).

Overall, most participants reported that the gardening experience motivated them to eat a healthier diet, eat more vegetables, or try new vegetables (median scores of 7 or 8 out of 10; Figure 1). Most participants also reported motivation to be more physically active (median score of 8 out of 10); additional activities reported included yard work and walking, with a few participants reporting yoga and other exercises. There was also a positive impact of the gardening experience on well-being (Figure 1). Average scores were highest for feeling connected to nature when gardening and mindfulness, i.e., being better able to stay in the present moment (median scores of 9.5 and 8.5 out of 10, respectively). Average scores were lowest for being more socially active (likely due to COVID-19). Sixty percent of cancer survivors reported that gardening helped them cope with pain, anxiety about test results, either fear of or having received a diagnosis of cancer recurrence or a second cancer, or dealing with cancer in a family member.

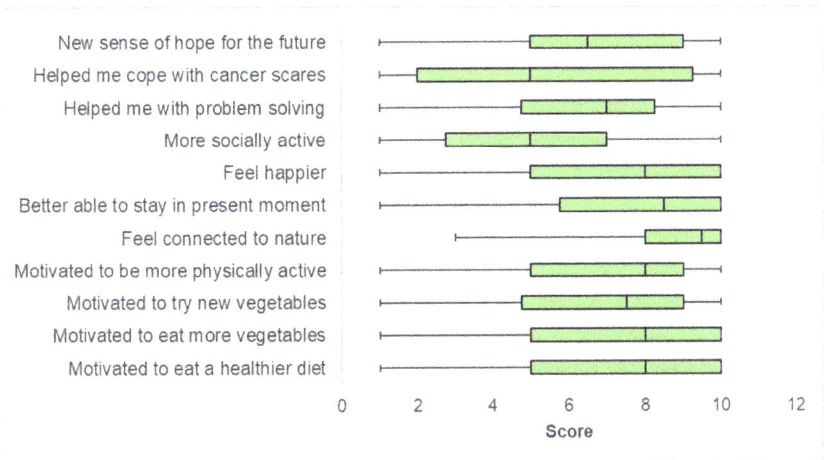

Figure 1. Impact of a home-based vegetable gardening intervention on diet, physical activity, and well-being. Shaded areas represent the interquartile range. Vertical lines within each shaded region represent the median score. Whiskers represent the minimum and maximum scores for the questions, ranging from 1 (not at all) to 10 (very much).

4. Discussion

This study explored the feasibility and acceptability of a home-based, mentored vegetable gardening intervention adapted for middle-aged and older cancer survivors living in the Southwest U.S. It represents the first systematic adaptation of the original Harvest for Health program for a different state/region of the U.S. Despite being conducted entirely during the COVID-19 pandemic, which necessitated remote instead of in-person mentoring, the gardening intervention was feasible. This was demonstrated by the high recruitment and retention rates, as well as the moderately high adherence rates. Satisfaction with the intervention was also high, based on the percentage of cancer survivors rating their experience, being willing to "do it again", and the likelihood of recommending the program to other cancer survivors.

The feasibility metrics of Southwest Harvest for Health compare favorably with the earlier pilot studies conducted in Alabama. The retention rates in this one-year (Alabama; AL) or nine-month (New Mexico; NM) vegetable gardening intervention have been very high (91–100%) [24–26], suggesting that the variety of gardening activities and benefits may help prevent satiation, which is more common with other lifestyle promotion programs. Satisfaction rates with Southwest Harvest for Health were also similar to those of the original Harvest for Health pilot programs: 93% vs. 100%, respectively, rating the experience as good to excellent; and 87% vs. 85–100%, respectively, stating they would "do it again" [24–26]. Another similarity between all these pilot studies is the expansion or plans to expand the garden space (NM: 84%; AL: 70–89%) as well as intention to continue vegetable gardening beyond the study (NM: 90%; AL: 85–100%) [24–26].

Although the current pilot study was a small, single-arm study, the health behavior outcomes compare favorably to the earlier pilot studies. We observed a meaningful overall increase of 1.2 servings per day of V and F, similar to the earlier pilot studies, which reported increases of 0.9 servings per day [25,26]. The magnitude of this change is particularly notable in view that increases of half a serving a day have been considered clinically important in previous work [48]. While maintaining a small, home vegetable garden likely represents light-intensity physical activities, these activities may serve as a gateway for additional physical activity. Similar to the earlier studies, participants in the current study reported being motivated by their garden to do more yard work and to walk more. In contrast to the earlier studies, but not surprisingly, few participants mentioned joining a fitness center during the pandemic. Engagement in these types of activities may have prevented more substantial declines in overall activity compared with other studies of adults during the COVID-19 pandemic [49]. While there are fewer reports of declines in HRQOL that have been systematically observed during COVID-19, Grajek et al. recently reported declines of 2% among 450 patients actively treated for cancer [50]. Therefore, the high satisfaction with the gardening intervention may have mitigated steeper reductions in HRQOL that may have been noted otherwise.

As with many research studies conducted during 2020, the potential effects of COVID-19 on this home-based gardening intervention must be taken into consideration. While the baseline assessment was conducted prior to the declaration of COVID-19 as a pandemic, both the intervention and follow-up assessment were conducted during the pandemic. Thus, diet, physical activity, and HRQOL could also have been affected by stay-at-home orders and recommendations, social distancing, and closed fitness centers, parks, and swimming pools. Additionally, a large proportion of the study participants were at higher risk for Sars-CoV-2 infection and complications, and thus greatly limited their time away from home or interactions with other people.

Limitations of this pilot study were the lack of a control group and the potential effect of seasonal variation on V and F intake and physical activity. Due to the colder winters in New Mexico (compared to Alabama), we shortened the one-year intervention to nine months. Some studies have shown that diet, especially consumption of fresh V and F, may be influenced by season [51,52]. Similarly, physical activities, especially outdoor activities, may be influenced by season [53–55]. In the Southwest US, we would expect more outdoor activity during the spring and fall seasons.

5. Conclusions and Future Directions

The adapted home-based vegetable gardening intervention was feasible, safe, and well-received by middle-aged and older cancer survivors living in the Southwest U.S. Future directions include moving from efficacy trials to effectiveness/pragmatic trials to observe the impact of this promising intervention under real-world conditions. The Cooperative Extension System is ideally situated for delivering health promotion programs in community settings, which can greatly expand their reach to a broader and more diverse population. Further research is needed to optimize the implementation of Harvest for Health within the Extension Master Gardener Programs within a state and ideally to other states across the nation.

Author Contributions: Conceptualization, C.K.B. and W.D.-W.; Data curation, C.K.B., E.M.H., J.R. and K.Q.; Formal analysis, C.K.B. and V.S.P.; Funding acquisition, C.K.B. and L.S.C.; Investigation, C.K.B., E.M.H., J.R., D.D., E.B. (Ellen Burgess), K.Q., T.V.K. and E.B. (Erika Baca); Methodology, C.K.B., P.A., D.D.G., A.L.S., E.M.H., D.D., S.D. and W.D.-W.; Project administration, C.K.B., E.M.H., J.R. and D.D.; Resources, D.D., U.B.-G., Z.D. and W.D.-W.; Supervision, C.K.B., D.D.G. and E.B. (Ellen Burgess); Writing—original draft, C.K.B.; Writing—review and editing, C.K.B., P.A., D.D.G., A.L.S., L.S.C., E.M.H., J.R., D.D., E.B. (Ellen Burgess), K.Q., U.B.-G., T.V.K., E.B. (Erika Baca), Z.D., V.S.P., S.D. and W.D.-W. All authors have read and agreed to the published version of the manuscript.

Funding: This research was supported by the UNM Comprehensive Cancer Center Support Grant NCI P30CA118100 and the Behavioral Measurement and Population Science Shared Resource and the Biostatistics Shared Resource. CKB is currently supported by an NIH K07 grant CA215937.

Institutional Review Board Statement: The study was conducted according to the guidelines of the Declaration of Helsinki, and approved by the Institutional Review Board of the University of New Mexico Health Sciences Center (protocol 19–551 approved 3 December 2019).

Informed Consent Statement: Informed consent was obtained from all subjects involved in the study.

Data Availability Statement: Aggregate data may be available for research purpose upon reasonable request to the corresponding author.

Acknowledgments: The authors thank the cancer survivors and the Extension Master Gardeners for their participation in the study. We are especially grateful for everyone's patience and understanding as we had to make several modifications to the study due to the COVID-19 pandemic.

Conflicts of Interest: The authors declare no conflict of interest. The funders had no role in the design of the study, in the collection, analyses, or interpretation of data, in the writing of the manuscript, or in the decision to publish the results.

References

1. Miller, K.D.; Nogueira, L.; Mariotto, A.B.; Rowland, J.H.; Yabroff, K.R.; Alfano, C.M.; Jemal, A.; Kramer, J.L.; Siegel, R.L. Cancer treatment and survivorship statistics, 2019. *CA Cancer J. Clin.* **2019**, *69*, 363–385. [CrossRef]
2. American Cancer Society. *Cancer Treatment & Survivorship Facts & Figures 2019–2021*; American Cancer Society: Atlanta, GA, USA, 2019.
3. Avis, N.E.; Deimling, G.T. Cancer survivorship and aging. *Cancer* **2008**, *113*, 3519–3529. [CrossRef]
4. Hewitt, M.; Rowland, J.H.; Yancik, R. Cancer survivors in the United States: Age, health, and disability. *J. Gerontol. A Biol. Sci. Med. Sci.* **2003**, *58*, 82–91. [CrossRef] [PubMed]
5. Morgans, A.K.; Fan, K.H.; Koyama, T.; Albertsen, P.C.; Goodman, M.; Hamilton, A.S.; Hoffman, R.M.; Stanford, J.L.; Stroup, A.M.; Resnick, M.J.; et al. Influence of age on incident diabetes and cardiovascular disease in prostate cancer survivors receiving androgen deprivation therapy. *J. Urol.* **2015**, *193*, 1226–1231. [CrossRef]
6. Singh, S.; Earle, C.C.; Bae, S.J.; Fischer, H.D.; Yun, L.; Austin, P.C.; Rochon, P.A.; Anderson, G.M.; Lipscombe, L. Incidence of Diabetes in Colorectal Cancer Survivors. *J. Natl. Cancer Inst.* **2016**, *108*, djv402. [CrossRef] [PubMed]
7. Armenian, S.H.; Xu, L.; Ky, B.; Sun, C.; Farol, L.T.; Pal, S.K.; Douglas, P.S.; Bhatia, S.; Chao, C. Cardiovascular Disease Among Survivors of Adult-Onset Cancer: A Community-Based Retrospective Cohort Study. *J. Clin. Oncol.* **2016**, *34*, 1122–1130. [CrossRef] [PubMed]
8. Strongman, H.; Gadd, S.; Matthews, A.; Mansfield, K.E.; Stanway, S.; Lyon, A.R.; Dos-Santos-Silva, I.; Smeeth, L.; Bhaskaran, K. Medium and long-term risks of specific cardiovascular diseases in survivors of 20 adult cancers: A population-based cohort study using multiple linked UK electronic health records databases. *Lancet* **2019**, *394*, 1041–1054. [CrossRef]
9. Schoormans, D.; Vissers, P.A.J.; van Herk-Sukel, M.P.P.; Denollet, J.; Pedersen, S.S.; Dalton, S.O.; Rottmann, N.; van de Poll-Franse, L. Incidence of cardiovascular disease up to 13 year after cancer diagnosis: A matched cohort study among 32,757 cancer survivors. *Cancer Med.* **2018**, *7*, 4952–4963. [CrossRef]
10. Overholser, L.S.; Callaway, C. Preventive Health in Cancer Survivors: What Should We Be Recommending? *J. Natl. Compr. Cancer Netw.* **2018**, *16*, 1251–1258. [CrossRef]
11. Denlinger, C.S.; Ligibel, J.A.; Are, M.; Baker, K.S.; Demark-Wahnefried, W.; Dizon, D.; Friedman, D.L.; Goldman, M.; Jones, L.; King, A.; et al. Survivorship: Healthy lifestyles, version 2.2014. *J. Natl. Compr. Cancer Netw.* **2014**, *12*, 1222–1237. [CrossRef]
12. Cancer.Net. What Is Survivorship? Available online: https://www.cancer.net/survivorship/what-survivorship#:~{}:text=A%20person%20who%20has%20had,reasons%20for%20this%20may%20vary (accessed on 3 July 2021).
13. Jacob, M.E.; Yee, L.M.; Diehr, P.H.; Arnold, A.M.; Thielke, S.M.; Chaves, P.H.; Gobbo, L.D.; Hirsch, C.; Siscovick, D.; Newman, A.B. Can a Healthy Lifestyle Compress the Disabled Period in Older Adults? *J. Am. Geriatr. Soc.* **2016**, *64*, 1952–1961. [CrossRef]
14. Li, Y.; Pan, A.; Wang, D.D.; Liu, X.; Dhana, K.; Franco, O.H.; Kaptoge, S.; Di Angelantonio, E.; Stampfer, M.; Willett, W.C.; et al. Impact of Healthy Lifestyle Factors on Life Expectancies in the US Population. *Circulation* **2018**, *138*, 345–355. [CrossRef] [PubMed]
15. McCullough, M.L.; Patel, A.V.; Kushi, L.H.; Patel, R.; Willett, W.C.; Doyle, C.; Thun, M.J.; Gapstur, S.M. Following cancer prevention guidelines reduces risk of cancer, cardiovascular disease, and all-cause mortality. *Cancer Epidemiol. Biomarkers Prev.* **2011**, *20*, 1089–1097. [CrossRef] [PubMed]
16. Heath, G.W.; Parra, D.C.; Sarmiento, O.L.; Andersen, L.B.; Owen, N.; Goenka, S.; Montes, F.; Brownson, R.C.; Lancet Physical Activity Series Working, G. Evidence-based intervention in physical activity: Lessons from around the world. *Lancet* **2012**, *380*, 272–281. [CrossRef]
17. Morey, M.C.; Snyder, D.C.; Sloane, R.; Cohen, H.J.; Peterson, B.; Hartman, T.J.; Miller, P.; Mitchell, D.C.; Demark-Wahnefried, W. Effects of home-based diet and exercise on functional outcomes among older, overweight long-term cancer survivors: RENEW: A randomized controlled trial. *JAMA* **2009**, *301*, 1883–1891. [CrossRef] [PubMed]
18. Demark-Wahnefried, W.; Rogers, L.Q.; Alfano, C.M.; Thomson, C.A.; Courneya, K.S.; Meyerhardt, J.A.; Stout, N.L.; Kvale, E.; Ganzer, H.; Ligibel, J.A. Practical clinical interventions for diet, physical activity, and weight control in cancer survivors. *CA Cancer J. Clin.* **2015**, *65*, 167–189. [CrossRef]
19. Basen-Engquist, K.; Alfano, C.M.; Maitin-Shepard, M.; Thomson, C.A.; Stein, K.; Syrjala, K.L.; Fallon, E.; Pinto, B.M.; Schmitz, K.H.; Zucker, D.S.; et al. *Moving Research into Practice: Physical Activity, Nutrition, and Weight Management for Cancer Patients and Survivors. NAM Perspectives*; Discussion Paper; National Academy of Medicine: Washington, DC, USA, 2018. [CrossRef]
20. Reis, R.S.; Salvo, D.; Ogilvie, D.; Lambert, E.V.; Goenka, S.; Brownson, R.C.; Lancet Physical Activity Series 2 Executive, C. Scaling up physical activity interventions worldwide: Stepping up to larger and smarter approaches to get people moving. *Lancet* **2016**, *388*, 1337–1348. [CrossRef]
21. USDA National Institute of Food and Agriculture: Cooperative Extension System. Available online: https://nifa.usda.gov/cooperative-extension-system (accessed on 3 July 2021).

MDPI

Article

Healthy Moves to Improve Lifestyle Behaviors of Cancer Survivors and Their Spouses: Feasibility and Preliminary Results of Intervention Efficacy

Cindy L. Carmack [1,*], Nathan H. Parker [2], Wendy Demark-Wahnefried [3], Laura Shely [4], George Baum [5], Ying Yuan [6], Sharon H. Giordano [7], Miguel Rodriguez-Bigas [8], Curtis Pettaway [9] and Karen Basen-Engquist [5]

1 Department of Palliative Care, Rehabilitation & Integrative Medicine, The University of Texas MD Anderson Cancer Center, Houston, TX 77030, USA
2 Moffitt Cancer Center, Tampa, FL 33612, USA; Nathan.Parker@moffitt.org
3 Department of Nutrition Sciences, The University of Alabama at Birmingham, Birmingham, AL 35294, USA; demark@uab.edu
4 Mental Health Care Line, Michael E. DeBakey Veterans Affairs Medical Center, Houston, TX 77030, USA; Laura.Shely@va.gov
5 Department of Behavioral Science, The University of Texas MD Anderson Cancer Center, Houston, TX 77030, USA; gpbaum@mdanderson.org (G.B.); kbasenen@mdanderson.org (K.B.-E.)
6 Department of Biostatistics, The University of Texas MD Anderson Cancer Center, Houston, TX 77030, USA; yyuan@mdanderson.org
7 Department of Health Services Research, The University of Texas MD Anderson Cancer Center, Houston, TX 77030, USA; sgiordan@mdanderson.org
8 Department of Colon & Rectal Surgery, The University of Texas MD Anderson Cancer Center, Houston, TX 77030, USA; mrodbig@mdanderson.org
9 Department of Urology, The University of Texas MD Anderson Cancer Center, Houston, TX 77030, USA; cpettawa@mdanderson.org
* Correspondence: ccarmack@mdanderson.org

Citation: Carmack, C.L.; Parker, N.H.; Demark-Wahnefried, W.; Shely, L.; Baum, G.; Yuan, Y.; Giordano, S.H.; Rodriguez-Bigas, M.; Pettaway, C.; Basen-Engquist, K. Healthy Moves to Improve Lifestyle Behaviors of Cancer Survivors and Their Spouses: Feasibility and Preliminary Results of Intervention Efficacy. *Nutrients* **2021**, *13*, 4460. https://doi.org/10.3390/nu13124460

Academic Editor: Sara Gandini

Received: 25 October 2021
Accepted: 9 December 2021
Published: 14 December 2021

Abstract: Spouses offer a primary source of support and may provide critical assistance for behavior change. A diet-exercise intervention previously found efficacious in improving cancer survivors' lifestyle behaviors was adapted to utilize a couples-based approach. The aims were to test the feasibility of this couples-based (CB) intervention and compare its efficacy to the same program delivered to the survivor-only (SO). Twenty-two survivor-spouse couples completed baseline assessments and were randomized to the CB or SO interventions. The study surpassed feasibility benchmarks; 91% of survivors and 86% of spouses completed a 6-month follow-up. Survivors and spouses attended 94% and 91% of sessions, respectively. The SO survivors showed significant improvements on the 30-s chair stand and arm curl tests, weight, and fruit and vegetable (F and V) consumption. The CB survivors showed significant improvements on the 6-min walk and 2-min step tests, body weight, and fat and F and V consumption. Improvement in the 30-s chair stand and arm curl tests was significantly better for SO survivors. The SO spouses showed no significant changes in outcome measures, but the CB spouses showed significant improvements in moderate-to-strenuous physical activity, weight, and fat and F and V consumption. Weight loss was significantly greater in CB spouses compared to SO spouses. Findings demonstrate feasibility, warranting further investigation of CB approaches to promote lifestyle change among cancer survivors and spouses.

Keywords: behavior change; diet; physical activity; couples; telehealth counseling

1. Introduction

Roughly half a century ago, the nation's War on Cancer was launched and has resulted in major increases in survival through improvements in early detection and treatment [1]. There are now over 17 million cancer survivors in the US alone, comprising roughly 4% of the population [2]. While many survivors have been definitively treated for cancer, they remain at risk for recurrence and are at an increased risk for second cancers, cardiovascular disease, and diabetes [3]. Many survivors also experience lingering effects of cancer and its treatment, including fatigue, psychological distress, and accelerated functional decline [3,4]. Collectively, these health conditions impose considerable costs. The total economic burden of cancer was previously estimated at $263.8 billion with $20.9 billion and $140.1 billion from indirect morbidity costs (lost productivity due to illness) and indirect mortality costs (lost productivity due to premature death), respectively [5]. Given the burgeoning number of survivors and their potential impact on the health care system, improving their health status is a national priority [6]. The proposed study targets survivors of breast, prostate, and colorectal cancers because they are the largest segment of cancer survivors where survival rates currently exceed 90% [7].

Research in cancer survivors has shown that interventions promoting a healthy weight, a healthy diet, and increased physical activity improve quality of life (QOL), physical functioning and overall health status, and reduce the risk of chronic disease [8–10] and possibly recurrence [11], and improve survival [12,13]. It is recommended that cancer survivors achieve and maintain a healthy weight, accumulate at least 150 min of moderate physical activity per week, and consume a healthy plant-based diet [10,13,14]. However, a study of 3367 racially- and ethnically-diverse cancer survivors identified through the National Health Interview Survey indicates that roughly 70% of survivors are overweight or obese, and over 80% do not meet the guidelines for physical activity or fruit and vegetable (F and V) consumption [15]. While cancer survivors in general have similar and equally high prevalence rates for physical inactivity, poor diets, and obesity as those without cancer [3,16], their risk for developing co-morbidities and downstream costly events that result from interactions between their cancer, its treatment, and these lifestyle factors results in a significant burden. As such, interventions that target diet, physical activity, and weight management are essential.

Research is ongoing to explore how best to deliver lifestyle interventions for cancer survivors [17–19]. Successful behavior change interventions integrate theory to maximize effectiveness. Social Cognitive Theory (SCT) [20], one of the most robust theories of behavior change, posits that behavior is influenced by expectations formed through direct and observed experiences, which includes expectations about the confidence (self-efficacy) in performing these tasks successfully. Behavior also is influenced by goals, both proximal and distal, and by barriers to performance [21]. In the behavior change process, change is more likely when: (1) Behaviors are successfully performed independently; (2) Support is received from others who express confidence in that behavior change and provide feedback on performance; and (3) Desired behaviors are then modelled by others [20,22]. Thus, an integral part of SCT is the role that social relationships have on behavior change.

A recent scoping review by Ellis et al. [23] calls for interventions that capitalize on existing support networks that are conducted within the family context to promote healthy behaviors not only among cancer survivors, but also among their family members. A couples-based (CB) intervention is consistent with this call and also the tenants of SCT. This format encourages couples to model healthy eating and physical activity for each other; observing the other's success can increase one's confidence. In the process, couples learn to provide one another with support and feedback regarding goal setting and to work together to overcome barriers. Indeed, couples report that having their partner perform and model goal behaviors, join in health discussions, and provide emotional support encourages their own behavior change [24]. Unfortunately, enlisting the spouse only as a supporter may result in negative consequences, because even well-intentioned spouses may offer assistance in ways that appear controlling or over-protective, rather than supportive. As such, interventions addressing behavior change for both members of the couple and promoting shared goals, conjoint coping, and mutual support may be more effective than those targeting the individual cancer survivor [22]. This approach capitalizes on the strength of the spousal bond and embraces the recommendations that family members also follow the American Cancer Society guidelines for nutrition and physical activity [13]. Ultimately, both partners may benefit by reducing disease burden for survivors (tertiary prevention) [13] and the risk of cancer and other diseases in spouses (primary prevention) [25]. While recent pilot trials have included spouses in such interventions [26,27], none have focused on changing multiple health behaviors and none have compared their efficacy to a survivor only multiple health behavior change approach in which the health behavior outcomes of both the survivor and their spouse are examined.

The aims of the present study were to conduct a pilot trial to test the feasibility of a CB multi-behavior change program (diet and physical activity) and to compare its efficacy to the same program delivered to the survivor only (SO) in 22 survivors of breast, prostate, and colorectal cancer and their spousal partners, both of whom were identified as having poor health behaviors. The hypothesis was that survivors randomized to the CB format compared to the SO format would show favorable changes in physical activity, physical performance, body weight status, body composition, and diet. Likewise, and additionally, spouses would show greater changes in outcomes with the CB format compared to the SO format.

2. Materials and Methods

2.1. Study Overview

This study employs a 2-arm, single-blinded, randomized controlled trial (RCT) that evaluated a 6-month diet and exercise intervention delivered in either a CB or SO format. All participants completed assessments at baseline and 6 months (post-intervention).

2.2. Participant Eligibility

Eligibility for the survivor included: (1) Diagnosis of loco-regional breast cancer (Stages 0-IIIA), prostate cancer (Stages I-II), or colorectal cancer (Stages I-II); (2) Completion of primary cancer treatment and at least 3 months from surgery; (3) No history of other cancers (excluding non-melanoma skin cancer); (4) <150 min of moderate-to-vigorous intense physical activity (PA) per week; (5) Fruit and vegetable (F and V) intake <7 servings/day for women or <9 servings/day for men; (6) Age 18 years or older; (7) Able to read and speak English; (8) Living within the Houston area (Harris or a contiguous county); (9) No pre-existing medical conditions that precluded adherence to an unsupervised PA program or high fruit and vegetable (F and V) diet [28]; (10) Having a spouse or significant other with whom they have resided for at least 1 year (includes heterosexual and same-sex couples); (11) Able to provide informed consent; and (12) Has access to a computer with high-speed internet.

Eligibility for the spouse included criteria 4–12 listed above. Exclusion criteria for survivor or spouse included using a walker or wheelchair/scooter, being pregnant, or reporting any conditions that are listed on the Physical Activity Readiness Questionnaire (PAR-Q) [29].

2.3. Recruitment and Screening Procedures

This research was approved by The University of Texas MD Anderson Institutional Review Board. Participants who participated in another MD Anderson approved protocol and who indicated they would like to be contacted for future lifestyle trials were recruited for this study. First, potential participants were contacted and verified for eligibility following verbal consent for screening, including cancer diagnosis and treatment status, and having lived with a spouse or significant other for at least 1 year. Next, they were screened for current physical activity using the Godin Leisure Time Exercise Questionnaire (GLTEQ) and F and V intake using the 2009 Texas Behavioral Risk Factor Surveillance System (BRFSS) dietary questions, as well as whether there were any pre-existing medical conditions that precluded their participation using the PAR-Q. Survivors endorsing any item on the PAR-Q were required to have a medical release from their physician to clear them for participation in the study. Following permission to contact their spouses, an identical process was then used to solicit spousal interest, gain verbal consent, and screen for eligibility. For this study, both survivors and their spouses had to be eligible and provide written consent for participation. For all survivors and/or spouses not interested in participating, information regarding reasons for refusal was collected.

2.4. Study Group Assignment

After completing baseline assessments, participants were assigned to 1 of the 2 study conditions using a form of adaptive allocation referred to as minimization [30]. The following survivor variables were used to ensure balance across study group assignment: baseline physical activity, baseline diet quality, age, race, gender, and marital quality. Spousal factors were not included because doing so would likely be redundant given the literature showing a strong concordance between spousal health behavior [31]. Group assignment was conducted separately by disease site. Minimization has been used successfully in previous trials resulting in a good group balance [32,33].

2.5. Study Conditions

The diet and exercise intervention, based on Social Cognitive Theory, was a tailored correspondence and web-based counseling regimen initially developed and proven efficacious for breast, prostate, and colorectal cancer survivors by Demark Wahnefried and colleagues in Reach out to EnhaNcE Wellness (RENEW) [34]. The behavioral goals were for participants to engage in 15 min of strength exercise every other day, $\geq$30 min of walking or other moderate-intensity exercise on 5 or more days per week, and consume a diet of $\geq$7 F and V servings/day for women or $\geq$9 F and V servings/day for men and $\leq$7% of total calories from saturated fat [35]. It also encouraged weight management; for those with a BMI $\geq$ 25 kg/m^2, a healthy weight loss goal of 1–2 lbs. per week (a loss of 5% body weight was used as a goal over the course of the 6-month study period) was encouraged. To adapt to the 6-month timeline, the RENEW intervention materials were modified such that they could be delivered within the study period and included materials directed toward the spouse. For survivors randomized to the survivor-only arm, the materials were similar; however, there was no reference to working with a spouse on behavior change efforts. Survivors (and their spouses if randomized to the CB arm) were provided with a tailored workbook and 3 tailored print newsletters over the 6-month study period. All print materials provided motivational messages tailored on stage of readiness [36] that accompanied illustrations of current behaviors in relation to national guidelines; progress reports depicted headway toward goals (which were incrementally set) and reinforced. The SO survivors and CB survivors and their spouses also received 9 web-based video counseling sessions; in the unforeseen event that there were problems connecting to the session online, participants had the option of receiving these sessions by telephone. The first 3 sessions were weekly; sessions changed to every other week after session 3, and then monthly after session 5. The counselor had a master's degree in Marriage and Family Counseling and was supervised by a licensed clinical psychologist with expertise in counseling and health behavior change. Each counseling session focused on specific cognitive-behavioral strategies for healthy behavior change. All sessions also emphasized SCT concepts, such as self-monitoring and incremental goal setting and specific session topics included the following: problem-solving; relapse prevention; goal-setting; cognitive restructuring; and time management [20]. Skills practice was assigned as homework to be reviewed in subsequent sessions. For both study conditions, assessments of adverse events were conducted at the start of each counseling session. Finally, participants in each arm were provided with the following materials: Therabands®; T-Factor 2006© Guide to the Fat Content of Foods; portion plate; pedometer; web camera; headset; and log books to track their exercise and diet behaviors. In addition to the intrapersonal (individual) cognitive-behavioral strategies oriented toward one's own behavior change, participants in the CB intervention learned interpersonal cognitive-behavioral skills, including communal coping, joint problem-solving, and healthy communication. They also received the counseling sessions together as a couple.

2.6. Assessment Procedures

All participants (survivors and spouses) were assessed in-person at MD Anderson at baseline and at a 6-month follow-up. Assessment personnel were blinded to the participant's study condition. Accrual, attrition, and patient satisfaction served as the primary endpoints of this feasibility study, and other outcomes, such as physical activity, physical performance, weight status, body composition, and dietary intake were also assessed. Each survivor and spouse who completed the baseline and 6-month assessments received compensation in the form of $25 gift cards: one following each completed assessment (up to $100 per couple). Participants also received relevant assessment results at the end of the study. If one member of the survivor-spouse dyad dropped out, the remaining member of the couple could continue.

2.7. Measures

To address the study aims, feasibility, and exploratory outcome measures were collected:

Feasibility measures included accrual, attrition, participant views on intervention acceptability, and the monitoring of adverse events. For recruitment, the number of participants contacted about the study, who were eligible and who consented to participate, were tracked. Retention was calculated as the percentage of participants assessed at baseline who completed the 6-month assessments. Drop outs were tracked by study condition. Session attendance was monitored to measure exposure. Intervention acceptability was assessed by asking participants, "Would you recommend this program to other cancer survivors?" at the 6-month follow-up. Possible responses included "Yes," "Maybe," and "No." Responses were compiled across study conditions.

Exploratory outcome measures included physical activity, physical performance, body composition, weight, and diet.

Physical activity was assessed with the 3-item modified version of the Godin Leisure Time Exercise Questionnaire to ascertain self-reported moderate and strenuous leisure time exercise [37]. To provide an objective measure of physical activity, for 1 week before the baseline and the 6-month assessments, the participants also wore a programmed Actigraph accelerometer (Fort Walton Beach, FL, USA). Accelerometers were downloaded according to the manufacturer's instructions and as per the previous studies of Basen-Engquist and colleagues [38].

Physical performance was measured using a variety of tests that assessed endurance, strength, and agility. The 6-min walk test and 2-min step test were used as measures of aerobic function. The 6-minute walk test has been validated in older adults by comparing it to a treadmill walking test measuring the time to get to 85% of an age-adjusted maximum heart rate [39]. The 2-min step test is self-paced and assesses the number of times within 2 min a participant can step in place raising the knees to a height halfway between the iliac crest and mid-patella. This test correlates moderately with common measures of aerobic capacity and is low risk [40]. For lower body strength, a 30-s chair-stand test was used [41]. For upper body strength and functionality, the timed arm curl task was used, taking into account that this test has been shown to be better tolerated than maximum-grip strength for participants with arthritis [41]. To assess agility and dynamic balance, an 8-foot up-and-go assessment was used. The task is a composite measure involving dynamic balance, power, and agility. The test is a modification of the 3-m time up-and-go test [42]. The modification to 8 feet is to increase the feasibility of administering the tests in areas with limited space, including home settings [41]. The height at baseline (for BMI calculation) and body weight were measured using a stadiometer and electronic scale, respectively.

Diet was assessed with the Automated Self-administered 24-h Dietary Recall (ASA24) to document the participant's food intake for a total of 24 h. Two interviews were obtained, 1 for a weekday and 1 for a weekend day [43]. The foods selected by the participants are from the USDA's Food and Nutrient Database for Dietary Studies' (FNDDS) most up-to-date database. Participants in both groups completed this assessment 1 week before their baseline and 6-month assessments. F and V intake and % of calories from dietary fat were extracted from the ASA24 output and averages for the 2-day recalls were taken at each of the time points.

Spouse exploratory outcome measures included the same measures of physical activity, physical performance, diet, and weight. These were assessed at the same time points as survivors.

Finally, demographic/medical questions for survivors were collected at baseline and included age, sex, race/ethnicity, education level, employment status, cancer type/stage, and treatment types. Demographic data for spouses included age, sex, and race/ethnicity.

2.8. Data Analysis

Summary statistics were calculated for demographic and clinical characteristics for the study population by study condition.

Feasibility was determined by 3 criteria: (1) The completion of accrual within 1 year; (2) An attrition rate of 20% or less; and (3) No occurrence of serious adverse events that are directly attributable to the intervention.

For the exploratory outcome measures, we calculated the means and standard deviations. Prior to computing the sum for moderate-to-strenuous physical activity, moderate and strenuous physical activity variables were each truncated at 420 min per week. Paired t-tests or Wilcoxon signed-rank tests were used to assess within group differences between the 6-month and baseline measurements. We also calculated the difference between the 6-month measurement and baseline and compared it between groups using a 2-sample t-test or Mann-Whitney U test. To estimate the effect of the study group assignment (CB arm relative to SO arm) on changes in exploratory outcome variables, multivariable linear regression models were fit for each outcome. The covariates included in multivariable linear regression models were selected based on the univariates analysis with $p < 0.05$.

3. Results

3.1. Participant Characteristics

Table 1 displays clinicodemographic characteristics of survivors and spouses randomized to the SO and CB groups. On average, survivors were in their early-to-mid 60s, and slightly more were female. Survivors were predominantly white, and the vast majority had at least some college-level education. Nearly one-third of survivors were employed full time, and nearly one-third classified themselves as a homemaker/volunteer, while slightly more than one-third were retired. The average BMI among survivors was in the overweight range. All female participants were breast cancer survivors, whereas most male participants were prostate cancer survivors. In terms of cancer treatment, most survivors had undergone surgery, radiation therapy, and hormonal therapy, but fewer than half of the survivors had undergone chemotherapy. There were no significant differences in clinicodemographic characteristics between survivors randomized to the SO and CB conditions (all p-values > 0.05). As published previously [44], survivors who enrolled in the study were younger and consumed less energy from fat than survivors who were screened but did not enroll.

Table 1. Clinic demographic characteristics of cancer survivor and spouse participants by study condition.

Characteristic	Cancer Survivors				Spouses			
	Overall (n = 22)	**Survivor-Only Condition** (n = 10)	**Couples-Based Condition** (n = 12)	p [a]	**Overall** (n = 22)	**Survivor-Only Condition** (n = 10)	**Couples-Based Condition** (n = 12)	p [a]
Age (years), mean (SD)	64.1 (10.8)	62.4 (11.4)	65.5 (10.5)	0.5	63.4 (8.2)	63.1 (9.0)	63.4 (7.8)	0.9
Sex				0.7				>0.9
Male, n (%)	10 (45.5)	4 (40.0)	6 (50.0)		13 (59.1)	6 (60.0)	7 (58.3)	
Female, n (%)	12 (54.5)	6 (60.0)	6 (50.0)		9 (40.9)	4 (40.0)	5 (41.7)	
Race/Ethnicity				0.1				0.6
Hispanic, n (%)	4 (18.2)	3 (30.0)	1 (8.3)		3 (13.6)	1 (10.0)	2 (16.7)	
Non-Hispanic Black, n (%)	1 (4.5)	1 (10.0)	0 (0.0)		2 (9.1)	1 (10.0)	1 (8.3)	
Non-Hispanic White, n (%)	16 (72.7)	5 (50.0)	11 (91.7)		16 (72.7)	7 (70.0)	9 (75.0)	
Other, n (%)	1 (4.5)	1 (10.0)	0 (0.0)		1 (4.5)	1 (10.0)	0 (0.0)	
Education, n (%)				0.7				-
High school diploma/GED	2 (9.1)	0 (0.0)	2 (16.7)		-	-	-	
Some college or 2-year degree	6 (27.3)	3 (30.0)	3 (25.0)		-	-	-	
Bachelor's degree	8 (36.4)	3 (30.0)	5 (41.7)		-	-	-	
Advanced degree	6 (27.2)	4 (40.0)	2 (16.6)		-	-	-	
Employment Status, n (%)				0.2				-
Full Time	6 (27.3)	1 (10.0)	5 (41.7)		-	-	-	
Part Time	2 (9.1)	1 (10.0)	1 (8.3)		-	-	-	
Retired	8 (36.4)	6 (60.0)	2 (16.7)		-	-	-	
Homemaker or volunteer	6 (27.3)	2 (20.0)	4 (33.3)		-	-	-	
Weight (kg), mean (SD)	76.4 (19.4)	70.4 (13.8)	81.4 (22.4)	0.2	85.7 (22.6)	80.8 (12.8)	89.8 (28.3)	0.4
BMI (kg/m^2), mean (SD)	27.7 (6.4)	25.4 (3.8)	29.7 (7.6)	0.1	29.6 (5.8)	28.2 (3.9)	30.7 (7.0)	0.3
Cancer type, n (%)				0.6				-
Breast	13 (59.1)	6 (60.0)	7 (58.3)		-	-	-	
Prostate	8 (36.3)	4 (40.0)	4 (33.3)		-	-	-	
Colorectal	1 (4.5)	0 (0.0)	1 (8.3)		-	-	-	
Surgery, n (%)				0.2				-
No	2 (10.5)	0 (0.0)	2 (22.2)		-	-	-	
Yes	17 (89.5)	10 (100.0)	7 (77.8)		-	-	-	
Chemotherapy				>0.9				-
No	11 (57.9)	6 (60.0)	5 (55.6)		-	-	-	
Yes	8 (42.1)	4 (40.0)	4 (44.4)			-	-	
Radiation therapy				>0.9				-
No	9 (45.0)	4 (40.0)	5 (50.0)		-	-	-	
Yes	11 (55.0)	6 (60.0)	5 (50.0)		-	-	-	
Hormonal therapy				>0.9				-
No	7 (38.9)	3 (33.3)	4 (44.4)		-	-	-	
Yes	11 (61.1)	6 (66.7)	5 (55.6)		-	-	-	
Other treatment				0.6				-
No	9 (75.0)	3 (60.0)	6 (85.7)		-	-	-	
Yes	3 (25.0)	2 (40.0)	1 (14.3)		-	-	-	

[a] For difference between individual and couple condition.

The average age of spouses was similar to that of survivors. The majority of spouses were male and white. The average BMI of spouses also was in the overweight range, though higher than survivors. There were no significant differences in demographic characteristics between spouses randomized to the SO and CB conditions (all p-values > 0.05).

3.2. Feasibility Measures

Recruitment spanned 15 months, with 22 survivors and 22 spouses enrolling between July 2011 and September 2012. One hundred ninety-seven survivors were contacted, and 22 survivors (11.2%) enrolled and completed baseline assessments (Figure 1). One couple enrolled but did not complete baseline assessments. The overall enrollment rate was 12.7%.

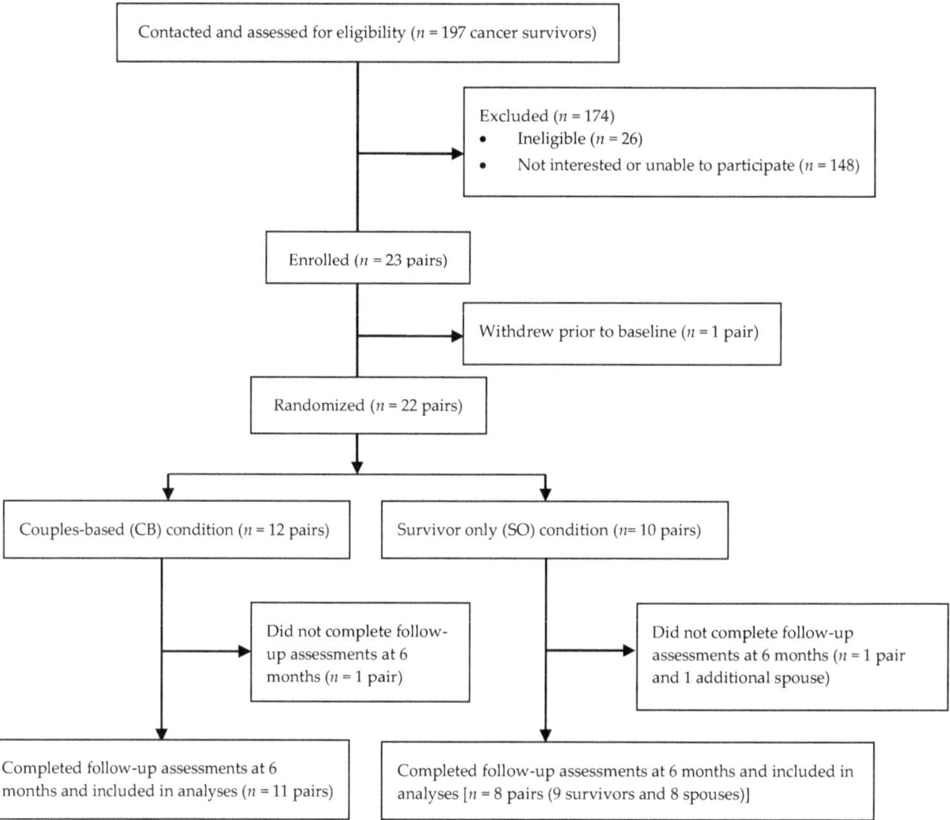

Figure 1. CONSORT flow diagram from recruitment to study completion.

Nine-of-ten survivors in the SO group (90.0%) completed the 6-month follow-up measures, and 11-of-12 survivors in the CB group (92.7%) completed the 6-month follow-up measures. Thus, the attrition rate among survivors was 9.1%. Eight-of-ten spouses in the SO condition (80%) and 11-of-12 spouses in the CB condition (92.7%) completed the 6-month follow-up measures, resulting in a 13.6% attrition rate among spouses. The overall attrition rate, regardless of survivor status or study condition, was 11.4%.

Survivors in the SO condition attended an average of 97% of sessions, and survivors in the CB condition attended an average of 92% of sessions. Combining both groups, survivors attended 94% of sessions with no significant differences between study conditions. Spouses (CB condition only) attended an average of 91% of sessions.

In terms of intervention acceptability, 12 survivors (6 CB and 6 SO) and 5 spouses (4 CB and 1 SO) completed the follow-up question asking whether they would recommend the program to other cancer survivors. Among CB survivors, 5 (83%) responded, "Yes," and 1 (17%) responded, "No." Among SO survivors, 4 (67%) responded, "Yes," and 2 (33%)

responded "Maybe." Among spouses, all 4 spouses from the CB group (100%) responded, "Yes" and the 1 spouse from the SO group responded "Maybe."

There were no intervention-related adverse events.

3.3. Exploratory Outcome Measures

Table 2 display secondary outcomes including self-reported and accelerometer-based physical activity, physical functioning, weight status, body composition, and diet at study time points for survivors and spouses. The p-values indicating the significance of change within the study arm and the significance of the difference in change between the study arms is also presented.

Table 2. (**a**) Summary statistics for outcome measures among cancer survivors by study condition; (**b**) Summary statistics for outcome measures among spouses by study condition.

(a)

		Cancer Survivors								
		Survivor-Only Condition				Couples-Based Condition				
Outcome	Assessment	n	Mean	SD	p [a]	n	Mean	SD	p [a]	p [b]
Self-reported moderate-to-strenuous PA (min/wk)	Baseline	10	176.0	138.7	0.3	12	96.0	116.6	0.8	0.6
	6 months	10	196.5	158.0		11	107.7	112.1		
MPA (min/wk)	Baseline	9	71.3	46.0	0.4	11	38.3	18.0	0.2	0.3
	6 months	8	90.0	28.8		7	57.0	25.3		
VPA (min/wk)	Baseline	9	6.0	7.7	0.9	11	0.3	0.7	0.1	0.7
	6 months	8	3.7	4.1		7	1.2	1.9		
MVPA (min/wk)	Baseline	9	77.3	48.7	0.5	11	38.6	18.2	0.2	0.7
	6 months	8	93.7	31.3		7	58.2	25.8		
6-min walk distance (m)	Baseline	10	514.7	68.4	0.2	12	443.0	87.2	<0.001	0.5
	6 months	9	580.8	100.1		11	531.2	81.9		
2-min step test (repetitions)	Baseline	10	79.8	27.9	0.3	12	82.0	16.3	0.01	0.5
	6 months	9	92.4	26.0		11	98.9	14.0		
30-Second Sit-to-Stand (repetitions)	Baseline	10	11.4	3.2	0.005	12	12.1	3.1	0.5	0.005
	6 months	9	15.0	5.1		11	12.8	3.1		
Arm Curls (repetitions)	Baseline	10	13.2	3.4	0.01	12	14.8	3.1	0.8	0.02
	6 months	9	17.6	5.8		11	15.6	4.4		
8 foot up-and-go time (s)	Baseline	10	6.3	1.9	0.3	12	7.0	1.5	0.7	0.5
	6 months	9	5.6	0.9		11	6.7	1.7		
Body weight (kg)	Baseline	10	70.4	13.8	0.02	12	81.4	22.4	0.01	0.5
	6 months	9	67.0	13.8		11	73.3	14.1		
Total fat consumption (g/day)	Baseline	10	72.9	19.0	0.5	12	76.7	24.6	0.07	0.5
	6 months	9	64.9	31.8		11	60.3	29.7		
Saturated fat consumption (g/day)	Baseline	10	21.1	7.1	0.6	12	24.5	10.5	0.03	0.2
	6 months	9	19.0	9.5		11	18.2	7.1		
Fruit and vegetable consumption (cups/day)	Baseline	10	2.5	1.3	0.02	12	2.6	1.3	<0.001	0.8
	6 months	9	4.4	2.0		11	4.5	1.6		

(b)

		Spouses								
		Survivor-Only Condition				Couples-Based Condition				
Outcome	Assessment	n	Mean	SD	p [a]	n	Mean	SD	p [a]	p [b]
Self-reported moderate-to-strenuous PA (min/wk)	Baseline	9	129.4	158.6	0.9	12	71.7	78.1	0.02	0.8
	6 months	9	120.0	157.9		10	124.5	47.9		
MPA (min/wk)	Baseline	9	80.3	47.8	0.6	12	52.0	36.1	0.2	0.06
	6 months	9	87.0	31.8		6	55.7	29.6		
VPA (min/wk)	Baseline	9	3.1	5.7	0.2	12	3.6	7.1	0.5	0.2
	6 months	9	6.1	12.3		6	5.5	13.0		
MVPA (min/wk)	Baseline	9	83.4	49.1	0.9	12	55.6	41.5	0.2	0.2
	6 months	9	92.9	30.8		6	61.2	40.1		

Table 2. *Cont.*

(b)

Outcome	Assessment	Spouses Survivor-Only Condition n	Mean	SD	p^a	Couples-Based Condition n	Mean	SD	p^a	p^b
6-min walk distance (m)	Baseline	10	504.5	112.4	>0.99	12	491.3	131.2	0.5	0.6
	6 months	8	533.8	96.9		11	508.0	86.3		
2-min step test (repetitions)	Baseline	10	86.9	22.9	0.8	12	80.3	32.7	0.6	0.6
	6 months	8	86.0	18.5		11	84.0	17.5		
30-Second Sit-to-Stand (repetitions)	Baseline	10	11.9	4.8	0.6	12	12.0	3.8	>0.99	0.5
	6 months	8	12.1	6.6		11	12.4	3.9		
Arm Curls (repetitions)	Baseline	10	16.3	4.8	0.8	12	16.9	5.4	0.2	0.3
	6 months	8	16.8	4.8		11	15.7	4.8		
8 foot up-and-go time (s)	Baseline	10	7.1	2.6	0.3	12	7.6	4.9	0.3	0.2
	6 months	8	6.7	1.6		11	6.3	1.6		
Body weight (kg)	Baseline	10	80.8	12.8	0.7	12	89.8	28.3	0.03	0.05
	6 months	8	83.6	11.3		11	80.2	20.1		
Total fat consumption (g/day)	Baseline	9	58.9	26.3	0.2	12	85.9	38.8	<0.001	0.8
	6 months	8	51.1	27.3		11	63.2	26.6		
Saturated fat consumption (g/day)	Baseline	9	20.1	10.1	0.4	12	28.5	13.0	0.002	0.4
	6 months	8	18.0	10.9		11	19.0	8.7		
Fruit and vegetable consumption (cups/day)	Baseline	9	2.8	1.5	0.6	12	2.4	1.3	0.01	0.2
	6 months	8	3.3	1.3		11	3.3	1.4		

Abbreviations: PA = physical activity, MPA = accelerometer-measured moderate physical activity, VPA = accelerometer-measured vigorous physical activity, MVPA = accelerometer-measured moderate-to-vigorous physical activity, [a] for within-group differences between baseline and 6 months; [b] for differences in change from baseline to 6-months between study conditions.

There were no significant changes in either self-reported or accelerometer measures or physical activity between study time points among survivors randomized to the SO vs. CB conditions, and there were no significant differences in physical activity change between these groups.

Despite no differences in physical activity, there were significant changes in physical performance from baseline to the 6-month follow-up. Survivors randomized to the CB condition showed significant improvement in both the 6-min walk test and the 2-min step test at 6 months, whereas survivors randomized to the SO condition showed no significant change in these measures. No significant between-arm differences were detected for either of these measures. Survivors randomized to the SO condition showed significant improvement in the 30-s sit-to-stand test and in the arm curl test, whereas survivors in the CB condition showed no significant change in either of these tests. Improvements in these tests were significantly better for the SO vs. the CB arms.

Survivors in both the SO and CB arms demonstrated significant weight loss over the 6-month period with no between-arm differences in weight loss noted.

The SO arm survivors reported significantly higher F&V consumption at 6 months compared to baseline, as did CB arm survivors. Survivors in the CB arm also had significant decreases in saturated fat consumption. There were no significant between-arm differences in change scores for any of the dietary variables.

Spouses randomized to the CB condition reported significantly higher strenuous + moderate physical activity at 6 months compared to baseline, but there was no significant change in this variable for spouses in the SO arm. There were no significant changes in either arm in accelerometer-measured physical activity. There were no significant differences between arms in the amount of change in either self-reported or accelerometer-measured physical activity.

There were no significant changes in physical performance measures from baseline to 6 months among spouses randomized to either study condition, and there were no significant differences in physical performance changes between groups. Spouses randomized to CB condition demonstrated significant weight loss at 6 months relative to baseline. Spouses randomized to the SO condition showed no significant change in weight, and there was no significant difference in weight change between groups.

Spouses randomized to the CB condition showed significantly reduced consumption of total fat and saturated fat and significantly increased consumption of F&Vs at 6 months relative to baseline. In contrast, spouses randomized to the SO condition showed no significant changes in total fat, saturated fat, or F&V consumption from baseline to 6 months. There were no significant differences in 6-month changes in dietary variables between spouses randomized to SO vs. CB study conditions.

Table 3 displays multivariable linear regression models estimating the effect of study groups on exploratory outcome measures for cancer survivors and spouses. Based on bivariate correlations with outcome variable change scores, the following variables were included as covariates in the models: baseline value of the outcome of interest, ethnicity (white vs. non-white), and BMI. With randomization to the SO condition as the comparison group, randomization to the CB condition showed significant, negative associations with change in 30-s chair stand repetitions (B = −2.7, $p = 0.04$) and arm curls (B = −4.5, $p = 0.02$) among survivors. Among spouses, randomization to the CB condition showed a significant, negative association with change in vigorous physical activity (B = −4.08, $p = 0.02$).

Table 3. Multiple linear regression models estimating treatment effects.

		Cancer Survivors				Spouses			
	Effect	βeta	95% LB	95% UB	*p*-Value	βeta	95% LB	95% UB	*p*-Value
Self-reported moderate-to-strenuous PA (min/wk)	Condition (SO vs. CB)	−12.08	−161.14	136.99	0.866	46.76	−51.05	144.56	0.323
MPA (min/wk)	Condition (SO vs. CB)	−33.86	−68.27	0.56	0.053	−15.47	−40.55	9.61	0.196
VPA (min/wk)	Condition (SO vs. CB)	−2.26	−7.16	2.65	0.320	−4.08	−7.32	−0.83	0.019
MVPA (min/wk)	Condition (SO vs. CB)	−36.30	−72.95	0.34	0.052	−15.76	−43.91	12.38	0.237
6-min walk distance (meters)	Condition (SO vs. CB)	23.57	−58.61	105.75	0.550	1.15	−46.03	48.34	0.959
2-min step test (repetitions)	Condition (SO vs. CB)	11.98	−6.67	30.62	0.191	1.93	−10.75	14.61	0.749
30-Second Sit-to-Stand (repetitions)	Condition (SO vs. CB)	−2.71	−5.30	−0.12	0.042	1.43	−2.28	5.15	0.421
Arm curls (repetitions)	Condition (SO vs. CB)	−4.46	−8.18	−0.73	0.022	−1.47	−4.32	1.37	0.285
8 foot up-and-go time (seconds)	Condition (SO vs. CB)	0.65	−0.62	1.92	0.294	−0.67	−1.84	0.50	0.237
Body weight (kg)	Condition (SO vs. CB)	−0.63	−3.58	2.33	0.658	−3.66	−7.90	0.57	0.085
Total fat consumption (g/day)	Condition (SO vs. CB)	−5.08	−35.10	24.94	0.723	4.10	−14.74	22.93	0.646
Saturated fat consumption (g/day)	Condition (SO vs. CB)	−2.45	−10.37	5.46	0.519	−1.62	−9.40	6.17	0.661
Fruit and vegetable consumption (cups/day)	Condition (SO vs. CB)	0.30	−1.37	1.98	0.704	0.69	−0.30	1.67	0.156

Abbreviations: PA = physical activity, MPA = accelerometer-measured moderate physical activity, VPA = accelerometer-measured vigorous physical activity, MVPA = accelerometer-measured moderate-to-vigorous physical activity.

4. Discussion

Few studies have examined the feasibility or outcomes of spouse-based interventions to improve lifestyle behaviors among cancer survivors [41]. This study examined the feasibility of a CB intervention to improve diet and physical activity among cancer survivors and their spouses, and also compared differences in exploratory outcomes between survivors and spouses randomized to the CB intervention and those randomized to an SO intervention.

As hypothesized, the intervention was indeed feasible, with survivors and spouses surpassing metrics for retention and intervention session attendance. The enrollment rate for this study was 12.7%, which is typical of interventions targeting diet and physical activity among cancer survivor-partner dyads [27,45], particularly when survivors are not referred directly by oncologists involved in the survivors' care. Furthermore, retention and adherence in the current trial were strong. The combined attrition rate of 11.4% among all participants in this study is similar to or exceeds those reported from other studies involving cancer survivor-caregiver or partner dyads [27,45,46]. Moreover, the high attendance, which ranged from 91–97%, exceed those reported in recent studies involving exercise for cancer survivor-partner dyads [27,46], further highlighting the feasibility of this intervention. Participants experienced no intervention-related adverse events thus safety, an important outcome to establish feasibility, was obviated as a concern. Both survivors and spouses tended to rate the program favorably, with 92% of survivors and 80% of spouses responding that they would recommend participating to other cancer survivors.

Cancer survivors and their partners tend to struggle to consume healthy diets and engage in sufficient exercise, [15] so establishing intervention feasibility, as we did in this study, is a critical first step. The high rates of retention and adherence observed in this study highlight the importance and benefits of couples embarking on paths to improve eating and physical activity habits together. Similarly, low rates of attrition and high rates of adherence between survivors randomized to both study arms suggest that the two groups may have inspired similar levels of motivation to participate and complete the intervention. Attending intervention sessions and following-up to measure progress set the stage for developing positive health behaviors as a couple, with survivors and spouses each taking active roles in supporting one another's efforts to improve health. The observed enrollment rate, though typical in the realm of behavior change interventions for cancer survivors, leaves significant room for improving intervention reach. Studies that rely on treating oncologists to refer cancer survivors to behavior change interventions tend to demonstrate higher enrollment rates [27]; this highlights the importance of integrating lifestyle improvement programming into standard care for cancer care and survivorship. Future trials involving dyadic interventions to improve diet and physical activity among cancer survivors and their spouses may benefit from involving oncology providers directly in referral pathways.

In addition to intervention feasibility, our study provides some evidence of intervention benefits for both cancer survivors and their partners. Survivors in both the CB and SO groups improved health behaviors and related outcomes with between-group comparisons demonstrating few differences. However, spouses in the CB intervention demonstrated significant improvements in health behaviors and related outcomes, while those examined as part of the SO group (i.e., did not receive an intervention) demonstrated none. Though samples were small, these findings suggest that CB interventions may help enhance delivery to some cancer survivors and may provide an important opportunity for behavior change among spouses.

Recently published studies involving lifestyle interventions for dyads featuring cancer survivors and their caregivers or family members have demonstrated mixed results. Kamen et al. found no significant improvements in physical activity (steps per day) among cancer survivors or caregivers (95% of whom were spouses/partners) following a 6-week exercise intervention, and there was no significant difference between participants who were randomized to engage with their caregivers and those randomized to individual intervention [47]. In contrast, Demark-Wahnefried et al. found significant improvements in physical activity, fitness, and anthropometrics among breast cancer survivors and their adult daughters enrolled in team and individual lifestyle interventions, but no significant differences between intervention formats [45]. To date, studies of dyadic lifestyle interventions for cancer survivors and their caregivers or partners, including the current study, have focused on understanding intervention feasibility. As such, they likely lack statistical power to detect true differences in behaviors or outcomes between groups of survivors and partners receiving the intervention as pairs and those receiving interventions individually.

This study has important strengths and limitations. Strengths include a strong, randomized study designed to compare intervention feasibility and primary and secondary outcomes between study groups. Primary outcomes included valid measures of diet, and both objective and self-reported physical activity, and secondary outcomes included valid and objective measures of physical performance, body composition, and anthropometrics. Intervention adherence and retention were very strong and similar between study groups, helping to limit concerns about intervention fidelity, attrition bias, or missing data. The enrollment rate of 12.7%, though on par with dyadic lifestyle interventions for cancer survivors, suggests that those who actually participated may have been particularly motivated to engage in a lifestyle intervention with their spouses. Future efforts to enroll cancer survivors in dyadic lifestyle interventions may benefit from directly involving clinicians in recruitment efforts. Though the randomized design helps ensure that there were no systematic differences in motivation between groups, the overall findings of the study may not generalize to the broader population of cancer survivors, many of whom may be less motivated to make healthy lifestyle changes. Generalizability may also be limited by the relatively homogeneous, sociodemographic profile of study participants, as most participants were non-Hispanic white and well-educated, and had an opposite sex partner/spouse. Future research should recruit more diverse samples; in particular the needs of couples who are not heterosexual or are from different racial/ethnic groups needs focused study. The study had a number of secondary outcomes, multiple testing, and potentially spurious significant differences, which is another important limitation. Finally, as the primary study purpose was to examine the feasibility of the lifestyle intervention, the lack of statistical power to detect differences between groups in outcome measures was a primary limitation. The promising feasibility metrics in adherence and retention we observed in this study, coupled with plans to enhance recruitment strategies to enroll a larger and potentially more generalizable population, lend promise to a large, impactful RCT examining differences in outcomes by intervention delivery strategies.

5. Conclusions

Few studies have incorporated spouses in behavioral interventions for cancer survivors, despite the importance of relationships between cancer survivors and their loved ones in survivorship and the potential to broaden the impact of positive behavior change [48]. Cancer diagnosis, treatment, and survivorship expose both parties of caregiving relationships to stressful events that can impact health and well-being. Given the many opportunities for important family decisions throughout cancer survivorship, these circumstances may be particularly opportune times for lifestyle interventions. The well-being of cancer survivors and their caregivers tend to covary over time throughout cancer treatment and survivorship, and the positive role modeling and social support for behavior change that may result from dyadic interventions can provide mutual benefits for both cancer survivors and their spouses in this context [27,49,50]. Spousal support plays an important role in diet, as couples generally rely on the same strategies for food procurement and preparation [51,52]. Improvements detected among both cancer survivors and spouses in the CB group may reflect this important dynamic of dietary habits for couples and highlights the window of opportunity to impact both members' diet with dyadic lifestyle interventions for cancer survivors. Study findings showing improvements in physical activity, physical performance, anthropometrics, and diet among cancer survivors in both groups and spouses in the CB group are promising, albeit preliminary. It will be important for a future trial to be powered to compare outcomes between couples randomized to receive the intervention together and separately. Overall, the findings from this study suggest that dyadic lifestyle multiple behavior change interventions are promising for both cancer survivors and their spouses, and they may provide a valuable strategy to broaden and deepen the impact of improving health during cancer survivorship.

Author Contributions: Conceptualization, C.L.C., W.D.-W. and K.B.-E.; methodology, C.L.C., W.D.-W. and K.B.-E.; software, G.B.; validation, N.H.P., Y.Y., K.B.-E. and G.B.; formal analysis, Y.Y., G.B. and K.B.-E.; investigation, C.L.C., W.D.-W., K.B.-E. and L.S.; resources, C.L.C., W.D.-W., K.B.-E., L.S., S.H.G., M.R.-B. and C.P.; data curation, N.H.P., G.B., L.S. and Y.Y.; writing—original draft preparation, C.L.C. and N.H.P.; writing—review and editing, C.L.C., N.H.P., W.D.-W., K.B.-E., L.S., S.H.G., M.R.-B. and C.P.; visualization, C.L.C., N.H.P., W.D.-W., K.B.-E., L.S., G.B. and Y.Y.; supervision, C.L.C.; project administration, C.L.C. and L.S.; funding acquisition, C.L.C., W.D.-W. and K.B.-E. All authors have read and agreed to the published version of the manuscript.

Funding: This research was supported by the National Cancer Institute (P30 CA016672), Multidisciplinary Research Program and Assessment, Intervention, and Measurement Shared Resource; and the Center for Energy Balance and Survivorship Research, Duncan Family Institute.

Institutional Review Board Statement: The study was conducted according to the guidelines of the Declaration of Helsinki, and approved by the Institutional Review Board of The University of Texas MD Anderson Cancer Center (protocol number 2009-0588 and approved on 25 February 2010).

Informed Consent Statement: Informed consent was obtained from all subjects involved in the study.

Data Availability Statement: Data supporting reported results can be accessed via the corresponding author.

Acknowledgments: We would like to thank all of the participants who generously devoted their time to this research.

Conflicts of Interest: The authors declare no conflict of interest.

References

1. National Academies of Sciences, Engineering, and Medicine. *Guiding Cancer Control: A Path to Transformation*; Nass, S.J., Amankwah, F.K., Madhavan, G., Johns, M.M.E., Eds.; National Academies Press: Washington, DC, USA, 2019.
2. National Cancer Institute. Statistics, Graphs and Definitions. Cancer Treatment & Survivorship Facts & Figures 2019–2021. Available online: https://cancercontrol.cancer.gov/ocs/statistics (accessed on 8 September 2021).
3. Han, X.; Robinson, L.A.; Jensen, R.E.; Smith, T.G.; Yabroff, K.R. Factors associated with health-related quality of life among cancer survivors in the United States. *JNCI Cancer Spectr.* **2021**, *5*, pkaa123. [CrossRef]
4. Office of Cancer Survivorship. Available online: http://cancercontrol.cancer.gov/ocs/prevalence (accessed on 29 May 2012).
5. American Cancer Society. *Cancer Facts and Figures 2011*; American Cancer Society: Atlanta, GA, USA, 2011.
6. de Moor, J.S.; Alfano, C.M.; Kent, E.E.; Norton, W.E.; Coughlan, D.; Roberts, M.C.; Grimes, M.; Bradley, C.J. Recommendations for research and practice to improve work outcomes among cancer survivors. *J. Natl Cancer Inst.* **2018**, *110*, 1041–1047. [CrossRef] [PubMed]
7. Siegel, R.L.; Miller, K.D.; Fuchs, H.E.; Jemal, A. Cancer Statistics, 2021. *CA Cancer J. Clin.* **2021**, *71*, 7–33. [CrossRef]
8. Solans, M.; Chan, D.S.M.; Mitrou, P.; Norat, T.; Romaguera, D. A systematic review and meta-analysis of the 2007 WCRF/AICR score in relation to cancer-related health outcomes. *Ann. Oncol.* **2020**, *31*, 352–368. [CrossRef]
9. Campbell, K.L.; Winters-Stone, K.M.; Wiskemann, J.; May, A.M.; Schwartz, A.L.; Courneya, K.S.; Zucker, D.S.; Matthews, C.E.; Ligibel, J.A.; Gerber, L.H.; et al. Exercise guidelines for cancer survivors: Consensus statement from international multidisciplinary roundtable. *Med. Sci. Sports Exerc.* **2019**, *51*, 2375–2390. [CrossRef]
10. Denlinger, C.S.; Ligibel, J.A.; Are, M.; Baker, K.S.; Demark-Wahnefried, W.; Dizon, D.; Friedman, D.L.; Goldman, M.; Jones, L.; King, A.; et al. Survivorship: Nutrition and weight management, Version 2.2014. Clinical practice guidelines in oncology. *J. Natl. Compr. Cancer Netw.* **2014**, *12*, 1396–1406. [CrossRef]
11. Meyerhardt, J.A.; Giovannucci, E.L.; Holmes, M.D.; Chan, A.T.; Chan, J.A.; Colditz, G.A.; Fuchs, C.S. Physical activity and survival after colorectal cancer diagnosis. *J. Clin. Oncol.* **2006**, *24*, 3527–3534. [CrossRef] [PubMed]
12. Swain, C.T.V.; Nguyen, N.H.; Eagles, T.; Vallance, J.K.; Boyle, T.; Lahart, I.M.; Lynch, B.M. Postdiagnosis sedentary behavior and health outcomes in cancer survivors: A systematic review and meta-analysis. *Cancer* **2020**, *126*, 861–869. [CrossRef]

13. Rock, C.L.; Doyle, C.; Demark-Wahnefried, W.; Meyerhardt, J.; Courneya, K.S.; Schwartz, A.L.; Bandera, E.V.; Hamilton, K.K.; Grant, B.; McCullough, M.; et al. Nutrition and physical activity guidelines for cancer survivors. *CA Cancer J. Clin.* **2012**, *62*, 243–274. [CrossRef]
14. Doyle, C.; Kushi, L.H.; Byers, T.; Courneya, K.S.; Demark-Wahnefried, W.; Grant, B.; McTiernan, A.; Rock, C.L.; Thompson, C.; Gansler, T.; et al. Nutrition and physical activity during and after cancer treatment: An American Cancer Society guide for informed choices. *CA Cancer J. Clin.* **2006**, *56*, 323–353. [CrossRef] [PubMed]
15. Byrd, D.A.; Agurs-Collins, T.; Berrigan, D.; Lee, R.; Thompson, F.E. Racial and ethnic differences in dietary intake, physical activity, and Body Mass Index (BMI) among cancer survivors: 2005 and 2010 National Health Interview Surveys (NHIS). *J. Racial Ethnic Health Dispar.* **2017**, *4*, 1138–1146. [CrossRef] [PubMed]
16. Coups, E.J.; Ostroff, J.S. A population-based estimate of the prevalence of behavioral risk factors among adult cancer survivors and noncancer controls. *Prev. Med.* **2005**, *40*, 702–711. [CrossRef] [PubMed]
17. Stacey, F.G.; James, E.L.; Chapman, K.; Courneya, K.S.; Lubans, D.R. A systematic review and meta-analysis of social cognitive theory-based physical activity and/or nutrition behavior change interventions for cancer survivors. *J. Cancer Surviv.* **2015**, *9*, 305–338. [CrossRef] [PubMed]
18. Goode, A.D.; Lawler, S.P.; Brakenridge, C.L.; Reeves, M.M.; Eakin, E.G. Telephone, print, and Web-based interventions for physical activity, diet, and weight control among cancer survivors: A systematic review. *J. Cancer Surviv.* **2015**, *9*, 660–682. [CrossRef] [PubMed]
19. Amireault, S.; Fong, A.J.; Sabiston, C.M. Promoting healthy eating and physical activity behaviors: A systematic review of multiple health behavior change interventions among cancer survivors. *Am. J. Lifestyle Med.* **2018**, *12*, 184–199. [CrossRef] [PubMed]
20. Bandura, A. *Social Foundations of thought and Action: A Social-Cognitive Theory*; Prentice-Hall: Englewood Cliffs, NJ, USA, 1986.
21. Bandura, A. *Self-Efficacy: The Exercise of Control*; W. H. Freeman and Company: New York, NY, USA, 1997.
22. Scott, J.L.; Halford, W.K.; Ward, B.G. United we stand? The effects of a couple-coping intervention on adjustment to early stage breast or gynecologic cancer. *J. Consult. Clin. Psychol.* **2004**, *72*, 1122–1135. [CrossRef] [PubMed]
23. Ellis, K.R.; Raji, D.; Olaniran, M.; Alick, C.; Nichols, D.; Allicock, M. A systematic scoping review of post-treatment lifestyle interventions for adult cancer survivors and family members. *J. Cancer Surviv.* **2021**. epub ahead of print. [CrossRef] [PubMed]
24. Tucker, J.S.; Mueller, J.S. Spouses' social control of health behaviors: Use and effectiveness of specific strategies. *Personal. Soc. Psychol. Bull.* **2000**, *26*, 1120–1130. [CrossRef]
25. Kushi, L.H.; Doyle, C.; McCullough, M.; Rock, C.L.; Demark-Wahnefried, W.; Bandera, E.V.; Gapstur, S.; Patel, A.V.; Andrews, K.; Gansler, T. American Cancer Society Guidelines on nutrition and physical activity for cancer prevention: Reducing the risk of cancer with healthy food choices and physical activity. *CA Cancer J. Clin.* **2012**, *62*, 30–67. [CrossRef] [PubMed]
26. Porter, L.S.; Gao, X.; Lyna, P.; Kraus, W.; Olsen, M.; Patterson, E.; Puleo, B.; Pollak, K.I. Pilot randomized trial of a couple-based physical activity videoconference intervention for sedentary cancer survivors. *Health Psychol.* **2018**, *37*, 861–865. [CrossRef]
27. Winters-Stone, K.M.; Lyons, K.S.; Dobek, J.; Dieckmann, N.F.; Bennett, J.A.; Nail, L.; Beer, T.M. Benefits of partnered strength training for prostate cancer survivors and spouses: Results from a randomized controlled trial of the Exercising Together project. *J. Cancer Surviv.* **2016**, *10*, 633–644. [CrossRef] [PubMed]
28. Snyder, D.C.; Morey, M.C.; Sloane, R.; Stull, V.; Cohen, H.J.; Peterson, B.; Pieper, C.; Hartman, T.J.; Miller, P.E.; Mitchell, D.C.; et al. Reach out to ENhancE Wellness in Older Cancer Survivors (RENEW): Design, methods and recruitment challenges of a home-based exercise and diet intervention to improve physical function among long-term survivors of breast, prostate, and colorectal cancer. *Psychooncology* **2009**, *18*, 429–439. [CrossRef] [PubMed]
29. Thomas, S.; Reading, J.; Shephard, R.J. Revision of the Physical Activity Readiness Questionnaire (PAR-Q). *Can. J. Sport Sci.* **1992**, *17*, 338–345.
30. Pocock, S.J. *Clinical Trials: A Practical Approach*; John Wiley & Sons: New York, NY, USA, 1983.
31. Meyler, D.; Stimpson, J.P.; Peek, M.K. Health concordance within couples: A systematic review. *Soc. Sci. Med.* **2007**, *64*, 2297–2310. [CrossRef] [PubMed]
32. Cox, M.; Basen-Engquist, K.; Carmack, C.L.; Blalock, J.; Li, Y.; Murray, J.; Pisters, L.; Rodriguez-Bigas, M.; Song, J.; Cox-Martin, E.; et al. Comparison of Internet and telephone interventions for weight loss among cancer survivors: Randomized controlled trial and feasibility study. *JMIR Cancer* **2017**, *3*, e16. [CrossRef]
33. Carmack Taylor, C.L.; Smith, M.A.; deMoor, C.; Dunn, A.L.; Pettaway, C.; Sellin, R.; Charnsangavej, C.; Hansen, M.; Gritz, E.R. Quality of life intervention for prostate cancer patients: Design and baseline characteristics of the *Active for Life* after cancer trial. *Control. Clin. Trials* **2004**, *25*, 265–285. [CrossRef]
34. Morey, M.; Sloane, R.; Snyder, D.C.; Cohen, H.J.; Peterson, B.; Miller, P.E.; Hartman, T.J.; Demark-Wahnefried, W. A home-based intervention among older long-term cancer survivors improves physical activity, diet quality, weight status, and physical functioning: Results of the RENEW (Reach-out to ENhancE Wellness) trial. In Proceedings of the American Association for Cancer Research; Frontiers in Cancer Prevention: Washington, DC, USA, 2008; p. 113.
35. Lichtenstein, A.H.; Brands, M.; Carnethon, M.; Daniels, S.; Franch, H.A.; Franklin, B.; Kris-Etherton, P.S.H.W.; Howard, B.; Karanja, N.; Howard, B.; et al. Diet and lifestyle recommendations revision 2006: A scientific statement for the American Heart Association Nutrition Committee. *Circulation* **2006**, *114*, 82–96. [CrossRef] [PubMed]

36. Prochaska, J.O.; Velicer, W.F.; Rossi, J.S.; Goldstein, M.G.; Marcus, B.H.; Rakowski, W.; Fiore, C.; Harlow, L.L.; Redding, C.A.; Rosenbloom, D.; et al. Stages of change and decisional balance for 12 problem behaviors. *Health Psychol.* **1994**, *13*, 39–46. [CrossRef]
37. Shephard, R. Godin leisure-time exercise questionnaire. *Med. Sci. Sports Exerc.* **1997**, *29*, S36–S38.
38. Basen-Engquist, K.; Carmack, C.L.; Li, Y.; Brown, J.; Jhingran, A.; Hughes, D.C.; Perkins, H.Y.; Scruggs, S.; Harrison, C.; Baum, G.; et al. Social-cognitive theory predictors of exercise behavior in endometrial cancer survivors. *Health Psychol.* **2013**, *32*, 1137–1148. [CrossRef] [PubMed]
39. Rikli, R.E.; Jones, C.J. The reliability and validity of a 6-minute walk test as a measure of physical endurance in older adults. *J. Aging Phys. Act.* **1999**, *6*, 363–375. [CrossRef]
40. Rikli, R.E.; Jones, J.C. Functional fitness normative scores for community-residing older adults, Ages 60–94. *J. Aging Phys. Act.* **1999**, *7*, 162–181. [CrossRef]
41. Rikli, R.E.; Jones, C.J. Development and validation of a functional fitness test for community-residing older adults. *J. Aging Phys. Act.* **1999**, *7*, 129–161. [CrossRef]
42. Podsiadlo, D.; Richardson, S. The timed "Up & Go": A test of basic functional mobility for frail elderly persons. *J. Am. Geriatr. Soc.* **1991**, *39*, 142–148.
43. National Cancer Institute. Risk Factor Monitoring and Methods. Available online: http://riskfactor.cancer.gov/tools/instruments/asa24 (accessed on 28 July 2009).
44. Cox-Martin, E.; Song, J.; Demark-Wahnefried, W.; Lyons, E.J.; Basen-Engquist, K. Predictors of enrollment in individual- and couple-based lifestyle intervention trials for cancer survivors. *Support Care Cancer* **2018**, *26*, 2387–2395. [CrossRef] [PubMed]
45. Demark-Wahnefried, W.; Jones, L.W.; Snyder, D.C.; Sloane, R.J.; Kimmick, G.G.; Hughes, D.C.; Badr, H.J.; Miller, P.E.; Burke, L.E.; Lipkus, I.M. Daughters and Mothers Against Breast Cancer (DAMES): Main outcomes of a randomized controlled trial of weight loss in overweight mothers with breast cancer and their overweight daughters. *Cancer* **2014**, *120*, 2522–2534. [CrossRef]
46. Milbury, K.; Mallaiah, S.; Lopez, G.; Liao, Z.; Yang, C.; Carmack, C.; Chaoul, A.; Spelman, A.; Cohen, L. Vivekananda yoga program for patients with advanced lung cancer and their family caregivers. *Integr. Cancer Ther.* **2015**, *14*, 446–451. [CrossRef] [PubMed]
47. Kamen, C.; Heckler, C.; Janelsins, M.C.; Peppone, L.J.; McMahon, J.M.; Morrow, G.R.; Bowen, D.; Mustian, K. A dyadic exercise intervention to reduce psychological distress among lesbian, gay, and heterosexual cancer survivors. *LGBT Health* **2016**, *3*, 57–64. [CrossRef]
48. Badr, H.; Bakhshaie, J.; Chhabria, K. Dyadic interventions for cancer survivors and caregivers: State of the science and new directions. *Semin. Oncol. Nurs.* **2019**, *35*, 337–341. [CrossRef]
49. Lyons, K.S.; Bennett, J.A.; Nail, L.M.; Fromme, E.K.; Dieckmann, N.; Sayer, A.G. The role of patient pain and physical function on depressive symptoms in couples with lung cancer: A longitudinal dyadic analysis. *J. Fam. Psychol.* **2014**, *28*, 692–700. [CrossRef]
50. Hagedoorn, M.; Sanderman, R.; Bolks, H.N.; Tuinstra, J.; Coyne, J.C. Distress in couples coping with cancer: A meta-analysis and criticial review of role and gender effects. *Psychol. Bull.* **2008**, *134*, 1–30. [CrossRef] [PubMed]
51. Beverly, E.A.; Miller, C.K.; Wray, L.A. Spousal support and food-related behavior change in middle-aged and older adults living with type 2 diabetes. *Health Educ. Behav.* **2008**, *35*, 707–720. [CrossRef] [PubMed]
52. Bovbjerg, V.E.; McCann, B.S.; Brief, D.J.; Follette, W.C.; Retzlaff, B.M.; Dowdy, A.A.; Walden, C.E.; Knopp, R.H. Spouse support and long-term adherence to lipid-lowering diets. *Am. J. Epidemiol.* **1995**, *141*, 451–460. [CrossRef] [PubMed]

MDPI

Article

WISER Survivor Trial: Combined Effect of Exercise and Weight Loss Interventions on Insulin and Insulin Resistance in Breast Cancer Survivors

Nicholas J. D'Alonzo [1], Lin Qiu [2], Dorothy D. Sears [3,4,5,6], Vernon Chinchilli [2], Justin C. Brown [7], David B. Sarwer [8], Kathryn H. Schmitz [2] and Kathleen M. Sturgeon [2,*]

1 College of Medicine, Pennsylvania State University, Hershey, PA 17033, USA; ndalonzo@pennstatehealth.psu.edu
2 Department of Public Health Sciences, College of Medicine, Pennsylvania State University, Hershey, PA 17033, USA; lqiu@pennstatehealth.psu.edu (L.Q.); vchinchilli@pennstatehealth.psu.edu (V.C.); kschmitz@phs.psu.edu (K.H.S.)
3 College of Health Solutions, Arizona State University, Phoenix, AZ 85004, USA; dorothy.sears@asu.edu
4 Department of Medicine, UC San Diego, La Jolla, CA 92093, USA
5 Department of Family Medicine, UC San Diego, La Jolla, CA 92093, USA
6 Moores Cancer Center, UC San Diego, La Jolla, CA 92037, USA
7 Cancer Metabolism Program, Pennington Biomedical Research Center, Louisiana State University, Baton Rouge, LA 70808, USA; justin.brown@pbrc.edu
8 Center for Obesity Research and Education, Temple University, Philadelphia, PA 19122, USA; dsarwer@temple.edu
* Correspondence: ksturgeon@phs.psu.edu; Tel.: +717-531-0003 (ext. 284676)

Citation: D'Alonzo, N.J.; Qiu, L.; Sears, D.D.; Chinchilli, V.; Brown, J.C.; Sarwer, D.B.; Schmitz, K.H.; Sturgeon, K.M. WISER Survivor Trial: Combined Effect of Exercise and Weight Loss Interventions on Insulin and Insulin Resistance in Breast Cancer Survivors. *Nutrients* **2021**, *13*, 3108. https://doi.org/10.3390/nu13093108

Academic Editors: Wendy Demark-Wahnefried and Christina Dieli-Conwright

Received: 16 August 2021
Accepted: 1 September 2021
Published: 4 September 2021

Abstract: Obesity-associated breast cancer recurrence is mechanistically linked with elevated insulin levels and insulin resistance. Exercise and weight loss are associated with decreased breast cancer recurrence, which may be mediated through reduced insulin levels and improved insulin sensitivity. This is a secondary analysis of the WISER Survivor clinical trial examining the relative effect of exercise, weight loss and combined exercise and weight loss interventions on insulin and insulin resistance. The weight loss and combined intervention groups showed significant reductions in levels of: insulin, C-peptide, homeostatic model assessment 2 (HOMA2) insulin resistance (IR), and HOMA2 beta-cell function (β) compared to the control group. Independent of intervention group, weight loss of ≥10% was associated with decreased levels of insulin, C-peptide, and HOMA2-IR compared to 0–5% weight loss. Further, the combination of exercise and weight loss was particularly important for breast cancer survivors with clinically abnormal levels of C-peptide.

Keywords: breast neoplasms; neoplasm recurrence; weight reduction program; resistance training; overweight; adiposity; caloric restriction; biomarkers

1. Introduction

Breast cancer mortality has significantly declined over the last two decades [1], leading to a growing population of breast cancer survivors. Breast cancer survivors comprise more than 50% of the 8.8 million female cancer survivors in the United States [2]. Unfortunately, the risk of breast cancer recurrence is 10–52% depending on tumor subtype and cancer stage [3]. As a result of the growing population of breast cancer survivors there has been an increase in focus on preventing breast cancer recurrence [4].

Obesity and physical inactivity are risk factors for breast cancer and breast cancer recurrence [5–7] with a 12% increase in risk of breast cancer diagnosis for every 5 unit increase in BMI [8]. At breast cancer diagnosis, 54–71% of women are overweight (BMI 25–30 kg/m^2) or have obesity (BMI > 30 kg/m^2) [9,10]. Further, weight gain after diagnosis is common, with the majority gaining weight during treatment [11]. Every five pounds gained after diagnosis is associated with a 12% and 13% increase of breast cancer-specific mortality and all-cause mortality, respectively [12]. One proposed mechanism linking obesity and breast cancer recurrence are elevated insulin levels and reduced insulin sensitivity [13,14]. Insulin resistance is associated with abdominal obesity, breast cancer specific and overall mortality [15,16], and increased breast cancer recurrence [17,18].

Exercise and weight loss reduces cancer and all-cause mortality in breast cancer patients [19–22]. It is currently unclear if exercise and weight loss interventions reliably improve metabolic markers and reduce insulin resistance in the breast cancer survivor population [23–27]. The WISER Survivor trial compared exercise training, weight loss and a combination of the two energetic interventions to a control group among breast cancer survivors with overweight or obesity [28,29]. In this secondary analysis, we have analyzed mechanistically relevant biomarkers. We hypothesized that the combined intervention group would lead to the greatest decrease, compared to control, in adiposity and concomitant beneficial changes in insulin, C-peptide, and insulin resistance.

2. Materials and Methods

2.1. Design

The WISER Survivor randomized controlled trial consisted of four intervention groups comparing the individual and combined effects of exercise and weight loss in breast cancer survivors with excess body weight and lymphedema. The complete study design, methods, and primary results have been published separately [28,29]. The 12-month intervention groups included: control (referred to American Cancer Society and/or their physician), exercise (weight training and aerobic exercise), weight loss (caloric restriction), and combined exercise and caloric restriction to promote weight loss.

2.2. Participants

Recruitment was conducted using local hospital and state tumor registries in the Philadelphia, PA, metropolitan area. All participants were breast cancer survivors with a BMI $\geq$ 25 kg/m^2 but <50 kg/m^2, younger than 80 years old, cancer free and having completed curative treatment more than 6 months before randomization, had breast cancer-related lymphedema, and sedentary lifestyle (assessed by self-report) prior to enrollment. Eligible participants had to be able to walk unaided for greater than 6 min. Exclusions included taking weight loss medication at the time of enrollment, weight loss greater than 4.5 kg in the previous 3 months, current engagement in moderate intensity exercise (e.g., bicycling or brisk walking) 3 or more times per week, weight training in the past year, and bariatric surgery. Additional recruitment methods and eligibility criteria have been published separately [30]. There were 351 women enrolled between 5 December 2011 and 21 April 2015 and all follow-up testing was performed by 28 May 2016. The primary aim was to evaluate the effects of these interventions on interlimb volume difference [29]. This report is a secondary analysis examining the effects of the trial on insulin-related biomarkers of breast cancer recurrence.

2.3. Exercise Intervention

All exercise sessions could be performed within the participant's home and weight-adjustable dumbbells were provided for the weight training component of the exercise intervention. Participants were asked to engage in two weight training sessions and 180 min of aerobic exercise per week. For the first six weeks, participants received in-person, on-site weekly instruction from certified fitness professionals that focused on the proper and safe execution and gradual increase of the prescribed resistance exercises. From weeks 7 to 52, participants received monthly in-person sessions at the study site, in addition to performing two weight training and 6 aerobic exercise sessions per week at home. Resistance for weight training exercises was gradually increased throughout the intervention. Aerobic exercise prescription remained constant the entire intervention. Behavioral counseling was included in the monthly sessions with the goal of maximizing adherence. All participants were asked to keep a log of their exercises performed. Exercise trainers called participants weekly to provide behavioral counseling, answer questions, and check on adherence.

2.4. Weight Loss Intervention

The first 24 weeks of the intervention included weekly group meetings and meals provided using NutriSystem® to promote weight loss. Meetings were designed to increase adherence through behavioral modification lessons with a different topic each week such as goal setting, problem solving, etc. During the first 20 weeks, daily caloric intake was restricted to 1200–1500 kcal/day through the provision of Nutrisystem® shelf stable meals and snacks. From weeks 20–24 participants were encouraged to transition to purchasing their own food from the grocery store while maintaining 1200–1500 kcal/day. During the following 28 weeks, participants increased their caloric intake to 1700–2000 kcal/day with the goal of maintaining the weight they had lost in the initial 24 weeks. In this 28-week period there were monthly group meetings and weekly individual calls with a registered dietitian.

2.5. Exercise and Weight Loss Intervention

Participants in this group engaged in the same exercise protocol as the exercise intervention group. After week 6, they continued the exercise protocol in addition to starting the weight loss program. Detailed study design has been published previously [28].

2.6. Control Group

Participants in the control group were directed to the American Cancer Society website for all diet-related questions. For exercise-related questions they were referred to their physician. Participants were asked to continue with the exercise regimen they had prior to enrollment in the study. However, all participants were sedentary (self-report) at study entry.

2.7. Biomarker Assays

Trained medical staff collected 12-h fasting blood samples at the baseline and 12-month follow-up clinic visit. EDTA plasma samples, aliquoted and stored at −80 °C until assay preparation were used for biomarker measures. Laboratory personnel were blinded to participants' study groups. Plasma samples were prepared and analyzed according to standard methods and quality control procedures. Fasting plasma glucose concentrations were measured using a glucose oxidase method (YSI 2900 Biochemistry Analyzer). Plasma insulin and C-peptide concentrations were determined using high-sensitivity immunoassays (Meso Scale Discovery, catalog #K15164C and # K151X5D, respectively). Intra-plate and inter-plate coefficients of variance (CV), respectively, were: insulin (3.5%, 6.5%), glucose (2.1%, 3.2%), and C-peptide (4.1%, 9.6%).

2.8. Measurements

Body weight was measured, for analytic purposes, at baseline and 12 months. Height was measured at baseline. Dual energy x-ray absorptiometry (DEXA) and blood draws were performed at baseline and 12 months. The treadmill exercise test was conducted according to the modified Bruce protocol [31]. Homeostatic model assessment (HOMA2) insulin resistance (IR) and beta-cell function (β) were calculated using the HOMA2 calculator released by the Diabetes Trials Unit, University of Oxford [32]. Insulin (pmol//L) and glucose (mmol/L) values were used to calculate HOMA2-IR and C-peptide values (nmol/L) and glucose (mmol/L) were used to calculate HOMA2-β.

2.9. Statistical Analysis

Two hundred and six of the 351 women who participated in the WISER Survivor trial were included in this analysis. Participants were excluded if baseline or follow-up biomarker data was not available (n = 127) (due to inadequate sample availability to assay, loss to follow up, or not meeting quality control), or self-reported as non-fasting prior to blood draw (n = 10), or if insulin, C-peptide or glucose fell outside of the usable range for the HOMA2 formulas (indicating a non-fasting blood draw, n = 8) (Figure 1). Of the 206 participants included, 199 completed baseline and follow-up DEXA scans, and 186 completed baseline and 12-month treadmill testing. The Shapiro–Wilk test was used to test for normality and logarithmic transformations were performed accordingly. The Kruskal–Wallis and Chi-Squared test were used to test for differences between groups at baseline. One-way ANOVA with post hoc Bonferroni-adjusted differences were compared between biomarkers at baseline. A multiple linear regression model was used to examine main effects of the intervention using log-transformed baseline biomarker of interest and use of glucose-related medication as covariates. A multiple linear regression model was used to examine differences between tertiles using log-transformed baseline biomarker of interest, age, intervention arm, glucose-related medication use, and race as covariates. A logistic regression model was used to determine odds ratios for impaired baseline C-peptide and glucose levels returning to normal at 12 months). Clinically impaired C-peptide values were defined as falling outside 0.78–1.89 ng/mL [33]. Clinically impaired fasting glucose levels were defined as ≥100 mg/dL [34]. The odds ratio represents the probability of returning to a normal C-peptide or glucose range at 12-months compared to the control group. Baseline biomarker of interest, age, change in fat mass, change in lean body mass and change in treadmill time were used as covariates. The datasets generated during and/or analyzed during the current study are available from the corresponding author on reasonable request.

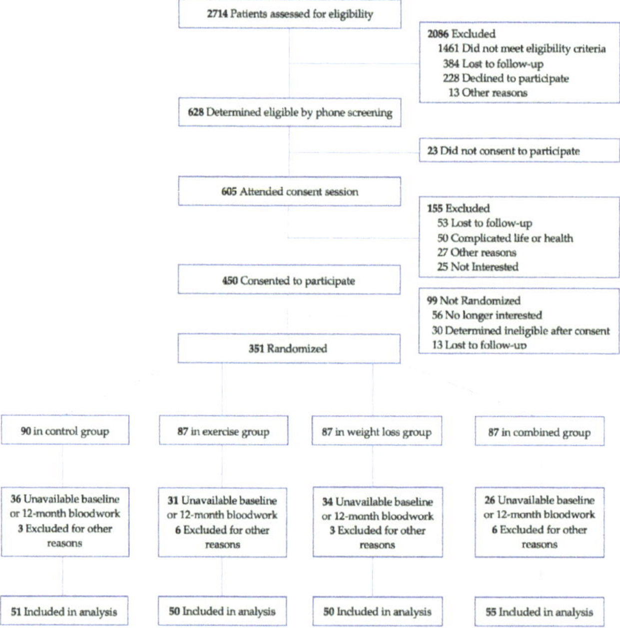

Figure 1. Two hundred and six of the 351 women who participated in the WISER Survivor trial were included in this analysis. Participants were excluded if baseline or follow-up biomarker data was not available (n = 127) (due to inadequate sample availability to assay, loss to follow up, or not meeting quality control), or self-reported as non-fasting prior to blood draw (n = 10), or if insulin, C-peptide or glucose fell outside of the usable range for the homeostatic model assessment 2 (HOMA2) formulas (indicating a non-fasting blood draw, n = 8).

3. Results

Participant characteristics are summarized in Table 1. The study population was 33% Black or other minority, on average with a BMI indicating overweight or obesity, greater than 5 years since diagnosis, and college educated. Forty-four percent of participants were on endocrine therapy and 8% were taking medication related to glucose management during the study. Adherence to aerobic exercise was reported at 145 ± 72 and 163 ± 94 min per week in the exercise group and combined group, respectively. Attendance at supervised exercise sessions was 85.8 ± 18% and 88.9 ± 29% in the exercise group and combined group, respectively.

Table 1. Demographic and clinical characteristics of the groups at baseline.

	Cohort *n* = 206	Control *n* = 51	Exercise *n* = 50	Weight Loss *n* = 50	Combined *n* = 55	*p* Value
Age, y	59.9 ± 8.9	60.5 ± 8.9	59.3 ± 8.8	59.7 ± 8.8	60.2 ± 9.2	0.90
BMI, kg/m^2	33.8 ± 5.9	33.3 ± 5.3	34.5 ± 7.0	34.1 ± 5.7	33.4 ± 5.8	0.86
Time since diagnosis, y	7.8 ± 5.3	8.8 ± 5.4	7.4 ± 5.3	7.6 ± 5.6	7.4 ± 4.9	0.41
Race						
Non-Hispanic White	138 (67)	37 (72.5)	33 (66.0)	34 (68.0)	34 (61.8)	
Black	61 (29.6)	12 (23.5)	16 (32.0)	16 (32.0)	17 (30.9)	
Other	7 (3.4)	2 (3.9)	1 (2.0)	0	4 (7.3)	0.45
Education						
High school diploma	36 (17.5)	11 (21.6)	7 (14.0)	5 (10.0)	13 (23.6)	
Some college	67 (32.5)	16 (31.4)	14 (28.0)	21 (42.0)	16 (29.1)	
College education	103 (50)	24 (47.1)	29 (58.0)	24 (48.0)	26 (47.3)	0.38
Endocrine therapy						
Aromatase inhibitors	70 (33)	17 (33.3)	16 (32.0)	19 (38.0)	18 (32.7)	0.92
Tamoxifen	20 (9.7)	4 (7.8)	8 (16.0)	5 (10.0)	3 (5.4)	0.31
Anti-Diabetes Medication	17 (8.2)	1 (2.0)	4 (8.0)	7 (14.0)	5 (9.1)	0.18

Data presented as mean ± SD or *n* (%).

3.1. Intervention Effects

Table 2 displays the baseline and follow up values of plasma biomarkers. There were no differences in levels between groups at baseline. At follow up, the weight loss and combined intervention groups experienced a significant decrease in insulin, C-peptide, HOMA2-IR, and HOMA2-β compared to the control group. Insulin decreased by 22.3% in the weight loss group and by 18.5% in the combined group. C-peptide decreased by 16.7% in the weight loss group, and by 13% in the combined group. HOMA2-IR decreased by 20% in the weight loss group, and by 16.7% in the combined group. HOMA2-β decreased by 3.2% in the exercise group, 2.2% in the weight loss group, and 5.6% in the combined group. The exercise group experienced a 4.5% increase in glucose.

Table 2. Main intervention effects on biomarkers of insulin sensitivity.

	Control	Exercise	Weight Loss	Combined
Insulin (uIU/mL)				
Baseline	17.8 ± 11.5	16.5 ± 10.2	17.5 ± 8.3	16.2 ± 11.4
12-Months	18.1 ± 10.6	16.8 ± 9.9	13.5 ± 7.3 [a]	13.2 ± 9.6 [a]
Change	0.3	0.3	−4.0	−3.0
C-peptide (ng/mL)				
Baseline	2.4 ± 1.0	2.3 ±1.0	2.4 ± 1.0	2.3 ± 1.2
12-Months	2.5 ± 0.9	2.5 ± 1.1	2.0 ± 0.9 [a]	2.0 ± 1.1 [a]
Change	0.1	0.2	−0.4	−0.3

Table 2. *Cont.*

	Control	Exercise	Weight Loss	Combined
Glucose (mg/dL)				
Baseline	103 ± 14.5	111 ± 31.7	110 ±14.9	102 ± 12.7
12-Months	99.7 ± 15.1	116 ± 40.6 [a]	106 ±17.5	101 ± 14.0
Change	−3.3	5.0	−4.0	−1.0
HOMA2-IR				
Baseline	2.0 ± 1.3	1.9 ± 1.2	2.0 ± 0.95	1.8 ± 1.3
12-Months	2.0 ± 1.2	2.0 ± 1.2	1.6 ± 0.8 [a]	1.5 ± 1.1 [a]
Change	0	0.1	−0.4	−0.3
HOMA2-β				
Baseline	110 ± 35.4	103 ± 33.8	97.2 ± 31.8	107 ± 38.1
12-Months	121 ± 38.1	99.7 ± 34.4 [a]	95.1 ± 32.6 [a]	101 ± 39.5 [a]
Change	11	−3.3	−2.1	−6.0

Data presented as mean ± SD. [a] $p < 0.05$. Homeostatic model assessment 2 (HOMA2) insulin resistance (IR), and HOMA2 beta-cell function (β).

3.1.1. Effects of Change in Body Composition and Treadmill Endurance

Table 3 displays the change in biomarker by categories of weight loss, and by tertiles of change in fat mass, lean mass, time on treadmill (i.e., fitness). Compared to those who lost between 0–5% of their weight, participants who lost ≥10% of their baseline weight experienced significant changes in insulin, C-peptide, and HOMA2-IR. Similarly, in tertiles 2 and 3 of change in fat mass (≥1.3 kg fat loss) had a significant change in insulin, C-peptide, and HOMA2-IR. Mean levels of lean mass change were 0.1 ± 1.8 kg, 0.41 ± 2.5 kg, −0.83 ± 3 kg, −1.2 ± 2.5 kg in the control, exercise, weight loss and combined intervention groups, respectively. The addition of the exercise intervention to caloric restriction did not mitigate the loss of lean mass. Participants in the upper tertile of change in lean mass (≥0.7 kg lean mass gained) experienced significantly less improvement in insulin, C-peptide, glucose, and HOMA2-IR, compared to participants that lost lean mass. Participants whom improved their treadmill test duration, "fitness capacity", by at least 31 s experienced significant decreases in insulin and HOMA-IR.

Table 3. Changes in biomarkers of insulin sensitivity among WISER Survivor participants stratified according to weight loss and tertiles of change in fat mass, change in lean mass, and change in aerobic fitness.

	Weight Loss (%) *		
	<0–5 (Reference)	≥5–10	≥10
n	70	30	47
Δ Insulin (uIU/mL)	−2.79 (8.5)	−3.16 (6.7)	−5.03 (7.7) [a]
Δ C-peptide (ng/mL)	−0.08 (0.7)	−0.17 (0.6)	−0.55 (0.7) [a]
Δ Glucose (mg/dL)	−0.20 (26.1)	0.07 (12.8)	−8.08 (11.6)
Δ HOMA2-IR	−0.31 (0.9)	−0.36 (0.8)	−0.60 (0.9) [a]
Δ HOMA2-β	1.20 (35.8)	−4.69 (23.4)	−4.35 (21.9)

Table 3. *Cont.*

	Δ Fat Mass (kg) **		
Tertile	1 (Reference)	2	3
Range (kg)	+15.5 to −1.2	−1.3 to −4.9	−5.0 to −30
Mean, [Median (SD)]	1.4, [0.95 (2.6)]	−2.7, [−2.6 (0.97)]	−9.1, [−7.7 (4.3)]
n	66	66	67
Δ Insulin (uIU/mL)	2.78 (7.9)	−2.39 (8.1) [a]	−5.52 (7.6) [a]
Δ C-peptide (ng/mL)	0.32 (0.7)	−0.08 (0.73) [a]	−0.53 (0.66) [a]
Δ Glucose (mg/dL)	3.82 (27.1)	0.15 (13.4)	−5.87 (12.3)
Δ HOMA2-IR	0.33 (0.91)	−0.26 (0.89) [a]	−0.65 (0.86) [a]
Δ HOMA2-β	5.45 (27.6)	−0.43 (35.1)	−6.33 (23.4) [a]
	Δ Lean Body Mass (kg) **		
Tertile	1 (Reference)	2	3
Range (kg)	−7.7 to −1.298	−1.297 to 0.68	0.7 to 13.3
Mean, [Median (SD)]	−3.1, [−2.8 (1.4)]	−0.4, [−0.4 (0.5)]	2.3, [1.8 (1.8)]
n	67	66	66
Δ Insulin (uIU/mL)	−3.18 (7.6)	−1.6 (9.1)	−0.38 (8.8) [a]
Δ C-peptide (ng/mL)	−0.26 (0.8)	−0.08 (0.8)	0.04 (0.7) [a]
Δ Glucose (mg/dL)	−6.95 (19.7)	2.16 (20.9) [a]	2.9 (15.1) [a]
Δ HOMA2-IR	−0.38 (0.9)	−0.17 (1.0)	−0.03 (1.0) [a]
Δ HOMA2-β	−0.26 (25.2)	−2.08 (25.3)	0.93 (36.6)
	Δ Time on Treadmill (s) ***		
Tertile	1 (Reference)	2	3
Range (sec)	−711 to −65	−64 to 28	31 to 752
Mean, [Median (SD)]	−194, [−166 (133)]	−13.4, [−0.5 (27.8)]	163, [117 (151)]
n	62	62	62
Δ Insulin (uIU/mL)	0.11 (8.6)	−1.67 (9.5)	−4.17 (7.3) [a]
Δ C-peptide (ng/mL)	−0.03 (0.7)	−0.07 (0.9)	−0.26 (0.6)
Δ Glucose (mg/dL)	−0.43 (14.7)	−0.76 (27.0)	−1.57 (13.4)
Δ HOMA2-IR	0.01 (1.0)	−0.19 (1.0)	−0.48 (0.1) [a]
Δ HOMA2-β	2.54 (26.7)	0.20 (34.9)	−4.65 (22.1)

Data presented as mean ± SD. [a] $p < 0.05$. * Participants (n = 59) who gained weight between baseline and 12-months were excluded. ** Participants (n = 7) without baseline and 12-month DEXA measurements were excluded. *** Participants (n = 20) without baseline and 12-month treadmill testing were excluded. (n = 20).

3.1.2. Normalization of C-Peptide

We observed that only the combined intervention group significantly increased odds of improving C-peptide levels from clinically impaired, to normal (Table 4). The odds (95% CI) of returning to normal C-peptide range for the exercise was 2.3 (0.8–6.6), 2.9 (1.0–9.0) in the weight loss group and 4.5 (1.4–14.1) for the combined group. Incidence of impaired C-peptide at follow up for participants who had normal baseline values was 40% for control, 25% for the exercise group, 12.5% for the weight loss group and 13.6% in the combined group. A regression to the mean analysis for impaired fasting glucose (>100 mg/dL) at baseline demonstrated that none of the intervention groups significantly increased the odds of returning to a healthy fasting glucose (<100 mg/dL) at 12 months.

Table 4. Odds ratio (95% CI) for C-peptide and glucose changing from clinically abnormal at baseline to normal at 12-month follow-up. Normal C-peptide and glucose range were 0.78–1.89 ng/mL and <100 mg/dL, respectively.

	Adjusted Model C-Peptide	Adjusted Model Glucose
Control	1.0 $n = 33^{b}, 37^{c}$	1.0 $n = 28^{b}, 25^{c}$
Exercise	2.3 (0.8–6.6) $n = 32^{b}, 29^{c}$	0.44 (0.2–1.2) $n = 27^{b}, 32^{c}$
Diet	2.9 (1.0–9.0) $n = 32^{b}, 23^{c}$	0.86 (0.3–2.4) $n = 39^{b}, 31^{c}$
Combined	4.5 (1.4–14.1) [a] $n = 31^{b}, 23^{c}$	0.75 (0.3–2.2) $n = 29^{b}, 28^{c}$

Data presented as odds ratio (95%, CI). [a] $p < 0.05$. [b] = number of participants abnormal at baseline. [c] = number of participants abnormal at 12-month.

4. Discussion

We observed that, compared to the control condition, weight loss with or without exercise led to significant reductions in insulin and insulin resistance. Insulin and associated metabolic pathways are associated with breast cancer recurrence and hypothesized to be a mechanistic driver of cancer. Obesity and lack of physical activity are common modifiable risk factors amongst breast cancer survivors and are strongly associated with hyperinsulinemia and insulin resistance. Using lifestyle modification in the form of exercise and weight loss, survivors can reduce insulin and insulin resistance through altered body composition. With an increasing number of breast cancer survivors, an increased emphasis on lifestyle modification to reduce recurrence and the sequela of breast cancer and breast cancer treatment is warranted. The results of this study demonstrate that reduction of insulin levels and increased insulin sensitivity is more effectively accomplished with a weight loss intervention than an exercise intervention when compared to the control. However, for participants with clinically impaired C-peptide levels, a combination of exercise and weight loss may be necessary to normalize C-peptide levels after taking into account changes in body composition and age.

We observed decreased insulin, C-peptide, HOMA2-IR, and HOMA2-β levels with a weight loss intervention and a combined weight loss and exercise intervention. Previous studies utilizing combined exercise and weight loss interventions in breast cancer survivors have observed similar results [26,27]. Unlike prior studies, this study isolates the individual and combined effects of exercise and weight loss to demonstrate that weight loss alone, or in combination with exercise, is more effective at improving biomarkers of insulin levels and insulin sensitivity than an exercise-only intervention.

The exercise prescription in the WISER Survivor trial was prescribed with specificity for lymphedema outcomes. We did not observe any change in insulin levels or insulin resistance in the exercise group. Yet, change has been observed in other trials using supervised aerobic and strength training interventions [25,35]. Thus, it is possible the exercise prescription was not appropriate for outcomes related to insulin levels and insulin resistance or the lack of supervision during the aerobic exercise portion may have led to bias in reporting the amount of aerobic exercise completed. Additionally, given that the upper tertile of fitness capacity in this study was set at an increase of 31 s on the Modified Bruce Treadmill protocol, it is likely that while participants may have increased their step count and physical activity minutes, they did not increase their exercise intensity and thus fitness capacity. Indeed, we observed that participants in the upper tertile for increased fitness capacity significantly decreased both fasting insulin and HOMA2-IR. Although not significant, C-peptide, fasting glucose and HOMA2-B did trend towards significant with increasing fitness capacity. This suggests that these biomarkers can be significantly decreased with an exercise intervention prescribed to increase fitness capacity via increased exercise intensity, or, a supervised exercise program. However, we cannot ignore that weight loss and increased fitness capacity are coupled and more work needs to be done to assess the ability of exercise, independent of weight loss, to improve insulin sensitivity in this patient population.

Independent of intervention group, we observed that a 10% or greater weight loss improves biomarkers of insulin sensitivity and insulin resistance. Our results align with Fabian et al. and their 6-month combined exercise and weight loss intervention that demonstrated weight loss of >10% resulted in improvements in serum and breast tissue biomarkers, including insulin levels, compared to <10% weight loss [36]. We demonstrate that exercise may not be necessary to achieve this 10% weight loss or significantly improve insulin levels or insulin sensitivity. Further assessment of body composition indicates that weight loss specific to fat mass (>1.3 kg) was sufficient to improve levels of insulin, C-peptide, and HOMA2-IR. However, 5.0 kg or more of fat mass loss was required for a significant decrease in HOMA2-β. The decrease in HOMA2-β indicates a decrease in beta cell function, however, it is likely that beta cell function is not decreased, rather, the decrease in insulin resistance is driving HOMA2-β down. Decreased adiposity lowers insulin resistance. This means that the beta cells of the pancreas are simply producing less insulin because the lowered insulin resistance has decreased the demand for insulin production.

Lifestyle interventions that improve body habitus alter biomarkers of insulin and insulin resistance, which are mechanistic contributors to breast cancer pathogenesis [37]. This can explain why women who engage in these interventions have better outcomes with respect to breast cancer survival [20,21], adverse sequelae of cancer/cancer treatment [19], and recurrence [18]. This study reinforces the current association between markers of insulin resistance and body composition.

When implementing lifestyle interventions, it is important to establish behavior change goals that are both attainable and have significant health benefits. Our results suggest that while any percentage of weight loss improves markers of insulin resistance, attaining a >10% weight loss is key. Unfortunately, weight loss of this size is at the upper range of what can be expected for most individuals treated with lifestyle modification or an anti-obesity medication. Additionally, the improvement in markers of insulin resistance in this patient population is tightly coupled with the magnitude of change in body composition. The most effective way to alter body composition and improve insulin resistance is through a combined intervention of caloric restriction and exercise. Indeed, for participants with impaired C-peptide levels, the combination intervention was the only intervention arm to return a significant number of participants to normal C-peptide levels. Fasting glucose levels did not respond to the interventions in the same way as insulin, C-peptide and HOMA-IR. Unlike the exercise, weight loss and combined interventions by Mason et al., our interventions were ineffective at returning impaired fasting glucose levels to normal levels [38]. The differences between the Mason et al. trial and the WISER Survivor study are that the Mason et al. trial diet intervention had a target weight loss of 10% and the exercise intervention had more supervision compared to the WISER Survivor trial exercise program. These differences may account for the lack of improvement in fasting glucose levels in our study population.

A strength of the WISER Survivor Trial was the diversity of the cohort. We were successful in recruiting a cohort with the largest number of Black breast cancer survivors reported to date [30]. The study sample is highly representative of the general US population at present. Additionally, the use of the 2 × 2 factorial design to examine relative effects of exercise and weight loss, or their combination on biomarkers of breast cancer recurrence (insulin, C-peptide, glucose, and clinically relevant indices such as HOMA2-IR and HOMA2-β) was novel. However, the WISER Survivor Trial was designed for a primary outcome associated with lymphedema. The exercise prescription was tailored specifically for slow progressive resistance training and a recommendation for 180 min per week of aerobic exercise training. The specificity of the WISER Survivor Trial exercise prescription is therefore a limitation for assessment of changes in insulin biomarkers. Yet, for many breast cancer survivors, lymphedema and insulin resistance may be co-occurring. Thus, the collective observations from the WISER Survivor Trial are important for both breast cancer survivors, clinicians, and exercise physiologists.

Author Contributions: Conceptualization, N.J.D. and K.M.S.; methodology, N.J.D., K.M.S., L.Q., V.C. and D.D.S.; formal analysis, L.Q. and V.C.; investigation, K.M.S. and K.H.S.; data curation, K.M.S. and N.J.D.; writing original draft, N.J.D., review and edits K.M.S., D.D.S., J.C.B., D.B.S. and K.H.S.; visualization, N.J.D. and K.M.S.; supervision, K.M.S.; project administration, K.M.S.; funding acquisition, K.H.S. and K.M.S. All authors have read and agreed to the published version of the manuscript.

Funding: This work was supported by, in part, by the National Cancer Institute of the National Institutes of Health under Award Numbers U54-CA155850, U54-CA155435, UL1-TR001878, P30-CA016520. Dr. Sturgeon is supported by, grant 5UL1TR002014 and 5KL2TR002015 from the National Center for Advancing Translational Sciences. Compression garments were donated by BSN Medical, and discounted meal replacements were provided by Nutrisystem, Inc. (Philadelphia, PA, USA).

Institutional Review Board Statement: The study was conducted according to the guidelines of the Declaration of Helsinki approved by the Institutional Review Board of The University of Pennsylvania. IRB project identification code is #812688 and was approved on 2 October 2011. This secondary analysis did not require additional IRB approval as all patient information was deidentified throughout the entire process. The randomized control trial registration number is NCT01515124.

Informed Consent Statement: Informed consent was obtained from all subjects involved in the study.

Data Availability Statement: All data used in this study can be made publicly available upon reasonable request.

Conflicts of Interest: Schmitz reports receiving nonfinancial support from BSN Medical, personal fees from Klose Training, and a licensed patent for a Strength after Breast Cancer course. No other disclosures were reported. Sarwer has consulting relationships with Ethicon and NovoNordisk which are unrelated to the study.

References

1. Ganz, P.A.; Goodwin, P.J. Breast Cancer Survivorship: Where Are We Today? *Stud. Biomark. New Targets Aging Res. Iran* **2015**, *862*, 1–8. [CrossRef]
2. Miller, K.D.; Nogueira, L.; Mariotto, A.B.; Rowland, J.H.; Yabroff, K.R.; Alfano, C.M.; Jemal, A.; Kramer, J.L.; Siegel, R.L. Cancer treatment and survivorship statistics, 2019. *CA Cancer J. Clin.* **2019**, *69*, 363–385. [CrossRef]
3. Colleoni, M.; Sun, Z.; Price, K.N.; Karlsson, P.; Forbes, J.F.; Thürlimann, B.; Gianni, L.; Castiglione, M.; Gelber, R.D.; Coates, A.S.; et al. Annual Hazard Rates of Recurrence for Breast Cancer During 24 Years of Follow-Up: Results From the Interna-tional Breast Cancer Study Group Trials I to V. *J. Clin. Oncol.* **2016**, *34*, 927–935. [CrossRef] [PubMed]
4. Chalasani, P.; Downey, L.; Stopeck, A.T. Caring for the Breast Cancer Survivor: A Guide for Primary Care Physicians. *Am. J. Med.* **2010**, *123*, 489–495. [CrossRef] [PubMed]
5. Ewertz, M.; Jensen, M.-B.; Gunnarsdóttir, K.Á.; Højris, I.; Jakobsen, E.H.; Nielsen, D.; Stenbygaard, L.E.; Tange, U.B.; Cold, S. Effect of Obesity on Prognosis After Early-Stage Breast Cancer. *J. Clin. Oncol.* **2011**, *29*, 25–31. [CrossRef]
6. Bergom, C.; Kelly, T.; Bedi, M.; Saeed, H.; Prior, P.; Rein, L.E.; Szabo, A.; Wilson, J.F.; Currey, A.D.; White, J. Association of Locoregional Control With High Body Mass Index in Women Undergoing Breast Conservation Thera-py for Early-Stage Breast Cancer. *Int. J. Radiat. Oncol. Biol. Phys.* **2016**, *96*, 65–71. [CrossRef]
7. Pizot, C.; Boniol, M.; Mullie, P.; Koechlin, A.; Boniol, M.; Boyle, P.; Autier, P. Physical activity, hormone replacement therapy and breast cancer risk: A meta-analysis of prospective studies. *Eur. J. Cancer* **2016**, *52*, 138–154. [CrossRef]
8. Renehan, A.G.; Tyson, M.; Egger, M.; Heller, R.F.; Zwahlen, M. Body-mass index and incidence of cancer: A systematic review and meta-analysis of prospective observational studies. *Lancet* **2008**, *371*, 569–578. [CrossRef]
9. Crozier, J.A.; Crozier, J.A.; Moreno-Aspitia, A.; Ballman, K.V.; Dueck, A.C.; Pockaj, B.A.; Perez, E.A. Effect of body mass index on tumor characteristics and disease-free survival in patients from the HER2-positive ad-juvant trastuzumab trial N9831. *Cancer* **2013**, *119*, 2447–2454. [CrossRef]
10. Cleveland, R.J.; Eng, S.M.; Abrahamson, P.E.; Britton, J.A.; Teitelbaum, S.L.; Neugut, A.I.; Gammon, M.D. Weight Gain Prior to Diagnosis and Survival from Breast Cancer. *Cancer Epidemiol. Biomark. Prev.* **2007**, *16*, 1803–1811. [CrossRef]
11. Vance, V.; Hanning, R.; Mourtzakis, M.; McCargar, L. Weight gain in breast cancer survivors: Prevalence, pattern and health consequences. *Obes. Rev.* **2010**, *12*, 282–294. [CrossRef] [PubMed]

12. Nichols, H.B.; Trentham-Dietz, A.; Egan, K.M.; Titus-Ernstoff, L.; Holmes, M.D.; Bersch, A.J.; Holick, C.N.; Hampton, J.M.; Stampfer, M.J.; Willett, W.C.; et al. Body Mass Index Before and After Breast Cancer Diagnosis: Associations with All-Cause, Breast Cancer, and Cardiovascular Disease Mortality. *Cancer Epidemiol. Biomark. Prev.* **2009**, *18*, 1403–1409. [CrossRef]
13. Van Kruijsdijk, R.C.M.; Van Der Wall, E.; Visseren, F. Obesity and Cancer: The Role of Dysfunctional Adipose Tissue. *Cancer Epidemiol. Biomark. Prev.* **2009**, *18*, 2569–2578. [CrossRef] [PubMed]
14. Rose, D.P.; Haffner, S.M.; Baillargeon, J. Adiposity, the Metabolic Syndrome, and Breast Cancer in African-American and White American Women. *Endocr. Rev.* **2007**, *28*, 763–777. [CrossRef] [PubMed]
15. Duggan, C.; Irwin, M.L.; Xiao, L.; Henderson, K.D.; Smith, A.W.; Baumgartner, R.N.; Baumgartner, K.B.; Bernstein, L.; Ballard-Barbash, R.; McTiernan, A. Associations of Insulin Resistance and Adiponectin With Mortality in Women With Breast Cancer. *J. Clin. Oncol.* **2011**, *29*, 32–39. [CrossRef]
16. Pan, K.; Chlebowski, R.T.; Mortimer, J.E.; Gunther, M.J.; Rohan, T.; Vitolins, M.Z.; Adams-Campbell, L.L.; Ho, G.Y.F.; Cheng, T.D.; Nelson, R.A. Insulin resistance and breast cancer incidence and mortality in postmenopausal women in the Women's Health Initiative. *Cancer* **2020**, *126*, 3638–3647. [CrossRef] [PubMed]
17. Pollak, M.N.; Chapman, J.W.; Shepherd, L.; Meng, D.; Richardson, P.; Orme, C.W.; Pritchard, K.I. Insulin resistance, estimated by serum C-peptide level, is associated with reduced event-free survival for postmeno-pausal women in NCIC CTG MA.14 adjuvant breast cancer trial. *J. Clin. Oncol.* **2006**, *24*, 524. [CrossRef]
18. Formica, V.; Tesauro, M.; Cardillo, C.; Roselli, M. Insulinemia and the risk of breast cancer and its relapse. *Diabetes Obes. Metab.* **2012**, *14*, 1073–1080. [CrossRef]
19. Kim, J.; Choi, W.J.; Jeong, S.H. The Effects of Physical Activity on Breast Cancer Survivors after Diagnosis. *J. Cancer Prev.* **2013**, *18*, 193–200. [CrossRef]
20. Holmes, M.D. Physical Activity and Survival after Breast Cancer Diagnosis. *JAMA* **2005**, *293*, 2479–2486. [CrossRef]
21. Schmitz, K.H.; Campbell, A.M.; Stuiver, M.M.; Pinto, B.M.; Schwartz, A.L.; Morris, G.S.; Ligibel, J.A.; Cheville, A.; Galvão, D.A.; Alfano, C.M.; et al. Exercise is medicine in oncology: Engaging clinicians to help patients move through cancer. *CA A Cancer J. Clin.* **2019**, *69*, 468–484. [CrossRef]
22. Maliniak, M.L.; Patel, A.V.; McCullough, M.L.; Campbell, P.T.; Leach, C.R.; Gapstur, S.M.; Gaudet, M.M. Obesity, physical activity, and breast cancer survival among older breast cancer survivors in the Cancer Preven-tion Study-II Nutrition Cohort. *Breast Cancer Res. Treat.* **2018**, *167*, 133–145. [CrossRef]
23. Shaikh, H.; Bradhurst, P.; Ma, L.X.; Tan, S.Y.; Egger, S.J.; Vardy, J.L. Body weight management in overweight and obese breast cancer survivors. *Cochrane Database Syst. Rev.* **2020**, *2020*, CD012110. [CrossRef]
24. Viskochil, R.; Blankenship, J.M.; Makari-Judson, G.; Staudenmayer, J.; Freedson, P.S.; Hankinson, S.E.; Braun, B. Metrics of Diabetes Risk Are Only Minimally Improved by Exercise Training in Postmenopausal Breast Cancer Survivors. *J. Clin. Endocrinol. Metab.* **2020**, *105*, e1958–e1966. [CrossRef] [PubMed]
25. Chang, J.S.; Kim, T.H.; Kong, I.D. Exercise intervention lowers aberrant serum WISP-1 levels with insulin resistance in breast cancer survivors: A randomized controlled trial. *Sci. Rep.* **2020**, *10*, 10898. [CrossRef]
26. Travier, N.; Buckland, G.; Vendrell, J.J.; Fernandez-Veledo, S.; Peiró, I.; Del Barco, S.; Pernas, S.; Zamora, E.; Bellet, M.; Margeli, M.; et al. Changes in metabolic risk, insulin resistance, leptin and adiponectin following a lifestyle intervention in overweight and obese breast cancer survivors. *Eur. J. Cancer Care* **2018**, *27*, e12861. [CrossRef]
27. Dittus, K.L.; Harvey, J.R.; Bunn, J.Y.; Kokinda, N.D.; Wilson, K.M.; Priest, J.; Pratley, R.E. Impact of a behaviorally-based weight loss intervention on parameters of insulin resistance in breast cancer survivors. *BMC Cancer* **2018**, *18*, 351. [CrossRef]
28. Winkels, R.M.; Sturgeon, K.M.; Kallan, M.J.; Dean, L.T.; Zhang, Z.; Evangelisti, M.; Brown, J.C.; Sarwer, D.B.; Troxel, A.B.; Denlinger, C.; et al. The women in steady exercise research (WISER) survivor trial: The innovative transdisciplinary design of a ran-domized controlled trial of exercise and weight-loss interventions among breast cancer survivors with lymphedema. *Contemp. Clin. Trials* **2017**, *61*, 63–72. [CrossRef]
29. Schmitz, S.H.; Troxel, A.B.; Dean, L.T.; DeMichele, A.; Brown, J.C.; Sturgeon, K.; Zhang, Z.; Evangelisti, M.; Spinelli, B.; Kallan, M.J.; et al. Effect of Home-Based Exercise and Weight Loss Programs on Breast Cancer-Related Lymphedema Outcomes Among Overweight Breast Cancer Survivors: The WISER Survivor Randomized Clinical Trial. *JAMA Oncol.* **2019**, *5*, 1605–1613. [CrossRef] [PubMed]
30. Sturgeon, K.M.; Bs, R.H.; Fornash, A.; Dean, L.T.; Laudermilk, M.; Brown, J.C.; Sarwer, D.B.; DeMichele, A.M.; Troxel, A.B.; Schmitz, K.H. Strategic recruitment of an ethnically diverse cohort of overweight survivors of breast cancer with lymphedema. *Cancer* **2017**, *124*, 95–104. [CrossRef] [PubMed]
31. Bruce, R.A.; Kusumi, F.; Hosmer, D. Maximal oxygen intake and nomographic assessment of functional aerobic impairment in car-diovascular disease. *Am. Heart J.* **1973**, *85*, 546–562. [CrossRef]
32. Wallace, T.M.; Levy, J.; Matthews, D.R. Use and Abuse of HOMA Modeling. *Diabetes Care* **2004**, *27*, 1487–1495. [CrossRef] [PubMed]
33. Pagana, K.D.; Pagana, T.J.; Pagana, T. *Mosby's Diagnostic & Laboratory Test Reference*, 14th ed.; Mosby: St. Louis, MO, USA, 2018.

34. The Expert Committee on the Diagnosis and Classification of Diabetes Mellitus. Follow-up report on the diagnosis of diabetes mellitus. *Diabetes Care* **2003**, *26*, 3160–3167. [CrossRef] [PubMed]
35. Ligibel, J.A.; Campbell, N.; Partridge, A.; Chen, W.Y.; Salinardi, T.; Chen, H.; Adloff, K.; Keshaviah, A.; Winer, E.P. Impact of a Mixed Strength and Endurance Exercise Intervention on Insulin Levels in Breast Cancer Survivors. *J. Clin. Oncol.* **2008**, *26*, 907–912. [CrossRef]
36. Fabian, C.J.; Kimler, B.F.; Donnelly, J.E.; Sullivan, D.K.; Klemp, J.R.; Petroff, B.K.; Phillips, T.A.; Metheny, T.; Aversman, S.; Yeh, H.-W.; et al. Favorable modulation of benign breast tissue and serum risk biomarkers is associated with >10% weight loss in postmenopausal women. *Breast Cancer Res. Treat.* **2013**, *142*, 119–132. [CrossRef] [PubMed]
37. Yee, L.D.; Mortimer, J.E.; Natarajan, R.; Dietze, E.C.; Seewaldt, V.L. Metabolic Health, Insulin, and Breast Cancer: Why Oncologists Should Care About Insulin. *Front. Endocrinol.* **2020**, *11*, 58. [CrossRef]
38. Mason, C.; Foster-Schubert, K.E.; Imayama, I.; Kong, A.; Xiao, L.; Bain, C.; Campbell, K.L.; Wang, C.-Y.; Duggan, C.; Ulrich, C.M.; et al. Dietary Weight Loss and Exercise Effects on Insulin Resistance in Postmenopausal Women. *Am. J. Prev. Med.* **2011**, *41*, 366–375. [CrossRef]

Article

Dietary Supplement Use and Interactions with Tamoxifen and Aromatase Inhibitors in Breast Cancer Survivors Enrolled in Lifestyle Interventions

Maura Harrigan [1,*], Courtney McGowan [1], Annette Hood [2], Leah M. Ferrucci [1,2], ThaiHien Nguyen [1], Brenda Cartmel [1,2], Fang-Yong Li [1], Melinda L. Irwin [1,2] and Tara Sanft [2,3]

1 Yale School of Public Health, Yale University, New Haven, CT 06510, USA; Courtney.McGowan@yale.edu (C.M.); leah.ferrucci@yale.edu (L.M.F.); Thaihien.nguyen@aya.yale.edu (T.N.); brenda.cartmel@yale.edu (B.C.); fangyong.li@yale.edu (F.-Y.L.); melinda.irwin@yale.edu (M.L.I.)
2 Yale Cancer Center, New Haven, CT 06510, USA; Annette.Hood@yale.edu (A.H.); tara.sanft@yale.edu (T.S.)
3 Yale School of Medicine, Yale University, New Haven, CT 06510, USA
* Correspondence: maura.harrigan@yale.edu

Abstract: The use of dietary supplements is common in the general population and even more prevalent among cancer survivors. The World Cancer Research Fund/American Institute for Cancer Research specifies that dietary supplements should not be used for cancer prevention. Several dietary supplements have potential pharmacokinetic and pharmacodynamic interactions that may change their clinical efficacy or potentiate adverse effects of the adjuvant endocrine therapy prescribed for breast cancer treatment. This analysis examined the prevalence of self-reported dietary supplement use and the potential interactions with tamoxifen and aromatase inhibitors (AIs) among breast cancer survivors enrolled in three randomized controlled trials of lifestyle interventions conducted between 2010 and 2017. The potential interactions with tamoxifen and AIs were identified using the Natural Medicine Database. Among 475 breast cancer survivors (2.9 (mean) or 2.5 (standard deviation) years from diagnosis), 393 (83%) reported using dietary supplements. A total of 108 different types of dietary supplements were reported and 36 potential adverse interactions with tamoxifen or AIs were identified. Among the 353 women taking tamoxifen or AIs, 38% were taking dietary supplements with a potential risk of interactions. We observed a high prevalence of dietary supplement use among breast cancer survivors and the potential for adverse interactions between the prescribed endocrine therapy and dietary supplements was common.

Keywords: dietary supplements; interactions; tamoxifen; aromatase inhibitors; breast cancer survivors; natural medicine

Citation: Harrigan, M.; McGowan, C.; Hood, A.; Ferrucci, L.M.; Nguyen, T.; Cartmel, B.; Li, F.-Y.; Irwin, M.L.; Sanft, T. Dietary Supplement Use and Interactions with Tamoxifen and Aromatase Inhibitors in Breast Cancer Survivors Enrolled in Lifestyle Interventions. *Nutrients* **2021**, *13*, 3730. https://doi.org/10.3390/nu13113730

Academic Editor: Tilman Kühn

Received: 6 September 2021
Accepted: 19 October 2021
Published: 22 October 2021

1. Introduction

Dietary supplements—defined by the Dietary Supplement Health and Education Act of 1994 [1] as herbal preparations, vitamins, and minerals—are commonly used, with 51% of U.S. adults using at least one dietary supplement [2]. Among cancer survivors, dietary supplement use is even more prevalent, with NHANES 2003–2016 data indicating a 70% use among this population [2]. Despite this high prevalence of use among cancer survivors, the joint World Cancer Research Fund (WCRF)/American Institute for Cancer Research (AICR) diet and exercise recommendations for cancer prevention state that "dietary supplements should not be used for cancer prevention" [3]. Supplement use for cancer prevention has not been shown to improve outcomes [4]. Furthermore, the use of dietary supplements is not associated with any improvement in the overall survival of cancer patients [5]. Additional cancer-specific nutrition guidelines recommend that supplements should not be used by cancer survivors for cancer prevention [6].

"Stacking" is a term used to describe a form of usage where multiple dietary supplements are consumed daily (for example, one patient may take vitamin D as a single nutrient, take a calcium plus vitamin D supplement (e.g., *Citracal*®), and also take a multivitamin that includes vitamin D). This makes assessing dietary supplement usage a challenge. because it can result in large combinations of nutrients from different products, especially when multivitamins and multiminerals are combined with single-nutrient dietary supplements. Assessing the nutrient exposures caused by a product can be complicated. For example, if a patient takes an herbal preparation with nine different nutrients in the ingredients, the potential for interactions would need to be evaluated for each nutrient [7].

Stacking and high nutrient exposure can lead to safety concerns. Many dietary supplements carry the potential risk of pharmacokinetic and pharmacodynamic interactions when taken with prescribed medications. For example, studies with tamoxifen taken alongside approved antidepressant medications that are CYP2D6 inhibitors demonstrate decreased levels of endoxifen, the active metabolite of tamoxifen [8]. These interactions may also be caused by dietary supplements, and until clinical trials can be performed to validate these supplements' safety, some researchers suggest that dietary supplement use by cancer patients should not be recommended [9].

While studies have looked at the prevalence of dietary supplement use among breast cancer patients receiving treatments including tamoxifen [10,11], to the best of our knowledge no research has investigated the frequency with which potential interactions may occur between dietary supplements and other endocrine therapies (i.e., aromatase inhibitors (AI)).

The purpose of this study was to evaluate the prevalence of dietary supplement use among breast cancer survivors enrolled in healthy lifestyle interventions and to identify any potentially harmful dietary supplement interactions between tamoxifen and AIs.

2. Materials and Methods

We conducted a cross-sectional analysis of the baseline prevalence of self-reported dietary supplement use among women treated for breast cancer who participated in several lifestyle intervention studies (NCT02109068, NCT02110641, NCT02681965, and NCT02056067) that were completed between 2010 and 2017 [12,13].

2.1. Participants and Recruitment

The eligibility criteria for the original studies were similar (Table 1). Eligible participants were breast cancer survivors diagnosed within the past 5 years with stage zero to three breast cancer, who had completed chemotherapy and/or radiation therapy at least 3 months before their enrollment. The women had to be physically able to exercise (i.e., be able to participate in a walking program), agree to be randomly assigned to a study group, and give informed consent to participate in all study activities. They also had to be reachable by telephone and able to communicate in English. Women were ineligible if they were pregnant or intending to become pregnant in the next year, had experienced a recent (during the past 6 months) stroke or myocardial infarction, or had any severe uncontrolled mental illness.

The breast cancer survivors were recruited between June 2010 and February 2017 via several approaches: (1) from five hospitals in Connecticut through the Rapid Case Ascertainment Shared Resource of the Yale Cancer Center, a field arm of the Connecticut Tumor Registry; (2) from self-referral via the study brochures in the Breast Center at the Smilow Cancer Hospital at Yale-New Haven; and (3) from the active recruitment of women attending the Yale Cancer Center Survivorship Clinic. The Connecticut Department of Public Health Human Investigation Committee (for NCT02056067 participants only) and the Yale School of Medicine Human Investigation Committee approved all of the procedures, including the written and verbal (via telephone) informed consent [13].

Table 1. Study eligibility criteria.

	Supervised Weight Loss Trial (NCT02109068 NCT02110641)	Self-Directed Weight Loss Trial (NCT02681965)	Supervised Exercise Trial (NCT02056067)
Study Description	6-month RCT	6-month RCT	12- month RCT
Number of Participants	151	205	119
Breast Cancer Stage	0–III	0–III	I–III
Endocrine Therapy	Tamoxifen, AI, or neither	Tamoxifen, AI, or neither	AI users only
BMI	$\geq$25.0 kg/m^2	$\geq$25.0 kg/m^2	Any BMI
Physical Activity	any amount	any amount	<90 min/week
Time Since Diagnosis	completed active treatment $\geq$ 3 months	completed active treatment $\geq$ 3 months	taking AI for 6 months to 4 years

2.2. Collection of Self-Reported Prescription Medication and Dietary Supplement Usage

All participants completed a self-reported frequency-based prescription medication and dietary supplement questionnaire. Regular use was defined as taking the agent at least 3 times a week for at least one month prior to the time of enrollment in the study. For the medications or supplements not listed on the questionnaire, open text fields allowed the participants to write the names of the prescription medications and dietary supplements they were taking (see the sample medication supplement form, Appendix A).

A registered dietitian (RD) with a certified specialty in oncology nutrition (CSO) (MH and CM) reviewed and standardized the self-reported dietary supplements using the Dietary Supplement Label Database (DSLD) developed by the Office of Dietary Supplements at the National Institutes of Health [14]. The generic and brand name formulas not found on the DSLD were reviewed on the manufacturer's website for each supplement fact label. Dietary supplements were then classified into categories: single nutrient, multivitamin, multimineral, and herbal preparations.

2.3. Collection of Other Lifestyle and Clinical Characteristics

Medical record review and self-report questionnaires were used to determine disease stage and endocrine therapy. The majority of height and weight measurements were taken during in-person baseline visits, though one study only collected self-reported height and weight measurements at the baseline (NCT02681965, n = 205).

2.4. Dietary Supplement Potential Interactions with Tamoxifen and AIs

The potential pharmacokinetic and pharmacodynamic interactions of all self-reported dietary supplements with tamoxifen and the AIs (anastrozole, letrozole, and exemestane) were identified using the Natural Medicines Database [15] by both a clinical pharmacy specialist (PharmD) specializing in oncology (AH) and a registered dietitian (RD) with a certified specialty in oncology nutrition (CSO) (MH and CM).

The stacking of nutrients from the use of multiple dietary supplements was enumerated, and proprietary formulas were broken down into their individual ingredients in order to more accurately report the nutrient exposures and assess potential interactions. Using the proprietary Natural Medicines Database interaction grading levels, only potential interaction levels of "moderate" (described as "a significant interaction or adverse outcome could occur") and "major" (described as "a serious adverse outcome could occur") grading were included. All research evidence grading levels were included: level A—a high-quality randomized control trial (RCT) or meta-analysis (a quantitative systematic review); level B—a nonrandomized clinical trial, non-quantitative systematic review, lower-quality RCT, clinical cohort study, case–control study, historical study, or epidemiological study; level

C—consensus or expert opinion; and level D—anecdotal evidence, an in vitro or animal study, or theory-based evidence from pharmacology.

2.5. Statistical Analysis

The patient characteristics were summarized using the means and standard deviations or the frequencies and percentages, as appropriate. A descriptive analysis was performed to describe the baseline dietary supplement use patterns in terms of the frequency and prevalence of each type of dietary supplement. The prevalence was also examined excluding women taking only vitamin D, calcium, and multivitamins, as these supplements are frequently recommended or prescribed by physicians to treat bone health or address other nutrient gaps in the customary diet. The number of dietary supplements taken at the baseline was classified by number of pills (e.g., *Hot Plants™ for Her* was counted as one pill, even though it has multiple active ingredients).

Using the Natural Medicines Database, we checked the individual nutrients for interactions with any of the endocrine therapy medications prescribed in our population (i.e., tamoxifen, anastrozole, letrozole, and exemestane). Patients not taking tamoxifen or AI were excluded from the analysis for supplement–drug interaction. Cross tabling was used to present the extent of the interaction.

3. Results

3.1. Baseline Characteristics

The baseline characteristics were similar for the women enrolled across the studies. The women were 58.6 (9.0) (mean (standard deviation)) years old, non-Hispanic white (86%), college-educated (68%), 2.9 (2.5) years from diagnosis, and had a BMI of 31.8 (5.9) kg/m^2. The women were diagnosed primarily with stage I breast cancer (47%) (Table 2).

3.2. Frequency of Dietary Supplement Usage

Among these 475 breast cancer survivors, 393 (83%) reported using dietary supplements at the baseline, with 51% of the women taking three or more individual dietary supplements and 23% taking five or more individual dietary supplements (range = 1–23) (Figure 1). Among all dietary supplement users, 108 (28%) reported taking either vitamin D, calcium, a multivitamin, or a combination of these supplements only, with 285 (73%) taking other supplements that are not traditionally prescribed or recommended by clinicians.

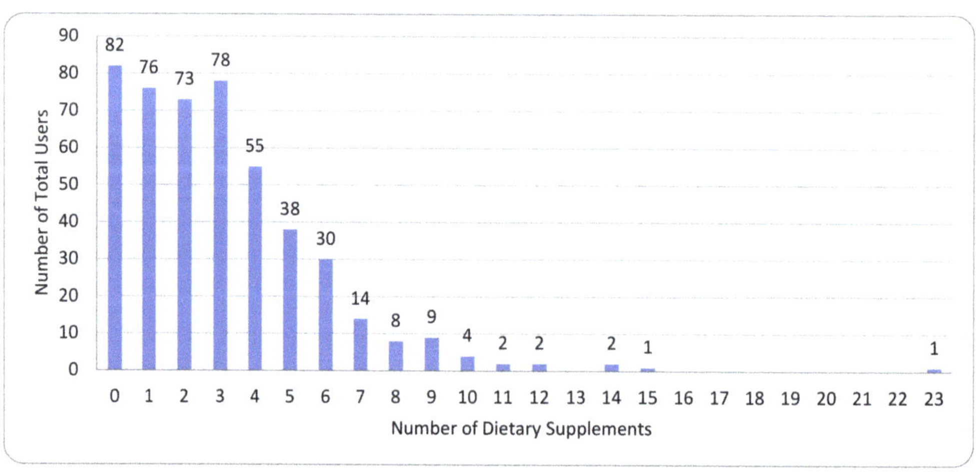

Figure 1. Dietary supplement use at baseline (n = 475).

Table 2. Participant baseline characteristics.

Variable	Total Sample n = 475 Mean (SD) or n (%)
Age (years) (mean (SD))	58.6 (9.0)
BMI (kg/m^2) (mean (SD)	31.8 (5.9)
Time since diagnosis (years)	2.9 (2.5)
Race/ethnicity	
Non-hispanic white	409 (86.1%)
Black	38 (8.0%)
Hispanic	16 (3.3%)
Other	12 (2.5)
Education	
≥College graduate	323 (68.0%)
Some school after high school	91 (19.2%)
High school graduate	57 (12.0%)
Refused to answer	4 (0.8%)
Stage	
0	52 (11.0%)
I	225 (47.4%)
II	132 (27.8%)
III	43 (9.1%)
Do not know	23 (4.8%)
Endocrine therapy usage (n = 475)	
None	122 (26%)
Tamoxifen	71 (15%)
Anastrozole	144 (30%)
Letrozole	110 (23%)
Exemestane	28 (6%)

SD: standard deviation.

The 393 dietary supplement users reported a total of 108 different types of dietary supplement. A total of 26 (24%) were single nutrients, 31 (29%) were paired nutrients (these include multivitamins and multiminerals), and 51 (47%) were herbal preparations. A total of 53 (14%) dietary supplement users took a combination of single nutrients, paired nutrients, and herbal preparations. The top 10 dietary supplements are presented in Table 3.

Table 3. Top 10 dietary supplements reported among the women reporting use of dietary supplements ($n = 393$).

Dietary Supplement	Participants Using Dietary Supplement $n = 393$
Vitamin D	238 (61%)
Calcium	200 (51%)
Multivitamin	198 (50%)
Omega 3	73 (19%)
Vitamin B12	68 (17%)
Vitamin C	52 (13%)
Glucosamine	42 (11%)
Fish oil	37 (9%)
Biotin	33 (9%)
Coenzyme Q10	31 (8%)

3.3. Dietary Supplement Interactions with Tamoxifen and AIs

When the nutrient exposures of all 108 self-reported dietary products were analyzed, 36 individual dietary supplement ingredients had potential interactions with either tamoxifen or any AI, as identified in the Natural Medicines Database (Table 4). The dietary supplements are classified by individual ingredient and are listed in alphabetical order, with the interacting endocrine medication and mechanism of potential interaction identified in superscript footnotes. We did not include the direction of metabolism (inducers or inhibitors) because the literature was inconsistent regarding the reporting of this information. Grapefruit extract was the only supplement that was considered a potential cause of major interactions; the remaining 35 were all considered potential causes of moderate interactions. Vitamin D was the most prevalent supplement: 191 women taking either tamoxifen or an AI reported taking vitamin D. The majority of interactions involved herbal preparations (89% versus 11% involving vitamins). The frequency of the 36 interactions varied with the type of endocrine therapy used: tamoxifen interacted with 100%, exemestane and letrozole both interacted with 72%, and anastrozole interacted with 36%.

Table 4. Potential Interactions with endocrine therapies.

Dietary Supplement *	Interactions with Endocrine Therapy
Astaxanthin	Tamoxifen [1] Exemestane [1] Letrozole [1]
Black Cohosh	Tamoxifen [2,3,4] Exemestane [4] Letrozole [4] Anastrozole [4]
Boswellia serrata extract	Tamoxifen [1,2,6] Exemestane [1] Letrozole [1]
Chamomile	Tamoxifen [1,2,4,6] Exemestane [1,4] Letrozole [1,4] Anastrozole [4]

Table 4. *Cont.*

Dietary Supplement *	Interactions with Endocrine Therapy
Cinnamon	Tamoxifen [3]
Cranberry extract	Tamoxifen [1] Exemestane [1] Letrozole [1]
Diindolylmethane	Tamoxifen [4] Exemestane [4] Letrozole [4] Anastrozole [4]
Diosmin	Tamoxifen [1,5,6] Exemestane [1] Letrozole [1]
Echinacea	Tamoxifen [1] Exemestane [1] Letrozole [1]
Eleuthero	Tamoxifen [4,5,6] Exemestane [4] Letrozole [4] Anastrozole [4]
Garlic extract	Tamoxifen [1] Exemestane [1] Letrozole [1]
Gingko biloba	Tamoxifen [1,5,6] Exemestane [1] Letrozole [1]
Ginseng	Tamoxifen [1,2,4,6,7] Exemestane [1,4] Letrozole [1,4] Anastrozole [4]
Glucomannan	Tamoxifen [9] Exemestane [9] Letrozole [9] Anastrozole [9]
Grapefruit extract**	Tamoxifen [1**,4,6,7**] Exemestane [1**,4] Letrozole [1**,4] Anastrozole [4]
Grapeseed	Tamoxifen [1,2] Exemestane [1] Letrozole [1]
Green tea extract	Tamoxifen [3]
Hesperidin	Tamoxifen [5]
Horny goat weed (Epimedium grandiflorum)	Tamoxifen [4] Exemestane [4] Letrozole [4] Anastrozole [4]
Jambolan (prune)	Tamoxifen [6]
Maca root	Tamoxifen [4] Exemestane [4] Letrozole [4] Anastrozole [4]

Table 4. *Cont.*

Dietary Supplement *	Interactions with Endocrine Therapy
Methoxylated flavones	Tamoxifen [1,5] Exemestane [1] Letrozole [1]
Milk thistle	Tamoxifen [1,4,10] Exemestane [4] Letrozole [4] Anastrozole [4]
Niacin	Tamoxifen [3]
Quercetin	Tamoxifen [1,2,6] Exemestane [1] Letrozole [1]
Red yeast rice	Tamoxifen [3]
Resveratrol	Tamoxifen [1,4] Exemestane [1,4] Letrozole [1,4] Anastrozole [4]
Rhodiola root	Tamoxifen [5,6]
Sesame seed	Tamoxifen [6,8]
Slippery elm bark	Tamoxifen [9] Exemestane [9] Letrozole [9] Anastrozole [9]
Sulforaphane	Tamoxifen [1] Exemestane [1] Letrozole [1]
Sweet orange	Tamoxifen [5]
Turmeric extract	Tamoxifen [1,3,4] Exemestane [1,4] Letrozole [1,4] Anastrozole [4]
Vitamin A	Tamoxifen [3]
Vitamin D	Tamoxifen [1] Exemestane [1] Letrozole [1]
Vitamin E	Tamoxifen [1] Exemestane [1] Letrozole [1]

* Classified by individual ingredient; ** indicates major interaction, otherwise all interactions listed below are moderate; [1] CYP3A4; [2] CYP2D6; [3] pharmacodynamic: hepatotoxic; [4] pharmacodynamic: estrogenic activity; [5] P-glycoprotein substrates; [6] CYP2C9; [7] may increase the effect of the drug; [8] pharmacodynamic: decreases the tumor inhibitory effect of tamoxifen; [9] decreases drug absorption; [10] inhibits UGT, causing decreased drug clearance.

Of the 353 women taking tamoxifen or AIs at the baseline, 38% were taking dietary supplements with the potential to produce major or moderate interactions. The highest interaction-to-use ratio was seen with exemestane and the lowest with anastrozole (exemestane 82% (23/28); letrozole 73% (80/110); tamoxifen 35% (25/71); anastrozole 4% (6/144)).

4. Discussion

We found a high dietary supplement usage among the breast cancer survivors, with 83% of the women taking at least one dietary supplement. Over half (51%) reported taking three or more supplements. Of those on endocrine therapy, 38% were taking supplements that had at least moderate potential for interactions.

Our study found higher dietary supplement usage compared to reports in the literature. For instance, different populations including women without cancer (51%) and cancer survivors of various disease types (76%) had a lower prevalence of dietary supplement use [16]. It should be noted that our study included all dietary supplements, both those recommended by clinicians and those initiated by patients without clinician involvement. While not all medical professionals recommend the use of vitamin D, calcium, or a daily multivitamin, these supplements are commonly recommended by clinicians, and we were unable to determine whether the use of these supplements had a medical indication. Even accounting for these sometimes-prescribed supplements, our study found a high use of "nontraditional" dietary supplements (60%).

The higher dietary supplement rate reported in our study could be explained by the high education level of our study participants, which fits with the profile of higher dietary supplement usage described by Cowan et al. In their analysis of NHANES data for 2011–2014, the overall dietary supplement use by healthy adults in the U.S. was found to be higher among women (59%) than men (45%), while higher-income and food-secure populations were more likely to consume one or more dietary supplements compared to less affluent participants [17].

The volume of supplement use per individual was also high in our study, with over half (51%) taking three or more and 23% taking five or more dietary supplements. There were 12 individuals taking 10 or more dietary supplements. To our knowledge, our study is the first to report on the broad spectrum of dietary supplement use including the volume (i.e., the total number of pills) and type of supplement by breast cancer survivors. The heterogenous, comprehensive list generated from our data collection included 108 unique supplements. Other studies have typically focused on a list of pre-specified supplements [9], a single class of nutrients (i.e., antioxidants) [11], or on non-cancer populations only [18]. In a study of healthy adults in the U.S. (2003–2006), most individuals reported taking one supplement daily, 10% reported taking three daily, and 10% reported taking five daily [19]. Du et al. found that adult cancer survivors had a higher prevalence of use of any dietary supplement compared to non-cancer survivors; however, the individual number of supplements was not reported [16].

Lee et al. looked at the potential interactions of all medications—including dietary supplements—taken by 67 prostate and breast cancer subjects before, during, and after chemotherapy. Dietary supplements were involved in 56% of the potential 1747 total interactions with chemotherapy that were identified. While there was a reported increased utilization of dietary supplements after chemotherapy (51% during vs. 66% after), the interactions of dietary supplements with tamoxifen and AIs were not evaluated [20].

Several studies have identified the health problems associated with synthetic xenoestrogens that are found in various materials, including additives or contaminants in food [21–23]. These endocrine-disrupting chemicals have become a part of everyday life, interfere with the natural cycle of the hormones in the body, and are thought to give rise to many endocrine-related disorders, including endocrine-related cancers. In our analysis, 11/36 (31%) of the reported dietary supplement ingredients caused estrogenic activity that could further potentiate this estrogenic exposure. This may reduce the effectiveness of hormone therapy and therefore worsen patients' prognoses.

Of the women in our study taking a prescribed endocrine therapy, 38% were taking supplements with potential moderate interaction with the endocrine therapy. It is notable that anastrozole produced the fewest interactions (36%), due to the fact that it is not metabolized through the CYP450 enzyme pathway. Anastrozole is metabolized through the N-dealkylation, hydroxylation and glucuronidation pathway, which is not a major pathway for drug interactions [24]. For women taking dietary supplements, clinicians may consider prescribing anastrozole, as it risks the least number of potential interactions.

Vitamin D was the most common dietary supplement reported in our population. This resonates with our own clinical experience, as many breast cancer survivors are taking this supplement for bone health or low vitamin D blood levels. The Natural Medicines Database lists vitamin D as risking potential moderate interactions with Level B evidence. However, the reference included in the database specifically studied vitamin D supplementation in relation to atorvastatin concentrations and cholesterol levels. This study was small (n = 16), and vitamin D was found to lower atorvastatin levels, which the authors concluded was a result of vitamin D inducing the CYP3A4 enzyme and increasing the clearance of drugs metabolized in this pathway [25]. Notably, the cholesterol levels were not adversely impacted. Given that vitamin D is commonly recommended in clinical practice, we conclude that more data on vitamin D and its potential to interact with endocrine therapy is needed.

This paper investigated the individual interactions of each dietary supplement, but it should be noted that stacking occurred frequently. While we were unable to calculate the total dose of dietary supplements per participant, clinicians should be more aware of stacking, as it can result in doses that are above the recommended daily allowances.

Cancer survivors do not readily discuss their dietary supplement usage. Du et al. reported that nearly half of 1355 cancer survivors used dietary supplements on their own without consulting health care providers [16]. In another study, Pouchieu et al. reported that only 2% of 1081 cancer survivors obtained advice on dietary supplement use from an RD [26]. In addition, fewer than one half of oncologists are initiating discussions with their patients about dietary supplement use, and many indicate that a lack of knowledge and education are barriers to such discussions [27].

One way to approach a review of dietary supplement usage is to begin with the premise of "first, do no harm". It is important for clinicians to remember that dietary supplements are legally defined as a food—and so are covered under food regulation laws—but may act like pharmacologic agents in the body [28]. Some common principles can be employed when addressing dietary supplement use with patients: (1) Meet the patient where they are at (i.e., accept that high-volume supplement users have many reasons and strong beliefs about the value added by dietary supplements). (2) Conversation starters can include questions such as, "Could you tell me about the foods you eat, and whether you take any supplements?". (3) The goals should be to maintain an ongoing assessment of usage and to reduce the use of dietary supplements that have potential interactions with cancer therapy.

This study has several limitations. Dietary supplement usage was self-reported and thus subject to recall bias. The dosages of the dietary supplements were not reported consistently, thus we could not take dose into account in our analyses. A selection bias may exist, as our participants were mostly from the northeast region and willing to enroll in lifestyle intervention studies focused on dietary-induced weight loss and exercise. Our study's strengths include its large study population, its comprehensive reporting of dietary supplement use, and its evaluation of all dietary supplements by ingredient. To the best of our knowledge, this is the first study to examine dietary supplement use and their potential interactions with the adjuvant endocrine therapy for breast cancer survivors.

5. Conclusions

We observed an 83% rate of dietary supplement use among breast cancer survivors enrolled in our study, and the potential for adverse interactions between the prescribed endocrine therapies and dietary supplements was common. These findings underscore the need for further research into the interactions between dietary supplements and endocrine therapies for breast cancer. Oncologists should be aware of dietary supplement use, understand their potential interactions with endocrine therapy, and discuss and/or refer patients to an RD and pharmacist in the multi-disciplinary team.

Author Contributions: Conceptualization, M.H., C.M., A.H., L.M.F., B.C., M.L.I., and T.S.; methodology, M.H., L.M.F., B.C., F.-Y.L., M.L.I., and T.S.; validation, F.-Y.L., M.L.I., and T.S.; formal analysis, M.H., L.M.F., B.C., F.-Y.L., M.L.I., and T.S.; investigation, M.H., C.M., A.H., T.N., B.C., M.L.I., and T.S.; data curation, M.H., B.C., F.-Y.L., and T.S.; writing—original draft preparation, M.H., C.M., A.H., T.N., F.-Y.L., M.L.I., and T.S.; writing—review and editing, M.H., C.M., A.H., L.M.F., T.N., B.C., F.-Y.L., M.L.I., and T.S.; visualization, F.-Y.L.; supervision, L.M.F., B.C., M.L.I., and T.S.; project administration, M.H., L.M.F., B.C., M.L.I., and T.S.; funding acquisition, M.L.I. All authors have read and agreed to the published version of the manuscript.

Funding: This research was funded by the American Institute for Cancer Research and in part by a grant from the Breast Cancer Research Foundation; NCT02056067: Supported by National Cancer Institute Grant No. R01 CA132931 and in part by a grant from the Breast Cancer Research Foundation (M.L.I.), a Yale Cancer Center Support Grant No. P30 CA016359, and a Clinical and Translational Science Award Grant No. UL1 TR000142 from the National Center for Advancing Translational Science, a component of the National Institutes of Health.

Institutional Review Board Statement: This study was conducted according to the guidelines of the Declaration of Helsinki, and approved by the Institutional Review Board (or Ethics Committee) of Yale School of Medicine (protocol codes 1012007780, 14100147716 and 0906005623).

Informed Consent Statement: Informed consent was obtained from all subjects involved in the study.

Acknowledgments: Certain data for the HOPE study (NCT02056067) used in this study were obtained from the Connecticut Tumor Registry located in the Connecticut Department of Public Health. The authors assume full responsibility for the analysis and interpretation of these data. We are indebted to the participants for their dedication to, and time with, the LEAN and HOPE studies.

Conflicts of Interest: The authors declare no conflict of interest. The funders had no role in the design of the study; in the collection, analysis, or interpretation of the data; in the writing of the manuscript; nor in the decision to publish the results.

Appendix A

Appendix A: sample medication supplement form.

Study ID: ______ Interviewer ID: ______ Today's Date: ___/___/___

month/day/year

Are you currently taking any <u>herbal remedies</u> or <u>nutritional supplements</u> (other than vitamins or minerals) on a <u>regular basis</u>? Regular is defined as <u>at least 3 times a week for at least 1 month</u>.

Herbal Preparation	√ box if taking	Date Started (mm/yyyy)
Ex: St. John's Wort		*03/1997*
Glucosamine		
Chondroitin		
Omega-3 fatty acids		
Coenzyme Q_{10}		
Black Cohosh		
Garlic		
Echinacea		
Sasparilla		
Cat's Claw		

Herbal Preparation	√ box if taking	Date Started (mm/yyyy)
Red Clover		
Fo Ti Teng		
Alfalfa		
Fenugreek		
Seaweed (kelp)		
Milk Thistle		
Astragalus		
Mushrooms (maitake, shitake, etc.)		
Turmeric		
Essiac		
Digestive enzymes		
Ensure or Boost		
Carnation Instant Breakfast		
Don quai		
Ginko biloba		
Ginseng		
Green Tea (tea or extract)		
Bee Pollen		
Royal Jelly		
Saw Palmetto		
Shark Cartilage		
Soy		
St. John's Wort		
Valerian		
Wild/Mexican yam		
Yerba Buena		
Acai Juice		
Flaxseed Oil		
Fish Oil		
Other:______________		
Other:______________		
Other:______________		

Are you currently taking any vitamins or minerals on a regular basis? Regular is defined as at least 3 times a week for at least 1 month. **NOTE: If you are taking a multivitamin please do NOT list its contents individually.**

Vitamin or Mineral	✓ Box if Taking	Date Started (mm/yyyy)	Dose	Unit	Frequency
Ex: Calcium	✓	*10/2008*	*1000*	*mg*	*once a day*
Multivitamin			N/A	N/A	
Vitamin A				IU	
Beta carotene				IU	
Vitamin B1 (thiamine)				mg	
Vitamin B2 (riboflavin)				mg	
Vitamin B6 (pyridoxine)				mg	
Vitamin B12 (cyanocobalamin)				mcg	
Biotin				mcg	
Vitamin C (ascorbic acid)				mg	
Vitamin D (calciferol)				IU	
Vitamin E (tocopherol)				IU	
Folic acid/folate (folacin)				mcg	
Niacin (niacinamide)				mg	
Pantothenic acid (pantothenate)				mg	
Calcium or Tums				mg	
Chromium				mcg	
Iron				mg	
Magnesium				mg	
Selenium				mcg	
Potassium				mg	
Zinc				mg	

Are you currently taking any medications (prescription or over the counter), other than those listed above, on a regular basis? Regular is defined as at least 3 times a week for at least 1 month.

Name of Medication	Dose	Unit of Dose	Frequency	Date Started (mm/yyyy)
Ex. Aspirin	81	mg	once per day	05/1998

References

1. Dietary Supplement Health and Education Act of 1994; Public Law 103-417, 103rd Congress.
2. National Institutes of Health, Office of Dietary Supplements. 2011 Dietary Supplements. *Background Information*. Available online: www.ods.od.nih.gov/factsheets/DietarySupplements-Consumer/ (accessed on 1 June 2021).
3. Diet, Nutrition, Physical Activity and Cancer: A Global Perspective. The 3rd Expert Report World Cancer Research Fund. 2018. Available online: www.wcrf.org/dietandcancer (accessed on 1 May 2021).
4. Manson, J.E.; Cook, N.R.; Lee, I.; Christen, W.; Bassuk, S.S.; Mora, S.; Gibson, H.; Gordon, D.; Copeland, T.; D'Agostino, D.; et al. Vitamin D Supplements and Prevention of Cancer and Cardiovascular Disease. *N. Engl. J. Med.* **2019**, *380*, 33–44. [CrossRef]
5. Abdel-Rahman, O. Dietary Supplements Use among Adults with Cancer in the United States: A Population-Based Study. *Nutr. Cancer* **2020**, 1–8. [CrossRef] [PubMed]
6. Rock, C.L.; Doyle, C.; Demark-Wahnefried, W.; Meyerhardt, J.; Courneya, K.S.; Schwartz, A.L.; Bandera, E.V.; Hamilton, K.K.; Grant, B.; McCullough, M.; et al. Nutrition and physical activity guidelines for cancer survivors. *CA Cancer J. Clin.* **2012**, *62*, 242–274. [CrossRef] [PubMed]
7. Bailey, R.L.; Dodd, K.W.; Jaime JGahche, J.J.; Dwyer, J.T.; Cowan, A.E.; Jun, S.; Eicher-Miller, H.A.; Guenther, P.M.; Bhadra, A.; Thomas, P.R.; et al. Best Practices for Dietary Supplement Assessment and Estimation of Total Usual Nutrient Intakes in Population-Level Research and Monitoring. *J. Nutr.* **2019**, *149*, 181–197. [CrossRef] [PubMed]
8. Jin, Y.; Desta, Z.; Stearns, V.; Ward, B.; Ho, H.; Lee, K.-H.; Skaar, T.; Storniolo, A.M.; Li, L.; Araba, A.; et al. CYP2D6 genotype, antidepressant use, and tamoxifen metabolism during adjuvant breast cancer treatment. *J. Natl. Cancer Inst.* **2005**, *97*, 30–39. [CrossRef]
9. Zeng, Z.; Mishuk, A.U.; Qian, J. Safety of dietary supplements use among patients with cancer: A systematic review. *Crit. Rev. Oncol. Hematol.* **2020**, *152*, 103013. [CrossRef]
10. Zirpoli, G.R.; Brennan, P.M.; Hong, C.C.; McCann, S.E.; Ciupak, G.; Davis, W.; Unger, J.M.; Budd, G.T.; Hershman, D.L.; Moore, H.C.F.; et al. Supplement use during an intergroup clinical trial for breast cancer (S0221). *Breast Cancer Res. Treat.* **2013**, *137*, 903–913. [CrossRef] [PubMed]
11. Greenlee, H.; Gammon, M.D.; Abrahamson, P.E.; Gaudet, M.M.; Terry, M.B.; Hershman, D.L.; Desai, M.; Teitelbaum, S.L.; Neugut, A.I.; Jacobson, J.S. Prevalence and Predictors of Antioxidant Supplement Use During Breast Cancer Treatment the Long Island Breast Cancer Study Project. *Cancer* **2009**, *115*, 3271–3282. [CrossRef] [PubMed]
12. Tara Sanft, T.; Usiskin, I.; Harrigan, M.; Cartmel, B.; Lu, L.; Li, F.Y.; Zhou, Y.; Chagpar, A.; Ferrucci, L.M.; Pusztai, L.; et al. Randomized controlled trial of weight loss versus usual care on telomere length in women with breast cancer: The lifestyle, exercise, and nutrition (LEAN) study. *Breast Cancer Res. Treat.* **2018**, *172*, 105–112. [CrossRef]
13. Irwin, M.L.; Cartmel, B.; Gross, C.; Ercolano, E.; Li, F.Y.; Yao, X.; Fiellin, M.; Capozza, S.; Rothbard, M.; Zhou, Y.; et al. Randomized Exercise Trial of Aromatase Inhibitor–Induced Arthralgia in Breast Cancer Survivors. *JCO* **2014**, *57*, 1547. [CrossRef]
14. Dwyer, J.T.; Bailen, R.A.; Saldanha, L.G.; Gahche, J.J.; Costello, R.B.; Betz, J.M.; Davis, C.D.; Bailey, R.L.; Potischman, N.; Ershow, A.G.; et al. The Dietary Supplement Label Database: Recent Developments and Applications. *J. Nutr.* **2018**, *148*, 1428S–1435S. [CrossRef] [PubMed]
15. TRC Natural Medicines™, Interaction Checker: Food, Herbs, Supplements, Drugs. Available online: naturalmedicines.therapeuticresearch.com (accessed on 1 June 2021).
16. Du, M.; Luo, H.; Blumberg, J.B.; Rogers, G.; Chen, F.; Ruan, M.; Shan, Z.; Biever, E.; Zhang, F.F. Dietary Supplement Use among Adult Cancer Survivors in the United States. *J. Nutr.* **2020**, *150*, 1499–1508. [CrossRef] [PubMed]
17. Cowan, A.E.; Jun, S.; Gahche, J.J.; Tooze, J.A.; Dwyer, J.T.; Eicher-Miller, H.A.; Bhadra, A.; Guenther, P.M.; Potischman, N.; Dodd, K.W.; et al. Dietary Supplement Use Differs by Socioeconomic and Health-Related Characteristics among U.S. Adults, NHANES 2011–2014. *Nutrients* **2018**, *10*, 1114. [CrossRef] [PubMed]
18. Kantor, E.D.; Colin, D.; Rehm, C.D.; Du, M.; White, E.; Giovannucci, E.L. Trends in Dietary Supplement Use Among US Adults From 1999-2012. *JAMA* **2016**, *316*, 1464–1474. [CrossRef]
19. Bailey, R.L.; Gahche, J.J.; Lentino, C.V.; Dwyer, J.T.; Engel, J.S.; Thomas, P.R.; Betz, J.M.; Sempos, C.T.; Piccano, M.F. Dietary Supplement Use in the United States, 2003–2006. *J. Nutr.* **2011**, *141*, 261–266. [CrossRef]
20. Lee, R.T.; Kwon, N.; Wu, J.; To, C.; To, S.; Szmulewitz, R.; Tchekmedyian, R.; Holmes, H.M.; Olopade, O.I.; Stadler, W.M.; et al. Prevalence of potential interactions of medications, including herbs and supplements, before, during, and after chemotherapy in patients with breast and prostate cancer. *Cancer* **2021**, *127*, 1827–1835. [CrossRef]
21. Rashid, H.; Alqahtani, S.S.; Alshahrani, S. Diet: A Source of Endocrine Disruptors. *Endocr. Metab. Immune Disord. Drug Targets* **2020**, *20*, 633–645. [CrossRef]
22. Paterni, I.; Granchi, C.; Minutolo, F. Risks and benefits related to alimentary exposure to xenoestrogens. *Crit. Rev. Food Sci. Nutr.* **2017**, *57*, 3384–3404. [CrossRef] [PubMed]
23. Qasem, R.J. The estrogenic activity of resveratrol: A comprehensive review of in vitro and in vivo evidence and the potential for endocrine disruption. *Crit. Rev. Toxicol.* **2020**, *50*, 439–462. [CrossRef]
24. *Arimidex [Package Insert]*; ANI Pharmaceuticals: Baudette, MN, USA, 2018.
25. Schwartz, J.B. Effects of vitamin D supplementation in atorvastatin-treated patients: A new drug interaction with an unexpected consequence. *Clin. Pharmacol. Ther.* **2009**, *85*, 198–203. [CrossRef]

26. Pouchieu, C.; Fassier, P.; Druesne-Pecollo, N.; Zelek, L.; Bachmann, P.; Touillaud, M.; Bairati, I.; Hercberg, S.; Galan, P.; Cohen, P.; et al. Dietary supplement use among cancer survivors of the NutriNet-Sante cohort study. *Br. J. Nutr.* **2015**, *113*, 1319–1329. [CrossRef] [PubMed]
27. Lee, R.T.; Barbo, A.; Lopez, G.; Melhem-Bertrandt, A.; Lin, H.; Olopade, O.I.; Curlin, F.A. National survey of US oncologists' knowledge, attitudes, and practice patterns regarding herb and supplement use by patients with cancer. *J. Clin. Oncol.* **2014**, *32*, 4095–4101. [CrossRef] [PubMed]
28. U.S. Food and Drug Administration. *Dietary Supplements*. Available online: www.fda.gov/food/dietary-supplements (accessed on 1 June 2021).

MDPI

Article

Effect of the Lifestyle, Exercise, and Nutrition (LEAN) Study on Long-Term Weight Loss Maintenance in Women with Breast Cancer

Alexa Lisevick [1,*], Brenda Cartmel [2,3], Maura Harrigan [2], Fangyong Li [2], Tara Sanft [3,4], Miklos Fogarasi [1], Melinda L. Irwin [2,3] and Leah M. Ferrucci [2,3]

[1] Frank H. Netter MD School of Medicine, Quinnipiac University, Hamden, CT 06518, USA; miklos.fogarasi@quinnipiac.edu

[2] Yale School of Public Health, Yale University, New Haven, CT 06510, USA; brenda.cartmel@yale.edu (B.C.); maura.harrigan@yale.edu (M.H.); fang-yong.li@yale.edu (F.L.); melinda.irwin@yale.edu (M.L.I.); leah.ferrucci@yale.edu (L.M.F.)

[3] Yale Cancer Center, New Haven, CT 06510, USA; tara.sanft@yale.edu

[4] Yale School of Medicine, Yale University, New Haven, CT 06510, USA

* Correspondence: Alexa.lisevick@quinnipiac.edu

Abstract: Lifestyle interventions among breast cancer survivors with obesity have demonstrated successful short-term weight loss, but data on long-term weight maintenance are limited. We evaluated long-term weight loss maintenance in 100 breast cancer survivors with overweight/obesity in the efficacious six-month Lifestyle, Exercise, and Nutrition (LEAN) Study (intervention = 67; usual care = 33). Measured baseline and six-month weights were available for 92 women. Long-term weight data were obtained from electronic health records. We assessed weight trajectories between study completion (2012–2013) and July 2019 using growth curve analyses. Over up to eight years (mean = 5.9, SD = 1.9) of post-intervention follow-up, both the intervention (n = 60) and usual care (n = 32) groups declined in body weight. Controlling for body weight at study completion, the yearly weight loss rate in the intervention and usual care groups was –0.20 kg (−0.2%/year) (95% CI: 0.06, 0.33, p = 0.004) and −0.32 kg (−0.4%/year) (95% CI: 0.12, 0.53, p = 0.002), respectively; mean weight change did not differ between groups (p = 0.31). It was encouraging that both groups maintained their original intervention period weight loss (6% intervention, 2% usual care) and had modest weight loss during long-term follow-up. Breast cancer survivors in the LEAN Study, regardless of randomization, avoided long-term weight gain following study completion.

Keywords: breast cancer; survivorship; weight loss maintenance; lifestyle intervention

Citation: Lisevick, A.; Cartmel, B.; Harrigan, M.; Li, F.; Sanft, T.; Fogarasi, M.; Irwin, M.L.; Ferrucci, L.M. Effect of the Lifestyle, Exercise, and Nutrition (LEAN) Study on Long-Term Weight Loss Maintenance in Women with Breast Cancer. *Nutrients* **2021**, *13*, 3265. https://doi.org/10.3390/nu13093265

Academic Editors: Wendy Demark-Wahnefried and Christina Dieli-Conwright

Received: 25 August 2021
Accepted: 17 September 2021
Published: 18 September 2021

Publisher's Note: MDPI stays neutral with regard to jurisdictional claims in published maps and institutional affiliations.

1. Introduction

In 2021, there will be an estimated 284,200 newly diagnosed cases of breast cancer and 44,130 breast cancer deaths in the United States, representing roughly 15% of all new cancer cases and 7% of all cancer deaths, respectively [1]. Currently, the five-year survival rate for breast cancer is 90% for all stages combined; nonetheless, mortality declines have slowed in recent years [1].

Obesity, defined as a body mass index (BMI) $\geq$30 kg/m^2, in the setting of breast cancer survivorship has been associated with increased risk of recurrence, therapy-related morbidity [2], poorer overall and breast cancer-specific survival [3–5], as well as reduced quality of life [2]. The molecular mechanisms linking obesity and breast cancer biology are not entirely understood; however, it is suggested that hormones, adipocytokines, inflammatory cytokines, and reactive oxygen species play important roles [6]. Obesity, weight gain, and physical inactivity during or following cancer treatment are highly prevalent in breast cancer survivors [2,7,8]. An analysis of the National Health Interview Survey found the prevalence of obesity increased more rapidly among cancer survivors, compared to the general population, from 1997 to 2014 [9]. The annual increase in obesity prevalence of 3.0% among breast cancer survivors was one of the highest rates of increasing obesity burden among all cancer survivors [9]. Data from the most recent years in this national sample indicated approximately 30–35% of breast cancer survivors were obese [9].

Although weight gain amongst adult women as they age is common [10,11], weight gain among breast cancer survivors may start during treatment and continue months to years after diagnosis [12]. Furthermore, it appears that women who are normal weight at diagnosis more commonly experience post-diagnosis weight gain than women with overweight or obesity at diagnosis [13]. Importantly, in comparison to women who maintain their weight following diagnosis, those that experience weight gain have increased all-cause mortality, especially when weight gain is 10% or higher [14]. Thus, there is a growing emphasis on finding efficacious interventions focused on preventing weight gain or promoting weight loss among breast cancer survivors with overweight or obesity (BMI > 25 kg/m^2) [2,15,16].

Lifestyle guidelines for breast cancer survivors highlight the importance of a healthy body weight with a focus on physical activity and diet [17]. The American Cancer Society and American Society of Clinical Oncology 2016 breast cancer survivorship care guidelines recommend physicians counsel survivors about consuming a diet high in vegetables, fruits, whole grains, and legumes, and low in saturated fats, as well as limiting alcohol intake. Further, recommendations for survivors include avoiding inactivity, completing at least 150 min of moderate or 75 min of vigorous aerobic exercise per week, and should include strength training exercises at least two days per week [18]. Some breast cancer survivors may have difficulty meeting these recommendations because of fatigue and therapy-related side effects, which may limit physical activity and achieving dietary goals. Data suggest only 18% and 37% of breast cancer survivors meet nutrition and physical activity guidelines, respectively [19].

A variety of short-term lifestyle interventions for breast cancer survivor populations have demonstrated successful weight loss [20,21]. While data on weight following the completion of these studies are limited, several studies have shown weight regain in the months following either the full intervention [22] or the intensive components of the intervention [23,24], prompting the question of whether measurable losses in body weight are sustainable for breast cancer survivors long-term.

The Lifestyle, Exercise, and Nutrition (LEAN) Study was a randomized-controlled weight-loss trial that compared the effect of in-person or telephone-based counseling versus usual care on changes in body composition, physical activity, diet, and serum biomarkers over six months in overweight or obese women with breast cancer [25]. The six-month trial led to a clinically meaningful 6% mean weight loss among the LEAN intervention group, compared to a 2% mean weight loss among the usual care group ($p < 0.05$) [25]. Given that the LEAN intervention provided a strong, clinically impactful short-term weight-loss benefit to breast cancer survivors, in the present analysis, we sought to determine the long-term impact of this intervention on weight change up to eight years post intervention. The primary aim of the current analysis was to evaluate long-term weight loss maintenance among breast cancer survivors enrolled in the LEAN Study. In an exploratory analysis, we also examined if weight change during the trial (weight loss, weight maintenance, weight gain) influenced weight change during long-term follow-up.

2. Materials and Methods

Women with BMI $\geq$ 25.0 kg/m^2 diagnosed with Stage 0 to III breast cancer within five years prior to study enrollment were eligible for the LEAN Study (clinicaltrials.gov registration number NCT02109068). Eligible participants had completed chemotherapy and/or radiation therapy, were physically able to exercise, accessible by telephone, and able to read and communicate in English. Women were excluded if they were pregnant, intending to become pregnant within a year, had a history of stroke or myocardial infarction within six months, or had a severe uncontrolled mental illness. Participants were self-referred or recruited between June 2011 and December 2012 through the Breast Center at Smilow Cancer Hospital at Yale-New Haven Hospital and the Yale Cancer Center Survivorship Clinic; a total of 100 women were enrolled. The study was approved by the Yale School of Medicine Human Investigation Committee. The detailed protocol and primary results of the trial related to the intervention's effect on 6-month change in body weight have been published previously [25].

Women were randomized to the LEAN intervention (either in-person or telephone-based counseling) or usual care group such that one-third of the participants were in each group. The weight loss intervention was centered around reduced caloric intake, increased physical activity, as well as behavioral therapy [25].

The intervention groups received 11 sessions of 30 min counseling led by a registered dietitian who was also a certified specialist in oncology nutrition, over the span of 6 months, either in person or via telephone, a breast cancer-specific healthy eating and exercise LEAN educational book, and a journal to guide counseling sessions. The in-person and telephone groups received the same lifestyle intervention. Participants received counseling sessions once per week throughout the first month, followed by every two weeks in the following two months, and then once per month in the final three months. The LEAN journal was used by participants to record all food and beverage intake, minutes of physical activity, and daily pedometer step counts, as well as their weight measured once a week on a scale provided by the study. Participants were provided with personalized energy intake goals based on baseline weight, such that they incurred an energy intake deficit of 500 kcal/day. The dietary fat goal was <25% total energy intake. Participants were encouraged to consume a plant-based diet and incorporate mindful eating practices alongside a home-based physical activity program with a goal of 150 moderate-intensity activity minutes per week and 10,000 steps per day [25].

The usual care group received one 30 min counseling session at the end of the six-month study period, in addition to the LEAN book and journal, American Institute for Cancer Research pamphlets on healthy eating and exercise, and referral to the Yale Cancer Center Survivorship Clinic, which offers a two-session weight management program [25].

Participant weights were measured by study staff in duplicate at baseline and the end of the six-month study period. Additional follow-up weight data assessed objectively via scales during patient visits at affiliated clinical sites through July 2019 was obtained retrospectively via patient electronic health records at Yale-New Haven Hospital.

The LEAN Study population comprised 100 breast cancer survivors; of those, 33 were randomized to usual care, 34 to intervention via telephone-based counseling, and 33 to intervention via in-person counseling. As there was no difference in weight loss between the two intervention groups, the telephone-based and in-person counseling groups were combined for this analysis [25]. After exclusion of participants who did not have six-month measured weights (end of LEAN Study), our analytic sample was 92 women (intervention = 60 (90%); usual care = 32 (97%)).

Participant weight trajectories were calculated between six months (end of LEAN Study) through July 2019 using the measured weight at the end of the LEAN Study and all available electronic health record data from thereon. Thus, up to 8 years of follow-up data were available.

Women's baseline characteristics were summarized using descriptive statistics. A growth curve analysis using mixed effect modeling was performed to compare the rate of body weight change from the six-month endpoint of the original LEAN Study through the follow-up period. A random intercept effect was included to account for within-subject correlation among repeated assessments. The difference in the slopes of weight change over time was examined by including time and group interaction as a fixed effect. The slope represents yearly weight gain (if positive) or weight loss (if negative) in kilograms.

In addition, in an exploratory analysis, we categorized changes in body weight during the six-month LEAN Study period from baseline to six months into three levels: weight loss was defined as losing greater than 1% of body weight, weight gain was defined as gaining greater than 1% of body weight, and weight maintenance was defined as weight remaining within 1% of body weight. The weight trajectories were compared by randomization group and weight change category. All analyses were performed using SAS 9.4 (Cary, NC, USA). Statistical significance was set at $p < 0.05$, two-sided.

3. Results

3.1. Baseline Characteristics of Study Participants

At the start of the LEAN Study, the mean participant age was 58.8 years (SD = 7.3) with a mean BMI of 33.1 kg/m^2 (SD = 6.6) (Table 1). The majority of participants were postmenopausal, identified as non-Hispanic whites, and had graduated from college. Among the 92 women in our analytic sample, the mean weight change over the six-month LEAN Study was significantly different between the intervention and usual care groups; on average, the intervention group lost 4.3 kg (5%), whereas the usual care group lost 1.6 kg (2%) ($p = 0.009$).

Table 1. Baseline characteristics of Lifestyle, Exercise and Nutrition (LEAN) Study participants with long-term weight data ($n = 92$).

Characteristic	Mean (SD) or *n* (%)			
	All *n* = 92	Intervention *n* = 60	Usual Care *n* = 32	*p*-Value
Age, years	58.8 (7.3)	59.4 (7.3)	57.6 (7.3)	0.28
BMI [a], kg/m^2	33.1 (6.6)	32.7 (6.2)	33.9 (7.6)	0.42
College graduate	47 (51%)	33 (55%)	14 (44%)	0.30
Non-Hispanic white	84 (91%)	55 (92%)	29 (91%)	0.49
Postmenopausal	75 (82%)	50 (83%)	25 (78%)	0.54
Time from diagnosis to study enrollment, years	2.7 (1.8)	2.7 (1.5)	2.8 (2.2)	0.72
Post study follow-up time, years	5.9 (1.9)	6.0 (1.8)	5.6 (2.1)	0.38
Disease Stage				0.94
0	15 (16%)	9 (15%)	6 (19%)	
I	48 (52%)	31 (52%)	17 (53%)	
II	21 (23%)	15 (25%)	6 (19%)	
III	6 (7%)	4 (7%)	2 (6%)	
Unknown	2 (2%)	1 (1%)	1 (3%)	
Treatment after surgery				0.67
None	14 (15%)	8 (13%)	6 (19%)	
Radiation only	34 (37%)	21 (35%)	13 (41%)	
Chemotherapy only	17 (18%)	13 (22%)	4 (13%)	
Radiation and Chemotherapy	27 (29%)	18 (30%)	9 (28%)	
Weight (kg)				
Baseline	87.5 (18.1)	86.1 (16.8)	90.4 (20.3)	0.27
Six-month	84.4 (19.3)	82.4 (18.0)	88.3 (21.2)	0.16
Weight change within study period (kg)	−3.4 (5.3)	−4.3(5.7)	−1.6 (3.7)	0.009

[a] BMI, body mass index.

3.2. Post-Intervention Weight Change

The median number of weight data points per participant was similar between groups, 20 (intervention) and 18 (usual care). Women were followed up to eight years post intervention (mean = 5.9 SD = 1.9), and the mean years of follow-up was similar between groups, 6.0 (intervention) and 5.6 (usual care) ($p = 0.38$). In the post LEAN follow-up period, both groups had a decline in body weight over time (intervention = −0.20 kg or −0.2% per year, SE 0.07, $p = 0.004$; usual care = −0.32 kg or −0.4% per year; SE 0.10, $p = 0.002$) (Table 2) (Figure 1a,b). There was no statistically significant difference in the yearly mean rates of weight change between the intervention and usual care groups ($p = 0.31$).

Table 2. Mean weight trajectories during the follow-up period by group.

	Yearly Mean Rate of Weight Change (kg)	SE [b]	95% CI [c]	*p*-Value
Intervention [a] (*n* = 60)	−0.20	0.07	[−0.06, −0.33]	0.004
Usual care (*n* = 32)	−0.32	0.10	[−0.12, −0.53]	0.002

[a] By-randomization group *p*-value was 0.31. [b] SE, standard error. [c] CI, confidence interval.

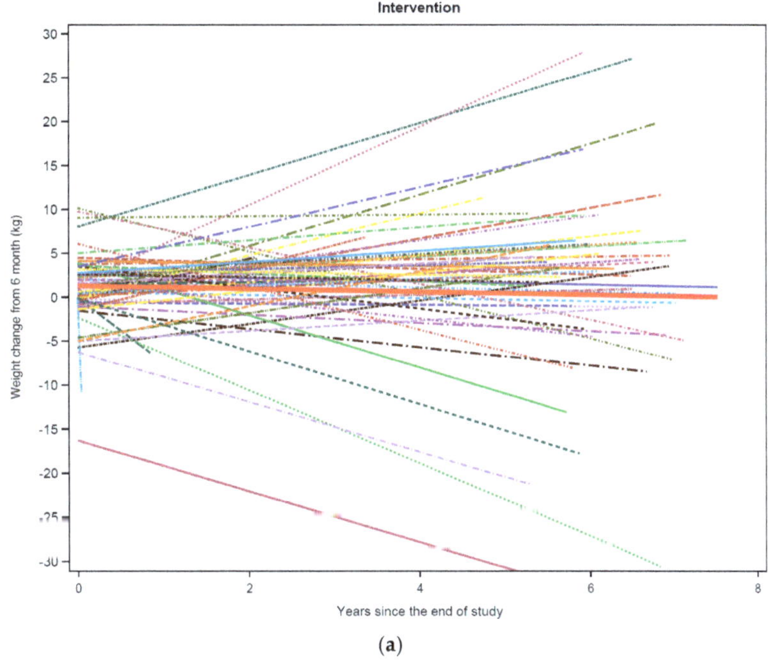

(a)

Figure 1. *Cont.*

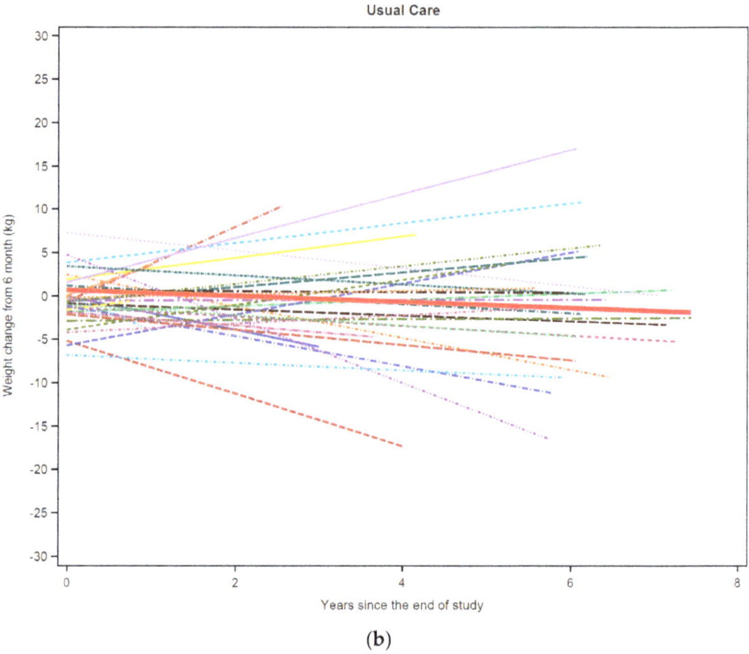

(b)

Figure 1. Weight trajectories by study group. Weight trajectories for each individual participant from LEAN Study completion through long-term follow-up (latest date July 2019) are shown: (**a**) intervention group ($n = 60$); (**b**) usual care ($n = 32$). The thick red line in (**a**) corresponds to the mean weight trajectory among intervention group participants, and the thick red line in (**b**) corresponds to the mean weight trajectory among usual care group participants.

3.3. Post-Intervention Weight Change by Weight Change during LEAN

The proportion of women who lost >1% body weight, gained >1% body weight, and maintained weight (>1% change) during the six-month LEAN Study was significantly different by study groups in our analytic sample ($p = 0.0005$) (Table 3). Considering weight-change categories during the LEAN Study, on average, women in the intervention group who lost weight during the LEAN Study, lost weight during follow-up (−0.09 kg per year), compared to those in the usual care group who had a non-significant gain in weight (0.20 kg per year) (group difference = −0.39 kg per year, $p = 0.02$) (Table 3). Women in both the usual care and intervention groups who gained weight during the six-month LEAN trial lost weight during follow-up; however, women in the usual care group had significantly greater weight loss (−1.32 kg per year) than those in the intervention group (−0.46 kg per year) (group difference = 0.85 kg per year, $p = 0.002$). Independent of the study group, women who maintained their weight during the study period did not experience significant weight change during follow-up (group difference = 0.08 kg per year, $p = 0.82$).

Table 3. Post-study weight change stratified by weight changes at the end of six-month study.

	Intervention ($n = 60$)	Usual Care ($n = 32$)	Chi-Square p-Value
Overall Weight Change			0.0005
Weight loss (>1% loss)	48 (80%)	16 (50%)	
Weight gain (>1% gain)	8 (13%)	7 (22%)	
Weight maintenance (>1% change)	4 (7%)	9 (28%)	

Table 3. *Cont.*

	Intervention (n = 60)	Usual Care (n = 32)	Chi-Square p-Value		
	Weight Change (kg/yr.)	**SE**	**Lower**	**Upper**	**p-Value**
Intervention					
Weight loss during LEAN	−0.09	0.04	−0.15	−0.02	0.02
Weight gain during LEAN	−0.46	0.18	−0.82	−0.10	0.01
Weight maintenance during LEAN	−0.07	0.25	−0.57	0.42	0.78
Usual Care					
Weight loss during LEAN	0.20	0.15	−0.08	0.49	0.16
Weight gain during LEAN	−1.32	0.20	−1.71	−0.92	<0.0001
Weight maintenance during LEAN	−0.15	0.25	−0.64	0.34	0.54
Intervention vs. Usual=Care Group Comparison					
Weight loss: intervention vs. usual care	−0.39	0.16	−0.71	−0.06	0.02
Weight gain: intervention vs. usual care	0.85	0.27	0.32	1.39	0.002
Weight maintenance: intervention vs. usual care	0.08	0.35	−0.62	0.78	0.82

4. Discussion

Breast cancer survivors participating in the LEAN Study, regardless of randomization to usual care or intervention, avoided long-term weight gain after the lifestyle intervention. Given there was modest weight loss overall during long-term follow-up (average of 5.9 years), it was encouraging that the women in the intervention and usual care groups were able to maintain their original LEAN trial weight loss (6% weight loss for intervention versus 2% weight loss for usual care). Participation in a lifestyle intervention led to the prevention of weight gain over time and maintenance of clinically meaningful weight loss among breast cancer survivors randomized to the intervention. Our results provide evidence of the benefits of lifestyle programs; thus, clinicians should consider recommending and referring patients to cancer survivorship and weight management programs following a diagnosis of breast cancer.

Our finding that both intervention and usual care women in our study lost weight over time differs from what is typically seen in the general female population. For example, amongst adult women without obesity or chronic disease, a mean weight increase of 2.33 lbs. (1.06 kg) to 5.24 lbs. (2.38 kg) per 4 years has been shown [10], suggesting that, on average, adult women gain weight over time. Another long-term study following adult women for a mean of 26 years, found an average BMI increase of 3.7 kg/m^2 and a mean weight change of 8.6 kg [26]. In this same study, baseline normal-weight women gained 2.4 kg more than obese women, and overweight women gained 3.3 kg more than obese women [26].

Although several lifestyle interventions similar to the LEAN intervention have demonstrated success in clinically meaningful weight loss among breast cancer survivors [20,21], several have observed weight regain in the months following either the full intervention [22] or the intensive components of the intervention [23,24]. A review of lifestyle interventions in female cancer survivors noted the challenge of maintaining participant motivation following the study conclusion and suggested that highly personalized approaches to weight loss may be more successful [27]. Thus far, the majority of existing studies have not reported long-term weight trajectories following the end of the intervention; therefore, additional research on this topic is vital. To our knowledge, this is the first study of long-term follow-up of a weight loss intervention in breast cancer survivors.

One possible explanation for our findings of modest weight loss during follow-up may be that all women enrolled in the LEAN Study had a BMI > 25 kg/m^2. Not all of the data for the general population has been stratified by BMI at baseline, which may be an important predictor of weight change. As noted above, there are limited follow-up weight data from other weight-loss interventions in breast cancer survivors with overweight/obesity to compare our results. It is also possible that the breast cancer survivors in the LEAN Study may differ in terms of weight patterns from breast cancer survivors not enrolled in a lifestyle intervention, as women enrolled in LEAN were willing to participate in a randomized weight-loss trial of diet and exercise. Therefore, regardless of the randomization group, they may have had greater readiness to adopt healthy behaviors, including behaviors resulting in weight loss, following breast cancer treatment. Additionally, recent studies have indicated that premenopausal women [28–31] and those with lower BMI at diagnosis [32,33] appear to be at increased risk of post-diagnosis weight gain, and the original LEAN study did not target these populations. One additional explanation for the loss of weight during the long-term follow-up among the usual care study participants specifically may be that the one weight-loss counseling session and the study material that the women in this group received at the end of the six-month intervention was efficacious for modest weight loss.

Since the intervention group lost significantly more weight (6% weight loss) during the six-month LEAN Study than the usual care group (2% weight loss), we examined long-term weight change by weight change during the trial. In these analyses, we found women in the intervention group who lost >1% body weight during the trial continued to lose weight in the follow-up period, compared to women in the usual care group, who lost >1% body weight during the six-month study period and, in contrast, did not continue to lose weight during the follow-up period. Moreover, in both the intervention and usual care groups, women who maintained their weight during the trial continued to experience weight maintenance during follow-up. Lastly, women who gained weight during the LEAN Study lost weight during the follow-up independent of the intervention group. However, these exploratory analyses should be interpreted with caution, as we could not investigate the more traditional 5% weight change cut-points due to our small sample size.

Our long-term weight data are from electronic health records, and therefore, we had rates of weight change per year (e.g., slopes) rather than change at a defined follow-up time (e.g., one year or two years post intervention). Therefore, we could not assess predictors of long-term weight change in our population. However, as demonstrated in Table 1, the baseline characteristics are fairly balanced; thus, predictors of the trajectory of weight change over time are not significant for a comparison between groups. Additional research is needed to elucidate the frequency of post-diagnosis weight gain, maintenance, and loss amongst breast cancer survivors, including those with overweight/obesity at diagnosis and those receiving modern anti-cancer therapies. Large prospective studies of women diagnosed with breast cancer could also help us understand which women are at the greatest risk of post-diagnosis weight gain, and how this impacts prognosis.

This study has several limitations. Due to the use of electronic health records for the collection of weight histories during the follow-up period, we had a variable number of weight measurements and lengths of follow-up for participants. There may also be some measurement error in weight data in the electronic health record, but we tried to eliminate recording errors through extensive data cleaning and hand review of abstracted data, and the data in the records were objective from scales in clinical settings. We also could not assess other measures of body composition, such as lean mass versus fat mass during our long-term follow-up. These data could be useful in future studies to fully understand body composition in relation to cancer outcomes, as evidence suggests an increased risk of mortality among early breast cancer patients with sarcopenia [34]. Additionally, we were unable to assess if participants sought additional resources to promote weight loss during post-intervention follow-up. Our participants were from one institution, and the majority were college educated, non-Hispanic White, post-menopausal breast cancer survivors diagnosed with stage I breast cancer, meaning that these findings may not be applicable to all breast cancer survivors. For instance, African American communities have both higher rates of obesity and breast cancer mortality rates, compared to non-

Hispanic white women; there are many factors that may contribute to these disparities, and further research to elucidate this relationship is necessary [35]. Important strengths of our study include the long-term follow-up period and multiple longitudinal-measured weights derived from the electronic health records, which eliminates the social desirability bias of self-reported weight.

5. Conclusions

In this sample of breast cancer survivors with overweight/obesity, we observed that overall women experienced modest long-term weight loss following the LEAN Study, even if randomized to usual care. Given our small size and the current lack of similar studies of long-term weight patterns following lifestyle interventions to which we could compare our results, additional research is necessary to understand long-term weight trajectories in breast cancer survivors with overweight/obesity both within and outside the context of lifestyle interventions.

Author Contributions: Conceptualization, B.C., M.H., T.S., M.L.I. and L.M.F.; methodology, A.L., B.C., M.H., F.L., T.S., M.L.I. and L.M.F.; validation, F.L., M.L.I. and L.M.F.; formal analysis, A.L., B.C., F.L., M.L.I. and L.M.F.; investigation, A.L., B.C., M.L.I. and L.M.F.; data curation, A.L., B.C., F.L. and L.M.F.; writing—original draft preparation, A.L., F.L., M.L.I. and L.M.F.; writing—review and editing, A.L., B.C., M.H., F.L., T.S., M.F., M.L.I. and L.M.F.; visualization, F.L.; supervision, B.C., M.L.I. and L.M.F.; project administration, B.C., M.H., T.S., M.L.I. and L.M.F.; funding acquisition, M.L.I. All authors have read and agreed to the published version of the manuscript.

Funding: Supported by the American Institute for Cancer Research and in part by a grant from the Breast Cancer Research Foundation; further supported in part by the Yale Cancer Center Support Grant P30 CA016359, the TREC Training Workshop R25 CA203650 (PI: Melinda Irwin), and the Clinical and Translational Science Award Grant Number UL1 TR000142 from the National Center for Advancing Translational Science, a component of the National Institutes of Health.

Institutional Review Board Statement: The study was conducted according to the guidelines of the Declaration of Helsinki and approved by the Institutional Review Board (or Ethics Committee) of Yale School of Medicine (Protocol Code 1012007780; 30 April 2019).

Informed Consent Statement: Informed consent was obtained from all subjects involved in the study.

Acknowledgments: We are indebted to the participants for their dedication to and time with the LEAN Study.

Conflicts of Interest: The authors declare no conflict of interest. The funders had no role in the design of the study; in the collection, analyses, or interpretation of data; in the writing of the manuscript, or in the decision to publish the results.

References

1. Siegel, R.L.; Miller, K.D.; Fuchs, H.E.; Jemal, A. Cancer Statistics. *CA Cancer J. Clin.* **2021**, *71*, 7–33. [CrossRef]
2. Sheng, J.; Sharma, D.; Jerome, G.; Santa-Maria, C.A. Obese Breast Cancer Patients and Survivors: Management Considerations. *Oncology* **2018**, *32*, 410–417.
3. Chan, D.S.M.; Vieira, A.R.; Aune, D.; Bandera, E.V.; Greenwood, D.C.; McTiernan, A.; Rosenblatt, D.N.; Thune, I.; Vieira, R.; Norat, T. Body mass index and survival in women with breast cancer—Systematic literature review and meta-analysis of 82 follow-up studies. *Ann. Oncol.* **2014**, *25*, 1901–1914. [CrossRef] [PubMed]
4. Bradshaw, P.T.; Ibrahim, J.G.; Stevens, J.; Cleveland, R.; Abrahamson, P.E.; Satia, J.A.; Teitelbaum, S.L.; Neugut, A.I.; Gammon, M.D. Postdiagnosis Change in Bodyweight and Survival after Breast Cancer Diagnosis. *Epidemiology* **2012**, *23*, 320–327. [CrossRef] [PubMed]
5. Maliniak, M.L.; Patel, A.V.; McCullough, M.L.; Campbell, P.T.; Leach, C.R.; Gapstur, S.M.; Gaudet, M.M. Obesity, physical activity, and breast cancer survival among older breast cancer survivors in the Cancer Prevention Study-II Nutrition Cohort. *Breast Cancer Res. Treat.* **2018**, *167*, 133–145. [CrossRef]
6. Simone, V.; D'Avenia, M.; Argentiero, A.; Felici, C.; Rizzo, F.M.; De Pergola, G.; Silvestris, F. Obesity and Breast Cancer: Molecular Interconnections and Potential Clinical Applications. *Oncology* **2016**, *21*, 404–417. [CrossRef] [PubMed]
7. Chen, X.; Lu, W.; Gu, K.; Chen, Z.; Zheng, Y.; Zheng, W.; Shu, X.O. Weight Change and Its Correlates Among Breast Cancer Survivors. *Nutr. Cancer* **2011**, *63*, 538–548. [CrossRef]

8. Lucas, A.R.; Levine, B.J.; Avis, N.E. Posttreatment trajectories of physical activity in breast cancer survivors. *Cancer* **2017**, *123*, 2773–2780. [CrossRef] [PubMed]
9. Greenlee, H.; Shi, Z.; Molmenti, C.L.S.; Rundle, A.; Tsai, W.Y. Trends in Obesity Prevalence in Adults with a History of Cancer: Results from the US National Health Interview Survey, 1997 to 2014. *J. Clin. Oncol.* **2016**, *34*, 3133–3140. [CrossRef]
10. Mozaffarian, D.; Hao, T.; Rimm, E.B.; Willett, W.C.; Hu, F.B. Changes in Diet and Lifestyle and Long-Term Weight Gain in Women and Men. *N. Engl. J. Med.* **2011**, *364*, 2392–2404. [CrossRef]
11. Malhotra, R.; Ostbye, T.; Riley, C.M.; Finkelstein, E.A. Young adult weight trajectories through midlife by body mass category. *Obesity* **2013**, *21*, 1923–1934. [CrossRef]
12. Vance, V.; Hanning, R.; Mourtzakis, M.; McCargar, L. Weight gain in breast cancer survivors: Prevalence, pattern and health consequences. *Obes. Rev.* **2010**, *12*, 282–294. [CrossRef] [PubMed]
13. Caan, B.J.; Kwan, M.L.; Shu, X.O.; Pierce, J.P.; Patterson, R.E.; Nechuta, S.; Poole, E.M.; Kroenke, C.H.; Weltzien, E.K.; Flatt, S.W.; et al. Weight Change and Survival after Breast Cancer in the After Breast Cancer Pooling Project. *Cancer Epidemiol. Biomark. Prev.* **2012**, *21*, 1260–1271. [CrossRef] [PubMed]
14. Playdon, M.C.; Bracken, M.B.; Sanft, T.B.; Ligibel, J.A.; Harrigan, M.; Irwin, M.L. Weight Gain After Breast Cancer Diagnosis and All-Cause Mortality: Systematic Review and Meta-Analysis. *J. Natl. Cancer Instig.* **2015**, *107*, djv275. [CrossRef] [PubMed]
15. Jiralerspong, S.; Goodwin, P. Obesity and Breast Cancer Prognosis: Evidence, Challenges, and Opportunities. *J. Clin. Oncol.* **2016**, *34*, 4203–4216. [CrossRef] [PubMed]
16. Nyrop, K.A.; Deal, A.M.; Lee, J.T.; Muss, H.B.; Choi, S.K.; Wheless, A.; Carey, L.A.; Shachar, S.S. Weight gain in hormone receptor-positive (HR+) early-stage breast cancer: Is it menopausal status or something else? *Breast Cancer Res. Treat.* **2018**, *167*, 235–248. [CrossRef] [PubMed]
17. Diet, Nutrition, Physical Activity, and Breast Cancer Survivors. 2014. Available online: www.wcrf.org/sites/default/files/Breast-Cancer-Survivors-2014-Report.pdf (accessed on 17 September 2021).
18. Runowicz, C.D.; Leach, C.R.; Henry, N.L.; Henry, K.S.; Mackey, H.T.; Cowens-Alvarado, R.L.; Cannady, R.S.; Pratt-Chapman, M.; Edge, S.B.; Jacobs, L.A.; et al. American Cancer Society/American Society of Clinical Oncology Breast Cancer Survivorship Care Guideline. *J. Clin. Oncol.* **2016**, *34*, 611–635. [CrossRef]
19. Blanchard, C.M.; Courneya, K.S.; Stein, K. Cancer survivors' adherence to lifestyle behavior recommendations and associations with health-related quality of life: Results from the American Cancer Society's SCS-II. *J. Clin. Oncol.* **2008**, *26*, 2198–2204. [CrossRef]
20. Shaikh, H.; Bradhurst, P.; Ma, L.X.; Tan, S.Y.C.; Egger, S.J.; Vardy, J.L. Body weight management in overweight and obese breast cancer survivors. *Cochrane Database Syst. Rev.* **2020**, *12*, Cd012110.
21. Playdon, M.; Thomas, G.; Sanft, T.; Harrigan, M.; Ligibel, J.A.; Irwin, M.L. Weight Loss Intervention for Breast Cancer Survivors: A Systematic Review. *Curr. Breast Cancer Rep.* **2013**, *5*, 222–246. [CrossRef] [PubMed]
22. Greenlee, H.A.; Crew, K.D.; Mata, J.M.; McKinley, P.S.; Rundle, A.G.; Zhang, W.; Liao, Y.; Tsai, W.Y.; Hershman, D.L. A Pilot Randomized Controlled Trial of a Commercial Diet and Exercise Weight Loss Program in Minority Breast Cancer Survivors. *Obesity* **2012**, *21*, 65–76. [CrossRef] [PubMed]
23. Harris, M.N.; Swift, D.L.; Myers, V.H.; Earnest, C.; Johannsen, N.M.; Champagne, C.M.; Parker, B.D.; Levy, E.; Cash, K.C.; Church, T.S. Cancer Survival Through Lifestyle Change (CASTLE): A Pilot Study of Weight Loss. *Int. J. Behav. Med.* **2012**, *20*, 403–412. [CrossRef]
24. Rock, C.L.; Flatt, S.W.; Byers, T.E.; Colditz, G.A.; Demark-Wahnefried, W.; Ganz, P.A.; Wolin, K.Y.; Elias, A.; Krontiras, H.; Liu, J.; et al. Results of the Exercise and Nutrition to Enhance Recovery and Good Health for You (ENERGY) Trial: A Behavioral Weight Loss Intervention in Overweight or Obese Breast Cancer Survivors. *J. Clin. Oncol.* **2015**, *33*, 3169–3176. [CrossRef]
25. Harrigan, M.; Cartmel, B.; Loftfield, E.; Sanft, T.; Chagpar, A.B.; Zhou, Y.; Irwin, M.L. Randomized Trial Comparing Telephone Versus In-Person Weight Loss Counseling on Body Composition and Circulating Biomarkers in Women Treated for Breast Cancer: The Lifestyle, Exercise, and Nutrition (LEAN) Study. *J. Clin. Oncol.* **2016**, *34*, 669–676. [CrossRef]
26. Kimokoti, R.W.; Newby, P.; Gona, P.; Zhu, L.; McKeon-O'Malley, C.; Guzman, J.P.; D'Agostino, R.B.; Millen, B.E. Patterns of weight change and progression to overweight and obesity differ in men and women: Implications for research and interventions. *Public Health Nutr.* **2012**, *16*, 1463–1475. [CrossRef] [PubMed]
27. Chlebowski, R.T.; Reeves, M.M. Weight Loss Randomized Intervention Trials in Female Cancer Survivors. *J. Clin. Oncol.* **2016**, *34*, 4238–4248. [CrossRef] [PubMed]
28. Nyrop, K.A.; Deal, A.M.; Shachar, S.S.; Park, J.; Choi, S.K.; Lee, J.T.; O'Hare, E.A.; Wheless, A.; Carey, L.A.; Muss, H.B. Weight trajectories in women receiving systemic adjuvant therapy for breast cancer. *Breast Cancer Res. Treat.* **2019**, *179*, 709–720. [CrossRef] [PubMed]
29. Vargas-Meza, A.; Chavez-Tostado, M.; Cortes-Flores, A.; Urias-Valdez, D.; Delgado-Gomez, M.; Morgan-Villela, G.; del Valle, C.J.Z.F.; Jimenez-Tornero, J.; Fuentes-Orozco, C.; García-Rentería, J.; et al. Body weight changes after adjuvant chemotherapy of patients with breast cancer: Results of a Mexican cohort study. *Eur. J. Cancer Care* **2016**, *26*, e12550. [CrossRef] [PubMed]
30. Yeo, W.; Mo, F.K.F.; Pang, E.; Suen, J.J.S.; Koh, J.; Loong, H.H.F.; Yip, C.C.H.; Ng, R.Y.W.; Yip, C.H.W.; Tang, N.L.S.; et al. Profiles of lipids, blood pressure and weight changes among premenopausal Chinese breast cancer patients after adjuvant chemotherapy. *BMC Women's Health* **2017**, *17*, 55. [CrossRef] [PubMed]

31. Chaudhary, L.; Wen, S.; Xiao, J.; Swisher, A.; Kurian, S.; Abraham, J. Weight change associated with third-generation adjuvant chemotherapy in breast cancer patients. *J. Community Support. Oncol.* **2014**, *12*, 355–360. [CrossRef]
32. Gu, K.; Chen, X.; Zheng, Y.; Chen, Z.; Zheng, W.; Lu, W.; Shu, X.O. Weight change patterns among breast cancer survivors: Results from the Shanghai breast cancer survival study. *Cancer Causes Control* **2010**, *21*, 621–629. [CrossRef] [PubMed]
33. Gandhi, A.; Copson, E.; Eccles, D.; Durcan, L.; Howell, A.; Morris, J.; Howell, S.; McDiarmid, S.; Sellers, K.; Evans, D.G.; et al. Predictors of weight gain in a cohort of premenopausal early breast cancer patients receiving chemotherapy. *Breast* **2019**, *45*, 1–6. [CrossRef] [PubMed]
34. Zhang, X.-M.; Dou, Q.-L.; Zeng, Y.; Yang, Y.; Cheng, A.S.K.; Zhang, W.-W. Sarcopenia as a predictor of mortality in women with breast cancer: A meta-analysis and systematic review. *BMC Cancer* **2020**, *20*, 172. [CrossRef] [PubMed]
35. Ford, M.E.; Magwood, G.; Brown, E.T.; Cannady, K.D.; Gregoski, M.; Knight, K.; Peterson, L.L.; Kramer, R.; Evans-Knowell, A.; Turner, D.P. Disparities in Obesity, Physical Activity Rates, and Breast Cancer Survival. *Adv. Cancer Res.* **2017**, *133*, 23–50. [CrossRef] [PubMed]

MDPI

Article

Comparing Outcomes of a Digital Commercial Weight Loss Program in Adult Cancer Survivors and Matched Controls with Overweight or Obesity: Retrospective Analysis

Christine N. May [1], Annabell Suh Ho [1], Qiuchen Yang [1], Meaghan McCallum [1], Neil M. Iyengar [2], Amy Comander [3], Ellen Siobhan Mitchell [1,*] and Andreas Michaelides [1]

[1] Academic Research, Noom Inc., 229 W. 28th St., New York, NY 10001, USA; christinem@noom.com (C.N.M.); annabell@noom.com (A.S.H.); qy2182@caa.columbia.edu (Q.Y.); meaghanm@noom.com (M.M.); andreas@noom.com (A.M.)

[2] Memorial Sloan Kettering Cancer Center, 300 East 66th St., New York, NY 10065, USA; iyengarn@mskcc.org

[3] Massachusetts General Hospital Cancer Center, Harvard Medical School, 55 Fruit St., Boston, MA 02114, USA; acomander@mgh.harvard.edu

* Correspondence: siobhan@noom.com

Abstract: Maintaining a healthy weight is beneficial for cancer survivors. However, weight loss program effectiveness studies have primarily been in highly controlled settings. This is a retrospective study exploring real-world outcomes (weight loss and program engagement) after use of a digital commercial weight loss program (Noom) in cancer survivors and matched controls. All participants had voluntarily self-enrolled in Noom. Weight and engagement data were extracted from the program. Cancer-related quality of life was secondarily assessed in a one-time cross-sectional survey for survivors. Controls were a sample of Noom users with overweight/obesity who had no history of cancer but 0–1 chronic conditions. Primary outcomes were weight change at 16 weeks and program engagement over 16 weeks. Engagement included frequency of weight, food, and physical activity logging, as well as number of coach messages. Multiple regression controlling for baseline age, gender, engagement, and BMI showed that survivors lost less weight than controls (B = −2.40, s.e. = 0.97, p = 0.01). Survivors also weighed in less (survivors: 5.4 [2.3]; controls: 5.7 [2.1], p = 0.01) and exercised less (survivors: 1.8 [3.2]; controls: 3.2 [4.1], $p < 0.001$) than controls. However, survivors sent more coach messages (survivors: 2.1 [2.4]; controls: 1.7 [2.0], $p < 0.001$). Despite controls losing more weight than cancer survivors (−7.0 kg vs. −5.3 kg), survivors lost significant weight in 4 months (M = −6.2%). Cancer survivors can have success on digital commercial programs available outside of a clinical trial. However, they may require additional support to engage in weight management behaviors.

Keywords: weight loss; obesity; cancer survivors; retrospective study

Citation: May, C.N.; Ho, A.S.; Yang, Q.; McCallum, M.; Iyengar, N.M.; Comander, A.; Mitchell, E.S.; Michaelides, A. Comparing Outcomes of a Digital Commercial Weight Loss Program in Adult Cancer Survivors and Matched Controls with Overweight or Obesity: Retrospective Analysis. *Nutrients* **2021**, *13*, 2908. https://doi.org/10.3390/nu13092908

Academic Editor: Maria Cristina Mele

Received: 7 July 2021
Accepted: 19 August 2021
Published: 24 August 2021

1. Introduction

The number of cancer survivors within the United States continues to increase rapidly as treatments improve and screening efforts expand. Over the next twenty years, the population of cancer survivors is expected to increase nearly two-fold, reaching 26.1 million individuals [1]. Given this projection, anticipating the complex health needs of cancer survivors represents a major public health concern [2]. Maintenance of a healthy body mass index (BMI) is one modifiable risk factor that has been associated with decreased risk for recurrence and mortality for survivors of certain types of cancers [3]. Current guidelines from the American Cancer Society and American Society of Clinical Oncology recommend maintaining a healthy weight after cancer treatment, but cancer survivors receive insufficient guidance on weight management from health providers [3,4]. Cancer survivors may try to manage their weight on their own or through commercial programs outside of a clinical trial. These self-management methods are estimated to be the most common weight management strategies [5].

In particular, commercial digital programs are rapidly proliferating due to widespread smartphone access in various geographical areas and among diverse populations [6,7]. These digital commercial programs raise important new empirical questions for cancer survivors. For instance, individuals use these programs in the comfort of their own home, which means they self-manage their participation in the absence of monitoring requirements in research protocols or in-person clinical sessions [5,8,9]. However, extant knowledge of weight management outcomes and behaviors is almost exclusively derived from formal study, clinical, or in-person settings. For the increasing number of cancer survivors who use digital commercial programs, the extent of their outcomes and behaviors in their real-world use of the program is entirely unknown.

Therefore, we conducted a retrospective analysis of weight loss and engagement in a commercially available digital program among self-enrolled cancer survivors with overweight or obesity compared with a group of matched controls. This question is particularly important for cancer survivors, who may have disease-related barriers to participation. Cancer survivors face post-treatment challenges, such as cancer-related fatigue and lack of energy, side effects, new health conditions, and physical limitations [10]. Survivors have also reported difficulty in self-sustaining weight-relevant behavioral changes [11]. In addition, weight gain is more common in breast cancer survivors than in non-cancer patients [12]. Moreover, commercially available weight loss programs are not typically designed specifically for cancer survivors. Thus, we hypothesized that matched controls would have greater weight loss and engagement than cancer survivors. We also conducted a subgroup analysis of breast cancer survivors since this was the most commonly reported cancer type and because obesity is associated with increased risk of breast cancer recurrence [13]. An additional aim of the study was to descriptively report survivors' cancer-related quality of life (QoL) after using this commercial digital weight loss program.

2. Materials and Methods

2.1. Participants

Only participants who had already signed up for the program were analyzed in this study. All participants provided consent for their program data to be used for research. Participants were also given the option to opt out. Participants were eligible if they signed up between July 2018 and August 2020, were still on the program (i.e., did one in-app action) in September 2020, had a BMI $\geq$ 25 kg/m^2, and had indicated a history of cancer during program sign-up (N = 363). A random sample of matched controls who had a similar BMI range (overweight or obese), 0–1 chronic health conditions (e.g., hypertension, type 2 diabetes), signed up for the program during the same time period, and were still on the program were selected (N = 2000). These criteria were selected so that controls were matched on key factors that could influence weight loss outcomes. Controls and survivors were contacted by email with a survey invitation at the time of data collection (September 2020). The survey measured self-reported cancer-related QoL (for survivors) and demographics (for survivors and controls). All participants were offered the chance to win one of three $100 gift cards for survey completion. 107 survivors and 150 controls completed the survey and were included in the study (see Figure 1 for a diagram of inclusion). For all participants, self-reported weight, engagement, and physical activity data were extracted from the program database from baseline through week 16 (the minimum length of the core weight loss program).

Because not all participants weighed in every week, weight and engagement analyses included only individuals who reported their weight at baseline and week 16 (43 survivors, 85 controls). Eligible survivors who responded to the survey, even if they did not report their weight at baseline and week 16, were included in descriptive QoL analysis (107 survivors).

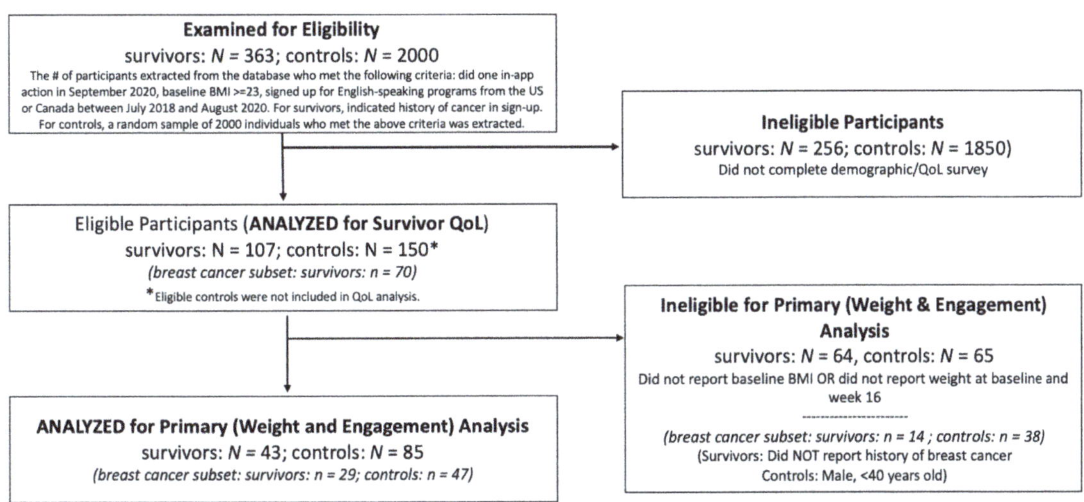

Figure 1. Flow diagram of participant eligibility. N refers to the main sample and n refers to the subsample of breast cancer survivors only.

In addition to these analyses, we analyzed a subset of breast cancer survivors. From the survivor samples described above, we included any participant who reported a history of breast cancer (n= 29 for weight and engagement analyses, n = 70 for QoL analyses). For weight and engagement analyses with this subset, controls were matched to breast cancer survivors on gender and age, such that from the original sample of 85 matched controls, only those who were female and ≥40 years old were selected (n = 47 controls).

2.2. Digital Platform

Noom is a mobile program that has been found to result in clinically significant weight loss in RCTs of a general population with overweight or obesity [14]. It is a publicly available program; individuals who elect to continue with the program after the free trial pay for a subscription. The program is based on cognitive behavioral therapy (CBT) and motivational interviewing techniques, which aid in weight control and increasing motivation to make behavioral changes [15]. Individuals are provided daily articles informed by federal guidelines and empirical work on healthy diet, physical activity, and the psychology of behavior change. The articles about nutrition are informed by MyPlate recommendations as well as empirical work on energy density [16,17]. Noom has a food color system that categorizes foods based on energy density, in terms of high (red), medium (yellow), and low (green) energy density. Previous work has shown that adherence to the food color system is associated with greater weight loss on Noom [18]. Individuals are guided through the entire program with behavior change principles derived from CBT, motivational interviewing, and third-wave CBT (e.g., dialectical behavior therapy) techniques, as well as behavior change techniques like self-monitoring and goal setting [15,19,20]. Individuals are provided with mobile logging features to self-monitor their weight and exercise, as well as a virtual group and the ability to exchange text messages with a 1:1 health coach. The health coach helps the individual to set individualized goals, recognize barriers, and identify individualized solutions to barriers. The coach also discusses awareness of behaviors and barriers (e.g., self-awareness), checks in on progress towards goals, and provides support to users [21]. A group coach oversees the group posts. The program does not have a required length and users can participate for as long as they would like.

2.3. Measures

Weight: Participants self-reported their weight on the program. Weight measurements from week 1 through 16 were extracted from the program database. Individuals are encouraged, but not required, to log their weight daily.

Engagement: We gathered data from the weight loss program database to assess differences in physical activity (steps) and program engagement. Steps were tracked by smartphone sensors, wearable devices connected to the program, or manually entered by participants. As in past work, engagement was measured by the number of times per week that participants self-reported their weight or exercises on the program, and the number of times they messaged their coach, which was tracked by the program [22]. Coaches reach out to users at least once a week, and individuals are encouraged, though not required, to log their exercise daily.

Quality of life: Survivors self-reported cancer history and cancer-related QoL via survey. Quality of life questions were adapted from the European Organization for Research and Treatment Core Quality of Life Questionnaire (EORTC QLQ-C30) [23] and transformed into scores ranging from 0 to 100, including overall quality of life rating with higher scores indicating better QoL; functioning scales with higher scores indicating better functioning role functioning, emotional function, cognitive functioning, social functioning; and symptom scales with higher scores indicating more severe symptoms including nausea, pain, dyspnea, insomnia, appetite loss, constipation, and diarrhea.

2.4. Statistical Analysis

Analyses were conducted in R (v 3.6.0) with α of 0.05. Descriptive statistics are expressed in means and standard deviations for normally distributed variables or median and interquartile range (IQR) for non-normally distributed variables. Weight loss constituted week 16 weight subtracted from baseline weight. Linear regressions were used to compare survivors and matched controls' weight loss while accounting for baseline BMI, age, gender, and engagement, since these are all factors that can influence the amount of weight lost [24–26]. *T*-tests, Mann-Whitney U test and chi-squared tests compared cancer survivors and matched controls on engagement and demographics. Cancer-related quality of life is presented descriptively with means and standard deviations.

3. Results

3.1. Demographics

Most cancer survivors had a history of breast cancer (n = 70). Other cancer types included melanoma (n = 9), cervical (n = 6), non-Hodgkin lymphoma (n = 5), renal (n = 5), skin (non-melanoma; n = 5), endometrial (n = 4), colon (n = 3), leukemia (n = 3), ovarian (n = 3), bladder (n = 2), rectal (n = 2), bone (n = 1), head and neck (n = 1), liver (n = 1), lung (n = 1), pancreatic (n = 1), prostate (n = 1), and other (n = 13). Survivors could indicate more than one cancer type. A majority of survivors reported having received chemotherapy (IV or pills) (63.5%), radiation (52.3%), and surgery (86.9%). Survivors could report more than one type of treatment. Most survivors received treatment 1 to less than 5 years ago (31.7%) or 5 to less than 10 years ago (24.3%), while 11.2% received treatment 10 or more years ago. For breast cancer survivors only, most reported receiving chemotherapy (IV or pills) (68.6%), radiation (68.6%), and surgery (94.3%). Controls had no history of cancer but had 0–1 chronic conditions. The most common chronic conditions were hypertension (13%) and depression (12%).

Demographic characteristics for all eligible cancer survivors and controls, as well as those included in primary analyses of weight and engagement, are displayed in Table 1. There were significant differences in employment status, where more cancer survivors were retired, and more controls worked 40+ hours per week. Eligible cancer survivors included significantly more females and were significantly older than controls. The same pattern for age but not gender emerged in the subset of breast cancer survivors and matched controls included in primary analyses.

Table 1. Demographic characteristics of eligible participants.

	All Eligible Participants			Participants Included in Primary Analyses		
	Cancer Survivors (*N* = 107), *N* (%) or Median (IQR)	Matched Controls (*N* = 150), *N* (%) or Median (IQR)	*p*-Value	Cancer Survivors (*N* = 43), *N* (%) or Median (IQR)	Matched Controls (*N* = 85), *N* (%) or Median (IQR)	*p*-Value
Hispanic/Latino			1			1
Yes	5 (4.7%)	7 (4.7%)		2 (4.7%)	3 (3.5%)	
No	102 (95.3%)	143 (95.3%)		41 (95.3%)	82 (96.5%)	
Race			0.10			0.20
Black or African American	4 (3.7%)	3 (2%)		2 (4.7%)	2 (2.4%)	
White	97 (90.7%)	141 (94%)		39 (90.7%)	81 (95.3%)	
Asian	0 (0%)	4 (2.7%)		0 (0%)	2 (2.4%)	
Other	6 (5.6%)	2 (1.4%)		2 (4.6%)	0 (0%)	
Employment status			<0.001			<0.001
Employed, 1–39 h per week	16 (15.0%)	43 (28.7%)		6 (14.0%)	25 (29.4%)	
Employed, 40+ hours per week	36 (33.6%)	84 (56%)		11 (25.6%)	50 (58.8%)	
Not employed	8 (7.4%)	9 (6.1%)		3 (7%)	4 (4.7%)	
Retired	36 (33.6%)	14 (9.4%)		17 (39.5%)	6 (7.1%)	
Disabled, not able to work	11 (10.3%)	0 (0%)		6 (14.0%)	0 (0%)	
Highest Education			0.95			0.85
High school degree or some high school	6 (5.6%)	7 (4.7%)		1 (2.3%)	4 (4.7%)	
Some college or vocational training	10 (9.3%)	20 (12.7%)		3 (7.0%)	16.4 (15.9%)	
2-year college degree	12 (11.2%)	12 (8%)		5 (11.6%)	6 (7.1%)	
4-year college degree	38 (35.5%)	48 (32%)		16 (37.2%)	30 (35.3%)	
Some graduate school	8 (7.5%)	13 (8.7%)		3 (7.0%)	7 (8.2%)	
Graduate degree	32 (29.9%)	49 (32.6%)		15 (34.8%)	23 (27.1%)	
I prefer not to answer	1 (0.9%)	1 (0.7%)		0 (0%)	1 (1.2%)	
Gender			<0.001			0.20
Female	100 (93.5%)	114 (76%)		39 (90.7%)	67 (78.8%)	
Male	7 (6.5%)	35 (23.3%)		4 (9.3%)	17 (20%)	
Other	0 (0%)	1 (0.7%)		0 (0%)	1 (1.2%)	
Current Age *	61 (53–67)	49 (38–58)	<0.001	62 (53.5–66.5)	49 (38–58)	<0.001

Note. * denotes variables that deviated from a normal distribution, so independent 2-group Mann-Whitney U tests were employed. Chi-squared tests were used for all other variables.

3.2. *Weight Loss*

Participants who had baseline and week 16 weight data were included in the weight loss outcomes (43 survivors, 85 controls). Among this subset, cancer survivors (M = 60.46, SD = 8.82) remained significantly older than matched controls (M = 47.51, SD = 13.07; t (115.69) = 6.64, $p < 0.001$) and had differing employment status. As seen in Table 2, matched controls lost significantly more weight (in kg) than cancer survivors [t (90.81) = −2.07, $p = 0.04$], but the percentage of body weight lost did not differ significantly. After controlling for gender, age, engagement, and baseline BMI, controls lost 1.9 kg more than cancer survivors (B = −1.90, S.E. = 0.95, $p = 0.05$). Overall, males lost more weight than females (B = 3.45, S.E. = 1.00, $p < 0.001$) and as participants aged, they lost more weight (B = −0.09, S.E. = 0.03, $p = 0.007$). The more participants engaged, the more weight they lost (B = −0.54, S.E. = 0.13, $p < 0.001$). Results did not change when controlling for engagement status, which was significantly different across groups but did not emerge as a significant predictor of weight (all *ps* > 0.60).

3.3. *Engagement*

There were differences between the groups in weekly engagement (Table 2). Cancer survivors sent more messages to their coaches compared to controls [W = 605862, $p < 0.001$]. However, controls logged more exercise sessions [W = 432425, $p < 0.001$], and took more steps [W = 415962, $p < 0.001$] compared to cancer survivors.

Table 2. Weight and engagement for cancer survivors and matched controls.

	Cancer Survivors (N = 43), Median (IQR) or Mean (SD)	Matched Controls (N = 85), Median (IQR) or Mean (SD)	p-Value
Baseline BMI *	32.78 (29.15–37.48)	31.82 (28.83–36.79)	0.50
Weight loss (kg)	−4.72 (4.34)	−6.52 (4.77)	0.04
Weight loss (%)	−6.20 (5.18)	−7.39 (4.67)	0.08
Engagement per week			
Coach messages *	2 (0–3)	1 (0–3)	<0.001
Weigh ins *	7 (4–7)	7 (5–7)	0.07
Exercises *	0 (0–2)	1 (0–7)	<0.001
Steps *	20321 (9386–39550)	35034 (16764–51781)	<0.001
Meals logged *	26 (19–33)	26 (21–31)	0.98
Articles read *	25 (11–28)	26 (12–28)	0.07

Note. * denotes variables that deviated from a normal distribution, so independent 2-group Mann-Whitney U tests were employed. *T*-tests were used for all other variables.

3.4. Quality of Life

Global QoL for all eligible survivors (N = 107) at the time surveys were administered was 72.0 on average (SD = 18.5). Scores on the functional subscales were as follows: role functioning: 78.3 (SD = 26.6), emotional functioning: 67.6 (SD = 22.1), cognitive functioning: 80.7 (SD = 19.3), and social functioning: 81.6 (SD = 25.5).

Average symptom scores were as follows: nausea: 4.4 (SD = 10.6), pain: 33.0 (SD = 29.0), dyspnoea: 11.8 (SD = 17.9), insomnia: 40.2 (SD = 28.5), appetite loss: 5.9 (SD = 15.7), constipation: 17.8 (SD = 26.8), diarrhoea: 11.5 (SD = 21.5), financial difficulties: 14.3 (SD = 25.9).

3.5. Subset Analysis

Due to the large proportion of breast cancer survivors in our sample, we conducted a subset analysis to compare breast cancer survivors to the controls. Compared to controls, breast cancer survivors wrote more coach messages (breast cancer survivors: Median = 2, IQR = 1–; controls: Median = 1, IQR = 0–3; p = 0.001), and logged fewer instances of exercise (breast cancer survivors: Median = 0, IRQ = 0–3; controls: Median = 1, IQR = 0–6; p < 0.001) and steps (breast cancer survivors: Median = 19996, IQR = 10181–38656; controls: Median = 33367, IQR = 12481–49324; p < 0.001). They logged their weight similarly to controls (breast cancer survivors: Median = 7, IQR = 5–7; controls: Median = 7, IQR = 5–7). When controlling for age, baseline BMI, and a composite score of overall engagement, breast cancer survivors lost significantly less weight than controls (B = −2.07, S.E. = −0.9, p = 0.03). Breast cancer survivors lost 5.37kg (SD = 4.39) on average, which was 6.5% body weight loss (SD = 5.4%). Controls lost 7.58kg (SD = 4.38) on average, which constituted 7.2% body weight loss (SD = 4.6%). Eligible breast cancer survivors had a global QoL of 74.5 (SD = 16.0), role functioning of 84.8 (SD = 21.6), emotional functioning of 67.4 (SD = 21.5), cognitive functioning of 80.9 (SD = 19.7), and social functioning of 87.4 (SD = 20.5). Their average symptom scores were as follows: nausea: 3.1 (SD = 8.2), pain: 30.5 (SD = 26.3), dyspnoea: 9.5 (SD = 18.1), insomnia: 40.9 (29.0), appetite loss: 5.2 (SD = 16.7), constipation: 13.8 (SD = 22.3), diarrhoea: 9.0 (SD = 14.9), financial difficulties: 13.8 (SD = 25.7).

4. Discussion

This retrospective study examined weight loss and engagement in cancer survivors compared to matched controls who were all trying to lose weight on a digital commercial weight loss program. For this population, real-world weight and engagement outcomes are unknown. This is a particularly pressing question for cancer survivors, who face post-treatment physical and mental health limitations which could impact their engagement and weight loss. It is therefore important to understand how weight loss outcomes for cancer survivors who signed up for a general weight loss program compare to those who do not have a history of cancer. Notably, previous investigations have only taken place in research study settings in which, at the very least, minimal participation requirements were salient. In McCarroll et al. [27], for example, participants were informed that they should provide baseline and follow-up measurements, received training on how to use the commercial program, and were contacted if they did not log food or exercise for more than 3 days in a row. To our knowledge, this is the first study to assess weight loss and engagement in a naturalistic environment where cancer survivors were using the program on their own initiative, without being reminded of participation requirements. We found that cancer survivors lost less weight by 16 weeks than matched controls, but still showed clinically significant weight loss (−5.3 kg or 6.2% body weight). In addition, cancer survivors had lower engagement for self-reported weight and exercise and objectively recorded steps throughout the program. However, when compared with the controls, the cancer survivors sent more messages to their coaches. When the analysis was limited to breast cancer survivors alone, this pattern still held. With regard to weight assessment, this behavior was similar in frequency between breast cancer survivors and controls.

In RCTs of digital commercial programs, cancer survivors lost on average 1.71 kg after 6 months and 2.3 kg after 4 weeks [27,28]. A systematic review found that body weight loss ranged from 2.4 to 6.8% in high-quality RCTs of generalized weight management interventions for survivors [29]. A systematic review of non-commercial weight loss interventions for breast cancer survivors found that survivors lost clinically significant amounts of weight (≥5%) in 14 out of 15 studies [30]. In the context of past studies, our results suggest that cancer survivors with overweight or obesity can lose significant and comparable weight on a digital commercial program, though they do not attain as much weight loss as individuals without a history of cancer.

We found that survivors showed less engagement in terms of logging or physical activity. This corroborates past studies showing that cancer survivors' engagement is relatively low in digital interventions, as well as work showing that cancer survivors experience fatigue and cognitive barriers to engaging as much as they would like [31–33]. We found for the first time to our knowledge that cancer survivors messaged their coaches more than matched controls. This could be because health coaches can provide additional motivation and trust [34]. Future studies should confirm that cancer survivors would benefit from amplified support from health coaches on a digital commercial program.

The study's additional aim was to describe survivors' cancer-related QoL at one time point during the program. One time point was chosen to minimize salient study requirements. These descriptive statistics provide rare data on survivors' QoL after real-world use of a self-managed commercial program and could inform future prospective trials, which are needed to directly compare QoL outcomes. Average global QoL for all survivors was 72.0 (SD = 18.5). In controlled trials of weight management interventions, survivors' scores were as follows: 56.4 (SD = n.a.) after a 12-week online weight loss intervention, 73.3 (S.E. = 3.7) after a 12-week stage-matched diet and exercise intervention, 79.5 (SD = 18.4) after a 12-week diet and exercise intervention, and 71.4 (SD = 18.8) after a 16-week physical activity and behavior change intervention [35–38]. A direct comparison cannot be made because of the difference between controlled and digital self-managed settings. Also, the study populations could have different demographic characteristics, with potentially higher socioeconomic status in a commercial program compared to other populations. Therefore, an aim of future work is to compare differences between QoL between cancer survivors and individuals without a history of cancer before and after using Noom. Future work should also compare long-term weight loss between survivors and controls on this type of program.

The study has a few limitations. In order to maximize ecological validity, quality of life was measured once rather than prospectively, weight loss was only analyzed with a subset of participants who provided baseline and 16-week weight measurements, and retrospective analyses were conducted. Because quality of life was not measured at baseline, it is unknown to what extent quality of life improved in cancer survivors over the course of the program. In addition, messaging, physical activity, and weight data were recorded throughout the program, which reduces recall bias, but causal interpretations cannot be made from a retrospective design. Another limitation is that the main outcome was self-reported weight, which can be prone to error or bias [39], and due to the study design, we could not assess its reliability compared to objective measurements. However, it should also be noted that self-reported weight can still be fairly accurate, and this type of observational design could decrease the opportunity for bias that stems from reporting weight directly to researchers (e.g., social desirability bias or from researchers' expectations) [40–44]. Still, future work should assess the reliability and validity of self-reported weight, and use other objective measurements (e.g., bioimpedance, plethysmography, or bone density measurement). Future research should also use accelerometers or other devices to objectively measure physical activity and calorie consumption. BMI also poses limitations. For instance, BMI does not account for weight variation due to changes in muscle mass (e.g., muscle mass loss from chemotherapy). Future studies should assess body composition specifically. Further, in addition to types of treatment, future work should also consider the duration of cancer treatments. Finally, only participants who did an in-app action in September 2020 were included in the study, since the goal was to investigate outcomes among those who actually participated in the program. This may limit generalizability of the findings and may represent a motivated sample that continued with the program and did not drop out early on.

5. Conclusions

This study contributes new knowledge with regard to the use of commercially available digital weight loss programs by cancer survivors outside the context of a clinical trial. Our findings highlight key differences in the experience of cancer survivors versus individuals without a history of cancer. While weight loss and engagement were lower in cancer survivors versus controls, cancer survivors interacted with coaches more than matched controls and lost a clinically significant amount of weight (>5%). Cancer-related QoL was also qualitatively comparable to previous post-weight loss intervention findings. Our results suggest that though cancer survivors can lose significant weight at 16 weeks on a digital commercial program, they may benefit from additional tailoring to improve their weight and engagement. Specifically, survivors may need cancer-specific support in terms of weight loss and motivation to engage in weight management behaviors. This could be done through support from health coaches, as survivors used this resource more than individuals with no history of cancer. These findings support future studies investigating the implementation of digital weight management platforms in oncology care.

Author Contributions: Conceptualization, C.N.M., E.S.M. and Q.Y.; methodology, C.N.M., Q.Y. and E.S.M.; formal analysis, Q.Y.; investigation, A.S.H.; Writing—Original draft preparation, A.S.H. and M.M.; Writing—Review and editing, A.S.H., C.N.M., N.M.I., A.C., E.S.M. and A.M.; supervision, A.M. All authors have read and agreed to the published version of the manuscript.

Funding: This research received no external funding.

Institutional Review Board Statement: The study was conducted according to the guidelines of the Declaration of Helsinki and approved by the Advarra Institutional Review Board.

Informed Consent Statement: Consent was obtained from all subjects involved in the study.

Data Availability Statement: Restrictions apply to the availability of these data. Data was obtained from Noom and are available by request from the corresponding author with the permission of Noom.

Conflicts of Interest: Authors C.N.M., A.S.H., Q.Y., M.M., E.S.M. and A.M. are employees of Noom Inc and have received salary and/or stock compensation from Noom, Inc. Author N.M.I. has the following relevant financial relationships to disclose: Consultant for: Novartis, Seattle Genetics; Grant support (to institution): Breast Cancer Research Foundation, American Cancer Society, National Cancer Institute, Conquer Cancer Foundation, Novartis. Author A.C. has no conflicts of interest to report.

References

1. Bluethmann, S.M.; Mariotto, A.B.; Rowland, J.H. Anticipating the "Silver Tsunami": Prevalence Trajectories and Comorbidity Burden among Older Cancer Survivors in the United States. *Cancer Epidemiol. Biomark. Prev.* **2016**, *25*, 1029–1036. [CrossRef] [PubMed]
2. Institute of Medicine; National Research Council. *From Cancer Patient to Cancer Survivor: Lost in Transition*; The National Academies Press: Washington, DC, USA, 2006; ISBN 978-0-309-09595-2.
3. Demark-Wahnefried, W.; Platz, E.A.; Ligibel, J.A.; Blair, C.; Courneya, K.S.; Meyerhardt, J.A.; Ganz, P.A.; Rock, C.L.; Schmitz, K.; Wadden, T.; et al. The Role of Obesity in Cancer Survival and Recurrence. *Cancer Epidemiol. Biomark. Prev.* **2012**, *21*, 1244–1259. [CrossRef] [PubMed]
4. Anderson, A.; Caswell, S.; Wells, M.; Steele, R.J.C. Obesity and lifestyle advice in colorectal cancer survivors-how well are clinicians prepared? *Color. Dis.* **2013**, *15*, 949–957. [CrossRef]
5. Stubbs, J.; Whybrow, S.; Teixeira, P.; Blundell, J.; Lawton, C.; Westenhoefer, J.; Engel, D.; Shepherd, R.; Mcconnon, Á.; Gilbert, P.; et al. Problems in Identifying Predictors and Correlates of Weight Loss and Maintenance: Implications for Weight Control Therapies Based on Behaviour Change: Predicting Weight Outcomes. *Obes. Rev.* **2011**, *12*, 688–708. [CrossRef] [PubMed]
6. Nikolaou, C.K.; Lean, M.E.J. Mobile applications for obesity and weight management: Current market characteristics. *Int. J. Obes.* **2017**, *41*, 200–202. [CrossRef]
7. Pew Research Center. Demographics of Mobile Device Ownership and Adoption in the United States. *Pew Research Center: Internet, Science & Tech.* Available online: https://www.pewresearch.org/internet/fact-sheet/mobile/ (accessed on 30 June 2021).
8. McCambridge, J.; Kypri, K.; Elbourne, D. Research participation effects: A skeleton in the methodological cupboard. *J. Clin. Epidemiol.* **2014**, *67*, 845–849. [CrossRef] [PubMed]
9. MacNeill, V.; Foley, M.; Quirk, A.; McCambridge, J. Shedding light on research participation effects in behaviour change trials: A qualitative study examining research participant experiences. *BMC Public Health* **2016**, *16*, 91. [CrossRef]
10. Valdivieso, M.; Kujawa, A.M.; Jones, T.; Baker, L.H. Cancer Survivors in the United States: A Review of the Literature and a Call to Action. *Int. J. Med. Sci.* **2012**, *9*, 163–173. [CrossRef]
11. Brunet, J.; Taran, S.; Burke, S.; Sabiston, C.M. A qualitative exploration of barriers and motivators to physical activity participation in women treated for breast cancer. *Disabil. Rehabil.* **2013**, *35*, 2038–2045. [CrossRef]
12. Gross, A.L.; May, B.J.; Axilbund, J.E.; Armstrong, D.K.; Roden, R.B.S.; Visvanathan, K. Weight Change in Breast Cancer Survivors Compared to Cancer-Free Women: A Prospective Study in Women at Familial Risk of Breast Cancer. *Cancer Epidemiol. Prev. Biomark.* **2015**, *24*, 1262–1269. [CrossRef]
13. Carmichael, A.R. Obesity and prognosis of breast cancer. *Obes. Rev.* **2006**, *7*, 333–340. [CrossRef]
14. Toro-Ramos, T.; Michaelides, A.; Anton, M.; Karim, Z.; Kang-Oh, L.; Argyrou, C.; Loukaidou, E.; Charitou, M.M.; Sze, W.; Miller, J.D. Mobile Delivery of the Diabetes Prevention Program in People With Prediabetes: Randomized Controlled Trial. *JMIR mHealth uHealth* **2020**, *8*, e17842. [CrossRef] [PubMed]
15. Van Dorsten, B.; Lindley, E.M. Cognitive and Behavioral Approaches in the Treatment of Obesity. *Endocrinol. Metab. Clin. N. Am.* **2008**, *37*, 905–922. [CrossRef] [PubMed]
16. United States Department of Agriculture. What Is MyPlate? Available online: https://www.myplate.gov/eat-healthy/what-is-myplate (accessed on 27 February 2021).
17. Rolls, B.J. Dietary energy density: Applying behavioural science to weight management. *Nutr. Bull.* **2017**, *42*, 246–253. [CrossRef]
18. Mitchell, E.; Yang, Q.; Ho, A.; Behr, H.; May, C.; DeLuca, L.; Michaelides, A. Self-Reported Nutritional Factors Are Associated with Weight Loss at 18 Months in a Self-Managed Commercial Program with Food Categorization System: Observational Study. *Nutrients* **2021**, *13*, 1733. [CrossRef] [PubMed]
19. Lawlor, E.R.; Islam, N.; Bates, S.; Griffin, S.J.; Hill, A.J.; Hughes, C.A.; Sharp, S.J.; Ahern, A.L. Third-wave cognitive behaviour therapies for weight management: A systematic review and network meta-analysis. *Obes. Rev.* **2020**, *21*. [CrossRef] [PubMed]
20. Michie, S.; Richardson, M.; Johnston, M.; Abraham, C.; Francis, J.; Hardeman, W.; Eccles, M.P.; Cane, J.; Wood, C.E. The Behavior Change Technique Taxonomy (v1) of 93 Hierarchically Clustered Techniques: Building an International Consensus for the Reporting of Behavior Change Interventions. *Ann. Behav. Med.* **2013**, *46*, 81–95. [CrossRef] [PubMed]
21. Kim, H.; Tietsort, C.; Posteher, K.; Michaelides, A.; Toro-Ramos, T. Enabling Self-management of a Chronic Condition through Patient-centered Coaching: A Case of an mHealth Diabetes Prevention Program for Older Adults. *Health Commun.* **2020**, *35*, 1791–1799. [CrossRef]
22. Madero, E.; Müller, A.; DeLuca, L.; Toro-Ramos, T.; Michaelides, A.; Seng, E.; Swencionis, C. Relationship Between Age and Weight Loss in Noom: Quasi-Experimental Study. *JMIR Diabetes* **2020**, *5*, e18363. [CrossRef]
23. Fayers, P.; Bottomley, A. EORTC Quality of Life Group Quality of Life Research within the EORTC—the EORTC QLQ-C30. *Eur. J. Cancer* **2002**, *38*, 125–133. [CrossRef]
24. Stubbs, R.J.; Pallister, C.; Whybrow, S.; Avery, A.; Lavin, J. Weight outcomes audit for 34,271 adults referred to a primary care/commercial weight management partnership scheme. *Obes. Facts* **2011**, *4*, 113–120. [CrossRef]
25. Carraça, E.V.; Santos, I.; Mata, J.; Teixeira, P.J. Psychosocial Pretreatment Predictors of Weight Control: A Systematic Review Update. *Obes. Facts* **2018**, *11*, 67–82. [CrossRef] [PubMed]
26. Perski, O.; Blandford, A.; West, R.; Michie, S. Conceptualising engagement with digital behaviour change interventions: A systematic review using principles from critical interpretive synthesis. *Transl. Behav. Med.* **2017**, *7*, 254–267. [CrossRef] [PubMed]

27. McCarroll, M.L.; Armbruster, S.; Pohle-Krauza, R.J.; Lyzen, A.M.; Min, S.; Nash, D.W.; Roulette, G.D.; Andrews, S.J.; von Gruenigen, V.E. Feasibility of a lifestyle intervention for overweight/obese endometrial and breast cancer survivors using an interactive mobile application. *Gynecol. Oncol.* **2015**, *137*, 508–515. [CrossRef] [PubMed]
28. Ferrante, J.M.; Devine, K.A.; Bator, A.; Rodgers, A.; Ohman-Strickland, P.A.; Bandera, E.V.; Hwang, K.O. Feasibility and potential efficacy of commercial mHealth/eHealth tools for weight loss in African American breast cancer survivors: Pilot randomized controlled trial. *Transl. Behav. Med.* **2020**, *10*, 938–948. [CrossRef]
29. Hoedjes, M.; van Stralen, M.M.; Joe, S.T.A.; Rookus, M.; van Leeuwen, F.; Michie, S.; Seidell, J.C.; Kampman, E. Toward the optimal strategy for sustained weight loss in overweight cancer survivors: A systematic review of the literature. *J. Cancer Surviv.* **2017**, *11*, 360–385. [CrossRef] [PubMed]
30. Playdon, M.; Thomas, G.; Sanft, T.; Harrigan, M.; Ligibel, J.; Irwin, M. Weight Loss Intervention for Breast Cancer Survivors: A Systematic Review. *Curr. Breast Cancer Rep.* **2013**, *5*, 222–246. [CrossRef]
31. Roberts, A.L.; Fisher, A.; Smith, L.; Heinrich, M.; Potts, H. Digital health behaviour change interventions targeting physical activity and diet in cancer survivors: A systematic review and meta-analysis. *J. Cancer Surviv.* **2017**, *11*, 704–719. [CrossRef]
32. Corbett, T.; Singh, K.; Payne, L.; Bradbury, K.; Foster, C.; Watson, S.E.; Richardson, A.; Little, P.; Yardley, L. Understanding acceptability of and engagement with Web-based interventions aiming to improve quality of life in cancer survivors: A synthesis of current research. *Psycho Oncol.* **2018**, *27*, 22–33. [CrossRef]
33. Balneaves, L.G.; Van Patten, C.; Truant, T.L.O.; Kelly, M.T.; Neil, S.E.; Campbell, K.L. Breast cancer survivors' perspectives on a weight loss and physical activity lifestyle intervention. *Support Care Cancer* **2014**, *22*, 2057–2065. [CrossRef]
34. Monteiro-Guerra, F.; Signorelli, G.R.; Rivera-Romero, O.; Dorronzoro-Zubiete, E.; Caulfield, B. Breast Cancer Survivors' Perspectives on Motivational and Personalization Strategies in Mobile App–Based Physical Activity Coaching Interventions: Qualitative Study. *JMIR mHealth uHealth* **2020**, *8*, e18867. [CrossRef]
35. Travier, N.; Guillamo, E.; Oviedo, G.R.; Valls, J.; Buckland, G.; Fonseca-Nunes, A.; Alamo, J.M.; Arribas, L.; Moreno, F.; Sanz, T.E.; et al. Is Quality of Life Related to Cardiorespiratory Fitness in Overweight and Obese Breast Cancer Survivors? *Women Health* **2015**, *55*, 1–20. [CrossRef]
36. Ligibel, J.A.; Meyerhardt, J.; Pierce, J.P.; Najita, J.; Shockro, L.; Campbell, N.; Newman, V.A.; Barbier, L.; Hacker, E.; Wood, M.; et al. Impact of a telephone-based physical activity intervention upon exercise behaviors and fitness in cancer survivors enrolled in a cooperative group setting. *Breast Cancer Res. Treat.* **2011**, *132*, 205–213. [CrossRef]
37. Lee, M.K.; Yun, Y.H.; Park, H.-A.; Lee, E.S.; Jung, K.H.; Noh, D.-Y. A Web-Based Self-Management exercise and Diet Intervention for Breast Cancer Survivors: Pilot Randomized Controlled Trial. *Int. J. Nurs. Stud.* **2014**, *51*, 1557–1567. [CrossRef] [PubMed]
38. Kim, S.H.; Shin, M.S.; Lee, H.S.; Lee, E.S.; Ro, J.S.; Kang, H.S.; Kim, S.W.; Lee, W.H.; Kim, H.S.; Kim, C.-J.; et al. Randomized Pilot Test of a Simultaneous Stage-Matched Exercise and Diet Intervention for Breast Cancer Survivors. *Oncol. Nurs. Forum* **2011**, *38*, E97–E106. [CrossRef]
39. Gorber, S.C.; Tremblay, M.; Moher, D.; Gorber, B. A comparison of direct vs. self-report measures for assessing height, weight and body mass index: A systematic review. *Obes. Rev.* **2007**, *8*, 307–326. [CrossRef]
40. Luo, J.; Thomson, C.A.; Hendryx, M.; Tinker, L.F.; Manson, J.E.; Li, Y.; Nelson, D.A.; Vitolins, M.Z.; Seguin, R.A.; Eaton, C.B.; et al. Accuracy of self-reported weight in the Women's Health Initiative. *Public Health Nutr.* **2018**, *22*, 1019–1028. [CrossRef]
41. Davies, A.; Wellard-Cole, L.; Rangan, A.; Allman-Farinelli, M. Validity of self-reported weight and height for BMI classification: A cross-sectional study among young adults. *Nutrition* **2020**, *71*, 110622. [CrossRef]
42. Hodge, J.M.; Shah, R.; McCullough, M.L.; Gapstur, S.M.; Patel, A.V. Validation of self-reported height and weight in a large, nationwide cohort of U.S. adults. *PLoS ONE* **2020**, *15*, e0231229. [CrossRef]
43. Kreuter, F.; Presser, S.; Tourangeau, R. Social Desirability Bias in CATI, IVR, and Web Surveys: The Effects of Mode and Question Sensitivity. *Public Opin. Q.* **2008**, *72*, 847–865. [CrossRef]
44. Béland, Y.; St-Pierre, M. *Mode Effects in the Canadian Community Health Survey: A Comparison of CATI and CAPI.*; American Statistical Association: Toronto, ON, Canada, 2004; pp. 297–314.

Article

Rationale and Methods for a Randomized Controlled Trial of a Dyadic, Web-Based, Weight Loss Intervention among Cancer Survivors and Partners: The DUET Study

Dorothy W. Pekmezi [1,2,*], Tracy E. Crane [3], Robert A. Oster [2,4], Laura Q. Rogers [2,4], Teri Hoenemeyer [5], David Farrell [6], William W. Cole [5], Kathleen Wolin [7], Hoda Badr [8] and Wendy Demark-Wahnefried [2,5]

1 Department of Health Behavior, University of Alabama at Birmingham (UAB), Birmingham, AL 35294, USA
2 O'Neal Comprehensive Cancer Center at University of Alabama at Birmingham (UAB), Birmingham, AL 35294, USA; roster@uabmc.edu (R.A.O.); lqrogers@uabmc.edu (L.Q.R.); demark@uab.edu (W.D.-W.)
3 Sylvester Comprehensive Cancer Center, Miami, FL 33136, USA; tecrane@med.miami.edu
4 Division of Preventive Medicine, Department of Medicine, Birmingham, AL 35294, USA
5 Department of Nutrition Sciences, University of Alabama at Birmingham (UAB), Birmingham, AL 35294, USA; tgw318@uab.edu (T.H.); colew14@uab.edu (W.W.C.)
6 People Designs, Inc., Durham, NC 27705, USA; dfarrell@peopledesigns.com
7 Coeus Health, LLC, Chicago, IL 60614, USA; kate@drkatewolin.com
8 Department of Medicine, Epidemiology and Population Sciences, Baylor College of Medicine, Houston, TX 77030, USA; hoda.badr@bcm.edu
* Correspondence: dpekmezi@uab.edu; Tel.: +1-205-975-8061

Citation: Pekmezi, D.W.; Crane, T.E.; Oster, R.A.; Rogers, L.Q.; Hoenemeyer, T.; Farrell, D.; Cole, W.W.; Wolin, K.; Badr, H.; Demark-Wahnefried, W. Rationale and Methods for a Randomized Controlled Trial of a Dyadic, Web-Based, Weight Loss Intervention among Cancer Survivors and Partners: The DUET Study. *Nutrients* **2021**, *13*, 3472. https://doi.org/10.3390/nu13103472

Academic Editor: William B. Grant

Received: 9 September 2021
Accepted: 24 September 2021
Published: 30 September 2021
Corrected: 29 July 2022

Abstract: Scalable, effective interventions are needed to address poor diet, insufficient physical activity, and obesity amongst rising numbers of cancer survivors. Interventions targeting survivors and their friends and family may promote both tertiary and primary prevention. The design, rationale, and enrollment of an ongoing randomized controlled trial (RCT) (NCT04132219) to test a web-based lifestyle intervention for cancer survivors and their supportive partners are described, along with the characteristics of the sample recruited. This two-arm, single-blinded RCT randomly assigns 56 dyads (cancer survivor and partner, both with obesity, poor diets, and physical inactivity) to the six-month DUET intervention vs. wait-list control. Intervention delivery and assessment are remotely performed with 0–6 month, between-arm tests comparing body weight status (primary outcome), and secondary outcomes (waist circumference, health indices, and biomarkers of glucose homeostasis, lipid regulation and inflammation). Despite COVID-19, targeted accrual was achieved within 9 months. Not having Internet access was a rare exclusion (<2%). Inability to identify a support partner precluded enrollment of 42% of interested/eligible survivors. The enrolled sample is diverse: ages 23–81 and 38% racial/ethnic minorities. Results support the accessibility and appeal of web-based lifestyle interventions for cancer survivors, though some cancer survivors struggled to enlist support partners and may require alternative strategies.

Keywords: diet; weight loss; exercise; physical activity; lifestyle; cancer survivors; Internet; dyads

1. Introduction

Given the high number of cancer survivors in the United States (over 16.9 million in 2019 [1]) due to improvements in early detection and treatment, new challenges emerge in terms of preventing second malignancies and common comorbidities and promoting quality of life (QoL) and healthy aging among survivors. Healthy diet, weight management, and physical activity can enhance the quality (and quantity) of life for cancer survivors and reduce their risk for developing secondary cancers; however, few cancer survivors meet the World Cancer Research Fund (WCRF) and American Institute of Cancer Research (AICR) recommendations for diet and physical activity [2]. Moreover, recent National

Health Interview Survey (NHIS) data indicate that 34% of adult cancer survivors report no leisure time physical activity and 33% have obesity [3], which clearly demonstrates the need for effective, scalable lifestyle interventions in this population.

Past studies have reported changes in diet, physical activity and/or weight loss among cancer survivors, with most interventions delivered face-to-face or via print and/or telephone [4]. For example, in the RENEW study, tailored mailed print materials and telephone counseling produced significant improvements in diet quality, physical activity, and body mass index at 12 months among 641 older survivors of breast, prostate, and colorectal cancer with overweight and obesity, which were sustained at two-year follow-up [5,6]. However, higher reach, more cost effective, technology-supported strategies might be required to address a public health problem of this magnitude.

Website intervention delivery requires less staff time and training than face-to-face and telephone-based approaches; furthermore, the incremental delivery costs per participant is minimal. Promising results were reported in a recent review of web-based lifestyle interventions for cancer survivors [7]; however, most of the programs were only 6–12 weeks in duration and did not assess long-term behavior change. As cancer survivors experience unique health concerns, more support and engagement may be required to achieve long-term maintenance of lifestyle changes.

Enrolling cancer survivors with support partners could lead to more sustainable gains in health behaviors. A dyad-based approach, in which 68 inactive breast cancer survivors with overweight or obesity partnered with their adult daughters who had similar characteristics, showed promise in the daughters and mothers (DAMES) study (n = 136) [8]. The dyads who received the tailored, self-help, print materials experienced significant improvements in body mass index (BMI), weight, waist circumference (WC), and physical activity at 12 months, compared to control arm dyads who received standard, publicly available brochures on diet, exercise, and weight status. Building on this research, the current study seeks to combine both the advantages of web-based platforms and partner support by developing and testing the first (to our knowledge) dyad- and web-based lifestyle intervention for cancer survivors.

The current paper describes the rationale, design, and recruited sample for an ongoing efficacy trial of DUET (daughters, dudes, mothers, and others together), a six-month web-based lifestyle intervention to promote weight loss among cancer survivors with overweight or obesity and their chosen supportive partners. The central hypothesis is that cancer survivor and support partner dyads that are assigned to the web-based intervention will lose significantly more weight at six months than dyads in the wait-list control. We also expect that the intervention will result in more favorable changes in other measures of adiposity (e.g., BMI and WC), diet quality, physical activity, QoL, and physical functioning and performance, as well as related biomarkers (e.g., insulin, glucose, total and high-density lipoprotein (HDL) cholesterol, triglycerides, leptin, adiponectin, interleukin 6(IL6), c-reactive protein (CRP), and tumor necrosis factor alpha).

2. Materials and Methods

2.1. Overall Design

DUET is a single-blinded, 2-arm randomized controlled trial (RCT) that will test a 6-month, web-based lifestyle intervention against a wait-list control among 56 dyads. Each dyad is comprised of a survivor of an obesity-related early-stage cancer and their supportive partner, both of whom have obesity or overweight, are insufficiently active, and consume suboptimal diets. The main outcome (body weight) is assessed at baseline and 6 months, along with several other secondary outcomes. This trial was approved by the University of Alabama at Birmingham (UAB) Institutional Review Board (300003882/Approval date: 10-28-2019) and registered with ClinicalTrials.gov (NCT04132219).

2.2. Participant Recruitment and Eligibility Screening

A two-step recruitment process was undertaken whereby initial enrollment efforts targeted cancer survivors, and once interest and eligibility were established, efforts were directed towards enrolling appropriate partners. Survivors of obesity-related cancers with 5-year cancer-free survival rates of at least 70% (i.e., localized renal cancer, and loco-regional ovarian, colorectal, prostatic, endometrial, and female breast cancers) [9]. Adult survivors of these cancers were identified through the UAB Cancer Registry, as well as through a wait-list of individuals who had previously expressed interest in lifestyle interventions and provided their contact email address or telephone numbers. Letters of invitation were posted to registry-ascertained cases, and telephone calls or email messages were placed to individuals on the wait-list. The Love Research Army (https://drsusanloveresearch.org/love-research-army, accessed on 27 September 2021) also initiated a series of email "blasts" to its members, and a recruitment website was established (https://duet4health.org, accessed on 27 September 2021). Finally, individuals not meeting eligibility criteria or disinterested in other ongoing cancer survivorship studies were apprised of DUET.

Study staff provided telephone follow-up on recruitment mailings and contacts by placing up to six calls at various days and times. The study was explained and interested survivors were screened for eligibility. Inclusion and exclusion criteria were established to target survivors who were most in need and who could best benefit from a web-based diet and exercise intervention. Inclusion criteria were as follows: (1) BMI > 25 kg/m^2 [10]; (2) vegetable and fruit intake <2.5 cups/day; (3) moderate-to-vigorous physical activity (MVPA) <150 min/week [11]; (4) completion of primary cancer treatment; (5) English speaking and writing; (6) educational attainment of 5th grade or higher; and (7) daily use of the Internet and mobile phone access. Exclusion criteria were few and limited to those who were already adhering to modified diets or enrolled in an exercise program, residing in assisted or skilled nursing facilities, or recently advised by their physician to limit physical activity and/or having health issues that might make participation in an unsupervised weight loss intervention unsafe (e.g., pregnancy, severe orthopedic conditions, end-stage renal disease, metastatic cancer or other cancers with poorer survival, paralysis, dementia, blindness, unstable angina, untreated stage 3 hypertension, recent history of heart attack, congestive heart failure or pulmonary conditions that required oxygen or hospitalization within 6 months) [12]. Once initial eligibility was established and cancer case status was verified by treating physicians of any self-referrals, the survivor was asked to identify a local support partner (preferably within a 10-min drive) and have them contact the research team for screening. Supportive partners were required to meet all inclusion/exclusion criteria, except for being a cancer survivor.

2.3. Study Protocol

The research team provided study overviews for eligible dyads and answered their questions via conference calls. Informed consent was obtained from all participants involved in the study and signed electronically (Adobe Sign®, San Jose, CA, USA). Participants completed baseline assessments and dyads were randomly and evenly assigned to study arms (DUET intervention or wait-list control) using a permuted block design (block size = 4). Participants complete assessments again at 6 months and then wait-list control dyads receive the DUET intervention.

2.4. DUET Intervention

The DUET web-based intervention was adapted from the previously mentioned tailored, mail-delivered dyadic DAMES intervention [8] and then expanded to meet the needs of a broader range of cancer survivors and support partners (i.e., not limited to post-menopausal breast cancer survivors and their biological daughters). Like DAMES, DUET was theoretically grounded and primarily based on the social cognitive theory (SCT) [13]; which posits that participation in health behaviors is determined by individual factors (e.g., self-efficacy, or confidence in the ability to exert control over one's own behavior) and the social and physical environment (e.g., barriers, social support from friends and family). The DUET intervention targets key SCT constructs by providing participants with resources (i.e., Fitbits and Aria Scales) to track diet, exercise, and weight, and provides guidance on setting incremental goals. Such strategies build upon small successes with lifestyle change and thereby enhance self-efficacy. The DUET weekly sessions also directly address weight loss barriers that are common for cancer survivors, such as fatigue and stress, as well as barriers common across populations, such as time constraints, to address the needs of both partners.

To further bolster dyadic interactions to enhance social support, concepts from interdependence theory and the theory of communal coping were incorporated [14]. DUET emphasizes relational factors such as joint problem solving, commitment to relationship quality and upholding mutual goals to promote adoption and maintenance of health behaviors. Moreover, dyads receive guidance on supporting their partners (e.g., how to ask for and provide help). In total, the DUET intervention draws upon 38 of the 40 behavioral change techniques that are categorized by the CALO-RE taxonomy purported by Michie and colleagues to promote adherence to healthful diet and physical activity patterns [15]. The two exceptions to the taxonomy are formal motivational interviewing (MI) and fear arousal. Although some MI elements were incorporated into the website design and information on cancer risk and recurrence and comorbidity are presented, emotionally evocative images were purposely avoided given the already high levels of anxiety related to such outcomes among cancer survivors and their loved ones [16].

The website includes the following sections: My Profile, Healthy Weight, Healthy Eating, Exercise, Weekly Sessions, Tools, News You Can Use, and Team Support. All users are given instructions on using the website features and encouraged to call the research team with any problems or questions. Once logged in, participants can update personal information (gender, age, height, weight, diet, physical activity, and cancer history) in the My Profile feature and then access tailored content in the Healthy Weight, Healthy Eating, and Exercise sections. Participants pursue and track their weight, diet, and exercise using study-provided equipment (Fitbit® Aria 2 digital scales and Inspire fitness trackers (San Francisco, CA, USA), Portion Doctor® tableware (Portion Health Products, St. Augustine Beach, FL, USA) and exercise bands (Theraband Academy, Akron, OH, USA)). Dyads also are encouraged to work as a team and use the commercially available MyFitnessPal (https://www.myfitnesspal.com/, accessed on 27 September 2021) app to set goals and view progress; log-ins and data are tracked to assess adherence. Participants are cued via text messages to complete 24 weekly interactive diet and exercise modules in the Sessions section; sessions range from 10–20 min. See Table 1 for session topics.

The Tools section includes tracking forms, online calculators, planning guides, tip sheets, and other healthy eating and exercise resources. Summarized updates on recent findings from salient research on diet, exercise, and/or weight loss for cancer prevention and control is provided in the News feature. The Team Support page offers practical tips on how dyads can support each other to promote lifestyle change (e.g., active listening). Regular website usage is encouraged via text messages (three per week) and tracked to assess intervention adherence. Short Message System (SMS) Text Messages also are a key component of the DUET intervention. After an initial welcome message, text messages are delivered at a frequency of three per week over the course of the intervention for a total of 72 messages for intervention.

Intervention adherence is evaluated using a variety of means. Completion of sessions, as well as website logins/duration and text message receipts/responses, are all tracked. Moreover, the dyad provides permission for the research office to access secure Fitbit wireless API generated by the Inspire tracker and Aria scale. These data are downloaded by study staff at intervention completion and stored on the study server by ID number until analysis.

Table 1. DUET diet and exercise sessions.

Week	Topic/Brief Description
1	What Can You Do to Lower Your Risk of Cancer? American Institute of Cancer Research's (AICR) dietary recommendations to lower cancer risk and why they are important.
2	Get on Track for Success! Using the Fitbit Aria weight scale each day and tracking tips and tools to promote weight loss.
3	Be Safe While Losing Weight! Importance of setting safe weekly weight loss goals that reduce the risk of sarcopenia.
4	Moving Towards Better Health Incremental goal setting to ultimately achieve 150 min of aerobic, resistance and flexibility exercises each week.
5	Let's Get Physical and Step It Up! Use of the Fitbit wrist monitor to track physical activity and making a safe exercise plan with incremental goals
6	Be S.M.A.R.T. About Physical Activity and Exercise. How to make exercise goals specific, measurable, achievable, relevant and time-based.
7	The Sweet 'n Low-down on Sugar and Fasting. Setting goals to limit of sugar intake and reviewing the concept and evidence for intermittent fasting.
8	Been Resisting "Resistance" Exercises? Importance of resistance exercise and instructions on how to perform them safely.
9	Yes, Portion Size Does Matter! Determining and tracking portions sizes; managing temptation while at the grocery store and dining out.
10	Why Are Bending Down and Touching Your Toes So Important for Good Health? How flexibility and balance exercises promote strength, and prevent falls and pain; instructions on how to begin safely.
11	Red and Processed Meats: How Can Something So Good Be So Bad? AICR recommendations on limiting red and processed meats. Harnessing social support to make dietary changes.
12	Did You Know that Your Surroundings Can Make You More Likely to Exercise? Managing environmental influences in support and promotion of good exercise habits.
13	Get the Skinny on Trimming the Fat. How high-fat foods contribute to risk of cancer and comorbidities; understanding different types of fat and food sources.
14	Reaping the Benefits of Whole Grains. Recommendations for whole grain daily intake and food sources.
15	Being Labeled is Not Always Bad. Importance of reading food labels and how.
16	Too pooped to Make Healthy Diet Choices? Recognizing and managing fatigue in support of healthy food preparation and making good choices.
17	Super Food Heroes: Fruits and Vegetables. Vegetables and fruits as sources of fiber, phytochemicals and antioxidants to reduce the risk for cancer and comorbidities; recommendation on sources, daily intake and serving sizes.
18	Problem Solving Strategies to Help You Get More Healthy Foods into Your Diet Identifying barriers/problems, brainstorming solutions, evaluating pros/cons, and developing an action plan.
19	Have Concerns About Pesticides Been Bugging You? Strategies to reduce pesticides in the diet and on foods; money and timesaving tips.
20	Want to Join the Party Without Blowing Your Diet? Recommendations on alcohol and cancer risk; making healthy choices when attending/hosting social gatherings.
21	Need a Break From Stress? Recognizing how stress influences physical and emotional wellbeing and strategies to manage it.
22	Why Am I Hungry All the Time? How to recognize hunger, and manage emotional or habitual eating.
23	Are Supplements Really Good for You? Information on safety and recommendations for supplement use
24	You did it! You Completed the DUET Program! Celebrating healthful eating and exercise behavior changes with positive rewards and planning for maintenance

2.5. Assessments

Baseline and 6-month follow-up assessments are largely identical except that some demographic and health characteristics that are likely to be time invariant for an adult sample over the study period (e.g., race/ethnicity, marital and educational status, height) are self-reported only at enrollment. Originally designed to include home-based assessments, the DUET protocol was modified prior to recruitment to virtual assessments via Zoom® (San Jose, CA, USA) in order to continue research activities during the COVID-19 pandemic. Assessors were trained and evaluated for accuracy prior to initiation; measures were evaluated for reliability, as well as validity with those collected in-person data that are featured in a separate report [17]. All Zoom® sessions are recorded to increase accuracy for timed performance testing to reduce discrepancies resulting from variable transmission of sight and sound and allow for periodic quality assurance evaluations among assessors. Once assessors review these files, time the tests, log the data, and quality assurance tests are completed, the recordings are deleted. Virtual assessments are scheduled during times when both dyad members can participate and occur in tandem with one member of the dyad undergoing the assessment first and the other recording the encounter on Zoom®, and then vice versa. Dyads are asked to prepare by viewing videos on performance testing (https://youtu.be/lbxctNuOgLk, accessed on 27 September 2021) and dried blood spot (DBS) collection (https://youtu.be/IBPLS4PoHv4, accessed on 27 September 2021).

Supplies also are sent to the home of the dyad member in which assessments will be performed. Mailed materials include an 8′ length of cord and two stickers (to mark the distance for the 8′ walk and up-and-go performance tests), two orange soccer cones (to enhance virtual visualization for walk testing), and a 36″ vinyl tape measure and two stickers (to measure and guide step height for 2-min step tests). The mailing also includes duplicate supplies to cover the assessment needs of each dyad member: (1) programmed Actigraphs (Walton Beach, FL, USA) with activity/sleep logs; (2) DBS kits (903TM Protein Saver Card, 2 lancets, 2 non-stick gauze pads, 2 small adhesive bandages, 2 alcohol prep wipes, 1–5 × 3″ foil biohazard envelop with desiccant) to self-collect fasting blood samples (12 h or more); and (3) Two ribbons (4–1″ × 55″) and a felt-tip marker (to perform repeated measures of WC). A digital scale is sent if participants do not have one. In addition to the virtual assessment and bio-specimen collection, each dyad member completes an on-line survey and a 2-day dietary recall conducted by telephone at each time point. Details of specific measures are provided in Table 2.

Table 2. Outcome measures.

PRIMARY OUTCOME
Body Weight: Weight is measured in light clothing without shoes. Zoom® images are captured of the "zeroed" scale display, the participant actively weighing, and final images of the display showing the participant's weight. The assessor verifies the weight with both the participant and partner. Weight is measured twice, and the average taken for analyses.
SECONDARY OUTCOMES
Waist circumference: The participant faces the camera and positions clothing to reveal midriff; as the partner is coached to encircle the waist with one of the ribbons at the level of the umbilicus [18]. As the participant rotates, the assessor checks to assure the ribbon is flat against the skin and parallel to the floor. Upon exhale, the partner uses a felt-tip marker to mark the ribbon at the point of overlap. The process is repeated with the second ribbon. Both ribbons are returned to the study office and measured in centimeters and the average taken for analyses. **Diet Quality:** A trained nutritionist conducts telephone-based dietary recalls of a non-consecutive weekday and weekend day using the National Cancer Institute (NCI)-developed Automated Self-Administered 24-h (ASA24) recall dietary assessment web-based tool (https://epi.grants.cancer.gov/asa24/, accessed on 27 September 2021). Calorie intake and nutrient density are averaged over the 2 days for each time point and Diet Quality is calculated using the Healthy Eating Index (HEI)-2015 [19]. **Physical Activity:** Programmed actigraphs (Fort Walton, FL, USA) objectively capture physical activity over a 7-day period and are then downloaded and processed using procedures and software supplied by the manufacturer and using methods similar to those we have reported previously [20,21]. Physical activity also will be measured by self-report using the Godin Leisure-Time Exercise Questionnaire, given its excellent reliability and validity with cancer survivors [22,23]. **Physical Performance Testing:** The Senior Fitness Battery assesses physical performance objectively across multiple domains, is sensitive to change, minimizes ceiling effects, and has normative scores [24]. Usually conducted in-person, tests were adapted to virtual use, refined, and then evaluated for validity and reliability [17]; arm curls and grip strength, were omitted because of excessive equipment and postage costs. - 30-s chair stand (lower body strength): A standard 18″ unpadded chair is used, though if the participant does not have one, the identical chair is used for both baseline and follow-up assessments. The participant sits in view of the camera and is instructed to cross arms with hands on shoulders. Upon the assessor's signal to start, the participant stands up and sits down as many times as possible during a 30-s timed period. - 8′ Get Up & Go (agility, dynamic balance): Participant begins seated with crossed arms and hands on shoulders while the partner places a sticker and the end of the 8′cord (from mailed supplies) beneath the toe and extends the cord fully in front of the chair. The endpoint is marked by a soccer cone and the cord removed. The camera is positioned to capture the full course with a focus on the chair (start and end points for this test). Upon the signal to start, the participant stands, walks as fast as possible (without running) around the cone, returns to the chair, and sits down. The test is timed using the video—starting from the sign of movement until seated again. - 8′ Walk (gait speed): The chair is removed, and the participant stands with their toe on the sticker (see test above). Upon the signal to start, they walk as fast as possible through the 8′point marked by 2 soccer cones (another cone is added to increase visibility of the finish line). This test also is timed using the video, starting from the sign of movement until the finish line is crossed. - Sit-and-reach (flexibility): Seated on the edge of the chair, the participant extends one leg with their heel on the floor, the knee straightened, and the toe pointed to the ceiling. The camera captures the side view, and the assessor guides the participant to overlap their hands and extend them towards the toe. The partner measures the distance from the middle finger to the big toe with a vinyl tape measure. Positive values are recorded for over-reaching, negative for under-reaching, and zero for touching. - Back scratch (flexibility): The camera captures a back view while the participant reaches over their same shoulder while at the same time reaching their other arm directly back in an attempt to their clasp hands. The partner measures the distance between the closest fingers. Positive for over-reach, negative for under-reaching, zero for touching. - 2-min step test (endurance): The partner is instructed to palpate the participant to locate their iliac crest and then uses the vinyl tape measure to record the distance to the top of the patella, which is called-out to the assessor. The assessor calculates the midpoint, which is denoted by a sticker. Then the partner is asked to measure the distance from the sticker to the floor and call-out the value to the assessor. The assessor records this value for future testing and instructs the partner to measure this distance against a wall and to mark it with another sticker. The camera captures the side view and upon the command to start, the participant is instructed to "march in place" for 2 min making sure to bring their knees up to point of the sticker. The participant is instructed not to talk, and to take breaks, and briefly reach out to the wall to regain balance as needed while timer continues (partners are instructed to "spot" the participant as needed). The assessor counts steps during the 2-min period (steps not reaching the mark are not counted). **Balance Testing:** Zoom® captures side-by-side, semi-tandem and tandem stance balance testing as advocated by the Centers for Disease Control [25]. To reduce ceiling effects, the latter test is extended for up to two minutes (or until the time the stance is broken). **Circulating Biomarkers:** DBS captured on the designated card are dried thoroughly (>4 h at room temperature), then inserted into a foil pouch with desiccant and frozen (0 F° or below) until analyzed. DBS eluents are batch-tested against known standards for insulin, glucose, leptin, adiponectin, high density lipoprotein (HDL) and total cholesterol, triglycerides, interleukin-6 (IL6), c-reactive protein (CRP) and tumor necrosis factor alpha (TNFα) at the University of Washington as described previously [26]. Values are expressed in plasma equivalent terms. **Quality of Life:** The PROMIS global health scale and the EuroQOL-5D-5L (EQ-5D-5L) will be used to measure QOL [27]. The EQ-5D-5L includes 5 dimensions (Mobility, Self-care, Pain/Discomfort, and Anxiety/Depression) and the scores are used to calculate Quality Adjusted Life Years. **Comorbidity:** The Older Americans Resources & Services (OARS) Comorbidity Index (43-items) will assess the number and severity of chronic medical conditions and symptoms. Since falls are a particular issue in this population, an item validated by Chen & Janke that assesses falls in the past year also will be included [28,29].

As indicated and in addition to comorbidity, other potential moderators of the intervention's effect on weight change, such as demographic factors, distance separating the dyad members (and dyad cohabitation vs. not), smoking status [30], and risk for depression (as measured by the PROMIS Cancer-Related Item Bank) will be explored [31]. Potential mediators of effect also will be studied and include specific SCT constructs directly targeted by the intervention (e.g., self-efficacy, social support, and barriers). Self-efficacy will be measured with a 20-item instrument (α = 0.70–0.88) for dietary weight management [32] and the 6-item Lifestyle Efficacy scale (α = 0.95) [33]. Social support for these lifestyle changes will be assessed using validated 5-point scales with acceptable test-retest reliabilities (r = 0.55–0.86) and internal consistencies (α = 0.61–0.91) [34]. Barriers will be captured using a list of 36 common barriers to a diet with reduced fat and sugar, and increased fruits and vegetable intake, whole grains, and exercise (cost, availability, time, etc.) [35–38].

Upon completion of the intervention both dyad members undergo separate telephone debriefings on the acceptability and satisfaction of the various intervention components (e.g., website, equipment, text messages) and their suggestions for improvement are solicited.

As with any lifestyle intervention trial conducted in a high-risk patient population, especially one that is home-based and unsupervised, adverse events are a key concern. Thus, changes in health status of both study arms are systematically ascertained at study midpoint (3 months), in addition to 6-month follow-up. Any hospitalizations are logged, and admittance to the hospital resulting in an overnight stay, as well as events that are permanently disabling or life threatening are categorized as "serious" with attribution of the intervention explored further. Furthermore, all study participants are encouraged to call a toll-free study number to report any adverse events that occur between assessments.

2.6. Statistical Power

This 2-arm RCT formally tests for differences in the loss of body weight from baseline to 6-months that occurs among 56 dyads (with each dyad comprised of a survivor and a partner) who are randomized to two study arms. All other analyses and tests are exploratory. Power calculations are based on the following assumptions: (1) the retention rate will be identical to the DAMES trial (i.e., 90%), which is conservative as the duration for DUET is only 6 months instead of 12 months; (2) the eHealth intervention will promote weight losses of ~3.46 kg during the 6-month study period (similar to those observed in the Healthy Moves tailored-print, web-based, spousal support intervention—see companion article by Carmack et al. in this Nutrients edition [39]), whereas the control arm will be weight stable—a conservative estimate, since the average American gains 0.5–1 kg/year [40]); and (3) two-sided tests at an alpha level of 0.05 will be used. Given these assumptions, along with those of a two-sided two-group *t*-test, a standard deviation of 4.6 kg (from Healthy Moves [39]), and a sample size of 25 dyads per arm (allowing for 90% retention of the initial 28 dyads per arm), there is at least 80% power to detect differences in weight loss of −3.72 kg or greater between the two arms.

2.7. Data Analyses

The primary analysis will be performed on an intent-to-treat basis using baseline to 6-month data. Arm differences in weight loss will be assessed using a mixed linear model which accounts for the covariance between the dyad members. Specifically, mixed models repeated measures analyses will be used to test differences between arms, the two time points, and the potential interaction between arm and time point simultaneously. An appropriate structure for the covariance matrix (e.g., unstructured) will be selected using the final data. Clinical and demographic covariates of interest will be included in these models. The Tukey-Kramer multiple comparisons test will be used to determine which pairs of means are significantly different. Overall cross-sectional comparisons of continuous variable at baseline will be performed using the two-group *t*-test to determine if there are any differences remaining between the arms after randomization. For categorical variables, comparisons between arms will be performed using the two-group chi-square test (or Fisher's exact test if the assumptions for the chi-square test are not tenable). The strength of the relationship between pairs of variables will be examined using Pearson (or Spearman, if needed) correlation analyses. Distributions of continuous study variables will be examined using stem-and-leaf, box, and normal probability plots and the Kolmogorov-Smirnov test; any of these variables that deviate from a normal distribution will be transformed prior to analysis or will be analyzed using non-parametric methods such as the Wilcoxon rank-sum test.

Statistical tests will use an alpha level of 0.05 and will be two-sided. SAS software (version 9.4 or later; SAS Institute, Inc., Cary, NC, USA) will be used to perform all statistical analyses. Analyses of secondary outcomes will occur similarly, though analyses are exploratory, and as such will not be controlled for multiple testing. To identify predictor variables associated with program efficacy, e.g., social support (type, amount), self-efficacy, and risk of depression, logistic regression analyses will be used. Odds ratios, along with their corresponding two-sided 95% confidence intervals, will be obtained for all variables included in these models.

For our initial statistical analyses that focus on determining potential demographic and cancer-type differences between enrolled DUET cancer survivors and partners, between cancer survivors who express interest in the DUET intervention versus those who refuse or are unresponsive, and those who enroll in the RCT versus those not enrolled, two-group t-tests are performed for continuous variables such as age, and chi-square tests are performed for categorical variables, such as gender, race/ethnicity, residence in a rural- or urban-classified county, and cancer-type. These analyses are now complete, and the results are presented in the next section—findings that are integral in assessing program interest and to appropriately generalize the main outcomes of this trial upon its completion.

3. Results

Recruitment for DUET spanned 8 October 2020 to 2 July 2021, and despite substantial overlap with the COVID-19 pandemic, met its accrual target of 56 partnered dyads within a 9-month period. To date, there have been no drop-outs; however, the trial is still in the field with completion of data collection anticipated within the next five months. Laboratory and statistical analyses will occur over the subsequent 6-month period.

Figure 1 details the study trajectory from self-referral or registry/wait-list ascertainment to randomization. Data suggest that for intervention RCTs like DUET, roughly 23 cancer survivors require contact for every participant enrolled. Granted, this number could be reduced to less than 20 if contact data in registries were current, but roughly 12% of cases were found to be deceased or had telephone and/or address information that was obsolete. Of those for whom contact is assumed, roughly 21% express interest in participating in the trial, but over one-third (37%) screen-out on various eligibility criteria with the leading causes for exclusion being normal or underweight status, already adhering to a healthful diet or regular exercise, or medical exclusions.

Of note, irregular computer use, or lack of Internet access was a rarely reported occurrence, with less than 2% being screened-out on this criterion. Moreover, while a substantial number of cancer survivors initially expressed interest, almost 20% were lost to follow-up afterward. This substantial loss to follow-up also occurred once enrollment became focused on partners; here loss to follow-up accounted for 28% of individuals identified either prior to consent or the baseline appointment. Additionally, the identification of a partner appeared to be a barrier since 42% of interested survivors were unable to enlist one.

DUET enrolled participants in Alabama, Illinois, Mississippi, North Carolina, and Tennessee. Table 3 provides the characteristics of the DUET cohort. The sample of survivors is comprised largely of individuals diagnosed with female breast cancer, though early-stage kidney, prostate, endometrial and ovarian cancer also are represented. Interestingly, 13% of supportive partners also reported cancer histories, again with most of these diagnoses being female breast cancer.

By and large, these are long-term cancer survivors more than five years (M = 67.5 months) out from diagnoses. Given the high representation of breast cancer (80%), it is unsurprising that most participants are female (86%). The age range is broad (i.e., 23–81 years) with a mean age of 58.4 years, and while most participants are employed (55%), one-third are retired. In addition, most (85%) acknowledged at least some college education, with a substantial proportion (43%) reporting annual incomes of at least $50,000 or refusing to answer this question. Minorities comprise almost 40% of the sample, with non-Hispanic Blacks (NHB) having the highest representation; however, very few participants are rural. While most dyads resided separately, over 40% cohabitated with their supportive partners (all of cohabitating partners were in spousal relationships). Both survivors and their supportive partners had average BMI's falling in the range of Class I obesity (M BMI = 32).

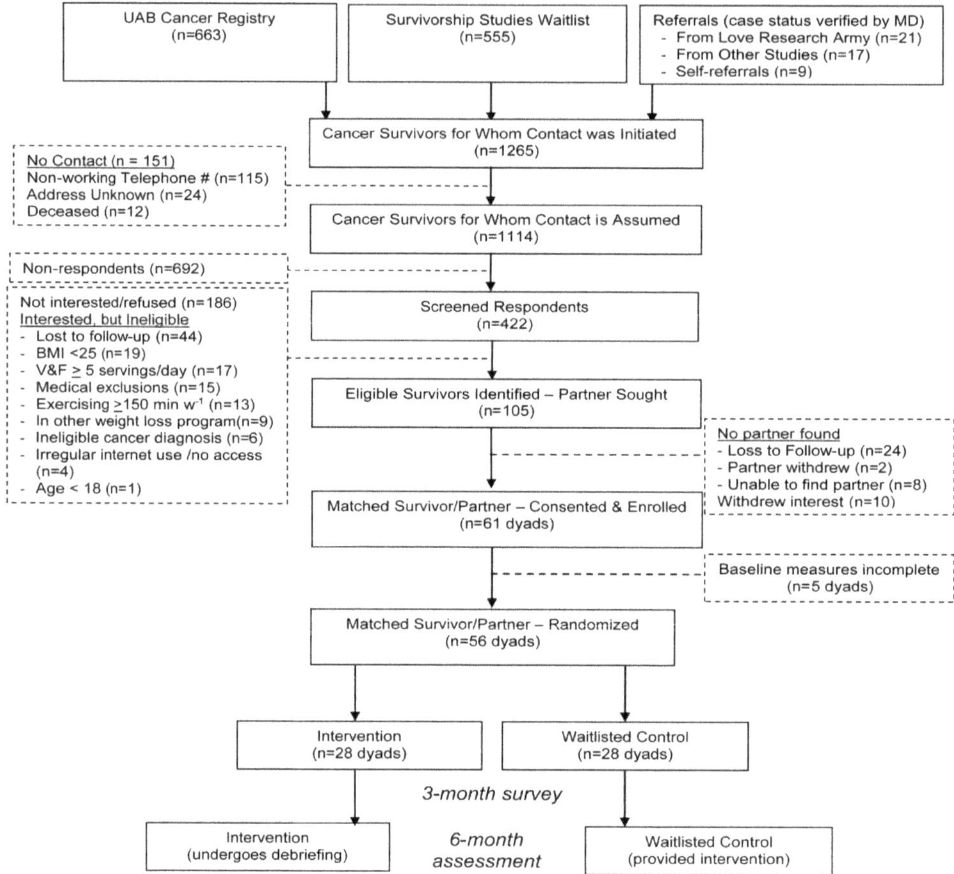

Figure 1. DUET study flow diagram that focuses on the enrollment trajectory.

Tests that compared the enrolled DUET cancer survivors ($n = 56$) to partners ($n = 56$) found significant differences for cancer diagnosis ($p < 0.001$), where the proportion of survivors with cancer (100%) was greater than the proportion of partners with cancer (13%), and for gender ($p = 0.025$), where the proportion of female survivors (86%) was greater than the proportion of female partners (68%).

Tests that compared the sample who responded with interest ($n = 236$) and ultimately enrolled in DUET ($n = 56$) to the larger pool who were unresponsive ($n = 1029$) and unenrolled ($n = 1209$) found no significant differences by rural (versus urban) residence or race/ethnicity ($p > 0.10$ for both); however, there were significantly higher response rates among survivors of breast cancer (and hence, females), as well as those who are younger ($p < 0.05$ for both). Though after screening, the only clear difference between enrolled participants versus the potential pool was the proportion of breast cancer survivors, which was significantly higher among enrollees ($p < 0.0001$).

Table 3. Study sample characteristics of DUET cancer survivors and partners *.

	Survivors (n = 56)	Partners (n = 56)	p-Value **
Cancer Diagnosis (n/%) *			
- Breast	45 (80%)	4 (7%)	
- Colorectal	1 (2%)	0	<0.001
- Gynecologic	2 (4%)	2 (4%)	
- Genitourinary	8 (14%)	1 (2%)	
Months elapsed since diagnosis			
- Mean (sd) (n = 53)	67.5 (72)		
- Range	10–303		
- Miles between Survivor and Partner (n/%)			
- 0 (cohabitate)	24 (43%)		
- Greater than 0, but less than 5	12 (21%)		
- 5 to 10	8 (14%)		
- More than 10	12 (21%)		
Race/Ethnicity			
- Non-Hispanic White	35 (63%)	34 (61%)	
- Hispanic White	0	1 (2%)	
- Non-Hispanic Black	19 (34%)	21 (38%)	0.8
- Hispanic Black	1 (2%)	0	
- Other	1 (2%)	0	
Gender (n/%)			
- Male	8 (14%)	18 (32%)	0.03
- Female	48 (86%)	38 (68%)	
Age (years)			
- Mean (sd)	60.3 (11)	56.5 (14.3)	0.1
- Range	32 79	23 81	
Educational Status			
- High School Graduate	7 (13%)	9 (16%)	
- Some College/Junior College/Trade School	18 (32%)	17 (30%)	0.9
- College Graduate/Post Graduate	30 (54%)	29 (52%)	
- Unknown	1 (2%)	1 (2%)	

Table 3. *Cont.*

	Survivors (n = 56)	Partners (n = 56)	p-Value **
Income			
- Less than $50k/year	11 (20%)	7 (13%)	
- $50k/year or more	24 (43%)	23 (41%)	0.5
- Unreported or Refused/Unknown	21 (38%)	26 (46%)	
Rural (n/%)	4 (7%)	5 (9%)	1
BMI (kg/m^2) Mean (sd)	31.8 (5.8)	32.9 (6.1)	0.3
Employment (n/%)			
- Employed	31 (55%)	31 (55%)	
- Retired	18 (32%)	18 (32%)	1
- Other	7 (13%)	7 (13%)	

* Information on cancer-type was verified for cancer survivors, but was self-reported for supportive partners. ** For cancer diagnosis, participants with cancer were compared to participants without cancer; for race/ethnicity, only Non-Hispanic Whites and Non-Hispanic Blacks were compared; for educational status, participants with a response of unknown were excluded from the analysis.

4. Discussion

To our knowledge, DUET is the first interactive web-based intervention aimed at improving body weight status, dietary intake, and physical activity among high-need survivors of obesity-related cancers and their supportive partners. As such, it represents a program that not only addresses the tertiary prevention needs of cancer survivors, but also serves to promote primary prevention among their friends and family members—a substantial proportion of whom are at higher risk due to common risk factors. In addition, DUET is unique from the perspective of capturing outcome data strictly using remote methodologies. Thus, it is among a new generation of trials in which the intervention is delivered, and the outcomes are assessed exclusively via remote means. The fact that we were able to meet our accrual target for this logistically challenging trial involving a 2-step process among both cancer survivors and partners within nine months, and during a pandemic when other cancer prevention and control trials are struggling (i.e., Unger et al. reports a decrease of 54% in enrollment during the same period [41]), is testimony to the fact that web-based approaches have appeal.

Remote delivered and assessed trials also have the ability to recruit participants broadly and thereby potentially increase the generalizability of findings. As such, DUET was able to engage participants residing in a broad swath of America, from Illinois to Alabama. Moreover, it attracted cancer survivors and supportive partners across a vast age range that extended from individuals in their second through eighth decades of life. It also enrolled a racial and ethnically diverse cohort, as supported by a minority accrual of almost 40%, thereby surpassing U.S. Bureau of Census statistics that suggest 31% for the mean age group of this sample (i.e., 58–59 years) [42]. Indeed, web-based trials remove several barriers that are commonly reported for both older and minority populations, such as transportation and time away from family or occupational commitments [43]. Furthermore, the concern that cancer survivors and their partners would not have Internet access nor adequate computer skills was unfounded based on our data that less than 2% of survivors screened-out on this criterion and there has been good uptake of the intervention to date. Albeit this percentage is far lower than the 18% computer-related exclusion that was recently reported by van der Hout and colleagues for a web-based supportive care intervention across a mixed sample of cancer survivors; however, recruitment for their Oncokompas trial occurred in 2016–2017 [44]. Given estimates indicating that there are on average 640,000 new users of the Internet each day globally (with sharp increases during the pandemic) [45], the concern that cancer survivors may be unreachable through web-based programs appears to be diminishing rapidly.

A much greater barrier to accrual was the identification of a partner in order to participate in this dyadic-based intervention. Forty-two percent of interested and eligible cancer survivors were unable to engage a supportive partner. While this proportion is lower than the 48% suggested by the DAMES trial, there was an expectation that expanding the criteria for a partner beyond just a biological child to include other family members, spouses, friends, and neighbors, would yield a far better response rate—it did not. Therefore, this is a key concern for dyadic-based interventions in the future, at least for those that are aimed at improving diet quality, physical activity and weight status among cancer survivors and their circles of friends and family members. That being said, the magnitude of change possible for dyadic interventions needs to be weighed against these logistical considerations. The fact that both the DAMES and Healthy Moves trials resulted significant improvements in vegetable and fruit consumption and/or weight loss with modest-sized samples suggests that although dyadic interventions are challenging, they still may be worth the effort [8,39]. The results for DUET will add substantially to this small body of research.

Of note, the considerable representation of dyad spouses within the DUET cohort and the relative ease with which the Healthy Moves cohort was assembled, suggests that the spousal relationship is perhaps the most fruitful to capitalize upon and engage potential participants [39]. Because the DUET sample is relatively evenly divided between survivor-spouse dyads versus dyads comprised of survivors and others, it will be one of the first (if not the first) to compare changes that occur in health behaviors and outcomes changes in these two different subgroups. While our study is likely to be underpowered in detecting significant differences, the descriptive data that result still will be helpful in supporting frameworks such as that proposed by Monterrosa et al. that identify and categorize the several different influences on food choices on various social and environmental levels [46]. Given that the spousal relationship and its inherent cohabitation affect food procurement and preparation, and other more far-reaching domains, we anticipate that intervention effects may be accentuated in this subgroup.

As stated, the outcomes of the DUET RCT are anticipated within the next calendar year. Given its potential to break new ground in the fairly small (yet growing) areas of dyadic interventions, as well as remote intervention delivery and evaluation, results should be of interest to researchers not only involved in cancer control, but also interventionists who implement diet and exercise interventions within a variety of patient populations and to prevent a multitude of chronic diseases. Strengths of DUET include its randomized controlled design, theoretically-grounded intervention, and attention to fidelity. As with all studies, DUET has limitations which include a significant overrepresentation of breast cancer survivors (many of whom are upper-socioeconomic), and few dyads with rural residence. These are common limitations that have been reported by other research teams [47–49]. To address these concerns, future studies might consider preferentially accruing from support groups that focus on other types of cancer and community cancer centers in smaller towns (rather than major tertiary oncologic care centers). Finally, and potentially as a last resort, future studies may need to stratify accrual by cancer type, socio-economic factors and/or rural/urban residence to assure adequate representation.

5. Conclusions

This paper provides an in-depth description of DUET—a theoretically grounded, dyadic-based lifestyle intervention for cancer prevention and control that is delivered and evaluated exclusively using remote technology. The report includes specific information on the overall protocol, intervention, and measures. Moreover, details are presented on the recruitment for this RCT which was successful in achieving its total accrual target with broad representation across the United States, and which also surpassed benchmarks for racial and ethnic representation.

Author Contributions: D.W.P.: Conceptualization, funding acquisition, investigation, methodology, project administration, supervision, visualization, writing—original draft, reviewing and editing. T.E.C.: conceptualization, investigation, methodology, visualization, writing—original draft, reviewing and editing. R.A.O.: conceptualization, formal analysis, funding acquisition, writing—original draft, reviewing and editing. L.Q.R.: conceptualization, investigation, methodology, supervision, visualization, writing—reviewing and editing. T.H.: data curation, investigation, project administration, supervision. Writing—original draft, reviewing and editing. D.F.: conceptualization, visualization, methodology, writing—reviewing and editing. W.W.C.: data curation, investigation, project administration, supervision. Writing—original draft, reviewing and editing. K.W.: conceptualization, writing—reviewing and editing. H.B.: conceptualization, writing—reviewing and editing. W.D.-W.: conceptualization, funding acquisition, investigation, methodology, project administration, resources, supervision, visualization, writing—original draft, reviewing and editing. All authors have read and agreed to the published version of the manuscript.

Funding: This research was funded by the American Institute for Cancer Research (585363), the American Cancer Society (CRP-19-175-06-COUN), the National Cancer Institute (T32 CA047888, P30 CA013148, R25 CA76023, P30 CA023074).

Institutional Review Board Statement: The study was conducted according to the guidelines of the Declaration of Helsinki, and approved by the Institutional Review Board of The University of Alabama at Birmingham (protocol code 300003882/Approval date: 28 October 2019) and is registered with ClinicalTrials.gov (NCT04132219).

Informed Consent Statement: Informed consent was obtained from all subjects involved in the study.

Data Availability Statement: Not applicable.

Acknowledgments: First: the authors thank all of the cancer survivors and their partners who have enrolled in DUET. Moreover, we acknowledge the tremendous contributions and dedication of our staff, pre- and post-doctoral fellows, student interns, and work study students.

Conflicts of Interest: The authors report no conflict of interest.

References

1. Siegel, R.L.; Miller, K.D.; Fuchs, H.E.; Jemal, A. Cancer Statistics, 2021. *CA Cancer J. Clin.* **2021**, *71*, 7–33. [CrossRef]
2. Blanchard, C.M.; Courneya, K.S.; Stein, K.; American Cancer Society's SCS-II. Cancer survivors' adherence to lifestyle behavior recommendations and associations with health-related quality of life: Results from the American Cancer Society's SCS-II. *J. Clin. Oncol.* **2008**, *26*, 2198–2204. [CrossRef]
3. Cancer Trends Progress Report. July 2021. Available online: https://progressreport.cancer.gov/ (accessed on 27 September 2021).
4. Pekmezi, D.W.; Demark-Wahnefried, W. Updated evidence in support of diet and exercise interventions in cancer survivors. *Acta Oncol.* **2011**, *50*, 167–178. [CrossRef]
5. Morey, M.C.; Snyder, D.C.; Sloane, R.; Cohen, H.J.; Peterson, B.; Hartman, T.J.; Miller, P.; Mitchell, D.C.; Demark-Wahnefried, W. Effects of home-based diet and exercise on functional outcomes among older, overweight long-term cancer survivors: RENEW: A randomized controlled trial. *JAMA* **2009**, *301*, 1883–1891. [CrossRef]
6. Demark-Wahnefried, W.; Morey, M.C.; Sloane, R.; Snyder, D.C.; Miller, P.E.; Hartman, T.J.; Cohen, H.J. Reach out to enhance wellness home-based diet-exercise intervention promotes reproducible and sustainable long-term improvements in health behaviors, body weight, and physical functioning in older, overweight/obese cancer survivors. *J. Clin. Oncol.* **2012**, *30*, 2354–2361. [CrossRef]
7. Williams, V.; Brown, N.; Becks, A.; Pekmezi, D.; Demark-Wahnefried, W. Narrative Review of Web-based Healthy Lifestyle Interventions for Cancer Survivors. *Ann. Rev. Res.* **2020**, *5*, 555670. [CrossRef]
8. Demark-Wahnefried, W.; Jones, L.W.; Snyder, D.C.; Sloane, R.J.; Kimmick, G.G.; Hughes, D.C.; Badr, H.J.; Miller, P.E.; Burke, L.E.; Lipkus, I.M. Daughters and Mothers Against Breast Cancer (DAMES): Main outcomes of a randomized controlled trial of weight loss in overweight mothers with breast cancer and their overweight daughters. *Cancer* **2014**, *120*, 2522–2534. [CrossRef] [PubMed]
9. Howlader, N.; Noone, A.M.; Krapcho, M.; Miller, D.; Brest, A.; Yu, M.; Ruhl, J.; Tatalovich, Z.; Mariotto, A.; Lewis, D.R.; et al. (Eds.) *SEER Cancer Statistics Review, 1975–2018*; Based on November 2020 SEER Data Submission, Posted to the SEER Web Site; National Cancer Institute: Bethesda, MD, USA, 2021. Available online: https://seer.cancer.gov/csr/1975_2018/ (accessed on 27 September 2021).
10. Center for Disease Control, Division of Nutrition, Physical Activity and Obesity. National Center for Chronic Disease Prevention and Health Promotion. Updated 7 June 2021. Available online: https://www.cdc.gov/obesity/adult/defining.html (accessed on 27 September 2021).

11. Rock, C.L.; Doyle, C.; Demark-Wahnefried, W.; Meyerhardt, J.; Courneya, K.S.; Schwartz, A.L.; Bandera, E.V.; Hamilton, K.K.; Grant, B.; McCullough, M.; et al. Nutrition and physical activity guidelines for cancer survivors. *Cancer* **2020**, *63*, 215. [CrossRef] [PubMed]
12. Canadian Society for Exercise Medicine. *PAR-Q and You*; Canadian Society for Exercise Physiology: Gloucester, ON, Canada, 1994.
13. Bandura, A. *Social Foundations of Thought and Action: A Social Cognitive Theory*; Prentice-Hall, Inc.: Englewood Cliffs, NJ, USA, 1986.
14. Kelley, H.H. The "stimulus field" for interpersonal phenomena: The source of language and thought about interpersonal events. *Pers. Soc. Psychol. Rev.* **1997**, *1*, 140–169. [CrossRef] [PubMed]
15. Michie, S.; Ashford, S.; Sniehotta, F.F.; Dombrowski, S.U.; Bishop, A.; French, D.P. A refined taxonomy of behaviour change techniques to help people change their physical activity and healthy eating behaviours: The CALO-RE taxonomy. *Psychol. Health* **2011**, *26*, 1479–1498. [CrossRef] [PubMed]
16. De Padova, S.; Grassi, L.; Vagheggini, A.; Belvederi Murri, M.; Folesani, F.; Rossi, L.; Farolfi, A.; Bertelli, T.; Passardi, A.; Berardi, A.; et al. Post-traumatic stress symptoms in long-term disease-free cancer survivors and their family caregivers. *Cancer Med.* **2021**, *10*, 3974–3985. [CrossRef]
17. Hoenemeyer, T.; Cole, W.W.; Pye, A.; Pekmezi Dorothy, W.; Oster, R.A.; Demark-Wahnefried, W. Remote Assessment of Anthropometric and Physical Function for a Randomized Controlled Trial of a Web-Based Lifestyle Intervention for Cancer Survivors and Their Support Partners during COVID-19: Validity and Reliability Analyses. (in preparation)
18. Lohman, T.G.; Roche, A.F.; Martorell, R. *Anthropometric Standardization Reference Manual*; Abridged, Ed.; Human Kinetics Books: Champaign, IL, USA, 1991; Volume vi, 90p.
19. National Cancer Institute, Division of Cancer Control and Population Sciences (DCCPS). The Healthy Eating Index. 2017. Available online: https://epi.grants.cancer.gov/hei/ (accessed on 27 September 2021).
20. Rogers, L.Q.; McAuley, E.; Anton, P.M.; Courneya, K.S.; Vicari, S.; Hopkins-Price, P.; Verhulst, S.; Mocharnuk, R.; Hoelzer, K. Better exercise adherence after treatment for cancer (BEAT Cancer) study: Rationale, design, and methods. *Contemp. Clin. Trials* **2012**, *33*, 124–137. [CrossRef]
21. Sloane, R.; Snyder, D.C.; Demark-Wahnefried, W.; Lobach, D.; Kraus, W.E. Comparing the 7-day physical activity recall with a triaxial accelerometer for measuring time in exercise. *Med. Sci. Sports Exerc.* **2009**, *41*, 1334–1340. [CrossRef]
22. Godin, G.; Shephard, R.J. A simple method to assess exercise behavior in the community. *Can. J. Appl. Sport Sci.* **1985**, *10*, 141–146.
23. Amireault, S.; Godin, G. The Godin-Shephard leisure-time physical activity questionnaire: Validity evidence supporting its use for classifying healthy adults into active and insufficiently active categories. *Percept. Mot. Skills* **2015**, *120*, 604–622. [CrossRef] [PubMed]
24. Rikli, R.; Jones, C.J. Development and Validation of a Functional Fitness Test for Community-Residing Older Adults. *J. Aging Phys. Act.* **1999**, *7*, 129–161. [CrossRef]
25. The 4-Stage Balance Test. Available online: https://www.cdc.gov/steadi/pdf/4-Stage_Balance_Test-print.pdf (accessed on 27 September 2021).
26. Borsch-Supan, A.; Weiss, L.M.; Borsch-Supan, M.; Potter, A.J.; Cofferen, J.; Kerschner, E. Dried blood spot collection, sample quality, and fieldwork conditions: Structural validations for conversion into standard values. *Am. J. Hum. Biol.* **2021**, *33*, e23517. [CrossRef] [PubMed]
27. Janssen, M.F.; Pickard, A.S.; Golicki, D.; Gudex, C.; Niewada, M.; Scalone, L.; Swinburn, P.; Busschbach, J. Measurement properties of the EQ-5D-5L compared to the EQ-5D-3L across eight patient groups: a multi-country study. *Qual. Life Res.* **2013**, *22*, 1717–1727. [CrossRef] [PubMed]
28. Fillenbaum, G.G.; Smyer, M.A. The development, validity, and reliability of the OARS multidimensional functional assessment questionnaire. *J. Gerontol.* **1981**, *36*, 428–434. [CrossRef]
29. Chen, T.Y.; Janke, M.C. Gardening as a potential activity to reduce falls in older adults. *J. Aging Phys. Act.* **2012**, *20*, 15–31. [CrossRef]
30. Ossip-Klein, D.J.; Bigelow, G., Parker, S.R.; Curry, S.; Hall, S.; Kirkland, S. Classification and assessment of smoking behavior. *Health Psychol.* **1986**, *5*, 3–11. [CrossRef]
31. Williams, M.S.; Snyder, D.C.; Sloane, R.; Levens, J.; Flynn, K.E.; Dombeck, C.B.; Demark-Wahnefried, W.; Weinfurt, K.P. A comparison of cancer survivors from the PROMIS study selecting telephone versus online questionnaires. *Psychooncology* **2013**, *22*, 2632–2635. [CrossRef]
32. Clark, M.M.; Abrams, D.B.; Niaura, R.S.; Eaton, C.A.; Rossi, J.S. Self-efficacy in weight management. *J. Consult. Clin. Psychol.* **1991**, *59*, 739–744. [CrossRef]
33. McAuley, E.; Hall, K.S.; Motl, R.W.; White, S.M.; Wojcicki, T.R.; Hu, L.; Doerksen, S.E. Trajectory of declines in physical activity in community-dwelling older women: Social cognitive influences. *J. Gerontol. B Psychol. Sci. Soc. Sci.* **2009**, *64*, 543–550. [CrossRef] [PubMed]
34. Sallis, J.F.; Grossman, R.M.; Pinski, R.B.; Patterson, T.L.; Nader, P.R. The development of scales to measure social support for diet and exercise behaviors. *Prev. Med.* **1987**, *16*, 825–836. [CrossRef]
35. Demark-Wahnefried, W.; Clipp, E.C.; Lipkus, I.M.; Lobach, D.; Snyder, D.C.; Sloane, R.; Peterson, B.; Macri, J.M.; Rock, C.L.; McBride, C.M.; et al. Main outcomes of the FRESH START trial: A sequentially tailored, diet and exercise mailed print intervention among breast and prostate cancer survivors. *J. Clin. Oncol.* **2007**, *25*, 2709–2718. [CrossRef] [PubMed]

36. Demark-Wahnefried, W. Print-to-Practice: Designing Tailored Print Materials to Improve Cancer Survivors' Dietary and Exercise Practices in the FRESH START Trial. *Nutr. Today* **2007**, *42*, 131–138. [CrossRef] [PubMed]
37. Arroyave, W.D.; Clipp, E.C.; Miller, P.E.; Jones, L.W.; Ward, D.S.; Bonner, M.J.; Rosoff, P.M.; Snyder, D.C.; Demark-Wahnefried, W. Childhood cancer survivors' perceived barriers to improving exercise and dietary behaviors. *Oncol. Nurs. Forum* **2008**, *35*, 121–130. [CrossRef] [PubMed]
38. Vijan, S.; Stuart, N.S.; Fitzgerald, J.T.; Ronis, D.L.; Hayward, R.A.; Slater, S.; Hofer, T.P. Barriers to following dietary recommendations in Type 2 diabetes. *Diabet. Med.* **2005**, *22*, 32–38. [CrossRef] [PubMed]
39. Carmack, C.L.; Basen-Engquist, K.; Shely, L.; Baum, G.; Giordano, S.; Rodriguez-Bigas, M.; Pettaway, C.; Demark-Wahnefried, W. Healthy Moves to Improve Lifestyle Behaviors of Cancer Survivors and Their Spouses: Feasibility and Preliminary Results of Intervention Efficacy. *Nutrients* **2021**.
40. Hutfless, S.; Maruthur, N.M.; Wilson, R.F.; Gudzune, K.A.; Brown, R.; Lau, B.; Fawole, O.A.; Chaudhry, Z.W.; Anderson CA, M.; Segal, J.B. *AHRQ Comparative Effectiveness Reviews, in Strategies to Prevent Weight Gain Among Adults*; Agency for Healthcare Research and Quality (US): Rockville, MD, USA, 2013.
41. Unger, J.M.; Xiao, H.; LeBlanc, M.; Hershman, D.L.; Blanke, C.D. Cancer Clinical Trial Participation at the 1-Year Anniversary of the Outbreak of the COVID-19 Pandemic. *JAMA Netw. Open* **2021**, *4*, e2118433. [CrossRef]
42. US Bureau of the Census. America Community Survey, 2008–2012. Available online: http://statecancerprofiles.cancer.gov/demographics/index.php?stateFIPS=01&topic=pop&demo=00022&type=manyareacensus&sortVariableName=value&sortOrder=default (accessed on 27 September 2021).
43. Arriens, C.; Aberle, T.; Carthen, F.; Kamp, S.; Thanou, A.; Chakravarty, E.; James, J.A.; Merrill, J.T.; Ogunsanya, M.E. Lupus patient decisions about clinical trial participation: A qualitative evaluation of perceptions, facilitators and barriers. *Lupus Sci. Med.* **2020**, *7*, e000360. [CrossRef]
44. van der Hout, A.; van Uden-Kraan, C.F.; Holtmaat, K.; Jansen, F.; Lissenberg-Witte, B.I.; Nieuwenhuijzen GA, P.; Hardillo, J.A.; Baatenburg de Jong, R.J.; Tiren-Verbeet, N.L.; Sommeijer, D.W.; et al. Reasons for not reaching or using web-based self-management applications, and the use and evaluation of Oncokompas among cancer survivors, in the context of a randomised controlled trial. *Internet Interv.* **2021**, *25*, 100429. [CrossRef] [PubMed]
45. Johnson, J. Worldwide Digital Population as of January 2021. Available online: https://www.statista.com/statistics/617136/digital-population-worldwide/ (accessed on 27 September 2021).
46. Monterrosa, E.C.; Frongillo, E.A.; Drewnowski, A.; de Pee, S.; Vandevijvere, S. Sociocultural Influences on Food Choices and Implications for Sustainable Healthy Diets. *Food Nutr. Bull.* **2020**, *41* (Suppl. S2), 59S–73S. [CrossRef] [PubMed]
47. Addison, S.; Shirima, D.; Aboagye-Mensah, E.B.; Dunovan, S.G.; Pascal, E.Y.; Lustberg, M.B.; Arthur, E.K.; Nolan, T.S. Effects of tandem cognitive behavioral therapy and healthy lifestyle interventions on health-related outcomes in cancer survivors: A systematic review. *J. Cancer Surviv.* **2021**, 1–24. [CrossRef]
48. Moinpour, C.M.; Lovato, L.C.; Thompson, I.M., Jr.; Ware, J.E., Jr.; Ganz, P.A.; Patrick, D.L.; Shumaker, S.A.; Donaldson, G.W.; Ryan, A.; Coltman, C.A., Jr. Profile of men randomized to the prostate cancer prevention trial: Baseline health-related quality of life, urinary and sexual functioning, and health behaviors. *J. Clin. Oncol.* **2000**, *18*, 1942–1953. [CrossRef]
49. Virani, S.; Burke, L.; Remick, S.C.; Abraham, J. Barriers to recruitment of rural patients in cancer clinical trials. *J. Oncol. Pract.* **2011**, *7*, 172–177. [CrossRef]

 MDPI

Correction

Correction: Pekmezi et al. Rationale and Methods for a Randomized Controlled Trial of a Dyadic, Web-Based, Weight Loss Intervention among Cancer Survivors and Partners: The DUET Study. *Nutrients* 2021, *13*, 3472

Dorothy W. Pekmezi [1,2,*], Tracy E. Crane [3], Robert A. Oster [2,4], Laura Q. Rogers [2,4], Teri Hoenemeyer [5], David Farrell [6], William W. Cole [5], Kathleen Wolin [7], Hoda Badr [8] and Wendy Demark-Wahnefried [2,5]

1 Department of Health Behavior, University of Alabama at Birmingham (UAB), Birmingham, AL 35294, USA
2 O'Neal Comprehensive Cancer Center at University of Alabama at Birmingham (UAB), Birmingham, AL 35294, USA; roster@uabmc.edu (R.A.O.); lqrogers@uabmc.edu (L.Q.R.); demark@uab.edu (W.D.-W.)
3 Sylvester Comprehensive Cancer Center, Miami, FL 33136, USA; tecrane@med.miami.edu
4 Division of Preventive Medicine, Department of Medicine, Birmingham, AL 35294, USA
5 Department of Nutrition Sciences, University of Alabama at Birmingham (UAB), Birmingham, AL 35294, USA; tgw318@uab.edu (T.H.); colew14@uab.edu (W.W.C.)
6 People Designs, Inc., Durham, NC 27705, USA; dfarrell@peopledesigns.com
7 Coeus Health, LLC, Chicago, IL 60614, USA; kate@drkatewolin.com
8 Department of Medicine, Epidemiology and Population Sciences, Baylor College of Medicine, Houston, TX 77030, USA; hoda.badr@bcm.edu
* Correspondence: dpekmezi@uab.edu; Tel.: +1-205-975-8061

Citation: Pekmezi, D.W.; Crane, T.E.; Oster, R.A.; Rogers, L.Q.; Hoenemeyer, T.; Farrell, D.; Cole, W.W.; Wolin, K.; Badr, H.; Demark-Wahnefried, W. Correction: Pekmezi et al. Rationale and Methods for a Randomized Controlled Trial of a Dyadic, Web-Based, Weight Loss Intervention among Cancer Survivors and Partners: The DUET Study. *Nutrients* 2021, *13*, 3472. *Nutrients* **2022**, *14*, 3141. https://doi.org/10.3390/nu14153141

Received: 2 February 2022
Accepted: 24 February 2022
Published: 29 July 2022

Publisher's Note: MDPI stays neutral with regard to jurisdictional claims in published maps and institutional affiliations.

Modifications to the Main Text and Table 2

The authors would like to correct errors in their prior publication [1]. In the original article, there were mistakes regarding the measures used for comorbidity and quality of life (Table 2) and depression (in the text, Materials and Methods, Section 2.5 Assessments). The correct measures were the Older Americans Resources & Services (OARS) Comorbidity Index [2,3], the EQ-5D-5L [4], and the Patient Reported Outcomes Measurement Information System (PROMIS) Global Health and Emotional Distress/Depression subscales [5], not the Charlson comorbidity index [6,7], RAND36 [8], and Center for Epidemiologic Studies of Depression [9].

The authors apologize for any inconvenience caused and state that the scientific conclusions are unaffected.

References

1. Pekmezi, D.W.; Crane, T.E.; Oster, R.A.; Rogers, L.Q.; Hoenemeyer, T.; Farrell, D.; Cole, W.W.; Wolin, K.; Badr, H.; Demark-Wahnefried, W. Rationale and Methods for a Randomized Controlled Trial of a Dyadic, Web-Based, Weight Loss Intervention among Cancer Survivors and Partners: The DUET Study. *Nutrients* **2021**, *13*, 3472. [CrossRef] [PubMed]
2. Fillenbaum, G.G.; Smyer, M.A. The development, validity, and reliability of the OARS multidimensional functional assessment questionnaire. *J. Gerontol.* **1981**, *36*, 428–434. [CrossRef] [PubMed]
3. Chen, T.Y.; Janke, M.C. Gardening as a potential activity to reduce falls in older adults. *J. Aging Phys. Act.* **2012**, *20*, 15–31. [CrossRef] [PubMed]
4. Janssen, M.F.; Pickard, A.S.; Golicki, D.; Gudex, C.; Niewada, M.; Scalone, L.; Swinburn, P.; Busschbach, J. Measurement properties of the EQ-5D-5L compared to the EQ-5D-3L across eight patient groups: A multi-country study. *Qual. Life Res.* **2013**, *22*, 1717–1727. [CrossRef] [PubMed]
5. Williams, M.S.; Snyder, D.C.; Sloane, R.; Levens, J.; Flynn, K.E.; Dombeck, C.B.; Demark-Wahnefried, W.; Weinfurt, K.P. A comparison of cancer survivors from the PROMIS study selecting telephone versus online questionnaires. *Psychooncology* **2013**, *22*, 2632–2635. [CrossRef] [PubMed]
6. Charlson, M.E.; Pompei, P.; Ales, K.L.; MacKenzie, C.R. A new method of classifying prognostic comorbidity in longitudinal studies: Development and validation. *J. Chronic Dis.* **1987**, *40*, 373–383. [CrossRef]
7. Charlson, M.; Szatrowski, T.P.; Peterson, J.; Gold, J. Validation of a combined comorbidity index. *J. Clin. Epidemiol.* **1994**, *47*, 1245–1251. [CrossRef]
8. Hays, R.D.; Sherbourne, C.D.; Mazel, R.M. The RAND 36-Item Health Survey 1.0. *Health Econ.* **1993**, *2*, 217–227. [CrossRef] [PubMed]
9. Kohout, F.J.; Berkman, L.F.; Evans, D.A.; Cornoni-Huntley, J. Two shorter forms of the CES-D (Center for Epidemiological Studies Depression) depression symptoms index. *J. Aging Health* **1993**, *5*, 179–193. [CrossRef] [PubMed]

MDPI

Article

Effect of a Remotely Delivered Weight Loss Intervention in Early-Stage Breast Cancer: Randomized Controlled Trial

Marina M. Reeves [1,*], Caroline O. Terranova [1], Elisabeth A. H. Winkler [1], Nicole McCarthy [2], Ingrid J. Hickman [3], Robert S. Ware [4], Sheleigh P. Lawler [1], Elizabeth G. Eakin [1] and Wendy Demark-Wahnefried [5]

1 School of Public Health, The University of Queensland, Brisbane 4006, Australia; caroline.terranova@qut.edu.au (C.O.T.); e.winkler@sph.uq.edu.au (E.A.H.W.); s.lawler@sph.uq.edu.au (S.P.L.); e.eakin@uq.edu.au (E.G.E.)

2 Wesley Centre, Icon Cancer Care, Brisbane 4066, Australia; Nicole.McCarthy@icon.team

3 Department of Nutrition and Dietetics, Princess Alexandra Hospital, Brisbane 4102, Australia; i.hickman@uq.edu.au

4 Menzies Health Institute Queensland, Griffith University, Brisbane 4111, Australia; r.ware@griffith.edu.au

5 O'Neal Comprehensive Cancer Center, University of Alabama at Birmingham, Birmingham, AL 35294, USA; demark@uab.edu

* Correspondence: marina.reeves@uq.edu.au; Tel.: +617-3346-4692

Abstract: Limited evidence exists on the effects of weight loss on chronic disease risk and patient-reported outcomes in breast cancer survivors. Breast cancer survivors (stage I–III; body mass index 25–45 kg/m^2) were randomized to a 12-month, remotely delivered (22 telephone calls, mailed material, optional text messages) weight loss (diet and physical activity) intervention (n = 79) or usual care (n = 80). Weight loss (primary outcome), body composition, metabolic syndrome risk score and components, quality of life, fatigue, musculoskeletal pain, menopausal symptoms, fear of recurrence, and body image were assessed at baseline, 6 months, 12 months (primary endpoint), and 18 months. Participants were 55 ± 9 years and 10.7 ± 5.0 months post-diagnosis; retention was 81.8% (12 months) and 80.5% (18 months). At 12-months, intervention participants had significantly greater improvements in weight (−4.5% [95%CI: −6.5, −2.5]; $p < 0.001$), fat mass (−3.3 kg [−4.8, −1.9]; $p < 0.001$), metabolic syndrome risk score (−0.19 [−0.32, −0.05]; $p = 0.006$), waist circumference (−3.2 cm [−5.5, −0.9]; $p = 0.007$), fasting plasma glucose (−0.23 mmol/L [−0.44, −0.02]; $p = 0.032$), physical quality of life (2.7 [0.7, 4.6]; $p = 0.007$; Cohen's effect size (d) = 0.40), musculoskeletal pain (−0.5 [−0.8, −0.2]; $p = 0.003$; $d = 0.49$), and body image (−0.2 [−0.4, −0.0]; $p = 0.030$; $d = 0.31$) than usual care. At 18 months, effects on weight, adiposity, and metabolic syndrome risk scores were sustained; however, significant reductions in lean mass were observed (−1.1 kg [−1.7, −0.4]; $p < 0.001$). This intervention led to sustained improvements in adiposity and metabolic syndrome risk.

Keywords: obesity; exercise; nutrition; supportive care; survivorship; telehealth

Citation: Reeves, M.M.; Terranova, C.O.; Winkler, E.A.H.; McCarthy, N.; Hickman, I.J.; Ware, R.S.; Lawler, S.P.; Eakin, E.G.; Demark-Wahnefried, W. Effect of a Remotely Delivered Weight Loss Intervention in Early-Stage Breast Cancer: Randomized Controlled Trial. *Nutrients* **2021**, *13*, 4091. https://doi.org/10.3390/nu13114091

Academic Editor: Vicente Martinez Vizcaino

Received: 19 October 2021
Accepted: 12 November 2021
Published: 15 November 2021

1. Introduction

Attention has been focused on modifiable risk factors (diet, obesity, physical activity) as a means to improve breast cancer outcomes [1,2]. Physical activity has been associated with reduced breast cancer recurrence risk and increased survival [3,4], with exercise interventions producing improvements in quality of life, physical function, and fatigue [5–7]. Breast cancer survivors who maintain a healthful weight (body mass index (BMI) = 18.5–24.9 kg/m^2) have 30–40% reduced mortality risk compared to those with obesity (BMI ≥ 30 kg/m^2) [8]. Consequently, weight management, physical activity, and dietary changes are encouraged for breast cancer survivors [1,2,9,10].

Weight loss trials in early-stage breast cancer have shown that modest weight loss is safe and feasible [11,12], with ongoing trials assessing effects on survival [13–15]. With limited evidence on prognostic benefit, there remains a need to understand the broader effects of weight loss on outcomes such as body composition [16], chronic disease risk (given that cardiovascular disease deaths surpass cancer-specific mortality for the majority of breast cancer survivors) [17], and patient-reported outcomes, i.e., quality of life and treatment-related side effects. Treatment-related side effects such as fatigue, arthralgia, and menopausal symptoms can persist long-term [18] and are exacerbated by excess body weight [19–23]. With the exception of quality of life, few trials have examined the effect of weight loss on patient-reported outcomes [11,12]. Further, weight loss trials to date have only evaluated effects on individual metabolic biomarkers and not broader measures of chronic disease risk [11,12]. A recent exercise-only trial reported large improvements in metabolic syndrome risk, following a short-term, supervised exercise intervention [24]. Metabolic syndrome, a cluster of risk factors that increases cardiovascular disease and type 2 diabetes risk [25], has also been associated with increased breast cancer mortality and recurrence risk [26,27].

Importantly, of relevance in the current COVID-19 environment is the need to understand the benefits that can be achieved with remotely delivered interventions (no face-to-face contact). The 'Living Well after Breast Cancer' trial aimed to evaluate the effectiveness of a 12-month, remotely delivered weight loss intervention versus usual care in women following treatment for early-stage breast cancer. This paper reports on the effects of the intervention on the primary outcome (percent weight loss), body composition, metabolic syndrome risk, and patient-reported outcomes [28], including whether intervention effects were sustained 6 months after intervention completion.

2. Materials and Methods

This two-arm, parallel, randomized trial was registered with the Australian and New Zealand Clinical Trial Registry (ACTRN12612000997853), with the trial protocol previously published [28]. The human research ethics committees of the Royal Brisbane & Women's Hospital, Greenslopes Private Hospital, St. Vincent's Health & Aged Care, and the University of Queensland granted approval. Signed, informed consent was obtained prior to participation.

2.1. Participants and Recruitment

Participants were recruited from seven hospitals in Brisbane, Australia, and the state-based cancer registry between October 2012 and December 2014. Women aged 18–75 years were eligible if they had: a diagnosis of stage I–III breast cancer in the previous two years, a BMI 25–45 kg/m^2, and completed primary cancer treatment (excluding endocrine treatment). Exclusions included pregnancy, contraindications to unsupervised exercise, >5% weight loss within the previous six months, insufficient English, or self-reported anxiety and/or depression that would interfere with participation [28]. Following baseline assessment, an off-site staff member randomized participants (1:1) into intervention or usual care arms using a computer-generated randomization program with uneven block sizes.

2.2. Usual Care

Participants in both arms received materials after each assessment, including a study newsletter and assessment feedback. Participants allocated to usual care received brief feedback on their assessment results, whereas for intervention participants, assessment results were compared to guidelines.

2.3. Weight Loss Intervention

The intervention was based on clinical practice guidelines for overweight and obesity (consistent with recommendations for cancer survivors) [9,10,29], piloted in a feasibility study [30,31], and described previously [28]. The intervention was remotely delivered via telephone by accredited dietitians (with optional text messages) and aimed for weight loss of 5–10%, by reducing energy intake (1200–1500 kcal/day) [32] and saturated fat (<7% total energy), increasing vegetables and fruit (5 and 2 servings/day, respectively), and limiting alcohol (≤1 serving/day). Additionally, incremental increases in moderate-to-vigorous intensity aerobic activity to 210 min/week and 2–3 resistance exercise sessions/week were encouraged. The intervention was grounded in social cognitive theory [33], emphasizing self-monitoring, goal setting, social support, problem solving, stimulus control, positive self-talk, and self-reward.

Intervention participants received a workbook, scale, measuring tape, pedometer, calorie-counter book, and self-monitoring diary. During the first 6 months, participants received up to 16 calls (six weekly then 10 bi-weekly calls) and optional text messages. During the second 6 months, participants received six monthly calls and tailored text messages. Dietitians used a semi-structured approach and motivational interviewing for each call.

2.4. Data Collection

Data were collected at baseline, 6 months, 12 months (primary endpoint), and 18 months by staff blinded to arm assignment. Methods and reliability/validity of measures have been reported previously [28].

2.4.1. Primary Outcome

Weight was measured without heavy clothing or shoes to the nearest 0.1 kg (Tanita BWB−600 Wedderburn Scales, Sydney, Australia), in duplicate, with the mean used, and expressed as percent weight change from baseline.

2.4.2. Secondary Outcomes

Secondary outcomes (all continuous) were body composition (total fat and lean mass), biomarkers of metabolic syndrome (risk score, waist circumference, triglycerides, high density lipoprotein (HDL) cholesterol, systolic and diastolic blood pressure, fasting plasma glucose), quality of life, fatigue, arthralgia, menopausal symptoms, fear of cancer recurrence, and body image. Detailed regional body composition outcomes were also explored.

Body composition was measured by a trained technician using Dual-Energy X-ray Absorptiometry (Lunar Prodigy, GE Medical Systems, Madison, WI, USA). Waist circumference was measured at the iliac crest in duplicate, with the mean used. Blood pressure was measured seated using an automated sphygmomanometer (300 Series Vital Signs Monitor, Welch Allyn, Beaverton, OR, USA) in duplicate, with the mean used. Lipids and glucose were determined through an overnight fasting (≥10 h) blood draw analyzed via a standard enzymatic colorimetric assay (c16000 Clinical Chemistry Analyzer, Abbott Diagnostics, Abbott Park, IL, USA). Lipid-lowering medication use (yes/no) was self-reported. Metabolic syndrome was classified using the harmonized definition [34] as outlined in Table S1, and a unitless continuous metabolic syndrome risk score was calculated, consistent with previous scoring (lower values being desirable) [35,36]. Each of the five metabolic syndrome components were log10-transformed, then standardized as z-scores—(value − population mean)/SD and (population mean − value)/SD for HDL—then averaged to yield a final unitless score (see Table S1). The z-scores used population means such that, for each biomarker, z > 0 indicates levels that are worse than average for the population of Australian women [37].

Quality of life was assessed using the Patient-Reported Outcome Measurement Information System (PROMIS) Global Health Scale, which solicits information across physical function, fatigue, pain, emotional distress, and social health, and provides summary scores for global physical and mental health components, with higher scores indicating better functioning [38]. Fatigue was assessed using the Functional Assessment of Chronic Illness Therapy Fatigue Scale, with higher scores indicating lower fatigue [39]. Arthralgia was measured using the Musculoskeletal Pain subscale from the Breast Cancer Prevention Trial Symptom Scale, with higher scores indicating worse pain [40]. Menopausal symptoms were assessed using the Greene Climacteric Scale [41]—psychological, somatic, and vasomotor symptoms subscales—with higher scores indicating more severe symptoms. Fear of cancer recurrence was measured using the Concerns About Recurrence Questionnaire, with higher scores indicating greater fear [42]. Body image was assessed using the Body Image and Relationships Scale (total score), with higher scores indicating greater impairment [43].

2.4.3. Adverse Events

At each follow-up assessment, participants self-reported any adverse events (AEs), with severity categorized according to the Common Terminology Criteria for Adverse Events (CTC-AE; v4.0) from Grade 1 'mild' to Grade 5 'fatal/death'. The 'relatedness' of the AE to the intervention was also recorded on a 5-point scale from 'clearly not related' to 'clearly related'.

2.5. Sample Size

The sample size was calculated to provide at least 90% power (5% two-tailed significance) to detect a between-arm minimum difference of 5% body weight [9] and at least 80% power to detect effects of 0.5 SD in secondary outcomes [28].

2.6. Data Analysis

Multivariable linear mixed models were used to evaluate primary and secondary outcomes. Marginal means evaluated at mean values were used to report within-arm changes and between-arm differences (intervention effects). Transformed outcomes were back-transformed prior to reporting. Standardized intervention effect sizes were reported using Cohen's *d* statistic. Based on a priori criteria [28], no potentially confounding variable met criteria for inclusion in models. Accordingly, models included fixed effects for the treatment arm, timepoint (6/12/18 months), and their interaction, along with the baseline value of the outcome [28]. To account for repeated measures from participants, models used restricted maximum likelihood estimation with an unstructured within-subjects covariance structure and no random intercept. The association between treatment arm and adverse events was assessed using Poisson regression.

Analyses followed intention-to-treat principles. Missing data were handled both using evaluable-case analysis and by multiple imputation (chained equations with m = 50 imputations) as sensitivity analyses since data were not missing completely at random. Variables included in imputation models are shown in Table S2. Due to the potential influence of medication use (endocrine treatment, lipid-lowering, and blood pressure medications), further sensitivity analyses were performed that adjusted for baseline and concurrent use of these medications on related outcomes (metabolic syndrome risk score, HDL-cholesterol, triglycerides, blood pressure, musculoskeletal pain, and menopausal symptoms). Statistical significance was set at $p < 0.05$ (two-tailed). Analyses were performed in Stata v16 (StataCorp LLC, College Station, TX, USA).

3. Results

Of the 394 women contacted, 170 were ineligible, 65 declined to participate, and 159 women (71% of those eligible) consented and were randomized (Figure 1). Baseline characteristics were similar between arms (Table 1), with the only noteworthy differences ($\geq$10%) being a greater proportion of post-menopausal and fewer peri-menopausal women at diagnosis, and a greater proportion with multi-comorbidities in the intervention versus usual care arm. Otherwise, women were, on average, 55 years old, approximately 11 months post-diagnosis, and half had obesity.

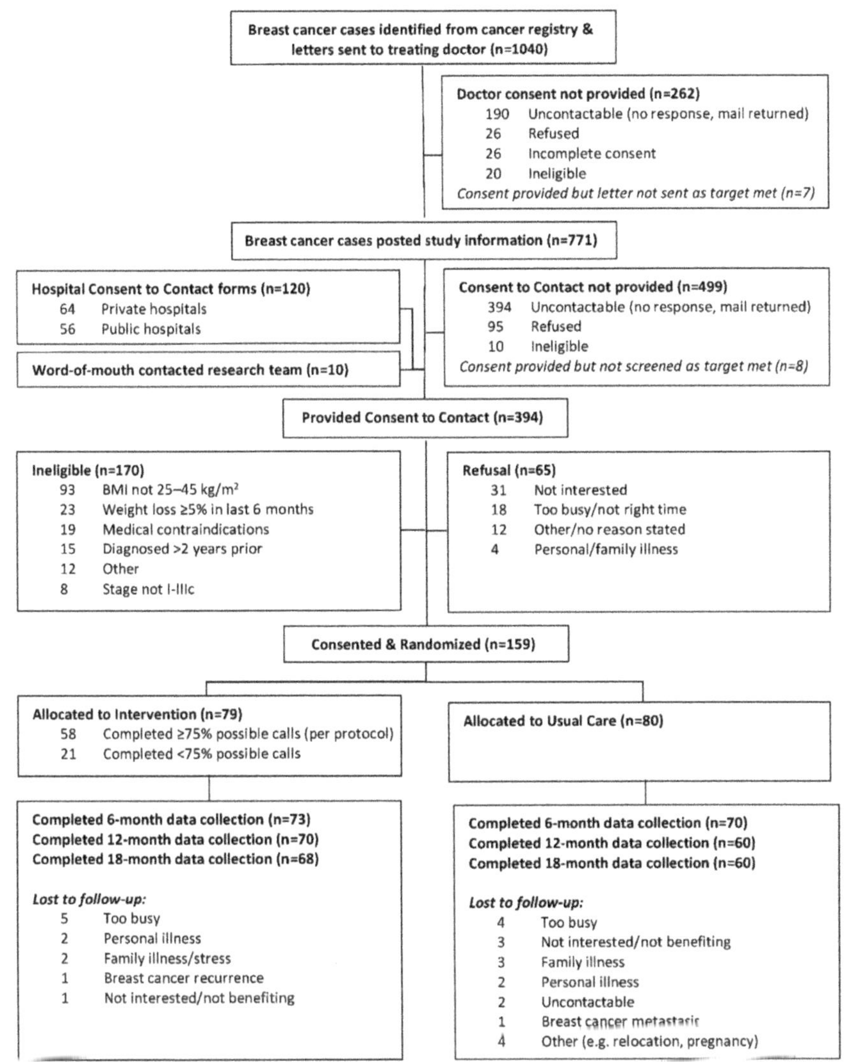

Figure 1. Participant flow diagram for Living Well after Breast Cancer trial.

Retention was 89.9% at 6 months, 81.8% at 12 months, and 80.5% at 18 months, with 124 (78.0%) participants completing all four assessments. Drop-out differed by arm, with 13.9% (n = 11) in intervention versus 30.0% (n = 24) in usual care (p = 0.02). Relative to those completing all assessments, drop-outs were younger, reported lower physical quality of life, and were more likely to have children at home, lower income, non-English speaking background, and received both chemotherapy and radiotherapy (see Table S3). Of the intervention participants, 73.4% (n = 58) received at least 75% ($\geq$17 out of 22) of intended calls, defined a priori.

Table 1. Baseline participant characteristics in the Living Well after Breast Cancer trial (n = 159).

Characteristic	Usual Care (n = 80)	Intervention (n = 79)
	Mean (SD)	
Age (years)	54.9 (9.3)	55.9 (9.1)
BMI (kg/m^2)	31.3 (5.2)	31.4 (4.9)
Months since diagnosis	10.8 (5.3)	10.7 (4.8)
Months since treatment completion	4.9 (4.6)	5.2 (4.7)
	n (%)	
Menopausal status at diagnosis		
Premenopausal	31 (39%)	28 (35%)
Perimenopausal [a]	15 (19%)	6 (8%)
Postmenopausal [a]	34 (42%)	45 (57%)
Breast cancer stage [b]		
Stage 1	46 (58%)	40 (51%)
Stage 2	24 (30%)	30 (38%)
Stage 3	9 (11%)	9 (11%)
Estrogen receptor status [b]		
Positive	72 (91%)	67 (85%)
Negative	7 (9%)	12 (15%)
HER2 [b]		
Positive	9 (11%)	11 (14%)
Negative	68 (86%)	68 (86%)
Equivocal	2 (3%)	0 (0%)
Chemotherapy treatment	51 (64%)	48 (61%)
Radiotherapy treatment	63 (79%)	63 (80%)
Endocrine treatment		
None	35 (44%)	32 (41%)
SERM	19 (24%)	22 (28%)
Aromatase inhibitor	26 (32%)	24 (30%)
GnRH agonist	0 (0%)	1 (1%)
Metabolic syndrome present	37 (48%)	37 (47%)
Charlson Comorbidity Index [c]		
0	51 (64%)	47 (60%)
1	18 (22%)	13 (16%)
≥2 [a]	11 (14%)	19 (24%)
Married or stable union	56 (70.0%)	54 (68.4%)
Caucasian	78 (97.5%)	78 (98.7%)
Employment status		
Paid work	44 (55%)	50 (63%)
Retired, home duties, unable to work, other	36 (45%)	29 (37%)
	n (%)	
Highest education level		
High school or less	32 (40%)	32 (41%)
Technical/trade/diploma	21 (26%)	16 (20%)
University or higher	27 (34%)	31 (39%)
Gross household income (AUD) [d]		
<$82,056 per year	34 (42%)	37 (47%)
≥$82,056 per year	38 (48%)	33 (42%)
Not reported/not known	8 (10%)	9 (11%)

Abbreviations: AUD, Australian dollar; BMI, body mass index; GnRH, gonadotropin releasing hormone; HER2, human epidermal growth receptor 2; SERM, selective estrogen receptor modulators. [a] Noteworthy difference (≥10%) between arms. [b] Percentages exclude missing data (n = 1, usual care arm). [c] Charlson Comorbidity Index was based on self-reported diagnosis of 13 conditions [17], with the addition of hypertension. [d] Threshold indicates 60th percentile of Australian population household income based on 2007–2008 census.

3.1. Weight and Body Composition

Significantly greater weight loss was observed in the intervention versus usual care arms at 12 months (−4.5% [95%CI: −6.5, −2.5], $p < 0.001$), which was largely maintained at 18 months (−3.1% [−5.3, −0.9], $p = 0.007$) (Table 2). Significant intervention effects on fat mass were observed at each assessment, with greater loss of lean mass observed in the intervention versus usual care at all follow-up assessments, being statistically significant at 6 and 18 months. Sensitivity analyses accounting for missing data (see Table S4) led to similar intervention effects (±20%) and conclusions regarding clinical relevance and statistical significance. Analysis of regional body composition showed that across each region, fat mass decreased primarily with small decreases in lean mass, and small to no change in bone mass, leading to lower proportions of body fat, higher proportions of lean mass, and slightly higher or similar percentages of bone mass within each body region (see Tables S5–S7).

3.2. Metabolic Syndrome

The intervention arm demonstrated statistically significant and more favorable metabolic syndrome risk scores across all follow-up assessments, which were statistically significant compared to usual care (Table 2). For individual metabolic syndrome components, significant intervention effects were observed for waist circumference at all follow-ups, systolic and diastolic blood pressure at 6 months only, and fasting plasma glucose at 12 months only. Sensitivity analyses adjusting for medication use (see Table S8) and accounting for missing data (see Table S4) yielded similar effect sizes and the same conclusions.

3.3. Patient-Reported Outcomes

Overall, the patient-reported outcomes (Table 3) favored intervention over usual care at most or all follow-up assessments. Significant intervention effects favoring intervention were seen at 12 months for physical quality of life ($d = 0.40$), musculoskeletal pain ($d = -0.49$), and body image ($d = -0.31$), with non-significant, small ($d \approx 0.2$–0.3) improvements observed for mental quality of life and psychological menopausal symptoms. At 18 months, most effects were attenuated. Changes in endocrine treatment medications did not account for observed effects on musculoskeletal pain or menopausal symptoms (see Table S8). After multiple imputation, effects on musculoskeletal pain were attenuated slightly and no longer significant at 12 months, while effects for quality of life and fatigue were of similar magnitude but no longer significant for physical quality of life at 6 months (see Table S4). For the remaining patient-reported outcomes, the magnitude of effects changed slightly, but with no change to overall conclusions.

Table 2. Within-arm and between-arm changes for weight, body composition, and metabolic syndrome risk biomarkers: Living Well after Breast Cancer trial.

Outcome	Timepoint	Intervention		Usual Care		Intervention Effect (Intervention—Usual Care)		
		n	Mean Change (95% CI)	*n*	Mean Change (95% CI)	Mean Difference (95% CI)	*p*	d [a]
Weight (% of baseline value)	Baseline M (SD)	79	83.9 (14.2)	80	83.6 (13.6)			
	6 months	73	−4.61 (−5.77, −3.44)	70	−0.52 (−1.70, 0.67)	−4.09 (−5.75, −2.43)	<0.001	−0.30
	12 months [b]	70	−5.06 (−6.46, −3.66)	60	−0.58 (−2.02, 0.85)	−4.48 (−6.48, −2.47)	<0.001	−0.32
	18 months	68	−3.69 (−5.23, −2.16)	60	−0.62 (−2.22, 0.97)	−3.07 (−5.28, −0.86)	0.007	−0.22
Weight (kg)	Baseline M (SD)	79	83.9 (14.2)	80	83.6 (13.6)			
	6 months	73	−3.74 (−4.71, −2.76)	70	−0.43 (−1.42, 0.56)	−3.31 (−4.70, −1.92)	<0.001	−0.24
	12 months [b]	70	−4.12 (−5.28, −2.96)	60	−0.52 (−1.71, 0.67)	−3.60 (−5.26, −1.94)	<0.001	−0.26
	18 months	68	−3.03 (−4.34, −1.73)	60	−0.56 (−1.91, 0.80)	−2.48 (−4.36, −0.59)	0.010	−0.18
Total fat mass (kg)	Baseline M (SD)	73	38.8 (10.4)	70	37.5 (10.2)			
	6 months	67	−3.13 (−3.97, −2.30)	62	0.13 (−0.73, 1.00)	−3.26 (−4.47, −2.06)	<0.001	−0.32
	12 months [b]	64	−3.27 (−4.26, −2.29)	54	0.05 (−0.98, 1.08)	−3.32 (−4.75, −1.90)	<0.001	−0.32
	18 months	63	−2.11 (−3.19, −1.03)	54	−0.29 (−1.43, 0.85)	−1.82 (−3.39, −0.25)	0.023	−0.18
Total lean mass (kg)	Baseline M (SD)	73	42.8 (5.0)	70	43.6 (5.2)			
	6 months	67	−0.96 (−1.28, −0.63)	62	−0.24 (−0.57, 0.09)	−0.71 (−1.18, −0.25)	0.002	−0.14
	12 months [b]	64	−1.07 (−1.46, −0.68)	54	−0.52 (−0.93, −0.10)	−0.55 (−1.12, 0.02)	0.059	−0.11
	18 months	63	−1.20 (−1.63, −0.77)	54	−0.14 (−0.59, 0.32)	−1.06 (−1.68, −0.43)	<0.001	−0.21
Metabolic syndrome risk score	Baseline M (SD)	78	0.65 (0.60)	77	0.63 (0.59)			
	6 months	69	−0.19 (−0.27, −0.11)	65	0.03 (−0.05, 0.12)	−0.22 (−0.34, −0.10)	<0.001	−0.37
	12 months [b]	67	−0.18 (−0.27, −0.08)	56	0.01 (−0.09, 0.11)	−0.19 (−0.32, −0.05)	0.006	−0.32
	18 months	66	−0.15 (−0.24, −0.06)	57	0.01 (−0.08, 0.11)	−0.16 (−0.29, −0.03)	0.014	−0.27
Waist circumference (cm)	Baseline M (SD)	79	106.7 (11.7)	80	104.9 (10.4)			
	6 months	73	−3.47 (−4.95, −1.99)	70	−0.64 (−2.14, 0.87)	−2.83 (−4.94, −0.71)	0.009	−0.26
	12 months [b]	70	−5.50 (−7.11, −3.89)	60	−2.30 (−3.98, −0.62)	−3.20 (−5.53, −0.87)	0.007	−0.29
	18 months	68	−5.29 (−6.81, −3.78)	60	−2.50 (−4.08, −0.91)	−2.80 (−4.99, −0.61)	0.012	−0.25
Triglycerides (mmol/L) [c]	Baseline M (SD)	78	1.4 (0.7)	78	1.5 (0.9)			
	6 months	71	−0.03 (−0.12, 0.05)	67	0.08 (−0.01, 0.18)	−0.11 (−0.24, 0.01)	0.081	−0.14
	12 months [b]	67	−0.08 (−0.18, 0.02)	57	0.04 (−0.08, 0.16)	−0.12 (−0.28, 0.03)	0.125	−0.15
	18 months	66	−0.11 (−0.20, −0.02)	59	−0.01 (−0.11, 0.09)	−0.10 (−0.24, 0.03)	0.124	−0.13
HDL-cholesterol (mmol/L)	Baseline M (SD)	78	1.4 (0.3)	78	1.4 (0.4)			
	6 months	71	0.02 (−0.02, 0.07)	67	−0.02 (−0.06, 0.03)	0.04 (−0.02, 0.10)	0.182	0.13
	12 months [b]	67	0.05 (0.00, 0.09)	57	−0.01 (−0.06, 0.04)	0.06 (−0.01, 0.12)	0.110	0.17
	18 months	66	0.06 (0.01, 0.11)	59	0.00 (−0.04, 0.05)	0.06 (−0.01, 0.12)	0.097	0.17

Table 2. *Cont.*

Outcome	Timepoint	Intervention		Usual Care		Intervention Effect (Intervention—Usual Care)		
		n	Mean Change (95% CI)	*n*	Mean Change (95% CI)	Mean Difference (95% CI)	*p*	d [a]
Systolic blood pressure (mmHg)	Baseline M (SD)	79	125.3 (12.2)	79	123.4 (11.3)			
	6 months	71	−1.69 (−4.21, 0.83)	68	3.44 (0.86, 6.02)	−5.13 (−8.73, −1.52)	0.005	−0.44
	12 months [b]	70	1.05 (−1.70, 3.80)	59	2.20 (−0.77, 5.17)	−1.15 (−5.20, 2.90)	0.577	−0.10
	18 months	68	3.07 (−0.37, 6.52)	59	5.54 (1.88, 9.19)	−2.46 (−7.48, 2.56)	0.336	−0.21
Diastolic blood pressure (mmHg)	Baseline M (SD)	79	78.7 (9.4)	79	77.9 (7.3)			
	6 months	71	−0.35 (−2.03, 1.32)	68	2.40 (0.69, 4.12)	−2.76 (−5.15, −0.36)	0.024	−0.33
	12 months [b]	70	0.60 (−1.07, 2.27)	59	1.51 (−0.29, 3.30)	−0.90 (−3.36, 1.55)	0.470	−0.11
	18 months	68	1.06 (−0.80, 2.92)	59	3.39 (1.41, 5.36)	−2.33 (−5.04, 0.39)	0.093	−0.28
Fasting plasma glucose (mmol/L)	Baseline M (SD)	78	5.5 (1.2)	78	5.6 (1.1)			
	6 months	71	−0.34 (−0.49, −0.19)	67	−0.21 (−0.36, −0.05)	−0.13 (−0.35, 0.08)	0.230	−0.11
	12 months [b]	67	−0.17 (−0.32, −0.03)	57	0.06 (−0.10, 0.21)	−0.23 (−0.44, −0.02)	0.032	−0.20
	18 months	66	−0.12 (−0.29, 0.04)	59	−0.10 (−0.27, 0.08)	−0.02 (−0.26, 0.22)	0.844	−0.02

Abbreviations: HDL, high-density lipoprotein. [a] Standardized effect: mean intervention effect divided by pooled baseline standard deviation of the outcome. [b] End-of-intervention contact; primary endpoint. [c] Modeled as log outcome adjusted for log outcome at baseline, with results back-transformed to change (follow-up minus baseline) in original units using the relevant expression of marginal means.

Table 3. Within-arm and between-arm changes for patient-reported outcomes: Living Well after Breast Cancer trial.

Outcome	Timepoint	Intervention		Usual Care		Intervention Effect (Intervention—Usual Care)		
		n	Mean Change (95% CI)	*n*	Mean Change (95% CI)	Mean Difference (95% CI)	*p*	d [a]
QOL Physical Health component (T score) [b]	Baseline M (SD)	78	44.8 (6.9)	77	45.9 (6.6)			
	6 months	69	2.43 (1.31, 3.56)	65	0.46 (−0.70, 1.62)	1.98 (0.36, 3.59)	0.017	0.29
	12 months [c]	65	3.16 (1.82, 4.50)	58	0.50 (−0.90, 1.91)	2.66 (0.71, 4.60)	0.007	0.39
	18 months	65	1.56 (0.13, 2.99)	57	0.38 (−1.13, 1.89)	1.18 (−0.90, 3.26)	0.266	0.18
QOL Mental Health component (T score) [b]	Baseline M (SD)	78	46.1 (7.2)	77	45.5 (6.3)			
	6 months	69	1.34 (0.10, 2.57)	65	−0.17 (−1.44, 1.10)	1.51 (−0.27, 3.28)	0.097	0.22
	12 months [c]	65	1.98 (0.68, 3.28)	58	0.21 (−1.15, 1.58)	1.77 (−0.12, 3.66)	0.067	0.26
	18 months	65	−0.36 (−1.98, 1.26)	57	0.17 (−1.54, 1.88)	−0.53 (−2.89, 1.83)	0.659	−0.08

Table 3. *Cont.*

Outcome	Timepoint	Intervention		Usual Care		Intervention Effect (Intervention—Usual Care)		
		n	Mean Change (95% CI)	*n*	Mean Change (95% CI)	Mean Difference (95% CI)	*p*	d [a]
Fatigue [b]	Baseline M (SD)	77	35.5 (9.7)	77	37.6 (9.5)			
	6 months	68	3.21 (1.57, 4.86)	65	0.75 (−0.94, 2.43)	2.47 (0.11, 4.83)	0.040	0.26
	12 months [c]	64	4.29 (2.57, 6.01)	58	2.26 (0.46, 4.05)	2.03 (−0.46, 4.53)	0.110	0.21
	18 months	63	2.63 (0.81, 4.46)	57	1.99 (0.09, 3.89)	0.65 (−2.00, 3.29)	0.632	0.07
Musculoskeletal Pain	Baseline M (SD)	63	1.5 (1.1)	59	1.6 (1.0)			
	6 months	56	−0.21 (−0.44, 0.02)	50	0.40 (0.16, 0.64)	−0.61 (−0.94, −0.28)	<0.001	−0.58
	12 months [c]	52	−0.19 (−0.42, 0.04)	46	0.32 (0.08, 0.56)	−0.51 (−0.84, −0.18)	0.003	−0.49
	18 months	53	−0.07 (−0.32, 0.18)	44	0.26 (−0.01, 0.52)	−0.33 (−0.69, 0.04)	0.079	−0.31
Menopausal Symptoms—Psychological subscale	Baseline M (SD)	75	10.0 (6.2)	76	9.6 (5.7)			
	6 months	66	−1.40 (−2.51, −0.30)	65	−0.14 (−1.25, 0.98)	−1.27 (−2.84, 0.30)	0.113	−0.21
	12 months [c]	62	−1.90 (−3.20, −0.60)	58	−0.62 (−1.96, 0.71)	−1.28 (−3.14, 0.59)	0.179	−0.22
	18 months	61	−1.24 (−2.54, 0.06)	56	−0.22 (−1.56, 1.12)	−1.02 (−2.88, 0.85)	0.286	−0.17
Menopausal Symptoms—Somatic subscale	Baseline M (SD)	76	5.5 (4.3)	76	5.1 (4.0)			
	6 months	67	−0.67 (−1.31, −0.04)	65	0.58 (−0.07, 1.22)	−1.25 (−2.15, −0.34)	0.007	−0.30
	12 months [c]	62	−0.77 (−1.53, −0.01)	58	−0.07 (−0.85, 0.71)	−0.70 (−1.79, 0.39)	0.206	−0.17
	18 months	63	−0.67 (−1.47, 0.13)	56	0.27 (−0.56, 1.10)	−0.94 (−2.10, 0.21)	0.108	−0.23
Menopausal Symptoms—Vasomotor subscale	Baseline M (SD)	76	2.6 (2.2)	76	2.4 (2.1)			
	6 months	67	0.24 (−0.13, 0.62)	65	0.49 (0.11, 0.87)	−0.25 (−0.78, 0.29)	0.367	−0.12
	12 months [c]	63	0.06 (−0.36, 0.48)	58	0.42 (−0.02, 0.85)	−0.35 (−0.96, 0.25)	0.250	−0.17
	18 months	63	−0.22 (−0.65, 0.21)	56	0.32 (−0.13, 0.77)	−0.54 (−1.16, 0.08)	0.089	−0.25
Fear of Cancer Recurrence	Baseline M (SD)	77	14.5 (9.6)	76	15.5 (9.7)			
	6 months	68	−2.20 (−3.74, −0.65)	64	−1.08 (−2.67, 0.51)	−1.12 (−3.34, 1.10)	0.321	−0.12
	12 months [c]	64	−2.24 (−3.83, −0.65)	57	−3.35 (−5.02, −1.67)	1.11 (−1.21, 3.42)	0.348	0.12
	18 months	64	−2.45 (−4.24, −0.65)	55	−1.99 (−3.91, −0.07)	−0.46 (−3.08, 2.17)	0.734	−0.05
Body Image—Total score	Baseline M (SD)	78	2.8 (0.6)	77	2.7 (0.6)			
	6 months	69	−0.35 (−0.46, −0.24)	65	−0.14 (−0.25, −0.03)	−0.21 (−0.37, −0.05)	0.010	−0.36
	12 months [c]	65	−0.43 (−0.54, −0.32)	58	−0.25 (−0.36, −0.13)	−0.18 (−0.35, −0.02)	0.030	−0.31
	18 months	65	−0.30 (−0.42, −0.17)	57	−0.21 (−0.35, −0.08)	−0.08 (−0.27, 0.10)	0.380	−0.14

Abbreviations: QOL, quality of life. [a] Standardized effect: mean intervention effect divided by pooled baseline standard deviation of the outcome. [b] Higher scores are preferable. [c] End-of-intervention contact; primary endpoint.

3.4. Adverse Events

Twenty-five serious AEs (SAE; CTC-AE grade 3–5) from 21 participants were observed over the trial (intervention: n = 13; usual care: n = 12) (Table 4). Only two of the SAEs were considered possibly related to the intervention (knee and foot injuries), neither of which was permanently disabling or life-threatening. Additionally, 180 moderate (grade 2) AEs were reported (intervention: n = 96 events, 53 participants; usual care: n = 84 events, 40 participants)—of these, 18 in the intervention arm were considered possibly related to the intervention and one was considered probably related, and all were primarily musculoskeletal injuries. There were no significant between-arm differences in the rate of either serious AEs (incidence rate ratio; IRR = 0.98 [95%CI: 0.41, 2.34], p = 0.95) or moderate AEs (IRR = 1.03 [95%CI: 0.76, 1.40], p = 0.85).

Table 4. Serious adverse events reported within each arm over the 18-month study period: Living Well after Breast Cancer trial.

	Intervention (n = 11 Participants)	No. of Events	Usual Care (n = 10 Participants)	No. of Events
Life-threatening (n = 4) [a]	Stage IV breast cancer (bone metastasis)	1	Heart episode during surgery	1
			Stage IV breast cancer (i.e., bone metastasis, site unknown)	2
Severe/undesirable (n = 21) [b]	Musculoskeletal events requiring hospitalization or surgery	6 [c]	Musculoskeletal events requiring hospitalization or surgery	1
	Genitourinary events requiring hospitalization or surgery	4	Gastrointestinal events requiring hospitalization or surgery	1
	Other events requiring hospitalization or surgery	1	Genitourinary events requiring hospitalization or surgery	2
	Local breast cancer recurrence	1	Respiratory events requiring hospitalization or surgery	3
			Other events requiring hospitalization or surgery	2

[a] Life-threatening symptoms. [b] Significant symptoms requiring hospitalization or invasive intervention. [c] Includes two adverse events possibly related to the intervention (i.e., knee injury, n = 1; and foot injury, n = 1).

4. Discussion

Intervention participants achieved statistically significant and clinically meaningful weight loss and improvement in metabolic syndrome risk at 12 months compared with usual care participants. Importantly, these improvements were largely sustained six months after intervention contact ceased, highlighting the durable effects of the intervention. Further, beneficial effects on patient-reported outcomes were observed. The magnitude of weight loss achieved is comparable to that observed in previous weight loss trials in breast cancer survivors [44–47], with the intervention effect on weight observed (−4.5% [−6.5, −2.5]) encompassing the clinically meaningful difference of 5% weight loss [9]. These findings provide further support for the use of remotely delivered interventions to successfully achieve weight loss [44,46,47], as well as the feasibility and acceptability of offering such interventions soon after diagnosis and treatment completion.

At study baseline, almost 50% of women were classified as having metabolic syndrome, putting them at considerably increased health risk [26,27]. Those allocated to the intervention observed significant and sustained improvements in metabolic syndrome risk score, with an effect size (Cohen's d) of approximately −0.3. This effect on metabolic syndrome risk is smaller than that observed by Dieli-Conwright et al. [24]; however, the baseline prevalence of metabolic syndrome (77%) was considerably higher in this previous trial of highly sedentary and largely Hispanic breast cancer survivors—in addition, the trial evaluated an intensive supervised exercise intervention. The outcomes observed in the present trial likely reflect the more realistic magnitude of effect achievable with a scalable, telehealth intervention.

Notably, though, we did not observe any significant or clinically meaningful intervention effects on lipids or blood pressure at the end of the intervention. Previous exercise-only intervention trials in breast cancer survivors with good adherence to exercise prescription have demonstrated small–large effects on triglycerides and HDL-cholesterol [24,48], but not in trials with lower adherence [49,50]. Dietary intervention studies have reported differing effects on lipids depending on the macronutrient composition of the diet—significant improvements in triglycerides with the low-carbohydrate diet group only, and a significant, albeit small, improvement in HDL-cholesterol in the low-fat group only [51]. A more specific focus on macronutrient composition or dietary patterns and more frequent assessment of adherence to exercise and dietary prescriptions may be necessary to improve lipid profiles and ultimately metabolic risk in breast cancer survivors. The Mediterranean diet is a dietary pattern that has shown consistent beneficial metabolic and cardiovascular effects in a number of populations [52,53]; however, there has been limited investigation of its benefits in breast cancer survivors [11]. Given the large burden of cardiovascular mortality in breast cancer survivors [17], the benefit of a Mediterranean-style diet and exercise intervention warrants further investigation.

Recent evidence suggests that body composition, defined by low muscle mass (sarcopenia) and adiposity, is more strongly associated with poorer survival in breast cancer than BMI [16]. In this trial, a significant reduction in total fat mass and central adiposity was observed with the intervention; however, reductions in lean mass of $\approx$1 kg were also observed, consistent with what is typically observed with weight loss [54]. When examined across body region, loss of lean mass occurred in every region, though not necessarily to equal degrees. Being much less than the loss of fat mass, the relative body composition shifted towards a higher percentage of lean mass. Although resistance exercise was encouraged, many women chose a less intensive resistance exercise program. Interventions emphasising supervised progressive resistance training, and perhaps gym-based sessions using specialized equipment [55,56], may be needed to minimize muscle loss. Further evidence on how to effectively achieve similar benefits via remotely delivered interventions, such as with telehealth, is needed [57].

This trial also examined the effect of the intervention on key patient-reported outcomes, including quality of life, and treatment-related side effects. A significant intervention effect on physical quality of life was observed, similar to improvements observed in previous trials [44,45], where significant short-term intervention effects were observed [44]. However, a particularly novel and important finding is the significant medium–large intervention effects observed for musculoskeletal pain, which was used to assess arthralgia. Arthralgia is common in breast cancer survivors, particularly those treated with aromatase inhibitors (AIs) [19], and can often lead to poor adherence or discontinuation of AI treatment [58–60]. Recent studies of exercise interventions have shown improvements in arthralgia and pain scores following intervention [61,62]; however, these trials exclusively recruited women on AIs reporting arthralgia/joint pain. Given the magnitude of intervention effects observed in our sample, where only a third were treated with AIs, and the very limited evidence to date [63], this finding warrants further investigation, as does the potential beneficial effect on menopausal symptoms (neither of which were attenuated following adjustment for changes in endocrine treatment).

Several ongoing trials are evaluating the effect of weight loss interventions on breast cancer-specific outcomes, including survival [13–15]. Preliminary findings from the SUCCESS-C trial [64] showed no significant effect on disease-free survival in the lifestyle intervention arm (vs. non-lifestyle intervention) in intention-to-treat analyses; however, weight loss in their telephone-delivered lifestyle arm at two-year follow-up was very modest (mean: 1.0 kg) and attrition was exceptionally high (51.8%) [64]. Post-hoc analyses in lifestyle intervention arm completers (vs. non-intervention) suggest a significant benefit for disease-free survival (HR: 0.51 [95%CI: 0.33, 0.78]) [64]. These results show promise but highlight the challenges of achieving and maintaining clinically meaningful weight loss (≥5%) and participant retention.

Strengths of this trial include the evaluation of a remotely delivered intervention, with potential for wider implementation; assessment of the durability of intervention effects; and the inclusion and retention of a broadly representative sample of breast cancer survivors. However, the sample included an over-representation of younger breast cancer survivors, with almost 40% reporting being pre-menopausal at diagnosis. Given that 22% of breast cancer cases are diagnosed in women <50 years in Australia [65], this likely reflects a particular interest and need for such interventions among younger survivors, who have been excluded from many of the previous trials [11,12]. Limitations of the trial include the primarily Caucasian sample, which limits generalizability to non-Caucasian populations. Although there was differential attrition, multiple imputation models showed that this did not affect the main study findings. The study was sufficiently powered to detect clinically important differences in the primary outcome and effect sizes in secondary outcomes of 0.5 SD—for some secondary outcomes, smaller differences were observed, which may still be clinically meaningful, but for which we were underpowered to detect between-arm effects. These should be examined in future trials or via pooling of trial findings in meta-analyses.

5. Conclusions

The COVID-19 pandemic has highlighted the need for quality cancer care that can be remotely delivered—an already advanced area of research in the field of cancer survivorship and lifestyle intervention. This trial adds to this evidence as both clinically meaningful and durable weight loss and improvement in metabolic syndrome risk were achieved, in women following a breast cancer diagnosis. Future research should further explore strategies for maximizing the health benefits achievable with remotely delivered interventions, particularly in relation to minimizing loss in lean mass, improvements in arthralgia and menopausal symptoms, and for achieving improvements across all metabolic syndrome components.

Supplementary Materials: The following are available online at https://www.mdpi.com/article/10.3390/nu13114091/s1, Table S1. Criteria for the clinical diagnosis of metabolic syndrome in women and calculation of continuous metabolic syndrome risk score in Living Well after Breast Cancer. Table S2. Variables included in multiple imputation models (outcomes imputed separately by chained equations, STATA 16, m = 50 imputations). Table S3. Predictors of drop-out. Table S4. Changes in primary and secondary outcomes in the Living well after Breast Cancer Trial (multiple imputation analysis). Table S5. Baseline mean regional body composition measured by dual X-ray absorptiometry in the Living Well after Breast Cancer intervention (n = 79) and usual care (n = 80) participants. Table S6. Within-arm and between-arm changes in regional body composition: Living Well after Breast Cancer trial (evaluable case analyses). Table S7. Within-arm and between-arm changes in regional body composition: Living Well after Breast Cancer trial (multiple imputation analyses). Table S8. Changes within each arm and intervention effects on medication-sensitive outcomes, adjusted for baseline and concurrent use of relevant medications in the Living Well after Breast Cancer Trial.

Author Contributions: Conceptualization, M.M.R., E.G.E. and W.D.-W.; Formal analysis, E.A.H.W., C.O.T. and R.S.W.; Funding acquisition, M.M.R., E.G.E., N.M. and I.J.H.; Investigation, M.M.R., C.O.T., E.A.H.W. and I.J.H.; Methodology, M.M.R., E.G.E., W.D.-W., E.A.H.W., I.J.H. and S.P.L.; Project administration, M.M.R. and C.O.T.; Supervision, M.M.R., E.G.E., N.M. and S.P.L.; Visualization, M.M.R., C.O.T. and E.A.H.W.; Writing—original draft, M.M.R., C.O.T. and E.A.H.W.; Writing—review and editing: All authors. All authors have read and agreed to the published version of the manuscript.

Funding: This work was supported by the Australian National Health and Medical Research Council (NHMRC; GNT1024739); a National Breast Cancer Foundation Fellowship (ECF-13-09 to M.M.R.); a Research Training Package Scholarship from the University of Queensland to C.O.T.; a NHMRC Centre for Research Excellence Grant (GNT1057608 to E.A.H.W.); a NHMRC Senior Research Fellowship (GNT1041789 to E.G.E.); and an ACS Clinical Research Professor Award (CRP-19-175-06-COUN to W.D.W.).

Institutional Review Board Statement: The study was conducted according to the guidelines of the Declaration of Helsinki and approved by the human research ethics committees of the Royal Brisbane & Women's Hospital (HREC/12/QRBW/149; 17 August 2012), Greenslopes Private Hospital (12/26; 19 June 2012), and St. Vincent's Health & Aged Care (13/02; 18 June 2013), and The University of Queensland (2012000944; 22 August 2012).

Informed Consent Statement: Informed consent was obtained from all participants involved in the study.

Data Availability Statement: Data available upon request to corresponding author with ethics approval.

Acknowledgments: The authors wish to acknowledge the support of the Breast Cancer Network Australia and the study consumer advocates, Lorraine Woods and Fiona Evans. The authors thank all the study participants and referring physicians and acknowledge the dedication and contribution of the project staff—Jane Erickson, Jennifer Job, Fiona Heys, Jodie Jetann, Zoe Thomson, Denise Brookes, and Charani Kiriwandeniya.

Conflicts of Interest: The authors declare no conflict of interest. The sponsors had no role in the design, execution, interpretation, or writing of the study.

References

1. Demark-Wahnefried, W.; Schmitz, K.H.; Alfano, C.M.; Bail, J.R.; Goodwin, P.J.; Thomson, C.A.; Bradley, D.W.; Courneya, K.S.; Befort, C.A.; Denlinger, C.S.; et al. Weight management and physical activity throughout the cancer care continuum. *CA Cancer J. Clin.* **2018**, *68*, 64–89. [CrossRef] [PubMed]
2. Ligibel, J.A.; Alfano, C.M.; Courneya, K.S.; Demark-Wahnefried, W.; Burger, R.A.; Chlebowski, R.T.; Fabian, C.J.; Gucalp, A.; Hershman, D.L.; Hudson, M.M.; et al. American society of clinical oncology position statement on obesity and cancer. *J. Clin. Oncol.* **2014**, *32*, 3568–3574. [CrossRef] [PubMed]
3. Lahart, I.M.; Metsios, G.S.; Nevill, A.M.; Carmichael, A.R. Physical activity, risk of death and recurrence in breast cancer survivors: A systematic review and meta-analysis of epidemiological studies. *Acta Oncol.* **2015**, *54*, 635–654. [CrossRef] [PubMed]
4. Ibrahim, E.M.; Al-Homaidh, A. Physical activity and survival after breast cancer diagnosis: Meta-analysis of published studies. *Med. Oncol.* **2011**, *28*, 753–765. [CrossRef]
5. Juvet, L.K.; Thune, I.; Elvsaas, I.K.O.; Fors, E.A.; Lundgren, S.; Bertheussen, G.; Leivseth, G.; Oldervoll, L.M. The effect of exercise on fatigue and physical functioning in breast cancer patients during and after treatment and at 6 months follow-up: A meta-analysis. *Breast* **2017**, *33*, 166–177. [CrossRef]
6. McNeely, M.L.; Campbell, K.L.; Rowe, B.H.; Klassen, T.P.; Mackey, J.R.; Courneya, K.S. Effects of exercise on breast cancer patients and survivors: A systematic review and meta-analysis. *CMAJ* **2006**, *175*, 34–41. [CrossRef]
7. Meneses-Echavez, J.F.; Gonzalez-Jimenez, E.; Ramirez-Velez, R. Effects of supervised exercise on cancer-related fatigue in breast cancer survivors: A systematic review and meta-analysis. *BMC Cancer* **2015**, *15*, 77. [CrossRef]
8. Chan, D.S.; Vieira, A.R.; Aune, D.; Bandera, E.V.; Greenwood, D.C.; McTiernan, A.; Navarro Rosenblatt, D.; Thune, I.; Vieira, R.; Norat, T. Body mass index and survival in women with breast cancer-systematic literature review and meta-analysis of 82 follow-up studies. *Ann. Oncol.* **2014**, *25*, 1901–1914. [CrossRef]
9. Rock, C.L.; Doyle, C.; Demark-Wahnefried, W.; Meyerhardt, J.; Courneya, K.S.; Schwartz, A.L.; Bandera, E.V.; Hamilton, K.K.; Grant, B.; McCullough, M.; et al. Nutrition and physical activity guidelines for cancer survivors. *CA Cancer J. Clin.* **2012**, *62*, 243–274. [CrossRef]
10. World Cancer Research Fund/American Institute for Cancer Research. *Diet, Nutrition, Physical Activity and Cancer: A Global Perspective*; The Third Expert Report; World Cancer Research Fund/American Institute for Cancer Research: London, UK, 2018.
11. Chlebowski, R.T.; Reeves, M.M. Weight Loss Randomized Intervention Trials in Female Cancer Survivors. *J. Clin. Oncol.* **2016**, *34*, 4238–4248. [CrossRef]
12. Reeves, M.M.; Terranova, C.O.; Eakin, E.G.; Demark-Wahnefried, W. Weight loss intervention trials in women with breast cancer: A systematic review. *Obes. Rev.* **2014**, *15*, 749–768. [CrossRef] [PubMed]
13. Ligibel, J.A.; Barry, W.T.; Alfano, C.; Hershman, D.L.; Irwin, M.; Neuhouser, M.; Thomson, C.A.; Delahanty, L.; Frank, E.; Spears, P.; et al. Randomized phase III trial evaluating the role of weight loss in adjuvant treatment of overweight and obese women with early breast cancer (Alliance A011401): Study design. *NPJ Breast Cancer* **2017**, *3*, 37. [CrossRef] [PubMed]
14. Pasanisi, P.; Villarini, A.; Gargano, G.; Bruno, E.; Raimondi, M.; Bellegotti, M.; Curtosi, P.; Berrino, F. A randomized controlled trial of diet, physical activity and breast cancer recurrences—The Diana-5 study. *Eur. J. Cancer* **2012**, *48*, S283–S284. [CrossRef]

15. Rack, B.; Andergassen, U.; Neugebauer, J.; Salmen, J.; Hepp, P.; Sommer, H.; Lichtenegger, W.; Friese, K.; Beckmann, M.W.; Hauner, D.; et al. The German SUCCESS C study—The first European lifestyle study on breast cancer. *Breast Care* **2010**, *5*, 395–400. [CrossRef]
16. Aleixo, G.F.P.; Williams, G.R.; Nyrop, K.A.; Muss, H.B.; Shachar, S.S. Muscle composition and outcomes in patients with breast cancer: Meta-analysis and systematic review. *Breast Cancer Res. Treat.* **2019**, *177*, 569–579. [CrossRef]
17. Patnaik, J.L.; Byers, T.; Diguiseppi, C.; Denberg, T.D.; Dabelea, D. The influence of comorbidities on overall survival among older women diagnosed with breast cancer. *J. Natl. Cancer Inst.* **2011**, *103*, 1101–1111. [CrossRef]
18. Schmitz, K.H.; Speck, R.M.; Rye, S.A.; DiSipio, T.; Hayes, S.C. Prevalence of breast cancer treatment sequelae over 6 years of follow-up: The Pulling Through Study. *Cancer* **2012**, *118*, S2217–S2225. [CrossRef]
19. Beckwee, D.; Leysen, L.; Meuwis, K.; Adriaenssens, N. Prevalence of aromatase inhibitor-induced arthralgia in breast cancer: A systematic review and meta-analysis. *Support. Care Cancer* **2017**, *25*, 1673–1686. [CrossRef]
20. Caan, B.J.; Emond, J.A.; Su, H.I.; Patterson, R.E.; Flatt, S.W.; Gold, E.B.; Newman, V.A.; Rock, C.L.; Thomson, C.A.; Pierce, J.P. Effect of postdiagnosis weight change on hot flash status among early-stage breast cancer survivors. *J. Clin. Oncol.* **2012**, *30*, 1492–1497. [CrossRef]
21. Schmidt, M.E.; Chang-Claude, J.; Seibold, P.; Vrieling, A.; Heinz, J.; Flesch-Janys, D.; Steindorf, K. Determinants of long-term fatigue in breast cancer survivors: Results of a prospective patient cohort study. *Psychooncology* **2015**, *24*, 40–46. [CrossRef]
22. Demark-Wahnefried, W.; Campbell, K.L.; Hayes, S.C. Weight management and its role in breast cancer rehabilitation. *Cancer* **2012**, *118*, S2277–S2287. [CrossRef]
23. Imayama, I.; Alfano, C.M.; Neuhouser, M.L.; George, S.M.; Wilder Smith, A.; Baumgartner, R.N.; Baumgartner, K.B.; Bernstein, L.; Wang, C.Y.; Duggan, C.; et al. Weight, inflammation, cancer-related symptoms and health related quality of life among breast cancer survivors. *Breast Cancer Res. Treat.* **2013**, *140*, 159–176. [CrossRef]
24. Dieli-Conwright, C.M.; Courneya, K.S.; Demark-Wahnefried, W.; Sami, N.; Lee, K.; Buchanan, T.A.; Spicer, D.V.; Tripathy, D.; Bernstein, L.; Mortimer, J.E. Effects of Aerobic and Resistance Exercise on Metabolic Syndrome, Sarcopenic Obesity, and Circulating Biomarkers in Overweight or Obese Survivors of Breast Cancer: A Randomized Controlled Trial. *J. Clin. Oncol.* **2018**, *36*, 875–883. [CrossRef]
25. Wilson, P.W.; D'Agostino, R.B.; Parise, H.; Sullivan, L.; Meigs, J.B. Metabolic syndrome as a precursor of cardiovascular disease and type 2 diabetes mellitus. *Circulation* **2005**, *112*, 3066–3072. [CrossRef]
26. Berrino, F.; Villarini, A.; Traina, A.; Bonanni, B.; Panico, S.; Mano, M.P.; Mercandino, A.; Galasso, R.; Barbero, M.; Simeoni, M.; et al. Metabolic syndrome and breast cancer prognosis. *Breast Cancer Res. Treat.* **2014**, *147*, 159–165. [CrossRef]
27. Calip, G.S.; Malone, K.E.; Gralow, J.R.; Stergachis, A.; Hubbard, R.A.; Boudreau, D.M. Metabolic syndrome and outcomes following early-stage breast cancer. *Breast Cancer Res. Treat.* **2014**, *148*, 363–377. [CrossRef]
28. Reeves, M.M.; Terranova, C.O.; Erickson, J.M.; Job, J.R.; Brookes, D.S.; McCarthy, N.; Hickman, I.J.; Lawler, S.P.; Fjeldsoe, B.S.; Healy, G.N.; et al. Living well after breast cancer randomized controlled trial protocol: Evaluating a telephone-delivered weight loss intervention versus usual care in women following treatment for breast cancer. *BMC Cancer* **2016**, *16*, 830. [CrossRef]
29. Jensen, M.D.; Ryan, D.H.; Apovian, C.M.; Ard, J.D.; Comuzzie, A.G.; Donato, K.A.; Hu, F.B.; Hubbard, V.S.; Jakicic, J.M.; Kushner, R.F.; et al. 2013 AHA/ACC/TOS Guideline for the Management of Overweight and Obesity in Adults: A Report of the American College of Cardiology/American Heart Association Task Force on Practice Guidelines and The Obesity Society. *Circulation* **2014**, *129*, S102–S138. [CrossRef]
30. Reeves, M.M.; Winkler, E.A.H.; McCarthy, N.; Lawler, S.P.; Terranova, C.O.; Hayes, S.C.; Janda, M.; Demark-Wahnefried, W.; Eakin, E.G. The Living Well after Breast Cancer™ Pilot Trial: A weight loss intervention for women following treatment for breast cancer. *Asia Pac. J. Clin. Oncol.* **2017**, *13*, 125–136. [CrossRef]
31. Spark, L.C.; Fjeldsoe, B.S.; Eakin, E.G.; Reeves, M.M. Efficacy of a Text Message-Delivered Extended Contact Intervention on Maintenance of Weight Loss, Physical Activity, and Dietary Behavior Change. *JMIR mHealth uHealth* **2015**, *3*, e88. [CrossRef]
32. Institute of Medicine of the National Academies. Dietary Reference Intakes for Energy, Carbohydrate, Fiber, Fat, Fatty Acids, Cholesterol, Protein, and Amino Acids (Macronutrients). Available online: http://www.nap.edu/openbook.php?record_id=10490&page=R1 (accessed on 2 November 2021).
33. Bandura, A. *Social Foundations of Thought and Action: A Social Cognitive Theory*; Prentice Hall: Englewood Cliffs, NJ, USA, 1986.
34. Alberti, K.G.; Eckel, R.H.; Grundy, S.M.; Zimmet, P.Z.; Cleeman, J.I.; Donato, K.A.; Fruchart, J.C.; James, W.P.; Loria, C.M.; Smith, S.C., Jr. Harmonizing the metabolic syndrome: A joint interim statement of the International Diabetes Federation Task Force on Epidemiology and Prevention; National Heart, Lung, and Blood Institute; American Heart Association; World Heart Federation; International Atherosclerosis Society; and International Association for the Study of Obesity. *Circulation* **2009**, *120*, 1640–1645. [CrossRef]
35. Wijndaele, K.; Healy, G.N.; Dunstan, D.W.; Barnett, A.G.; Salmon, J.; Shaw, J.E.; Zimmet, P.Z.; Owen, N. Increased cardiometabolic risk is associated with increased TV viewing time. *Med. Sci. Sports Exerc.* **2010**, *42*, 1511–1518. [CrossRef]
36. Ekelund, U.; Griffin, S.J.; Wareham, N.J. Physical activity and metabolic risk in individuals with a family history of type 2 diabetes. *Diabetes Care* **2007**, *30*, 337–342. [CrossRef]
37. Healy, G.N.; Winkler, E.A.H.; Owen, N.; Anuradha, S.; Dunstan, D.W. Replacing sitting time with standing or stepping: Associations with cardio-metabolic risk biomarkers. *Eur. Heart J.* **2015**, *36*, 2643–2649. [CrossRef]

38. Hays, R.D.; Bjorner, J.B.; Revicki, D.A.; Spritzer, K.L.; Cella, D. Development of physical and mental health summary scores from the patient-reported outcomes measurement information system (PROMIS) global items. *Qual. Life Res.* **2009**, *18*, 873–880. [CrossRef]
39. Yellen, S.B.; Cella, D.F.; Webster, K.; Blendowski, C.; Kaplan, E. Measuring fatigue and other anemia-related symptoms with the Functional Assessment of Cancer Therapy (FACT) measurement system. *J. Pain Symptom Manag.* **1997**, *13*, 63–74. [CrossRef]
40. Swenson, K.K.; Nissen, M.J.; Henly, S.J.; Maybon, L.; Pupkes, J.; Zwicky, K.; Tsai, M.L.; Shapiro, A.C. Identification of tools to measure changes in musculoskeletal symptoms and physical functioning in women with breast cancer receiving aromatase inhibitors. *Oncol. Nurs. Forum* **2013**, *40*, 549–557. [CrossRef]
41. Greene, J.G. Constructing a standard climacteric scale. *Maturitas* **1998**, *29*, 25–31. [CrossRef]
42. Thewes, B.; Zachariae, R.; Christensen, S.; Nielsen, T.; Butow, P. The Concerns About Recurrence Questionnaire: Validation of a brief measure of fear of cancer recurrence amongst Danish and Australian breast cancer survivors. *J. Cancer Surviv.* **2015**, *9*, 68–79. [CrossRef]
43. Hormes, J.M.; Lytle, L.A.; Gross, C.R.; Ahmed, R.L.; Troxel, A.B.; Schmitz, K.H. The body image and relationships scale: Development and validation of a measure of body image in female breast cancer survivors. *J. Clin. Oncol.* **2008**, *26*, 1269–1274. [CrossRef]
44. Goodwin, P.J.; Segal, R.J.; Vallis, M.; Ligibel, J.A.; Pond, G.R.; Robidoux, A.; Blackburn, G.L.; Findlay, B.; Gralow, J.R.; Mukherjee, S.; et al. Randomized trial of a telephone-based weight loss intervention in postmenopausal women with breast cancer receiving letrozole: The LISA trial. *J. Clin. Oncol.* **2014**, *32*, 2231–2239. [CrossRef] [PubMed]
45. Rock, C.L.; Flatt, S.W.; Byers, T.E.; Colditz, G.A.; Demark-Wahnefried, W.; Ganz, P.A.; Wolin, K.Y.; Elias, A.; Krontiras, H.; Liu, J.; et al. Results of the Exercise and Nutrition to Enhance Recovery and Good Health for You (ENERGY) Trial: A behavioral weight loss intervention in overweight or obese breast cancer survivors. *J. Clin. Oncol.* **2015**, *33*, 3169–3176. [CrossRef] [PubMed]
46. Patterson, R.E.; Marinac, C.R.; Sears, D.D.; Kerr, J.; Hartman, S.J.; Cadmus-Bertram, L.; Villaseñor, A.; Flatt, S.W.; Godbole, S.; Li, H.; et al. The Effects of Metformin and Weight Loss on Biomarkers Associated with Breast Cancer Outcomes. *J. Natl. Cancer Inst.* **2018**, *110*, 1239–1247. [CrossRef] [PubMed]
47. Harrigan, M.; Cartmel, B.; Loftfield, E.; Sanft, T.; Chagpar, A.B.; Zhou, Y.; Playdon, M.; Li, F.; Irwin, M.L. Randomized trial comparing telephone versus in-person weight loss counseling on body composition and circulating biomarkers in women treated for breast cancer: The Lifestyle, Exercise, and Nutrition (LEAN) study. *J. Clin. Oncol.* **2016**, *34*, 669–676. [CrossRef]
48. Nuri, R.; Kordi, M.R.; Moghaddasi, M.; Rahnama, N.; Damirchi, A.; Rahmani-Nia, F.; Emami, H. Effect of combination exercise training on metabolic syndrome parameters in postmenopausal women with breast cancer. *J. Cancer Res. Ther.* **2012**, *8*, 238–242. [CrossRef]
49. Thomas, G.A.; Alvarez-Reeves, M.; Lu, L.; Yu, H.; Irwin, M.L. Effect of exercise on metabolic syndrome variables in breast cancer survivors. *Int. J. Endocrinol.* **2013**, *2013*, 168797. [CrossRef]
50. Guinan, E.; Hussey, J.; Broderick, J.M.; Lithander, F.E.; O'Donnell, D.; Kennedy, M.J.; Connolly, E.M. The effect of aerobic exercise on metabolic and inflammatory markers in breast cancer survivors—A pilot study. *Support. Care Cancer* **2013**, *21*, 1983–1992. [CrossRef]
51. Thomson, C.A.; Stopeck, A.T.; Bea, J.W.; Cussler, E.; Nardi, E.; Frey, G.; Thompson, P.A. Changes in body weight and metabolic indexes in overweight breast cancer survivors enrolled in a randomized trial of low-fat vs. reduced carbohydrate diets. *Nutr. Cancer* **2010**, *62*, 1142–1152. [CrossRef]
52. Dinu, M.; Pagliai, G.; Casini, A.; Sofi, F. Mediterranean diet and multiple health outcomes: An umbrella review of meta-analyses of observational studies and randomised trials. *Eur. J. Clin. Nutr.* **2018**, *72*, 30–43. [CrossRef]
53. Garcia, M.; Bihuniak, J.D.; Shook, J.; Kenny, A.; Kerstetter, J.; Huedo-Medina, T.B. The Effect of the Traditional Mediterranean-Style Diet on Metabolic Risk Factors: A Meta-Analysis. *Nutrients* **2016**, *8*, 168. [CrossRef]
54. Heymsfield, S.B.; Gonzalez, M.C.; Shen, W.; Redman, L.; Thomas, D. Weight loss composition is one-fourth fat-free mass: A critical review and critique of this widely cited rule. *Obes. Rev.* **2014**, *15*, 310–321. [CrossRef]
55. Buffart, L.M.; Kalter, J.; Sweegers, M.G.; Courneya, K.S.; Newton, R.U.; Aaronson, N.K.; Jacobsen, P.B.; May, A.M.; Galvao, D.A.; Chinapaw, M.J.; et al. Effects and moderators of exercise on quality of life and physical function in patients with cancer: An individual patient data meta-analysis of 34 RCTs. *Cancer Treat. Rev.* **2017**, *52*, 91–104. [CrossRef]
56. Lacroix, A.; Hortobagyi, T.; Beurskens, R.; Granacher, U. Effects of Supervised vs. Unsupervised Training Programs on Balance and Muscle Strength in Older Adults: A Systematic Review and Meta-Analysis. *Sports Med.* **2017**, *47*, 2341–2361. [CrossRef]
57. Schmitz, K.H.; Troxel, A.B.; Dean, L.T.; DeMichele, A.; Brown, J.C.; Sturgeon, K.; Zhang, Z.; Evangelisti, M.; Spinelli, B.; Kallan, M.J.; et al. Effect of Home-Based Exercise and Weight Loss Programs on Breast Cancer-Related Lymphedema Outcomes Among Overweight Breast Cancer Survivors: The WISER Survivor Randomized Clinical Trial. *JAMA Oncol.* **2019**, *5*, 1605–1613. [CrossRef]
58. Kwan, M.L.; Roh, J.M.; Laurent, C.A.; Lee, J.; Tang, L.; Hershman, D.; Kushi, L.H.; Yao, S. Patterns and reasons for switching classes of hormonal therapy among women with early-stage breast cancer. *Cancer Causes Control* **2017**, *28*, 557–562. [CrossRef]
59. Brier, M.J.; Chambless, D.L.; Chen, J.; Mao, J.J. Ageing perceptions and non-adherence to aromatase inhibitors among breast cancer survivors. *Eur. J. Cancer* **2018**, *91*, 145–152. [CrossRef]
60. Pineda-Moncusi, M.; Servitja, S.; Tusquets, I.; Diez-Perez, A.; Rial, A.; Cos, M.L.; Campodarve, I.; Rodriguez-Morera, J.; Garcia-Giralt, N.; Nogues, X. Assessment of early therapy discontinuation and health-related quality of life in breast cancer patients treated with aromatase inhibitors: B-ABLE cohort study. *Breast Cancer Res. Treat.* **2019**, *177*, 53–60. [CrossRef]

61. Irwin, M.L.; Cartmel, B.; Gross, C.P.; Ercolano, E.; Li, F.; Yao, X.; Fiellin, M.; Capozza, S.; Rothbard, M.; Zhou, Y.; et al. Randomized exercise trial of aromatase inhibitor-induced arthralgia in breast cancer survivors. *J. Clin. Oncol.* **2015**, *33*, 1104–1111. [CrossRef]
62. Nyrop, K.A.; Callahan, L.F.; Cleveland, R.J.; Arbeeva, L.L.; Hackney, B.S.; Muss, H.B. Randomized Controlled Trial of a Home-Based Walking Program to Reduce Moderate to Severe Aromatase Inhibitor-Associated Arthralgia in Breast Cancer Survivors. *Oncologist* **2017**, *22*, 1238–1249. [CrossRef]
63. Befort, C.A.; Klemp, J.R.; Austin, H.L.; Perri, M.G.; Schmitz, K.H.; Sullivan, D.K.; Fabian, C.J. Outcomes of a weight loss intervention among rural breast cancer survivors. *Breast Cancer Res. Treat.* **2012**, *132*, 631–639. [CrossRef]
64. Janni, W.; Rack, B.; Friedl, T.; Müller, V.; Lorenz, R.; Rezai, M.; Tesch, H.; Heinrich, G.; Andergassen, U.; Harbeck, N.; et al. Lifestyle Intervention and Effect on Disease-free Survival in Early Breast Cancer Pts: Interim Analysis from the Randomized SUCCESS C Study. *Cancer Res.* **2019**, *79*, GS5-03. [CrossRef]
65. Australian Institute of Health and Welfare. *Australian Cancer Incidence and Mortality (ACIM) Books: Breast Cancer*; Australian Institute of Health and Welfare: Canberra, ACT, Australia, 2015.

 MDPI

Article

'Energy-Dense, High-SFA and Low-Fiber' Dietary Pattern Lowered Adiponectin but Not Leptin Concentration of Breast Cancer Survivors

Mohd Razif Shahril [1,*], Nor Syamimi Zakarai [2], Geeta Appannah [3], Ali Nurnazahiah [2], Hamid Jan Jan Mohamed [4], Aryati Ahmad [2], Pei Lin Lua [5] and Michael Fenech [1,6]

1 Centre for Healthy Ageing and Wellness (H-CARE), Faculty of Health Sciences, Universiti Kebangsaan Malaysia, Jalan Raja Muda Abdul Aziz, Kuala Lumpur 50300, Selangor, Malaysia; mf.ghf@outlook.com
2 School of Nutrition and Dietetics, Faculty of Health Sciences, University Sultan Zainal Abidin, Gong Badak Campus, Kuala Nerus 21300, Terengganu, Malaysia; norsyamimizakarai@gmail.com (N.S.Z.); nurnazahiahali@unisza.edu.my (A.N.); aryatiahmad@unisza.edu.my (A.A.)
3 Department of Nutrition, Faculty of Medicine and Health Sciences, Universiti Putra Malaysia, Seri Kembangan 43400, Selangor, Malaysia; geeta@upm.edu.my
4 Nutrition and Dietetics Programme, School of Health Sciences, Universiti Sains Malaysia, Kubang Kerian 16150, Kelantan, Malaysia; hamidjan@usm.my
5 Faculty of Pharmacy, Universiti Sultan Zainal Abidin, Besut Campus, Besut 22200, Terengganu, Malaysia; peilinlua@unisza.edu.my
6 Health and Biomedical Innovation, UniSA Clinical and Health Sciences, University of South Australia, Adelaide, SA 5000, Australia
* Correspondence: razifshahril@ukm.edu.my

Citation: Shahril, M.R.; Zakarai, N.S.; Appannah, G.; Nurnazahiah, A.; Mohamed, H.J.J.; Ahmad, A.; Lua, P.L.; Fenech, M. 'Energy-Dense, High-SFA and Low-Fiber' Dietary Pattern Lowered Adiponectin but Not Leptin Concentration of Breast Cancer Survivors. *Nutrients* **2021**, *13*, 3339. https://doi.org/10.3390/nu13103339

Academic Editors: Wendy Demark-Wahnefried and Christina Dieli-Conwright

Received: 19 August 2021
Accepted: 21 September 2021
Published: 24 September 2021

Abstract: Dietary pattern (DP) and its relationship with disease biomarkers have received recognition in nutritional epidemiology investigations. However, DP relationships with adipokines (i.e., adiponectin and leptin) among breast cancer survivors remain unclear. Therefore, we assessed relationships between DP and high-molecular weight (HMW) adiponectin and leptin concentration among breast cancer survivors. This cross-sectional study involved 128 breast cancer survivors who attended the oncology outpatient clinic at two main government hospitals in the East Coast of Peninsular Malaysia. The serum concentration of HMW adiponectin and leptin were measured using enzyme-linked immunosorbent assay (ELISA) kits. A reduced rank regression method was used to analyze DP. Relationships between DP with HMW adiponectin and leptin were examined using regression models. The findings show that with every 1-unit increase in the 'energy-dense, high-SFA, low-fiber' DP z-score, there was a reduction by 0.41 µg/mL in HMW adiponectin which was independent of age, BMI, education level, occupation status, cancer stage, and duration since diagnosis. A similar relationship with leptin concentration was not observed. In conclusion, the 'energy-dense, high-saturated fat and low-fiber' DP, which is characterized by high intake levels of sugar-sweetened drinks and fat-based spreads but low intake of fruits and vegetables, is an unhealthy dietary pattern and unfavorable for HMW adiponectin concentration, but not for leptin. These findings could serve as a basis in developing specific preventive strategies that are tailored to the growing population of breast cancer survivors.

Keywords: HMW adiponectin; leptin; dietary patterns; breast cancer survivors

1. Introduction

With the increase in the number and life expectancy of breast cancer survivors, the focus of cancer care has shifted towards better survivor care, including nutritional interventions and lifestyle changes. Dietary pattern has received considerable critical attention in nutritional epidemiology as one of the potential factors in modifying cancer risk, recurrence, or mortality [1,2]. Sotos-Prieto et al. [3] found that the risk of all-cause and cause-specific

in the present study. Each of the breast cancer survivors received a z-score for the DP identified, discriminating how strongly their dietary intakes linked with the DP.

2.4. Adipokines

Approximately, five milliliters of fasting blood from breast cancer survivors were drawn by venipuncture and transferred in a red-top tube, a BD Vacutainer® Plus Plastic Serum Tubes containing no anticoagulant, during the data collection. The tube was centrifuged at 3500 rpm, for 10 min at 4 degrees Celsius. Serum was transferred into the 1.5 L tube and stored at −80 degrees Celsius. The present study measured two target proteins: HMW adiponectin and leptin by using the enzyme-linked immunosorbent assay (ELISA) method following a typical two-step capture or 'sandwich' type assay for the detection of the target protein. The ELISA kits used were Human Adiponectin Immunoassay Kit Cat.No.47-ADPHU-E01 and Human Leptin Immunoassay Kit Cat.No.11-LEPHU-E01 (American Laboratory Products Company (ALPCO) Diagnostics, Salem, NH, USA).

2.5. Covariates Assessment

2.5.1. Socio-Demographic and Clinical Characteristics

The present study used a set of questionnaires consisting of sociodemographic and clinical characteristics and was interviewer-administered on a one-to-one basis. Sociodemographic questions consisted of age, home address, monthly income, ethnicity, marital status, and education level, as well as job status. Clinical characteristics questionnaires included the year cancer was diagnosed, stage of cancer, treatments and medications, other health problems faced by the respondents as well as complications experienced by the respondents after their cancer's treatment.

2.5.2. Anthropometric and Body Compositions Assessments

Anthropometric and body composition assessments such as body weight and per centage of body fat were measured using a body composition analyzer (Tanita BC-587, TANITA Corporation, Tokyo, Japan). Height was measured to the nearest 0.1 cm by using a mobile stadiometer (Seca 217, Seca, Hamburg, Germany) and waist measurement was taken by using a measuring tape (Seca 201, Seca, Hamburg, Germany) at the smallest waist area.

2.6. Statistical Analysis

All data were analyzed using IBM SPSS Statistics for Windows Version 22.0 software (IBM Corporation, Armonk, NY, USA), except for the dietary pattern, for which the partial least squares procedure with reduced rank regression option was used, analyzed using SAS Software Version 9.4 (SAS Institute, Cary, NC, USA). Descriptive statistics including the mean, standard deviation and range were used to present the respondent's serum adipokines concentration. Simple linear regression was performed to identify the possible independent factors related to adipokines concentration without considering any confounder. Next, multivariate regression analysis was conducted to analyze the relationship between the mean DP z-scores of breast cancer survivors and their serum adipokines (HMW adiponectin and leptin) concentrations including the adjusted variables which could be biologically important during model development were included. Overall, age, BMI, cancer stage, duration since diagnosis, education level and occupation were the selected confounders for the link between DP and adipokines concentration in this study.

3. Results

Table 1 describes the characteristics of breast cancer survivors, including the sociodemographic, anthropometric measurement and adipokine (HMW adiponectin and leptin) profile. In summary, the majority of the respondents were Malay (94.5%), married (77.3%) and had secondary education (59.4%). The majority of the respondents had a range of monthly income from MYR 500 to 2000 (45.3%), and the mean income was MYR

2409.80 ± 2325.85. The majority of the respondents in this study also had a long period of survivorship, in which 61.7% of them had survived for more than five years after they had been diagnosed.

Table 1. Characteristics of breast cancer survivors.

	n (%)	Mean ± SD	Range
Age		52.7 ± 7.9	37–72
Ethnic			
Malay	121 (94.5)		
Chinese	7 (5.5)		
Marital Status			
Single	5 (3.9)		
Married	99 (77.3)		
Widowed	20 (15.6)		
Divorced	4 (3.1)		
Education level			
None	1 (0.8)		
Primary	11 (8.6)		
Secondary	76 (59.4)		
College/University	40 (31.2)		
Occupational Status			
Working	66 (51.6)		
Not Working	62 (48.4)		
Monthly income (MYR)		2409.80 ± 2325.85	100–12,000
≤500	22 (17.2)		
500–2000	58 (45.3)		
≥2000	48 (37.5)		
Duration since diagnosis (years)		7.14 ± 3.92	2–33
≤5 year	49 (38.3)		
>5 year	79 (61.7)		
Cancer stage			
Stage I	23 (18.0)		
Stage II	71 (55.5)		
Stage III	34 (26.5)		
Body Weight (kg)		66.48 ± 12.52	38–115
Body mass index (kg/m^2)		27.72 ± 5.03	15–50
Underweight	3 (2.3)		
Normal	29 (22.7)		
Overweight	58 (45.3)		
Obese	38 (29.7)		
Waist circumference (cm)		87.98 ± 11.30	56–125
≤80 cm	28 (21.9)		
>80 cm	100 (78.1)		
HMW Adiponectin (µg/mL) [a]		3.69 ± 2.65	0.17–14.73
Leptin (ng/mL) [b]		45.85 ± 19.45	2.29–88.44

[a] Intra-assay coefficients variation, CV HMW (high-molecular weight) adiponectin = 13.19%; [b] Intra-assay coefficients variation, CV Leptin = 9.75%.

As three dietary factors (response variables), i.e., dietary energy density, saturated fat and dietary fiber were included in the RRR analysis, three dietary patterns were identified based on the combined dietary factors. The characteristics of the three dietary patterns are displayed in Table 2. The first dietary pattern presented with the maximum percent of variation explained in all response variables, 34.6%, compared to only 16.3% and 9.1% for dietary patterns 2 and 3, respectively. In addition, the first dietary pattern, which was positively correlated with DED (energy-dense; r = 0.67), SFA (high-SFA; r = 0.36) but negatively correlated with DF (low-fiber; r = −0.65), was shown to be the most pragmatic dietary pattern to be interpreted and in line with the hypothesized link with breast cancer risk. Dietary patterns that explain more than 20% of variation in all response variables were usually retained for further analysis [40]. The other two dietary patterns in the current study were not easily interpretable and are not hypothesized to be associated with the risk of breast cancer risk and mortality. Hence, only the first dietary pattern, the 'energy-dense, high-SFA and low-fiber' DP was highlighted for further analysis in this study.

Table 2. Characteristics of dietary patterns by reduced rank regression.

	Explained Variation (%)					Correlation Coefficient		
	All Food Intakes (Current)	All Responses (Current)	DED (kcal/g)	SFA (%E)	DF (g/kcal)	DED (kcal/g)	SFA (%E)	DF (g/kcal)
DP 1	5.3	34.6	46.7	13.6	43.4	0.67	0.36	−0.65
DP 2	3.5	16.3	52.1	55.7	44.9	−0.33	0.93	0.17
DP 3	4.0	9.1	64.1	55.9	60.0	0.66	0.10	0.74

DP: dietary pattern; DED: dietary energy density; SFA: saturated fatty acid; DF: dietary fibre; %E: percentage of energy intake.

Figure 1 presented the factor loading of the 'energy-dense, high-SFA and low-fiber' DP. Intake of foods with a positive factor loading increased the DP z-score, whilst the intake of foods with a negative factor loading decreased the DP z-score. According to previous studies by Jacobs et al. [41] and Kim, Shin, and Song [42], food groups with factor loading $\geq$0.20 and ≤-0.20 were significant and considered as the largest positive or negative contribution to the dietary pattern z-scores, respectively. In this present study, the 'energy-dense, high-SFA, low-fiber' DP was strongly characterized by sugar-sweetened beverages and fat-based spreads ($\geq$0.20 factors loadings) but negatively characterized by fruits, total vegetables, and green vegetables (≤-0.20 factors loadings). Therefore, these five food groups were considered as key foods and were further investigated with the biomarker of interest (HMW adiponectin and leptin) in this present study. The factor loadings for the other two neglected DPs are presented as Supplementary Materials (Figures S1 and S2).

Table 3 summarizes the multiple linear regression analysis of the relationship between the identified DP, key five food groups (those with high factor loadings) and adipokines (HMW adiponectin and leptin) concentration. Only HMW adiponectin had a significant inverse relation with the 'energy-dense, high-SFA, low-fiber' DP ($\beta = -0.410$; 95% CI = −0.806, −0.014; $p = 0.043$), but no relationship was observed with leptin, independent of age, BMI, cancer stage, duration since diagnosis, education level and occupation status. The findings show that for every 1-unit increase in the 'energy-dense, high-SFA, low-fiber' DP z-score, there is a reduction by 0.41 µg/mL in HMW adiponectin. Meanwhile, no significant findings were observed for the other two rejected DP (Supplementary Materials Table S1).

In addition, regression analysis between the key food groups with adipokines concentration showed no significant relationship with HMW adiponectin, but for leptin, green leafy vegetables showed a negative association with leptin concentration even after adjusting for confounding variables ($\beta = -0.079$; 95% CI = −0.151, −0.007; $p = 0.032$). This could be interpreted as for every 1 g per day increase in the green leafy vegetable intake, there is a reduction by 0.079 ng/mL in leptin.

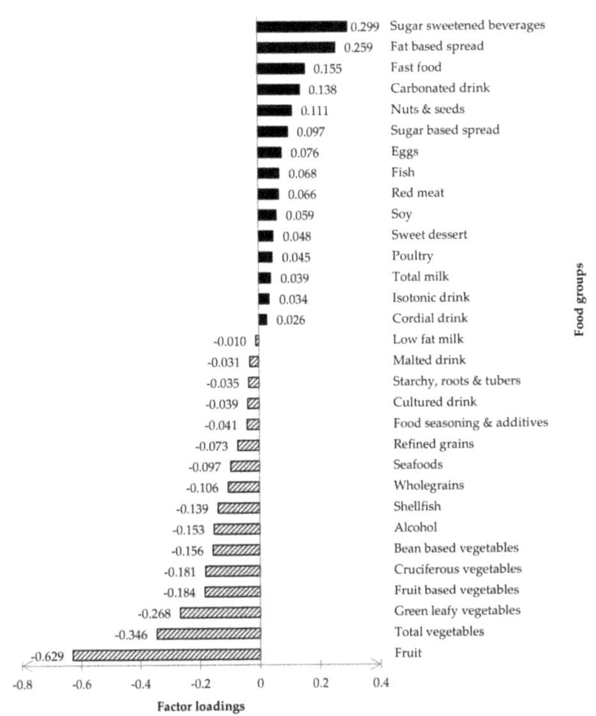

Figure 1. Factor loadings for the 'energy-dense, High-SFA and Low-Fiber' DP.

Table 3. Relationship between dietary pattern and food groups with adipokines.

	HMW Adiponectin		Leptin	
	β (95% CI)	*p*-Value	β (95% CI)	*p*-Value
	'Energy dense, High-SFA and low-fiber' DP			
[a] Unadjusted	−0.369 (−0.777,0.039)	0.075	1.701 (−1.350,4.752)	0.271
[b] Adjusted	−0.410 (−0.806,−0.014)	0.043 *	0.815 (−2.110,3.740)	0.581
	Sugar sweetened beverages (mL/day)			
[a] Unadjusted	0.000 (−0.003,0.002)	0.765	−0.012 (−0.031,0.008)	0.235
[b] Adjusted	0.000 (−0.003,0.002)	0.826	−0.009 (−0.028,0.010)	0.357
	Fat based spread (g/day)			
[a] Unadjusted	−0.071 (−0.261,0.119)	0.459	−0.290 (−1.692,1.112)	0.682
[b] Adjusted	−0.076 (−0.267,0.114)	0.428	−0.286 (−1.650,1.077)	0.677
	Fruits (g/day)			
[a] Unadjusted	0.003 (0.000,0.006)	0.050 *	0.002 (−0.019,0.023)	0.861
[b] Adjusted	0.003 (0.000,0.006)	0.055	0.002 (−0.019,0.023)	0.842
	Total vegetables (g/day)			
[a] Unadjusted	−0.004 (−0.014,0.006)	0.419	−0.080 (−0.152,−0.009)	0.027 *
[b] Adjusted	−0.008 (−0.017,0.002)	0.119	−0.067 (−0.137,0.002)	0.057
	Green leafy vegetables (g/day)			
[a] Unadjusted	−0.006 (−0.016,0.004)	0.216	−0.092 (−0.165,−0.018)	0.016 *
[b] Adjusted	−0.009 (−0.020,0.001)	0.068	−0.079 (−0.151,−0.007)	0.032 *

β Regression coefficient; [a] crude regression coefficient by simple linear regression; [b] adjusted regression coefficient by multiple linear regression, controlled for energy intake (kcal/d), age (years), BMI (kg/m^2), cancer stage (stage of cancer upon diagnosed either stage I/II/III), duration since diagnosis (years), education level (primary, secondary, college/university) and occupation status; only food groups with strongest positive factor loadings (≥0.2) and strongest negative factor loadings (≤−0.2) are shown. * statistically significant difference ($p < 0.05$).

4. Discussion

The present study identified three dietary factors, i.e., DED, SFA and DF, which were hypothesized to be associated with breast cancer survival and mortality. This was the first study to date that utilized DED, SFA and DF as response variables in RRR analysis for breast cancer survival outcomes. However, an 'energy-dense, high-SFA and low-fiber' DP in this present study seemed to be similar to the dietary pattern characterized in another study among breast cancer survivors which was named as 'Western' DP [43]. This Western DP showed high factor loading for dessert, high-fat dairy, processed and red meat, whereas low factor loadings were observed for fruit, vegetables, and whole grain. Nonetheless, similar characteristics of DP were found by Vrieling et al. [32]; however, the DP was named as an 'unhealthy' DP with a high factor loading of red meat, processed meat, deep-frying, and low factor loading for fruits and vegetables. These previous studies concluded that lower intake of the 'Western' DP may protect against mortality from causes unrelated to breast cancer and an increasing intake of an 'unhealthy' dietary pattern may increase the risk of non-breast cancer mortality.

The 'energy-dense, high-SFA and low-fiber' DP in this study was significantly and inversely related to HMW adiponectin concentration after adjusting for the potential confounding factors. This inverse relationship appeared to be consistent with other research which also found a negative relationship between the Traditional English pattern [20], the "Izakaya" pattern [21] and the Western pattern [44,45] with adiponectin concentration. As compared to DP in this current study, all dietary patterns observed earlier shared similar characteristics of high consumption levels of energy-dense food such as fried foods, fast foods, processed meat, sugar, refined grains intake and have low consumptions of vegetables, fruits, wholegrain, and low-fat dairies.

This significant negative relationship between unhealthy DPs and adiponectin concentration might be explained by the role of adiponectin in regulating food intake and energy expenditure [46]. In terms of energy metabolism, adiponectin acts as a starvation hormone that enhances energy storage by stimulating food intake and suppressing energy expenditure. Therefore, a decline in HMW adiponectin might explain the fact that the energy storage in the body has exceeded and preceded the development of insulin resistance. Previous studies have found positive relationships between dietary patterns that were characterized by healthy food consumption and adiponectin concentration [8]. This further supported the explanation of the effects of dietary pattern towards circulating adiponectin [20–22,45]. Furthermore, breast cancer patients showed that a reduction in the concentration of HMW adiponectin had an important effect on insulin resistance and metabolic syndrome, and this was associated with an increased risk of breast cancer mortality [47]. Hence, it could conceivably be suggested that a healthy DP has benefits in improving adiponectin concentration, while an unhealthy DP lowers the serum adiponectin levels.

Nonetheless, no significant relationship was found between leptin concentration and the 'energy-dense, high-SFA and low-fiber' DP among breast cancer survivors in this present study. This result was similar to earlier studies' observations, in which leptin concentration was not independently associated with the 'Western' DP (unhealthy DP characterized by red and processed meats, high-energy drinks, refined grains, pizza/lasagna, eggs, fats, and snacks/sweets) [45,48]. In contrast, several previous studies reported a positive association between the 'Western' DP with serum leptin concentration [19,49]. Different sample characteristics from different studies, in terms of sex, age, ethnicity, culture, food habits and potential confounding factors, may show different outcomes.

Although our study did not show a significant association between serum concentration of leptin and DP, it did provide evidence that circulating leptin had a significant negative relationship with green leafy vegetables. This finding mirrored those of the previous studies that have examined reductions in circulating leptin levels following healthy hypocaloric diets and regular physical activity [50,51]. Leptin sensitivity was increased with a high amount of fiber intake and has led to control in the secretion of leptin [52].

According to the studies by Harris et al. [53] and Khan et al. [15], a decline in leptin concentration was directly associated with reduced breast cancer recurrence and mortality. Thus, to improve survival and prevent the recurrence of breast cancer, a healthy diet particularly diets high in vegetables is recommended. This coincides with the central recommendation by WCRF/AICR to "Eat mostly food of plant origin, with a variety of non-starchy vegetables and of fruit every day with unprocessed cereals and/or pulses within every meal" [30,54]. It is also known that leptin-induced mammalian target of rapamycin (mTOR) activation may have implications for obesity-related pathophysiological conditions such as breast cancer [55]. Whether the findings from the current study depend on the altered mTOR pathway warrants further investigation.

Overall, this study looks into the DPs practiced among a group of breast cancer survivors in East Coast of Peninsular Malaysia by using RRR analysis, a hybrid method that had used a priori information to identify a nutrient-specific DP. Additionally, this study had contributed significant evidence concerning the health status of breast cancer survivor in Malaysia. However, the present study should be interpreted with caution due to the established limitations attributed to the cross-sectional study design. The current study also did not assess the physical activity level of breast cancer survivors which is one of the important factors that may influence adipokines concentration [14]. Therefore, there is a need to perform a prospective study to obtain stronger evidence that the converse of the identified DP does indeed predictably improve breast cancer survivorship.

5. Conclusions

HMW adiponectin concentration among breast cancer survivors in the East Coast of Peninsular Malaysia negatively associated with 'energy-dense, high-SFA and low-fiber' DP, which was characterized by the high level of consumption of sugar-sweetened drinks and fat-based spreads but low consumption levels of fruits, total vegetables, and green leafy vegetables. The present finding and those of some previous studies support the hypothesis that an 'energy-dense, high-SFA and low-fiber' DP or a similar unhealthy DP is associated with lower beneficial adipokines concentration. The high prevalence of both obesity and overweight might be fundamental to the alternating HMW adiponectin and leptin concentrations among respondents in this study. Future work should consider the long-term effects of adopting healthy dietary practices on breast cancer recurrence and other disease risks in this group of women as an important part of survival after cancer.

Supplementary Materials: The following are available online at https://www.mdpi.com/article/10.3390/nu13103339/s1, Figure S1: Factor loadings for DP 2, i.e., 'low energy-density, high-SFA and high-fiber', Figure S2: Factor loadings for DP 3, i.e., 'energy-dense, high-SFA and high-fiber', Table S1: Relationship between dietary patterns (DP 2 and DP 3) and adipokines.

Author Contributions: Conceptualization, M.R.S., A.A., P.L.L.; methodology, M.R.S., N.S.Z., G.A., A.N.; formal analysis, N.S.Z. and G.A.; investigation, N.S.Z., A.N.; resources, H.J.J.M., A.A., P.L.L., M.F.; data curation, M.R.S. and N.S.Z.; writing—original draft preparation, M.R.S. and N.S.Z.; writing—review and editing, G.A., A.N., H.J.J.M., A.A., P.L.L., M.F.; visualization, N.S.Z., G.A., M.F.; supervision, M.R.S.; project administration, M.R.S.; funding acquisition, M.R.S. All authors have read and agreed to the published version of the manuscript.

Funding: This research was funded by the Ministry of Higher Education Malaysia via Research Acculturation Grant Scheme, grant number RAGS/1/2014/SKK10/UniSZA/2 and by University Sultan Zainal Abidin via Dana Penyelidikan University, grant number UniSZA/2015/DPU/40.

Institutional Review Board Statement: The study was conducted according to the guidelines of the Declaration of Helsinki and approved by the Medical Research and Ethics Committee of the Ministry of Health Malaysia (Protocol number: NMRR-14-1618-23717; Approved: 18 May 2015).

Informed Consent Statement: Informed consent was obtained from all subjects involved in the study.

Data Availability Statement: The data presented in this study are available on request from the corresponding author (M.R.S). The data are not publicly available due to the privacy of research participants and ethical restrictions.

Acknowledgments: The authors thank the Director-General of Health Malaysia for his permission to publish this article. Thank you to all breast cancer survivors, clinicians and oncology nurses for your co-operation and participation in this research.

Conflicts of Interest: The authors declare no conflict of interest. The funders had no role in the design of the study; in the collection, analyses, or interpretation of data; in the writing of the manuscript, or in the decision to publish the results.

References

1. Sharif, R.; Shahar, S.; Rajab, N.F.; Fenech, M. Dietary pattern, genomic stability and relative cancer risk in Asian food landscape. *Nutr. Cancer* **2021**, in press. [CrossRef]
2. Zang, J.; Shen, M.; Du, S.; Chen, T.; Zou, S. The association between dairy intake and breast cancer in Western and Asian populations: A systematic review and meta-analysis. *J. Breast Cancer* **2015**, *18*, 313–322. [CrossRef]
3. Sotos-Prieto, M.; Bhupathiraju, S.N.; Mattei, J.; Fung, T.T.; Li, Y.; Pan, A.; Willett, W.C.; Rimm, E.B.; Hu, F.B. Association of Changes in Diet Quality with Total and Cause-Specific Mortality. *N. Engl. J. Med.* **2017**, *377*, 143–153. [CrossRef]
4. Coa, K.I.; Smith, K.C.; Klassen, A.C.; Caulfield, L.E.; Helzlsouer, K.J.; Peairs, K.; Shockney, L.D. Capitalizing on the "teachable moment" to promote healthy dietary changes among cancer survivors: The perspectives of health care providers. *Support. Care Cancer* **2015**, *23*, 679–686. [CrossRef]
5. Zainuddin, L.R.M.; Zakarai, N.S.; Yusoff, N.A.M.; Ahmad, A.; Sulaiman, S.; Shahril, M.R. Dietary intake among breast cancer survivors in East Coast of Peninsular Malaysia. *Malay. J. Public Health Med. Suppl.* **2017**, *2*, 59–65.
6. Greenlee, H.; Gaffney, A.O.; Aycinena, A.C.; Koch, P.; Contento, I.; Karmally, W.; Richardson, J.M.; Shi, Z.; Lim, E.; Tsai, W.-Y.; et al. Long-term diet and biomarker changes after a short-term intervention among Hispanic breast cancer survivors: The *¡Cocinar Para Su Salud!* randomized controlled trial. *Cancer Epidemiol. Biomark. Prev.* **2016**, *25*, 1491–1502. [CrossRef] [PubMed]
7. Zainordin, N.H.; Talib, R.A.; Shahril, M.R.; Sulaiman, S.; Karim, N.A. Dietary changes and its impact on quality of life among Malay breast and gynaecological cancer survivors in Malaysia. *Asian Pac. J. Cancer Prev.* **2020**, *21*, 3689–3696. [CrossRef] [PubMed]
8. Zakarai, N.S.; Shahril, M.R. Dietary patterns and its associations with adipokines (adiponectin and leptin) among adults: A narrative review. *Sains Malays.* **2017**, *46*, 1849–1857. [CrossRef]
9. Izadi, V.; Azadbakht, L. Specific dietary patterns and concentrations of adiponectin. *J. Res. Med. Sci.* **2015**, *20*, 178–184.
10. Alhazmi, A.; Stojanovski, E.; McEvoy, M.; Garg, M.L. The association between dietary patterns and type 2 diabetes: A systematic review and meta-analysis of cohort studies. *J. Hum. Nutr. Diet.* **2014**, *27*, 251–260. [CrossRef]
11. Quatromoni, P.A.; Copenhafer, D.L.; Demissie, S.; D'Agostino, R.B.; O'Horo, C.E.; Nam, B.-H.; Millen, B.E. The internal validity of a dietary pattern analysis. The Framingham Nutrition Studies. *J. Epidemiol. Community Health* **2002**, *56*, 381–388. [CrossRef]
12. Catsburg, C.; Miller, A.B.; Rohan, T.E. Adherence to cancer prevention guidelines and risk of breast cancer. *Int. J. Cancer* **2014**, *135*, 2444–2452. [CrossRef] [PubMed]
13. Swisher, A.K.; Abraham, J.; Bonner, D.; Gilleland, D.; Hobbs, G.; Kurian, S.; Yanosik, M.A.; Vona-Davis, L. Exercise and dietary advice intervention for survivors of triple-negative breast cancer: Effects on body fat, physical function, quality of life, and adipokine profile. *Support. Care Cancer* **2015**, *23*, 2995–3003. [CrossRef] [PubMed]
14. Patterson, R.E.; Cadmus, L.A.; Emond, J.A.; Pierce, J.P. Physical activity, diet, adiposity and female breast cancer prognosis: A review of the epidemiologic literature. *Maturitas* **2010**, *66*, 5–15. [CrossRef]
15. Khan, S.; Shukla, S.; Sinha, S.; Meeran, S.M. Role of adipokines and cytokines in obesity-associated breast cancer: Therapeutic targets. *Cytokine Growth Factor Rev.* **2013**, *24*, 503–513. [CrossRef]
16. Pellatt, A.J.; Lundgreen, A.; Wolff, R.K.; Hines, L.; John, E.M.; Slattery, M.L. Energy homeostasis genes and survival after breast cancer diagnosis: The Breast Cancer Health Disparities Study. *Cancer Causes Control* **2016**, *27*, 47–57. [CrossRef]
17. Kashino, I.; Nanri, A.; Kurotani, K.; Akter, S.; Yasuda, K.; Sato, M.; Hayabuchi, H.; Mizoue, T. Association of dietary patterns with serum adipokines among Japanese: A cross-sectional study. *Nutr. J.* **2015**, *14*, 58. [CrossRef]
18. Llanos, A.A.; Krok, J.L.; Peng, J.; Pennell, M.L.; Olivo-Marston, S.; Vitolins, M.Z.; DeGraffinreid, C.R.; Paskett, E.D. Favorable effects of low-fat and low-carbohydrate dietary patterns on serum leptin, but not adiponectin, among overweight and obese premenopausal women: A randomized trial. *SpringerPlus* **2014**, *3*, 175. [CrossRef]
19. Ko, B.-J.; Park, K.H.; Shin, S.; Zaichenko, L.; Davis, C.R.; Crowell, J.A.; Joung, H.; Mantzoros, C.S. Diet quality and diet patterns in relation to circulating cardiometabolic biomarkers. *Clin. Nutr.* **2016**, *35*, 484–490. [CrossRef] [PubMed]
20. Cassidy, A.; Skidmore, P.; Rimm, E.B.; Welch, A.; Fairweather-Tait, S.; Skinner, J.; Burling, K.; Richards, J.B.; Spector, T.D.; MacGregor, A.J. plasma adiponectin concentrations are associated with body composition and plant-based dietary factors in female twins. *J. Nutr.* **2009**, *139*, 353–358. [CrossRef] [PubMed]
21. Guo, H.; Niu, K.; Monma, H.; Kobayashi, Y.; Guan, L.; Sato, M.; Minamishima, D.; Nagatomi, R. Association of Japanese dietary pattern with serum adiponectin concentration in Japanese adult men. *Nutr. Metab. Cardiovasc. Dis.* **2012**, *22*, 277–284. [CrossRef]

22. Farhangi, M.A.; Jahangiry, L.; Jafarabadi, M.A.; Najafi, M. Association between dietary patterns and metabolic syndrome in a sample of Tehranian adults. *Obes. Res. Clin. Pract.* **2016**, *10*, S64–S73. [CrossRef] [PubMed]
23. Appannah, G.; Pot, G.K.; O'Sullivan, T.; Oddy, W.H.; Jebb, S.A.; Ambrosini, G.L. The reliability of an adolescent dietary pattern identified using reduced-rank regression: Comparison of a FFQ and 3 d food record. *Br. J. Nutr.* **2014**, *112*, 609–615. [CrossRef] [PubMed]
24. Institute for Public Health. *National Health and Morbidity Survey 2014: Malaysian Adult Nutrition Survey (MANS) Vol. II: Survey Findings*; Institute for Public Health, National Institutes of Health, Ministry of Health Malaysia: Kuala Lumpur, Malaysia, 2014.
25. Wen, V.N.J. Validity of Semi-Quantitative Food Frequency Questionnaires for Dietary Assessment among Adults. Bachelor's Thesis, Universiti Sultan Zainal Abidin, Kuala Terengganu, Malaysia, 31 July 2017.
26. Rhee, J.J.; Cho, E.; Willett, W.C. Energy adjustment of nutrient intakes is preferable to adjustment using body weight and physical activity in epidemiological analyses. *Public Health Nutr.* **2014**, *17*, 1054–1060. [CrossRef]
27. Ministry of Health Malaysia. Malaysian Food Composition Database (MyFCD). Available online: https://myfcd.moh.gov.my (accessed on 15 June 2019).
28. U.S. Department of Agriculture, Agricultural Research Service. Food Data Central. Available online: https://fdc.nal.usda.gov (accessed on 1 September 2019).
29. Shahar, S.; Shahril, M.R.; Abdullah, N.; Borhanuddin, B.; Kamaruddin, M.A.; Yusuf, N.A.M.; Dauni, A.; Rosli, H.; Zainuddin, N.S.; Jamal, R. Development and relative validity of a semiquantitative food frequency questionnaire to estimate dietary intake among a multi-ethnic population in the Malaysian Cohort Project. *Nutrients* **2021**, *13*, 1163. [CrossRef]
30. WCRF/AICR. *Diet, Nutrition, Physical Activity and Cancer: A Global Perspective*; Continuous Update Project Expert Report 2018; World Cancer Research Fund International: London, UK, 2018.
31. Pierce, J.P.; Natarajan, L.; Caan, B.; Parker, B.A.; Greenberg, E.R.; Flatt, S.W.; Rock, C.L.; Kealey, S.; Al-Delaimy, W.; Bardwell, W.A.; et al. Influence of a diet very high in vegetables, fruit, and fiber and low in fat on prognosis following treatment for breast cancer. *JAMA* **2007**, *298*, 289–298. [CrossRef]
32. Vrieling, A.; Buck, K.; Seibold, P.; Heinz, J.; Obi, N.; Flesch-Janys, D.; Chang-Claude, J. Dietary patterns and survival in German postmenopausal breast cancer survivors. *Br. J. Cancer* **2013**, *108*, 188–192. [CrossRef]
33. Mendoza, J.A.; Drewnowski, A.; Christakis, D.A. Dietary energy density is associated with obesity and the metabolic syndrome in U.S. adults. *Diabetes Care* **2007**, *30*, 974–979. [CrossRef]
34. Vernarelli, J.A.; Mitchell, D.C.; Rolls, B.J.; Hartman, T.J. Dietary energy density is associated with obesity and other biomarkers of chronic disease in US adults. *Eur. J. Nutr.* **2015**, *54*, 59–65. [CrossRef] [PubMed]
35. Howarth, N.C.; Murphy, S.P.; Wilkens, L.R.; Hankin, J.H.; Kolonel, L.N. Dietary energy density is associated with overweight status among 5 ethnic groups in the multiethnic cohort study. *J. Nutr.* **2006**, *136*, 2243–2248. [CrossRef]
36. Ledikwe, J.H.; Blanck, H.M.; Khan, L.K.; Serdula, M.K.; Seymour, J.D.; Tohill, B.C.; Rolls, B.J. Dietary energy density is associated with energy intake and weight status in US adults. *Am. J. Clin. Nutr.* **2006**, *83*, 1362–1368. [CrossRef] [PubMed]
37. De Pergola, G.; Silvestris, F. Obesity as a major risk factor for cancer. *J. Obes.* **2013**, *2013*, 291546. [CrossRef] [PubMed]
38. Lorincz, A.M.; Sukumar, S. Molecular links between obesity and breast cancer. *Endocr.-Relat. Cancer* **2006**, *13*, 279–292. [CrossRef]
39. Livingstone, K.; McNaughton, S. Dietary patterns by reduced rank regression are associated with obesity and hypertension in Australian adults. *Br. J. Nutr.* **2017**, *117*, 248–259. [CrossRef] [PubMed]
40. Gao, M.; Jebb, S.A.; Aveyard, P.; Ambrosini, G.L.; Perez-Cornago, A.; Carter, J.; Sun, X.; Piernas, C. Associations between dietary patterns and the incidence of total and fatal cardiovascular disease and all-cause mortality in 116,806 individuals from the UK Biobank: A prospective cohort study. *BMC Med.* **2021**, *19*, 83. [CrossRef] [PubMed]
41. Jacobs, S.; Kroeger, J.; Schulze, M.B.; Frank, L.K.; Franke, A.A.; Cheng, I.; Monroe, K.R.; Haiman, C.A.; Kolonel, L.N.; Wilkens, L.R.; et al. Dietary patterns derived by reduced rank regression are inversely associated with type 2 diabetes risk across 5 ethnic groups in the multiethnic cohort. *Curr. Dev. Nutr.* **2017**, *1*, e000620. [CrossRef]
42. Kim, W.K.; Shin, D.; Song, W.O. Are dietary patterns associated with depression in U.S. adults? *J. Med. Food* **2016**, *19*, 1074–1084. [CrossRef]
43. Kroenke, C.H.; Fung, T.T.; Hu, F.B.; Holmes, M.D. Dietary patterns and survival after breast cancer diagnosis. *J. Clin. Oncol.* **2005**, *23*, 9295–9303. [CrossRef]
44. Jafari-Vayghan, H.; Tarighat-Esfanjani, A.; Jafarabadi, M.A.; Ebrahimi-Mameghani, M.; Ghadimi, S.S.; Lalezadeh, Z.; Jafarabadi, M.A. Association between dietary patterns and serum leptin-to-adiponectin ratio in apparently healthy adults. *J. Am. Coll. Nutr.* **2015**, *34*, 49–55. [CrossRef]
45. Alves-Santos, N.H.; Cocate, P.G.; Eshriqui, I.; Benaim, C.; Barros, É.G.; Emmett, P.M.; Kac, G. Dietary patterns and their association with adiponectin and leptin concentrations throughout pregnancy: A prospective cohort. *Br. J. Nutr.* **2018**, *119*, 320–329. [CrossRef]
46. Wang, Y.; Li, J.; Fu, X.; Li, J.; Liu, L.; Alkohlani, A.; Tan, S.C.; Low, T.Y.; Hou, Y. Association of circulating leptin and adiponectin levels with colorectal cancer risk: A systematic review and meta-analysis of case-control studies. *Cancer Epidemiol.* **2021**, *73*, 101958. [CrossRef] [PubMed]
47. Duggan, C.; Irwin, M.L.; Xiao, L.; Henderson, K.D.; Smith, A.W.; Baumgartner, R.N.; Baumgartner, K.B.; Bernstein, L.; Ballard-Barbash, R.; McTiernan, A. Associations of insulin resistance and adiponectin with mortality in women with breast cancer. *J. Clin. Oncol.* **2011**, *29*, 32–39. [CrossRef] [PubMed]

48. Ganji, V.; Kafai, M.R.; McCarthy, E. Serum leptin concentrations are not related to dietary patterns but are related to sex, age, body mass index, serum triacylglycerol, serum insulin, and plasma glucose in the US population. *Nutr. Metab.* **2009**, *6*, 3. [CrossRef] [PubMed]
49. Zuo, H.; Shi, Z.; Dai, Y.; Yuan, B.; Wu, G.; Luo, Y.; Hussain, A. Serum leptin concentrations in relation to dietary patterns in Chinese men and women. *Public Health Nutr.* **2014**, *17*, 1524–1530. [CrossRef]
50. You, T.; Berman, D.M.; Ryan, A.S.; Nicklas, B.J. Effects of hypocaloric diet and exercise training on inflammation and adipocyte lipolysis in obese postmenopausal women. *J. Clin. Endocrinol. Metab.* **2004**, *89*, 1739–1746. [CrossRef] [PubMed]
51. Pierce, B.; Ballard-Barbash, R.; Bernstein, L.; Baumgartner, R.N.; Neuhouser, M.L.; Wener, M.H.; Baumgartner, K.B.; Gilliland, F.D.; Sorensen, B.E.; McTiernan, A.; et al. Elevated biomarkers of inflammation are associated with reduced survival among breast cancer patients. *J. Clin. Oncol.* **2009**, *27*, 3437–3444. [CrossRef]
52. Jensen, M.K.; Koh-Banerjee, P.; Franz, M.; Sampson, L.; Grønbaek, M.; Rimm, E.B. Whole grains, bran, and germ in relation to homocysteine and markers of glycemic control, lipids, and inflammation. *Am. J. Clin. Nutr.* **2006**, *83*, 275–283. [CrossRef]
53. Harris, H.R.; Tworoger, S.; Hankinson, S.E.; Rosner, B.A.; Michels, K.B. Plasma leptin levels and risk of breast cancer in premenopausal women. *Cancer Prev. Res.* **2011**, *4*, 1449–1456. [CrossRef] [PubMed]
54. Bruno, E.; Gargano, G.; Villarini, A.; Traina, A.; Johansson, H.; Mano, M.P.; De Magistris, M.S.; Simeoni, M.; Consolaro, E.; Mercandino, A.; et al. Adherence to WCRF/AICR cancer prevention recommendations and metabolic syndrome in breast cancer patients. *Int. J. Cancer* **2016**, *138*, 237–244. [CrossRef]
55. Magaway, C.; Kim, E.; Jacinto, E. Targeting mTOR and metabolism in cancer: Lessons and innovations. *Cells* **2019**, *8*, 1584. [CrossRef]

Article

Chronic Nutrition Impact Symptoms Are Associated with Decreased Functional Status, Quality of Life, and Diet Quality in a Pilot Study of Long-Term Post-Radiation Head and Neck Cancer Survivors

Sylvia L. Crowder [1,2,*], Zonggui Li [3], Kalika P. Sarma [4] and Anna E. Arthur [1]

1 Department of Food Science and Human Nutrition, University of Illinois at Urbana-Champaign, 386 Bevier Hall 905 S Goodwin Ave, Urbana, IL 61801, USA; aarthur@illinois.edu
2 Department of Health Outcomes and Behavior, Moffitt Cancer Center, 4117 E Fowler Ave, Tampa, FL 33617, USA
3 Department of Psychology and Neuroscience, Boston College, 140 Commonwealth Ave, Chestnut Hill, MA 02467, USA; il@bc.edu
4 Carle Cancer Center, Carle Foundation Hospital, 602 W University Ave, Urbana, IL 61801, USA; Kalika.sarma@carle.com
* Correspondence: sylvia.crowder@moffitt.org; Tel.: +1-217-244-4090

Abstract: Background: As a result of tumor location and treatment that is aggressive, head and neck cancer (HNC) survivors experience an array of symptoms impacting the ability and desire to eat termed nutrition impact symptoms (NISs). Despite increasing cancer survival time, the majority of research studies examining the impact of NISs have been based on clinical samples of HNC patients during the acute phase of treatment. NISs are often chronic and persist beyond the completion of treatment or may develop as late side effects. Therefore, our research team examined chronic NIS complications on HNC survivors' functional status, quality of life, and diet quality. Methods: This was a cross-sectional study of 42 HNC survivors who were at least 6 months post-radiation. Self-reported data on demographics, NISs, quality of life, and usual diet over the past year were obtained. Objective measures of functional status included the short physical performance battery and InBody© 270 body composition testing. NISs were coded so a lower score indicated lower symptom burden, (range 4–17) and dichotomized as $\leq$10 vs. >10, the median in the dataset. Wilcoxon rank sum tests were performed between the dichotomized NIS summary score and continuous quality of life and functional status outcomes. Diet quality for HNC survivors was calculated using the Healthy Eating Index 2015 (HEI-2015). Wilcoxon rank sum tests examined the difference between the HNC HEI-2015 as compared to the National Health and Nutrition Examination Survey (NHANES) data calculated using the population ratio method. Results: A lower NIS score was statistically associated with higher posttreatment lean muscle mass (p = 0.002). A lower NIS score was associated with higher functional (p = 0.0006), physical (p = 0.0007), emotional (p = 0.007), and total (p < 0.0001) quality of life. Compared to NHANES controls, HNC survivors reported a significantly lower HEI-2015 diet quality score (p = 0.0001). Conclusions: Lower NIS burden was associated with higher lean muscle mass and functional, physical, emotional, and total quality of life in post-radiation HNC survivors. HNC survivors reported a significantly lower total HEI-2015 as compared to healthy NHANES controls, providing support for the hypothesis that chronic NIS burden impacts the desire and ability to eat. The effects of this pilot study were strong enough to be detected by straight forward statistical approaches and warrant a larger longitudinal study. For survivors most impacted by NIS burden, multidisciplinary post-radiation exercise and nutrition-based interventions to manage NISs and improve functional status, quality of life, and diet quality in this survivor population are needed.

Keywords: survivorship; head and neck; diet; symptoms; quality of life

Citation: Crowder, S.L.; Li, Z.; Sarma, K.P.; Arthur, A.E. Chronic Nutrition Impact Symptoms Are Associated with Decreased Functional Status, Quality of Life, and Diet Quality in a Pilot Study of Long-Term Post-Radiation Head and Neck Cancer Survivors. *Nutrients* **2021**, *13*, 2886. https://doi.org/10.3390/nu13082886

Academic Editor: Henry J. Thompson

Received: 28 July 2021
Accepted: 20 August 2021
Published: 22 August 2021

1. Introduction

Head and neck cancer (HNC) is a heterogeneous disease including cancer of the oral cavity, oropharynx, hypopharynx, and larynx [1]. As a result of treatment and treatment that is aggressive, HNC survivors experience an array of symptoms impacting the desire and ability to eat termed nutrition impact symptoms (NISs) [2]. Common NISs include dysphagia, xerostomia, and difficulty chewing that lead to comprised food intake, malnutrition, and increased susceptibility to infection [2–4]. HNC survivors are living longer, thus increasing the survivorship period [2,5]. Despite increasing cancer survival time, the majority of research studies examining the impact of NISs have been based on clinical samples of HNC patients during the acute phase of treatment. NISs are often chronic and persist beyond the completion of treatment or may develop as late side effects [2]. Therefore, it is crucial for healthcare providers to examine chronic NIS complication on survivors' functional status, quality of life, and diet quality.

While it has previously been established that the time around formal diagnosis until approximately three months post-treatment has been associated with decreased quality of life, little is known regarding the chronic burden, defined as greater than 6 months, of treatment-related outcomes in HNC survivors [6]. Quality of life is a measure of survivors' overall well-being and often encompasses several domains—functional, physical, social, and emotional health. HNC is considered as one of the most emotionally traumatic cancer types [7] and impacts quality of life outcomes [8,9]. Studies have identified anxiety [10], depression [11], and psychological problems [12] in HNC survivors to be associated with poorer quality of life. However, few studies have examined the chronic impact of aggregated NIS burden on overall and domain-specific quality of life and functional status. This is of the upmost importance as one of the most important quality of life factors is nutrition [13] and adaptations to eating and psychological concerns regarding dysphagia and xerostomia may further reduce functional status in survivors [2,14,15]. The HNC population faces unique nutritional challenges as compared to other cancer types, likely decreasing functional status and quality of life.

Before interventions can be designed to enhance survivorship outcomes of long-term survivors, data is needed to inform researchers that provides evidence of reduced quality of life, functional status, and diet quality in long-term HNC survivors that identify critical targets for interventions. Therefore, the objective of this pilot study was to evaluate the relationship between an NIS summary score on quality of life and functional status and compare the diet quality of post-radiation head and neck cancer (HNC) survivors to age-matched National Health and Nutrition Examination Survey (NHANES) controls.

2. Subjects and Methods

2.1. Design

This was a cross-sectional study of 42 HNC survivors who were previously diagnosed or treated in a Midwestern hospital within six months to nine years post treatment. In the quality of life model, the dependent variables of interest were total, emotional, physical, functional, and social quality of life. In the functional status model, the dependent variables of interest were body mass index, functional status composite score, body fat percentage, and lean muscle mass. The independent variable of interest was aggregated NIS burden reflected by a dichotomized NIS summary score. Dietary intake for HNC survivors was assessed using the Healthy Eating Index-2015 (HEI-2015) [16] collected with the semi-quantitative Harvard food frequency questionnaires (FFQ) [17] as compared to NHANES controls using the population-ratio method [18]. All study activities were approved by the Institutional Review Boards of Carle Foundation Hospital and the University of Illinois at Urbana-Champaign (Project ID: 17088; UIUC number: 181933; Original Approval Date: 1 February 2018) and adhered to the principles of the Declaration of Helsinki. All participants were informed of the purpose and procedures of the study and informed written consent was obtained from all participants before data collection.

2.2. Study Population

Participant screening and recruitment occurred between March 2018 and May 2019. Criteria for eligibility included: (1) Previous diagnosis of stage I–IV primary cancer of the oropharynx, hypopharynx, larynx, or oral cavity; (2) within 6 months to 10 years posttreatment with radiation; (3) no evidence of disease, deemed by oncologist and/or surgeon; (4) ability to consume food orally; (5) ≥18 years of age; and (6) English-speaking. HNC survivors treated at the hospital were identified via the Hospital Cancer Registry and a letter was mailed to potential participants explaining the research study. Participants were called within 2–3 weeks of receiving the mailed letter. Medical records were searched to prevent calling deceased individuals. Interested participants were then scheduled for an in-person study visit. At the study visit, formal written consent was obtained.

2.3. Procedures

Participants completed a self-administered health survey that included data on demographics, behavioral characteristics, quality of life, and NIS burden. Dietary data were obtained using the self-administered 2007 Harvard FFQ [17,19,20]. Functional status data were obtained using the short physical performance battery [21] and InBody© body composition testing [22]. An electronic medical record (EMR) review was conducted to collect clinical data on cancer stage, treatment type, and time since diagnosis. Participants were compensated $50.

3. Measures

3.1. Predictor: Nutrition Impact Symptoms

The Functional Assessment of Cancer Therapy-Head and Neck (FACT-H&N) Additional Concerns (AC) Subscale was used to measure perceived NIS barriers [23]. The scale consists of 12 questions, six specifically referring to NIS including: (1) ability to eat any foods desired, (2) ability to eat as much as desired, (3) no presence of xerostomia, (4) ability to swallow naturally and easily, (5) ability to eat solid foods, and (6) no presence of pain in the mouth, throat, or neck, with answers ranked on a 5-point Likert scale ranging from 0 (not at all) to 4 (very much). The scale was coded so that a lower score indicated fewer disease- or treatment-related symptoms. The individual symptom scores were summed to create a total overall NIS summary score (range 4–17) and dichotomized as ≤10 vs. >10, the median in the dataset.

3.2. Outcome Variable: Functional Status

A study team member trained in anthropometrics collected the following functional status measures: anthropometrics, bioelectrical impedance analysis (BIA) [22], and the short physical performance battery [21].

Anthropometric measures for height were conducted in accordance with the Anthropometric Standardization Reference Manual [24]. Height was collected by a trained research staff member during the study visit and measured to the nearest 0.5 inch (without shoes).

Measures of body composition were determined by a vertical direct segmental multi-frequency BIA analyzer (InBody© 270, Cerritos, CA, USA). The InBody© 270 records a user's weight, lean muscle mass, body mass index, and percent body fat to the nearest 0.1 lb (without shoes and in light clothing with pockets emptied). The method of measuring body composition via a BIA has been previously validated and used in similar clinical studies [22]. Four participants were unable to complete InBody© measures as a result of other health conditions (pacemakers and physical impairments).

The short physical performance battery consists of three functional tests assessing performance. The physical performance battery has been previously used in the HNC population [21]. Tests of standing balance include tandem, semi-tandem, and side-by-side stands. Walking speed tests included an 8-foot walking course, with no obstructions for an additional two feet at either end. Participants were allowed to use assistive devices when needed, and each participant was timed for two walks [25]. To test the ability to rise from a chair (termed chair rise-and-sits), a straight-backed chair was placed next to a wall; participants were asked to fold their arms across their chest and to stand up and sit down five times as quickly as possible [25]. Participants were timed from the initial sitting position to the final sitting position at the end of the fifth stand. The short physical performance battery provides a summed composite score based on the number of seconds able to hold a semi-tandem, tandem, and/or side-by-side stance with feet together; 8-foot walk time; and time to complete 5 chair rise-and-sits [25].

3.3. Outcome Variable: Quality of Life

Overall quality of life was assessed using the FACT-H&N quality of life questionnaire (Cronbach's coefficient alpha = 0.86) [23]. The FACT-H&N assesses the impact of cancer diagnosis and therapy in four subdomains: physical, social, emotional, and functional. The FACT-H&N has 28 questions, with answers ranked on a 5-point Likert scale ranging from 0 (not at all) to 4 (very much). The scale was coded so that a higher score indicated higher quality of life. Questions were summed to create four subdomain scores and the four subdomains were summed to create a continuous total summary score.

3.4. Outcome Variable: Dietary Intake

HNC survivors' dietary intake information was collected using the validated 131 item semi-quantitative Harvard Food Frequency Questionnaire, which includes standard portions sizes for each item and the frequency of consumption over the past year [26]. The Harvard FFQ is a feasible instrument suited to assess associations of usual dietary intake and has been extensively used as a measure of dietary exposure in cancer studies [27]. The FFQ allows participants to choose the average frequency of consumption of food items over the past year from a Likert scale. Healthy eating of the HNC survivors was assessed using the HEI-2015 dietary measurement and compared to the NHANES controls using the population ratio method [16,28]. This method was employed because advice from the Dietary Guidelines for Americans (DGA) is designed to be met over time and this method best encompasses that intent for diet evaluation [29]. The HEI was developed to assess diet quality issued by the United States Department of Agriculture (USDA) based on the standards of a healthy lifestyle in association with health outcomes [30]. HEI is composed of 13 scored components and include 5 major food groups: fruit (total and whole), vegetable (total and greens/beans), grains (total and whole), dairy or alternative dairy and protein, oils, and nuts; in addition to limiting saturated fats, sodium, and empty calories [16]. Nine of the components focus on adequacy (dietary components to increase) and four focus on moderation (dietary components to decrease, including refined grains, sodium, added sugars, and saturated fats) [16]. The daily intakes for each component were standardized for energy by diving each study participant's daily component intake by his or her total daily energy intake in kilocalories and multiplying by 100 prior to applying the HEI-2015 scoring algorithm [16]. Each of the 13 components of the HEI-2015 had a minimum score of zero and a maximum score ranging from 5 to 10 that reflected a pre-established level of intake [16]. The total HEI score is the sum of the components, with a range of 0 to 100 [16]. A score between 0 and 50 indicates a poor diet; 51 and 80, a moderate diet quality that needs improvement; and a score greater than 80, a good diet [16].

4. Statistical Considerations and Analyses

Descriptive statistics (means and frequencies) were generated for demographic and clinical variables. ANOVA tests were computed to detect statistical difference between quality of life and demographic variables. The FACT-H&N scoring manual was used to calculate the mean score and standard deviations [31]. Summary scores and subscale scores were extracted into three tertiles of the actual range of scores and categorized as mild, moderate, or severe using methods described in similar studies [32,33]. For example, in the FACT-G, the actual range is 0 to 108; therefore, the three tertiles were 0–36 as severe impairment, 37–72 as moderate impairment, and 73–108 as mild impairment [33].

For the functional status model, Wilcoxon rank sum tests between lean muscle mass, body fat percentage, body mass index, and functional status and the dichotomized NIS summary score were computed. For the quality of life model, Wilcoxon rank sum tests between subdomain and total quality of life measures and the dichotomized NIS summary score were computed. Wilcoxon rank sum tests examined the difference among HNC HEI-2015 as compared to NHANES controls using the population ratio method. Statistical significance was set as an alpha level <0.05. Statistical analyses were performed using SAS software, version 9.4 or later [34].

5. Results

5.1. Participant Characteristics

The overall demographic characteristics of the study population are shown in Table 1. The mean age of the study population was 62 years old, and most participants were married (57%). Most participants were white males (59.5%), with at least some college education (62%). The most common tumor location was the oral cavity, and most participants were diagnosed with stage III-IV cancer (30%). The majority of participants were 1–4 years post treatment (64%). The majority were former smokers (57%) and current alcohol users (48%). Accrual was met within 14 months and required screening of 266 HNC cases. Of these, 79 were eligible for study participation and 187 were ineligible or excluded. Of the 79 eligible HNC cases, 42 agreed to participate for an enrollment rate of 52.2%. Reasons for ineligibility were distance (N = 15), timing (N = 13), too sick (N = 6), and too busy (N = 3).

Table 1. Demographic and clinical characteristics N = 42.

Characteristic	Total Participants
Age: Mean ± SD [range], years	62.7 ± 11.8 (32–81)
Body Mass Index: Mean ± SD [range], kg/m^2	26.2 ± 4.85 (16.5–38.3)
Under/normal weight N (%)	21 (50.0)
Overweight/obese N (%)	21 (50.0)
Lean muscle mass [a]: N (%)	
Under/normal	27 (71.0)
High	11 (29.0)
Body fat percentage [a]: N (%)	
Under/normal	13 (34.2)
High	25 (65.8)
Gender: N (%)	
Male	25 (59.5)
Female	17 (40.5)
Ethnicity: N (%)	
Non-Hispanic	42 (100)
Race: N (%)	
European American/White	39 (92.9)
Other	3 (7.1)
Education: N (%)	
High school or less	16 (38.1)
Some college or more	26 (61.9)
Annual household income: (dollars/year) N (%)	
Less than $54,999	24 (57.1)
$55,000 or more	18 (42.9)
Marital Status: N (%)	
Married	24 (57.1)
Not married	18 (42.9)

Table 1. *Cont.*

Characteristic	Total Participants
Smoking Status: *N* (%)	
Current	3 (7.1)
Former	24 (57.2)
Never	15 (35.7)
Alcohol Status: *N* (%)	
Current	20 (47.7)
Former	19 (45.2)
Never	3 (7.1)
Time since diagnosis: *N* (%)	
<1 to 4 years	27 (64.3)
≤4 to 9 years	15 (35.7)
Tumor site: *N* (%)	
Oral cavity	20 (47.6)
Pharynx/Larynx	22 (52.4)
Cancer stage: *N* (%)	
I–II	12 (28.6)
III–IV	30 (71.4)
Treatment: *N* (%)	
Concurrent chemoradiation	26 (61.9)
Radiation only	16 (38.1)
Nutrition Impact Symptom Score	
NIS ≤10	15 (35.7)
NIS > 10	27 (64.3)

[a] *N* = 38.

5.2. Functional Status and Nutrition Impact Symptoms

Table 2 reports the associations between functional status measures and the NIS summary score. A lower NIS summary score was statistically associated with higher post-treatment lean muscle mass (p = 0.002). A higher NIS summary score was non-statistically associated with a higher post-treatment body mass index and functional status composite score.

Table 2. Nutrition impact symptom burden and associated quality of life and functional status outcomes in head and neck cancer survivors *N* = 42.

Quality of Life Outcome	Mean (SD)	Median	*p*-Value [a]
Functional QOL			
NIS < 10	24.4 (3.7)	16.0	0.0006 [b]
NIS > 10	17.8 (6.9)	26.0	
Physical QOL			
NIS < 10	25.1 (2.6)	26.0	0.0007 [b]
NIS > 10	20.5 (5.8)	21.0	
Emotional QOL			
NIS < 10	21.2 (2.9)	22.0	0.007 [b]
NIS > 10	18.5 (3.8)	19.0	
Social QOL			
NIS < 10	22.4 (5.7)	23.5	0.09
NIS > 10	20.0 (7.2)	20.5	
Total QOL			
NIS < 10	93.0 (11.7)	95.5	0.0001 [b]
NIS > 10	76.8 (14.2)	73.5	

Table 2. *Cont.*

Functional Status Outcome	Mean (SD)	Median	*p*-Value [a]
Lean muscle mass [c]			
NIS < 10	75.7 (17.1)	76.4	0.002 [b]
NIS > 10	59.9 (15.0)	55.8	
Body fat percentage[c]			
NIS < 10	26.4 (9.0)	25.3	0.26
NIS > 10	28.8 (9.9)	25.5	
Body mass index			
NIS < 10	27.0 (5.1)	25.3	0.18
NIS > 10	25.2 (4.5)	24.9	
Functional Status Score			
NIS < 10	9.7 (2.5)	10.0	0.18
NIS > 10	9.4 (1.9)	9.0	

[a] Wilcoxon rank sum test; [b] Indicates statistical significance; [c] $N = 38$.

5.3. Quality of Life and Nutrition Impact Symptoms

Table 2 also reports associations between quality of life measures and the NIS summary score. A lower NIS summary score was statistically associated with higher functional ($p = 0.0006$), physical ($p = 0.0007$), emotional ($p = 0.007$), and total (<0.0001) quality of life.

5.4. Mean Quality of Life Summary Score

All FACT-H&N mean summary scores and FACT-H&N subscale scores were in the mild category (higher score), except for the Nutrition Impact Symptom Subscale (NIS) and Head and Neck Specific Concerns Subscale (HNCS) 6 item and 10 item, in which participants scored in the moderate category (Table 3).

Table 3. Functional Assessment Cancer Therapy (FACT) summary and subscale scores $N = 42$.

	# Items	Actual Range	Observed Range	Mean (SD)	Impairment Category
		FACT summary scores			
FACT-General	27	0–108	52–107	85.31 (15.21)	Mild
FACT-Head and neck	37	0–148	65–135	105.19 (19.59)	Mild
		FACT subscale scores			
Physical well-being	7	0–28	5–28	22.90 (4.91)	Mild
Social well-being	7	0–28	0–28	21.26 (6.47)	Mild
Emotional well-being	6	0–24	10–24	19.90 (3.58)	Mild
Functional well-being	7	0–28	7–28	21.24 (6.34)	Mild
Nutrition impact symptom questions	6	0–24	3–22	12.93 (5.26)	Moderate
Head and neck specific concerns	10	0–40	7–32	19.88 (6.39)	Moderate

5.5. Healthy Eating Index 2015 (HEI-2015) HNC Population vs. NHANES Data

As compared to NHANES controls, HNC survivors reported a significantly lower total HEI-2015 diet quality score ($p = 0.0001$) (Table 4). In the adequacy component, HNC survivors reported statistically significant lower consumption of total vegetables ($p < 0.0001$), whole grains ($p < 0.01$), total protein foods ($p < 0.0001$), seafood and plant proteins ($p < 0.0001$), and fatty acids ($p < 0.0001$). HNC survivors reported statistically significant higher consumption of dairy products ($p < 0.01$) and non-statistically significant higher consumption of total fruits and whole fruits. In the moderation component, HNC survivors consumed significantly more refined grains ($p < 0.0001$) and sodium ($p < 0.0001$), and significantly less added sugars ($p < 0.0001$) and saturated fats ($p < 0.01$) as compared to NHANES data.

Table 4. Healthy Eating Index-2015 (HEI-2015) Carle HNC survivors vs. NHANES data.

Component	Actual Max Points	Carle HNC Survivors N = 42	Age-Matched NHANES Population 2015–2016 Data [a]
Total HEI Score	100	54.3	60.6 [b]
	Adequacy		
Total Fruits	5	2.9	2.8
Whole Fruits	5	3.2	4.4
Total Vegetables	5	2.1	3.8 [b]
Greens and Beans	5	3.4	3.7
Whole Grains	10	2.5	3.0 [c]
Dairy	10	6.9	5.4 [b]
Total Protein Foods	5	3.0	5.0 [b]
Seafood and Plant Proteins	5	3.0	5.0 [b]
Fatty Acids	10	3.3	4.5 [b]
	Moderation		
Refined Grains	10	10.0	7.2 [b]
Sodium	10	8.9	3.7 [b]
Added Sugars	10	0.5	6.9 [b]
Saturated Fats	10	4.4	5.3 [c]

[a] Wilcoxon Rank Sum test to test difference; [b] $p < 0.0001$; [c] $p < 0.01$.

6. Discussion

To our knowledge, this study was among one of the first to explore the chronic complications of self-reported NIS burden on quality of life, objective measures of functional status, and diet quality in HNC survivors greater than 6 months post treatment. Notable findings were that higher post-treatment quality of life scores were associated with a lower NIS summary score (lower NIS burden). Furthermore, higher post-treatment lean muscle mass was associated with a lower NIS summary score, suggesting those who reported lower symptom burden had higher functional capacity. As compared to NHANES controls using the HEI-2015 population ratio method, HNC survivors in our study consumed a statistically significant lower total overall diet quality, which may be a result of NIS burden impacting the ability and desire to eat, though further longitudinal studies exploring this association are warranted.

Associations between quality of life measures and NIS were explored. Findings indicated that a lower NIS summary score was significantly associated with higher physical, functional, emotional, and total quality of life. Surprisingly, NIS burden was not associated with social quality of life for long-term survivors. A study by List et al. examining pre-treatment coping strategies, reported that social support-seeking was the most common coping strategy used by patients with HNC and commonly begins immediately following diagnosis [35]. Therefore, it may be possible that quality of life outcomes, such as social well-being, improve in the months and years after treatment as it is likely long-term HNC survivors have had sufficient time to adapt to their new normal and seek social support groups [36].

Previous research has suggested self-reported impairments in functional performance are common in HNC survivors [37]. Despite their high prevalence, few multidisciplinary rehabilitation programs designed for HNC exist in the peer-reviewed literature, and in those studies, survivors were assessed during and immediately after treatment [38,39]. Our study was among one of the first to objectively measure functional status in post-treatment HNC survivors. Despite our small sample size, higher post-treatment lean muscle mass was associated with a lower NIS summary score. Furthermore, while our findings were not statistically significant, higher body mass index and functional status were non-statistically associated with a lower NIS summary score. Power calculations suggest a sample size of N = 80 HNC survivors would detect a significant difference between the groups [40]. This work encourages larger, more robust clinical trials assessing risk factors for symptom development coupled with exercise and nutrition-based rehabilitation in long-term survivors to improve functional capacity.

Given the emphasis on the totality of the diet by national guidelines, our study team examined diet quality using the HEI-2015 among HNC cancer survivors as compared to NHANES data. HNC survivors reported a significantly lower total diet quality score as compared to NHANES controls, a possible consequence of treatment-related NIS burden. HNC survivors had higher diet quality scores among the adequacy component—dairy. As HNC survivors often experience difficulties with certain textures and flavors, we hypothesize this increase is likely due to HNC survivors' preference for soft foods, such as yogurt, which can be easily blended into smoothies and supplement drinks [41]. Additionally, the HNC population consumed nearly double the NHANES population for sodium and far less added sugars, likely a consequence of taste dysfunction resulting in food preference and aversion that is commonly reported in qualitative literature [2,4,42,43].

Limitations of this study should be noted. The cross-sectional design of the project is a limitation as quality of life and functional status measures were taken at only one time point. A prospective cohort study would be able to determine changes in different survivorship phases. Additionally, there is likely respondent bias inherent in the individuals in the individuals willing to complete the research study. The study population was selected from one Midwestern cancer center; consequently, the participants may not be fully representative of the total population of HNC survivors. The subjective bias of the FACT-H&N is a limitation of the study. Because the direction of the relationship between symptom burden and quality of life is unclear, objective measures are needed to determine if symptoms, such as swallowing dysfunction, result in quality of life declines or if declining quality of life emphasizes perceived symptom burden. Additionally, while the NISs examined were from a validated quality of life questionnaire, the NIS summary score was created for this specific analysis and is not validated.

This pilot study consisted of 42 HNC survivors. Although this study was severely underpowered, significant findings using simple Wilcoxon rank sum tests were noted. Furthermore, survivorship time ranged from 6 months to 10 years post treatment. Therefore, there was great heterogenicity in the population. A larger, longitudinal, and sufficiently powered study using multivariable regression models could further confirm directionality of quality of life, functional status, and diet quality changes post treatment based on this study's preliminary findings. Other strengths of the study include the use of validated questionnaires and objective measures of functional status in addition to the a priori approach to characterizing diet quality.

7. Conclusions

Self-reported NIS impairments were associated with lower quality of life and functional status outcomes among a population of long-term HNC survivors. As compared to an age-matched population from NHANES, HNC survivors reported lower overall diet quality, likely a result of symptoms impacting the ability and desire to eat. Multidisciplinary post-radiation exercise and nutrition-based interventions to manage NISs and improve quality of life, functional status, and dietary intake in this vulnerable survivor population are needed.

Author Contributions: S.L.C., K.P.S., and A.E.A. designed the study. S.L.C. and A.E.A. contributed to the development of the topic guide. S.L.C. collected the data. S.L.C. and Z.L. performed data analysis. S.L.C. wrote the first draft of the manuscript with contributions from all authors. All authors have read and agreed to the published version of the manuscript.

Funding: This study was supported by an Academy of Nutrition and Dietetics Colgate Palmolive Fellowship in Nutrition and Oral Health (PI: Crowder), USDA National Institute of Food and Agriculture, Hatch project 1011487 (PI: Arthur), a Division of Nutritional Sciences Vision 20/20 Grant PI: Arthur), and a Carle-Illinois Cancer Scholars for Translational and Applied Research Fellowship. SC was supported by NCI Cancer Prevention and Control Training Grant: 5T32CA090314-

17 (MPIs: Brandon/Vadaparampil) and Carle Illinois Cancer Scholars for Translational and Applied Research Fellowship.

Institutional Review Board Statement: The study was conducted according to the guidelines of the Declaration of Helsinki, and approved by the Institutional Review Board Institutional Review Boards of Carle Foundation Hospital and the University of Illinois at Urbana-Champaign (Project ID: 17088; UIUC number: 181933; Original Approval Date: 1 February 2018).

Informed Consent Statement: Informed consent was obtained from all subjects involved in the study.

Data Availability Statement: The dataset generated and analyzed in the current study are not publicly available due to the sensitive nature of responses, but deidentified data are available from the corresponding author on reasonable request.

Acknowledgments: The authors would like to thank Ozette Ostrow for assisting with data collection.

Conflicts of Interest: The authors declare no conflict of interest.

References

1. UM Head and Neck SPORE Program; Arthur, A.E.; Peterson, E.K.; Rozek, L.S.; Taylor, J.M.G.; Light, E.; Chepeha, D.; Hébert, J.R.; Terrell, J.E.; Wolf, G.T.; et al. Pretreatment dietary patterns, weight status, and head and neck squamous cell carcinoma prognosis. *Am. J. Clin. Nutr.* **2013**, *97*, 360–368.
2. Crowder, S.L.; Douglas, K.G.; Pepino, M.Y.; Sarma, K.P.; Arthur, A.E. Nutrition impact symptoms and associated outcomes in post-chemoradiotherapy head and neck cancer survivors: A systematic review. *J. Cancer Surviv.* **2018**, *12*, 479–494. [CrossRef]
3. Payakachat, N.; Ounpraseuth, S.; Suen, J.Y. Late complications and long-term quality of life for survivors (>5 years) with history of head and neck cancer. *Head Neck* **2013**, *35*, 819–825. [CrossRef] [PubMed]
4. Ganzer, H.; Rothpletz-Puglia, P.; Byham-Gray, L.; Murphy, B.A.; Touger-Decker, R. The eating experience in long-term survivors of head and neck cancer: A mixed-methods study. *Support. Care Cancer* **2015**, *23*, 3257–3268. [CrossRef]
5. Nguyen, N.-T.A.; Ringash, J. Head and Neck Cancer Survivorship Care: A Review of the Current Guidelines and Remaining Unmet Needs. *Curr. Treat. Options Oncol.* **2018**, *19*, 44. [CrossRef]
6. Howren, M.B.; Christensen, A.J.; Karnell, L.H.; Van Liew, J.R.; Funk, G.F. Influence of pretreatment social support on health-related quality of life in head and neck cancer survivors: Results from a prospective study. *Head Neck* **2012**, *35*, 779–787. [CrossRef]
7. Papadakos, J.; McQuestion, M.; Gokhale, A.; Damji, A.; Trang, A.; Abdelmutti, N.; Ringash, J. Informational Needs of Head and Neck Cancer Patients. *J. Cancer Educ.* **2018**, *33*, 847–856. [CrossRef]
8. Ringash, J. Survivorship and Quality of Life in Head and Neck Cancer. *J. Clin. Oncol.* **2015**, *33*, 3322–3327. [CrossRef] [PubMed]
9. Taylor, J.C.; Terrell, J.E.; Ronis, D.L.; Fowler, K.; Bishop, C.; Lambert, M.T.; Myers, L.L.; Duffy, S.A. Disability in Patients With Head and Neck Cancer. *Arch. Otolaryngol. Head Neck Surg.* **2004**, *130*, 764–769. [CrossRef] [PubMed]
10. Leeuw, I.M.V.-D.; Van Bleek, W.-J.; Leemans, C.R.; De Bree, R. Employment and return to work in head and neck cancer survivors. *Oral Oncol.* **2010**, *46*, 56–60. [CrossRef]
11. Koch, R.; Wittekindt, C.; Altendorf-Hofmann, A.; Singer, S.; Guntinas-Lichius, O. Employment pathways and work-related issues in head and neck cancer survivors. *Head Neck* **2014**, *37*, 585–593. [CrossRef]
12. Isaksson, J.; Wilms, T.; Laurell, G.; Fransson, P.; Ehrsson, Y.T. Meaning of work and the process of returning after head and neck cancer. *Support. Care Cancer* **2015**, *24*, 205–213. [CrossRef] [PubMed]
13. Nelke, K.H.; Pawlak, W.; Gerber, H.; Leszczyszyn, J. Head and neck cancer patients' quality of life. *Adv. Clin. Exp. Med.* **2014**, *23*, 1019–1027. [CrossRef] [PubMed]
14. Cartmill, B.; Cornwell, P.; Ward, E.; Davidson, W.; Porceddu, S. Long-term Functional Outcomes and Patient Perspective Following Altered Fractionation Radiotherapy with Concomitant Boost for Oropharyngeal Cancer. *Dysphagia* **2012**, *27*, 481–490. [CrossRef] [PubMed]
15. Crowder, S.L.; Najam, N.; Sarma, K.P.; Fiese, B.H.; Arthur, A.E. Head and Neck Cancer Survivors' Experiences with Chronic Nutrition Impact Symptom Burden after Radiation: A Qualitative Study. *J. Acad. Nutr. Diet.* **2020**, *120*, 1643–1653. [CrossRef]
16. Krebs-Smith, S.M.; Pannucci, T.E.; Subar, A.F.; Kirkpatrick, S.I.; Lerman, J.L.; Tooze, J.A.; Wilson, M.M.; Reedy, J. Update of the Healthy Eating Index: HEI-2015. *J. Acad. Nutr. Diet.* **2018**, *118*, 1591–1602. [CrossRef] [PubMed]
17. Willett, W. Monographs in epidemiology and biostatistics. In *Nutritional Epidemiology*, 3rd ed.; Oxford University Press: New York, NY, USA, 2013; 529p.
18. National Cancer Institute. Population Ratio Method. Available online: https://epi.grants.cancer.gov/hei/population-ratio-method.html (accessed on 3 March 2020).
19. Rimm, E.B.; Giovannucci, E.L.; Stampfer, M.J.; Colditz, G.A.; Litin, L.B.; Willett, W.C. Reproducibility and Validity of an Expanded Self-Administered Semiquantitative Food Frequency Questionnaire among Male Health Professionals. *Am. J. Epidemiol.* **1992**, *135*, 1114–1126, discussion 1127–1136. [CrossRef]
20. Willett, W.C.; Sampson, L.; Stampfer, M.J.; Rosner, B.; Bain, C.; Witschi, J.; Hennekens, C.H.; Speizer, F.E. Reproducibility and validity of a semiquantitative food frequency questionnaire. *Am. J. Epidemiol.* **1985**, *122*, 51–65. [CrossRef] [PubMed]

21. Rogers, L.Q.; Anton, P.M.; Bs, A.F.; Hopkins-Price, P.; Verhulst, S.; Rao, K.; Malone, J.; Robbs, R.; Courneya, K.S.; Nanavati, P.; et al. Pilot, randomized trial of resistance exercise during radiation therapy for head and neck cancer. *Head Neck* **2013**, *35*, 1178–1188. [CrossRef]
22. Böhm, A.; Heitmann, B.L. The use of bioelectrical impedance analysis for body composition in epidemiological studies. *Eur. J. Clin. Nutr.* **2013**, *67* (Suppl. S1), S79–S85. [CrossRef]
23. List, M.A.; D'Antonio, L.L.; Cella, D.F.; Siston, A.; Mumby, P.; Haraf, D.; Vokes, E. The performance status scale for head and neck cancer patients and the functional assessment of cancer therapy-head and neck scale: A study of utility and validity. *Cancer* **1996**, *77*, 2294–2301. [CrossRef]
24. Lohman, T.G.; Roche, A.F.; Martorell, R. *Anthropometric Standardization Reference Manual*; Human Kinetics Books: Champaign, IL, USA, 1988.
25. Guralnik, J.M.; Simonsick, E.M.; Ferrucci, L.; Glynn, R.J.; Berkman, L.F.; Blazer, D.G.; Scherr, P.A.; Wallace, R.B. A Short Physical Performance Battery Assessing Lower Extremity Function: Association with Self-Reported Disability and Prediction of Mortality and Nursing Home Admission. *J. Gerontol.* **1994**, *49*, M85–M94. [CrossRef]
26. Harvard school of public health. Harvard School of Public Health Nutrition Department's file download site. Available online: https://regepi.bwh.harvard.edu/health/FFQ/files (accessed on 21 August 2021).
27. Willett, W. *Nutrient Epidemiology*; Oxford Press: New York, NY, USA, 1998.
28. Data-National Center for Health Statistics. Healthy Eating Index-2015 Scores—U.S. Department of Agriculture, Center for Nutrition Policy and Promotion. Available online: https://www.fns.usda.gov/healthy-eating-index-hei (accessed on 11 February 2020).
29. National Cancer Institute. Healthy Eating Index: Overview of the Methods and Calculations. Available online: https://epi.grants.cancer.gov/hei/hei-methods-and-calculations.html (accessed on 14 February 2020).
30. Guenther, P.M.; Reedy, J.; Krebs-Smith, S.M. Development of the Healthy Eating Index—2005. *J. Am. Diet. Assoc.* **2008**, *108*, 1896–1901. [CrossRef]
31. Webster, K.; Cella, D.; Yost, K. The Functional Assessment of Chronic Illness Therapy (FACIT) Measurement System: Properties, applications, and interpretation. *Health Qual. Life Outcomes* **2003**, *1*, 79. [CrossRef]
32. Fisch, M.J.; Titzer, M.L.; Kristeller, J.L.; Shen, J.; Loehrer, P.J.; Jung, S.-H.; Passik, S.D.; Einhorn, L.H. Assessment of Quality of Life in Outpatients With Advanced Cancer: The Accuracy of Clinician Estimations and the Relevance of Spiritual Well-Being—A Hoosier Oncology Group Study. *J. Clin. Oncol.* **2003**, *21*, 2754–2759. [CrossRef]
33. Bilal, S.; Doss, J.G.; Cella, D.; Rogers, S.N. Quality of life associated factors in head and neck cancer patients in a developing country using the FACT-H&N. *J. Cranio Maxillofac. Surg.* **2015**, *43*, 274–280.
34. SAS. *9.4 [Computer Program]*; SAS: Cary, NC, USA, 2014.
35. List, M.A.; Rutherford, J.L.; Stracks, J.; Haraf, D.; Kies, M.S.; Vokes, E.E. An exploration of the pretreatment coping strategies of patients with carcinoma of the head and neck. *Cancer* **2002**, *95*, 98–104. [CrossRef] [PubMed]
36. Feinstein, A.R. Multi-item "instruments" vs Virginia Apgar's principles of clinimetrics. *Arch. Intern. Med.* **1999**, *159*, 125–128. [CrossRef]
37. van Deudekom, F.J.; Schimberg, A.S.; Kallenberg, M.H.; Slingerland, M.; van der Velden, L.-A.; Mooijaart, S.P. Functional and cognitive impairment, social environment, frailty and adverse health outcomes in older patients with head and neck cancer, a systematic review. *Oral Oncol.* **2017**, *64*, 27–36. [CrossRef] [PubMed]
38. Ringash, J.; Bernstein, L.; Devins, G.; Dunphy, C.; Giuliani, M.; Martino, R.; McEwen, S. Head and Neck Cancer Survivorship: Learning the Needs, Meeting the Needs. *Semin. Radiat. Oncol.* **2018**, *28*, 64–74. [CrossRef]
39. Capozzi, L.C.; Nishimura, K.C.; McNeely, M.; Lau, H.; Culos-Reed, S.N. The impact of physical activity on health-related fitness and quality of life for patients with head and neck cancer: A systematic review. *Br. J. Sports Med.* **2016**, *50*, 325–338. [CrossRef]
40. Faul, F.; Erdfelder, E.; Lang, A.-G.; Buchner, A. G*Power 3: A flexible statistical power analysis program for the social, behavioral, and biomedical sciences. *Behav. Res. Methods* **2007**, *39*, 175–191. [CrossRef] [PubMed]
41. Kristensen, M.B.; Mikkelsen, T.B.; Beck, A.M.; Zwisler, A.-D.; Wessel, I.; Dieperink, K.B. To eat is to practice—managing eating problems after head and neck cancer. *J. Cancer Surviv.* **2019**, *13*, 792–803. [CrossRef]
42. Larsson, M.; Hedelin, B.; Athlin, E. Lived experiences of eating problems for patients with head and neck cancer during radiotherapy. *J. Clin. Nurs.* **2003**, *12*, 562–570. [CrossRef] [PubMed]
43. Bressan, V.; Bagnasco, A.; Aleo, G.; Catania, G.; Zanini, M.; Timmins, F.; Sasso, L. The life experience of nutrition impact symptoms during treatment for head and neck cancer patients: A systematic review and meta-synthesis. *Support. Care Cancer* **2017**, *25*, 1699–1712. [CrossRef] [PubMed]

MDPI

Article

Plant-Based Dietary Patterns and Breast Cancer Recurrence and Survival in the Pathways Study

Ijeamaka C. Anyene [1], Isaac J. Ergas [1], Marilyn L. Kwan [1], Janise M. Roh [1], Christine B. Ambrosone [2], Lawrence H. Kushi [1] and Elizabeth M. Cespedes Feliciano [1,*]

[1] Division of Research, Kaiser Permanente Northern California, Oakland, CA 94612, USA; ijeamaka.c.anyene@kp.org (I.C.A.); Isaac.J.Ergas@kp.org (I.J.E.); Marilyn.L.Kwan@kp.org (M.L.K.); Janise.M.Roh@kp.org (J.M.R.); Larry.Kushi@kp.org (L.H.K.)

[2] Roswell Park Comprehensive Cancer Center, Buffalo, NY 14263, USA; christine.ambrosone@roswellpark.org

* Correspondence: elizabeth.m.cespedes@kp.org; Tel.: +1-510-891-5988

Abstract: Plant-based diets are recommended for cancer survivors, but their relationship with breast cancer outcomes has not been examined. We evaluated whether long-term concordance with plant-based diets reduced the risk of recurrence and mortality among a prospective cohort of 3646 women diagnosed with breast cancer from 2005 to 2013. Participants completed food frequency questionnaires at diagnosis and 6-, 25-, and 72-month follow-up, from which we derived plant-based diet indices, including overall (PDI), healthful (hPDI), and unhealthful (uPDI). We observed 461 recurrences and 653 deaths over a median follow-up of 9.51 years. Using multivariable-adjusted Cox proportional hazards models, we estimated hazard ratios (HR) and 95% confidence intervals for breast cancer recurrence and all-cause, breast-cancer-specific, and non-breast-cancer mortality. Increased concordance with hPDI was associated with a reduced hazard of all-cause (HR 0.93, 95% CI: 0.83–1.05) and non-breast-cancer mortality (HR 0.83, 95% CI: 0.71–0.98), whereas increased concordance with uPDI was associated with increased hazards (HR 1.07, 95% CI: 0.96–1.2 and HR 1.20, 95% CI: 1.02–1.41, respectively). No associations with recurrence or breast-cancer-specific mortality were observed. In conclusion, healthful vs. unhealthful plant-based dietary patterns had differing associations with mortality. To enhance overall survival, dietary recommendations for breast cancer patients should emphasize healthful plant foods.

Keywords: plant-based diet; dietary patterns; breast cancer; cancer survival; recurrence; lifestyle; survivorship

Citation: Anyene, I.C.; Ergas, I.J.; Kwan, M.L.; Roh, J.M.; Ambrosone, C.B.; Kushi, L.H.; Cespedes Feliciano, E.M. Plant-Based Dietary Patterns and Breast Cancer Recurrence and Survival in the Pathways Study. *Nutrients* **2021**, *13*, 3374. https://doi.org/10.3390/nu13103374

Academic Editors: Wendy Demark-Wahnefried and Christina Dieli-Conwright

Received: 21 July 2021
Accepted: 22 September 2021
Published: 25 September 2021

1. Introduction

There are an estimated 3.8 million female breast cancer survivors in the United States [1]. Due to insufficient evidence on whether dietary intake influences breast cancer survival and recurrence, breast cancer survivors are encouraged to observe general cancer prevention recommendations [2]. These recommendations include eating a healthy diet with an "emphasis on plant foods"; which is a diet "rich in whole grains, vegetables, fruits and beans" [2,3].

Plant-based diets in which individuals consume low amounts of animal-based foods, rather than completely excluding animal-based foods, have been adopted in nutritional research to reflect patterns of eating common in the population. A popular method of assessing plant-based diets is using a dietary pattern index, which is a numerical score measuring concordance to an overall pattern of eating [4]. Many plant-based diet indices, such as the original pro-vegetarian diet score, treat all plant foods the same regardless of quality [5]. To address this shortcoming, plant-based diet indices that differentiate between the consumption of healthful plants (e.g., fruits, whole grains, vegetables, legumes, nuts) and less healthful plants (e.g., refined grains, fruit juices, potatoes, sugar-sweetened beverages (SSBs)) were created [6]. A dietary pattern concordant with a healthful plant-based diet has been associated with reduced risk of coronary heart disease, type 2 diabetes, and breast cancer risk, but to the best of our knowledge, no prior study has examined breast cancer survival, as few cohorts collect dietary assessments both at and following diagnosis, and even fewer measure recurrence or breast-cancer-specific mortality [6–10].

Our study examined the relationship between repeated measures of an a priori plant-based diet index and its healthful and unhealthful variations at the time of diagnosis and in the post-diagnosis period, with breast cancer recurrence, all-cause mortality, breast-cancer-specific mortality, and non-breast-cancer mortality in a large cohort of breast cancer survivors.

2. Materials and Methods

2.1. Study Population

This analysis used data from the Pathways Study, a prospective cohort of 4505 female invasive breast cancer survivors diagnosed at Kaiser Permanente Northern California (KPNC) between the years 2005 and 2013. The protocol for this cohort has been previously published [11]. In brief, participants were enrolled on average 2.3 months (range: 0.7–18.7 months) post-diagnosis and completed an in-person baseline interview. Participants were eligible if they were current KPNC members, at least 21 years of age at the time of diagnosis, but had no previous diagnosis of cancer, spoke English, Spanish, Mandarin, or Cantonese, and lived within a 65-mile radius of a field staffer.

Dietary data were collected at baseline with a 139-item modified version of the Block 2005 Food Frequency Questionnaire (FFQ) [11]. During follow-up, repeated dietary intake data were collected via mailed questionnaires at 6, 24, and 72 months. In every questionnaire, participants reported how often, on average, did they eat each food in the past 6 months, and how much did they usually eat of the food. NutritionQuest scanned the questionnaires using a nutrient database developed from the USDA Food and Nutrient Database for Dietary Studies. Participants were excluded from this analysis for not completing a dietary assessment at baseline (n = 782, 17.4%), reporting a daily total energy intake of less than 400 or greater than 4000 kcal (n = 63, 1.4%). An additional 14 (0.

2.2. Plant-Based Diet Indices

Using a methodology defined by Satija et al., three plant-based diet indices were created based on the FFQ: an overall plant-based diet index (PDI), a healthful plant-based diet index (hPDI), and an unhealthful plant-based diet index (uPDI) [7]. Table S1 provides an example food item for each food group and indicates the scoring methodology by index. The indices were derived using 18 food groups, where a participant's total serving size consumption for each food group was broken into cohort-specific quintiles, and each quintile was given a score between 1 and 5. The 18 food groups are made up of healthful plant foods (whole grains, fruits, vegetables, nuts, legumes, vegetable oils, tea, and coffee), unhealthful plant foods (fruit juices, refined grains, potatoes, sugar-sweetened beverages (SSBs), sweets and desserts), and animal foods (dairy, animal fat, egg, meat, fish or seafood, and miscellaneous animal-based foods). Then, from these three main categories, the indices of PDI, hPDI, and uPDI are created. For all three indices, reverse scores are assigned to animal foods. For PDI, positive scores are assigned to all plant foods. For hPDI, positive scores are assigned to healthful plant foods, and reverse scores are assigned to unhealthful plant foods. For uPDI, positive scores are assigned to unhealthful plant foods, and reverse scores are assigned to healthful plant foods. Depending on the index, positive or reverse scores were given when a participant did not consume any foods within a group. For example, if an index used positive scoring for a food group (e.g., whole grains are assigned positive scores on the hPDI), participants received a score of 5 if they were in the highest quintile of consumption for the food group, and a score of 1 if they were in the lowest quintile of consumption (including no consumption). If an index used reverse scoring (e.g., whole grains are assigned reverse scores on the uPDI), participants received a score of 1 if they were in the highest quintile of consumption, and a score of 5 if they were in the lowest quintile or did not consume foods in that grouping. For each participant, scores for the 18 food groups were totaled to obtain their index-specific score. All the plant-based indices have a theoretical range from 18 to 90 with higher scores, indicating greater concordance with the dietary index of interest. The observed ranges for the PDI, hPDI, and uPDI scores were 32 to 79, 31 to 81, and 27 to 77, respectively.

The scores from the indices were operationalized in two approaches: a baseline score and a time-dependent cumulative average score. The baseline score used only the first dietary measurement. This approach allowed us to assess diet around the time of diagnosis. The cumulative average score used the time-updated average of all scores if a participant had repeat dietary measurements. Baseline dietary measurements were carried forward for participants missing follow-up questionnaires, thus assuming dietary intake to be constant over the study period. Using time-dependent cumulative average scores allowed us to account for long-term diet.

2.3. Ascertainment of Breast Cancer Recurrence and Survival

Breast cancer recurrences were identified using a combination of follow-up health status questionnaires and KPNC electronic medical record searches [11]. Breast cancer survival was defined by three outcomes: all-cause, breast-cancer-specific, and non-breast-cancer mortality. Mortality and causes of death were ascertained from KPNC's Virtual Data Warehouse (VDW) mortality files, which incorporate internal data from the KPNC health system, and external linkages with mortality information from the State of California, the Social Security Administration, and the National Death Index [12].

2.4. Covariate Selection

Demographic and behavioral covariates collected at the baseline interview included age at diagnosis, race/ethnicity (White, Hispanic, Black, Asian/Pacific Islander, and American Indian/Alaska Native), education (at least high school, some college, college graduate, postgraduate), menopausal status, smoking status (never, former, current), total energy in kcal/d, and physical activity as metabolic equivalent of task hours/week of moderate-vigorous activity (MET-hours/week). Breast cancer diagnosis characteristics including stage at diagnosis, estrogen receptor status (ER), and human epidermal growth factor receptor 2 status (HER2) were obtained from the VDW tumor file, which is based on data from the KPNC Cancer Registry [12]. The KPNC Cancer Registry meets the standards of the NCI SEER Program and reports to the San Francisco Bay Area and Greater California SEER Registries [13].

2.5. Statistical Analysis

Spearman correlation coefficients were estimated to compare the scores of PDI, hPDI, and uPDI. Cox proportional hazards models were used to estimate hazard ratios (HR) and 95% confidence intervals (CI) for recurrence, all-cause mortality, breast-cancer-specific mortality, and non-breast-cancer mortality. Separate models were fit for PDI, hPDI, and uPDI. Within the regression analyses, all scores were expressed as a 10-point continuous scale. Expressing the score as quintiles and restricted cubic splines was considered, but we found no evidence of non-linearity, and the 10-point continuous scale was the selected method using the Bayesian Information Criterion. Person-time was calculated as the years from baseline dietary assessment to the date of first confirmed recurrence or death, depending on the model. If no event occurred, participants were censored at the end of the study period, 31 December 2018.

Two models for each plant-based index were fit. Model 1 is a minimally adjusted model that adjusted for age at diagnosis, total energy intake in kcal, and physical activity in MET-hours/week. Model 2 is a stratified multivariable-adjusted model that adjusted for covariates in Model 1, and additionally for education, race/ethnicity, smoking status, menopausal status, HER2 status, and stratified by tumor stage and ER status (Model 2).

All statistical analyses were conducted using R [14]. Specifically, the survival analyses were enabled by the survival package, tables were generated using gtsummary and flextable packages, and figures were generated using ggplot2, gridExtra, and patchwork packages [15–20].

3. Results

3.1. Study Population

The mean age at diagnosis was 60 years (SD, 12 years). Baseline characteristics of the study population stratified by PDI quintiles are listed in Table 1. The baseline characteristics stratified by hPDI and uPDI are listed in Tables S2 and S3, respectively. Before the 10-point rescaling of the indices scores, the average index scores were 53.96 (PDI, SD: 6.72), 54.22 (hPDI, SD: 8.19), and 53.78 (uPDI, SD: 8.19). Cohort-specific quintile cut points are reported within Table 1 for PDI, Table S2 for hPDI, and Table S3 for uPDI, respectively. Due to the scoring methodology, hPDI and uPDI are perfectly inversely correlated (correlation coefficient of −1). Neither index was strongly correlated with PDI (hPDI r = 0.16; p-value < 0.001 and uPDI r = −0.16; p-value < 0.001). A total of 461 breast cancer recurrences and 653 deaths occurred during a median follow-up period of 9.2 years for recurrence and 9.51 years for deaths (range: 0.05 to 12.9 years). A higher proportion of participants with repeated measurements were white, had a post-graduate education, and were postmenopausal.

Table 1. Baseline characteristics of participants by plant-based diet index (PDI) quintiles, Pathways Study.

	Overall	**Quintiles of PDI Score**				
Characteristic	***n*** **= 3646** [1]	**Q1, *n* = 784** [1] **Scores: 32–48**	**Q2, *n* = 749** [1] **Scores: 49–52**	**Q3, *n* = 815** [1] **Scores: 53–56**	**Q4, *n* = 682** [1] **Scores: 57–60**	**Q5, *n* = 616** [1] **Scores: 61–79**
Age at diagnosis	60 (12)	60 (12)	60 (12)	60 (12)	59 (12)	58 (12)
BMI (kg/m^2)	28 (7)	29 (7)	28 (7)	28 (7)	28 (7)	28 (7)
Physical Activity (MET h/week)	54 (36)	46 (34)	50 (34)	54 (35)	58 (36)	63 (39)
Energy intake (kcal/day)	1465 (568)	1112 (432)	1296 (461)	1481 (505)	1679 (551)	1864 (580)
Race/Ethnicity						
White	2481 (68%)	552 (70%)	501 (67%)	563 (69%)	446 (65%)	419 (68%)
Black	237 (6.5%)	51 (6.5%)	53 (7.1%)	51 (6.3%)	45 (6.6%)	37 (6.0%)
Asian/Pacific Islander	474 (13%)	96 (12%)	96 (13%)	111 (14%)	93 (14%)	78 (13%)
Hispanic	378 (10%)	74 (9.4%)	85 (11%)	73 (9.0%)	77 (11%)	69 (11%)
American Indian/Alaska Native	76 (2.1%)	11 (1.4%)	14 (1.9%)	17 (2.1%)	21 (3.1%)	13 (2.1%)
Education						
High school or less	544 (15%)	138 (18%)	129 (17%)	118 (14%)	85 (12%)	74 (12%)
Some college	1241 (34%)	305 (39%)	239 (32%)	269 (33%)	241 (35%)	187 (30%)
College graduate	1022 (28%)	194 (25%)	223 (30%)	226 (28%)	199 (29%)	180 (29%)
Postgraduate	839 (23%)	147 (19%)	158 (21%)	202 (25%)	157 (23%)	175 (28%)

Table 1. *Cont.*

Characteristic	Overall	Quintiles of PDI Score				
	n = 3646 [1]	Q1, n = 784 [1] Scores: 32–48	Q2, n = 749 [1] Scores: 49–52	Q3, n = 815 [1] Scores: 53–56	Q4, n = 682 [1] Scores: 57–60	Q5, n = 616 [1] Scores: 61–79
Menopausal Status						
Premenopausal	1057 (29%)	207 (26%)	205 (27%)	236 (29%)	207 (30%)	202 (33%)
Postmenopausal	2589 (71%)	577 (74%)	544 (73%)	579 (71%)	475 (70%)	414 (67%)
Smoking status						
Never	2091 (57%)	422 (54%)	410 (55%)	469 (58%)	418 (61%)	372 (60%)
Former	1403 (38%)	325 (41%)	305 (41%)	317 (39%)	233 (34%)	223 (36%)
Current	152 (4.2%)	37 (4.7%)	34 (4.5%)	29 (3.6%)	31 (4.5%)	21 (3.4%)
AJCC Cancer Stage						
1	1998 (55%)	424 (54%)	411 (55%)	463 (57%)	369 (54%)	331 (54%)
2	1247 (34%)	279 (36%)	267 (36%)	268 (33%)	226 (33%)	207 (34%)
3	346 (9.5%)	72 (9.2%)	60 (8.0%)	69 (8.5%)	77 (11%)	68 (11%)
4	55 (1.5%)	9 (1.1%)	11 (1.5%)	15 (1.8%)	10 (1.5%)	10 (1.6%)
ER Status						
Positive	3063 (84%)	651 (83%)	631 (84%)	680 (83%)	573 (84%)	528 (86%)
Negative	583 (16%)	133 (17%)	118 (16%)	135 (17%)	109 (16%)	88 (14%)
HER2 Status						
Positive	469 (13%)	103 (13%)	92 (12%)	109 (13%)	86 (13%)	79 (13%)
Negative	3037 (83%)	654 (83%)	628 (84%)	673 (83%)	567 (83%)	515 (84%)
Missing	140 (3.8%)	27 (3.4%)	29 (3.9%)	33 (4.0%)	29 (4.3%)	22 (3.6%)

[1] Mean (SD); n (%). PDI = plant-based diet index; BMI = body mass index; MET (h/week) = metabolic equivalent of task hours/week; ER = estrogen receptor; HER2 = human epidermal growth factor receptor 2.

3.2. Food Consumption Patterns

In Figure 1, we used radar plots to qualitatively assess the overall food consumption patterns of participants with the greatest concordance (quintile 5) with hPDI (Figure 1a) and greatest concordance (quintile 5) with uPDI (Figure 1b). In both radar plots, the line represents the median score assigned for each food group for the quintile. If the line is pulled toward the outside of the circle, the median score is higher, indicating greater consumption of that food group. Overall, to achieve high concordance with hPDI, participants did not exclude animal foods or unhealthful plants from their diets but consumed low amounts in each of these food categories (median points 1–3). However, the preponderance of their dietary intake was healthful plants (e.g., whole grains, vegetables, and legumes). This trend is similar in the participants with a high concordance of uPDI, in which the median score reveals a low to moderate consumption of animal foods and healthful plants, but their diet was skewed toward unhealthful plants (e.g., sweets and desserts, potatoes, refined grains, fruit juices and sugar-sweetened beverages (SSBs)).

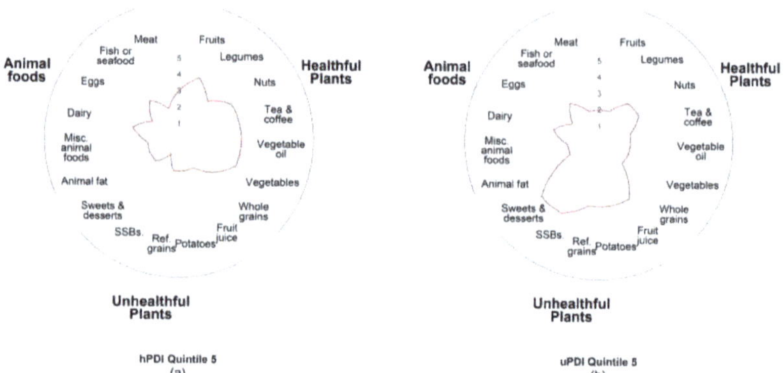

Figure 1. Food groups contributing to plant-based diet indices among breast cancer survivors in the Pathways Study: (**a**) radar plot showing the food consumption patterns of the highest quintile and greatest concordance with hPDI (healthful plant-based diet index). The line represents the median score for each food group within the quintile. A higher score indicates greater consumption of the food group; (**b**) radar plot showing the food consumption patterns of the highest quintile of uPDI (unhealthful plant-based diet index). The line represents the median score for each food group within the quintile. A higher score indicates greater consumption of the food group.

3.3. Plant-Based Indices and Breast Cancer Recurrence and Survival

Figure 2 shows the hazard ratios and 95% confidence intervals for a 10-unit increase in baseline scores. When using the baseline scores in Model 1, a 10-point increase in concordance with hPDI has a statistically significant inverse relationship with all-cause mortality (HR 0.84, 95% CI: 0.75–0.93) and non-breast-cancer mortality (HR 0.80, 95% CI: 0.69–0.93). Conversely, a 10-point increase in concordance with uPDI has a statistically significant positive relationship with all-cause mortality (HR 1.19, 95% CI 1.08–1.33) and non-breast-cancer mortality (HR 1.25, 95% CI: 1.08–1.45). When adjusting for additional demographic characteristics and stratifying on tumor characteristics in Model 2, the estimates are no longer statistically significant for all-cause mortality (hPDI HR 0.94, 95% CI: 0.85–1.05; uPDI HR 1.06, 95% CI: 0.96–1.18) and non-breast-cancer mortality (hPDI HR 0.88, 95% CI: 0.88, 0.76–1.02; uPDI HR 1.14, 95% CI: 0.98–1.32).

Figure 3 shows the hazard ratios and 95% confidence intervals for a 10-unit increase in the time-dependent cumulative average scores. When using the time-dependent cumulative average score in Model 1, models results were consistent with the baseline scores. hPDI had an inverse relationship and uPDI had a positive relationship with all-cause mortality and non-breast-cancer mortality. In Model 2, a 10-point increase in concordance with hPDI had no association with all-cause mortality (HR 0.93, 95% CI: 0.83, 1.05) and a reduced hazard of non-breast-cancer mortality (HR 0.83, 95% CI: 0.71–0.98). In contrast, a 10-point increase in concordance with uPDI had no association with all-cause mortality (HR 1.07, 95% CI: 0.96–1.2) and an increased hazard of non-breast-cancer mortality (HR 1.20, 95% CI: 1.02–1.41). A 10-point increase in concordance with PDI had no association with all-cause mortality (HR 0.96, 95% CI: 0.82–1.11) and non-breast-cancer mortality (HR 0.90, 95% CI: 0.73–1.11).

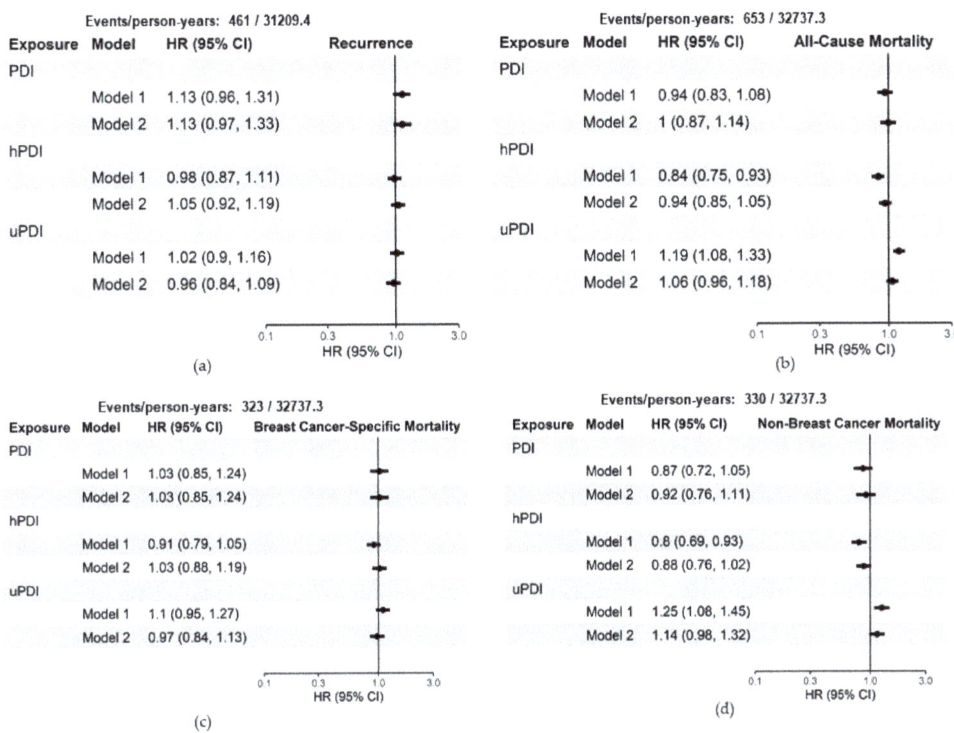

Figure 2. Hazard ratios and 95% confidence intervals for a 10-unit increase in baseline measurements of plant-based indices with breast cancer recurrence and survival among 3646 breast cancer survivors, Pathways Study. Model 1 adjusted for the following covariates: age at diagnosis, total energy intake (kcal/d), and physical activity (moderate-vigorous MET-hours/week). Model 2 adjusted for the following covariates: Model 1 covariates, race/ethnicity, education, menopausal status, smoking status, and stratified by tumor stage and ER status: (**a**) includes estimates of the hazard of recurrence; (**b**) estimates of the hazard of all-cause mortality; (**c**) estimates of the hazard of breast-cancer-specific mortality; (**d**) estimates of the hazard of non-breast-cancer mortality. CI = confidence interval; HR = hazard ratio; PDI = plant-based diet index; hPDI = healthful plant-based diet index; uPDI = unhealthful plant-based diet index.

Neither PDI, hPDI, and uPDI were associated with recurrence or breast cancer mortality, regardless of exposure method (baseline vs. time-dependent cumulative average) or model (Model 1 vs. Model 2) (Figures 2 and 3).

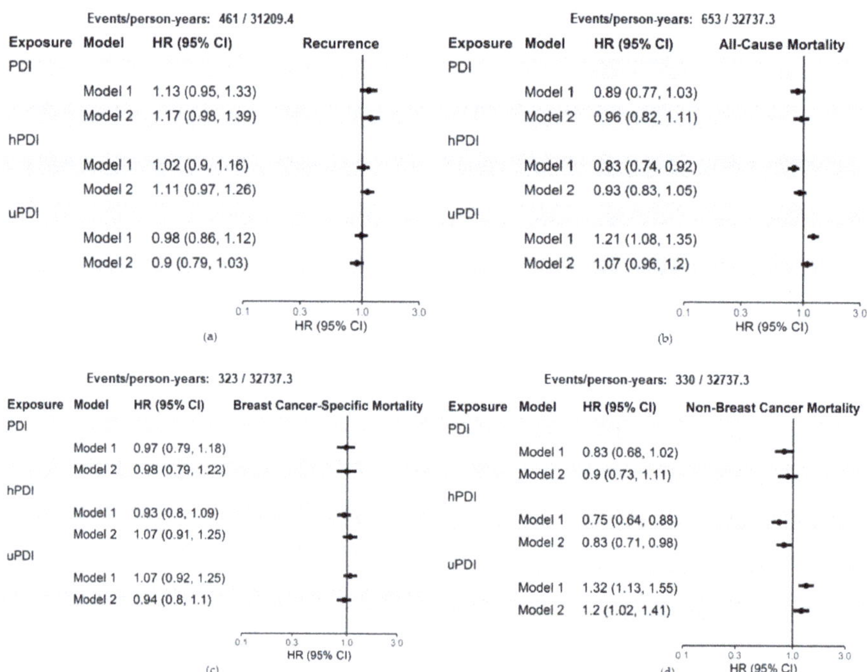

Figure 3. Hazard ratios and 95% confidence intervals for a 10-unit increase in time-dependent cumulative average plant-based indices with breast cancer recurrence and survival amongst 3646 breast cancer survivors, Pathways Study. Model 1 adjusted for the following covariates: age at diagnosis, total energy intake (kcal/d), and physical activity (moderate-vigorous MET-hours/week). Model 2 adjusted for the following covariates: Model 1 covariates, race/ethnicity, education, menopausal status, smoking status, and stratified by tumor stage and ER status: (**a**) includes estimates of the hazard of recurrence; (**b**) estimates of the hazard of all-cause mortality; (**c**) estimates of the hazard of breast-cancer-specific mortality; (**d**) estimates of the hazard of non-breast-cancer mortality. CI = confidence interval; HR = hazard ratio; PDI = plant-based diet index; hPDI = healthful plant-based diet index; uPDI = unhealthful plant-based diet index.

4. Discussion

In this study of 3646 breast cancer survivors, we found that concordance with healthful plant-based eating patterns (hPDI) in the postdiagnosis period (time-dependent cumulative measurements) reduced the risk of non-breast-cancer mortality. In contrast, greater concordance with unhealthful plant-based eating patterns (uPDI) increased non-breast-cancer mortality risk. We observed no associations between plant-based eating patterns and breast cancer recurrence or breast-cancer-specific mortality.

Few studies have addressed long-term dietary patterns and prognosis after breast cancer, and none have examined the hPDI, uPDI, or PDI. Even fewer studies have examined dietary patterns and breast cancer recurrence, and none emphasized the distinction between healthful and unhealthful plant foods or found associations with recurrence [21]. Several cohorts have examined other popular a priori dietary pattern indices that are not necessarily "plant based" but do emphasize healthful plant foods such as fruits, legumes, and whole grains, e.g., healthy eating index (HEI), dietary approaches to stop hypertension (DASH), and alternative eating index (AHEI) [21–24]. Consistent with our findings, a majority of these studies have found protective associations of dietary patterns emphasizing healthful plant foods with all-cause mortality and non-breast-cancer mortality [23,25].

Differentiating between healthful and unhealthful plants in plant-based diet indices is a relatively new concept [7]. Our findings of higher hPDI scores being associated with reduced all-cause mortality in the minimally adjusted model and associated with reduced non-breast-cancer mortality in the fully adjusted model and higher uPDI scores being associated with increased mortality are consistent with a study using National Health and Nutrition Examination Survey (NHANES) III [26]. In this study, a 10-unit increase above the median hPDI was associated with a reduced hazard of all-cause mortality in their study population [26]. However, this study observed no association between uPDI and all-cause mortality. In the Nurses' Health Study (NHS) and NHS2 cohorts, an inverse relationship between hPDI and coronary heart disease, and a positive relationship between uPDI and coronary heart disease were observed [7]. Our findings are consistent with these observations as cardiovascular disease is a major cause of death among breast cancer survivors [27].

Though prior research has shown healthful dietary patterns to be important for breast cancer prevention, our study did not observe an association between hPDI, uPDI, and PDI and recurrence or mortality after a breast cancer diagnosis. Potential explanations include that the effect of plant-based diets on breast-cancer-specific outcomes, if they exist, are small, and overwhelmed by the strong influence of breast tumor characteristics (e.g., stage and hormone receptor status) for which we controlled. Another possibility is that a precision nutrition approach with specific foods or nutrients, rather than overall concordance with a healthful, plant-based diet, would be influential in specific breast cancer subtypes [2]. Further, while our median follow-up was 9 years, many breast cancers recur more than a decade after initial diagnosis; it is possible long-term adherence to healthful, plant-based dietary patterns could impact later recurrences or second cancers [28].

No prior study has examined plant-based diets and breast-cancer-specific outcomes such as recurrence, making direct comparison difficult, and previous studies of breast cancer incidence have had inconsistent results [8–10]. In the Seguimiento Universidad de Navarra cohort (SUN), they observed an inverse association between a healthful version of the pro-vegetarian diet and breast cancer incidence and a positive association between an unhealthful pro-vegetarian diet and breast cancer incidence; however, neither finding was statistically significant [8]. Other case–control studies had differing results between hPDI, uPDI, and risk of breast cancer [9,10].

Our study assessed plant-based dietary indices using baseline and time-dependent cumulative average scores. We used both approaches to assess the effect of concordance with plant-based diets at diagnosis and in the postdiagnosis period. We observed that the cumulative average scores, as compared to the baseline scores, yielded stronger associations between hPDI and uPDI, and non-breast-cancer mortality. One explanation for these differences is that an individual's long-term maintenance of a dietary pattern matters more with regard to breast cancer survival than their diet at the time of diagnosis. However, it is also possible that leveraging repeated measures of a diet (rather than a one-time dietary measurement at baseline) mitigates measurement error, resulting in stronger associations. In addition, since we were observing usual consumption patterns, we could not assess the impact of maximum adherence to any variation of a plant-based diet, as no participant in our study scored at the theoretical maximum. A greater contrast in extremes of exposure could be achieved through a dietary intervention that coaches or provides food to help participants achieve high levels of concordance.

There are several strengths to this study, including its population being a large, representative prospective breast cancer survivor cohort recruited from a community setting. By combining active data collection via study questionnaires with KPNC electronic medical records and state and federal data sources, we were able to leverage a rich resource of covariates and outcomes data. With a long follow-up period and repeated measures, we were able to (1) have more robust measures of baseline and long-term dietary intake, (2) distinguish between the quality of different plant food patterns, and (3) assess, for the first time, the association between long-term concordance with healthful and unhealthful plant-based diets and breast cancer recurrence and survival.

5. Conclusions

In summary, healthful plant-based dietary patterns may reduce the risk of non-brea cancer mortality, whereas an unhealthful plant-based dietary pattern may increase t risk of this outcome. It is thus important to consider the quality of plant foods to achie a healthful dietary pattern. Healthful plant-based dietary patterns may improve overa survival in breast cancer survivors.

Supplementary Materials: The following are available online at https://www.mdpi.com/articl 10.3390/nu13103374/s1, Table S1: Scoring methods and examples of food items comprising th 18 food groups that contribute to PDI, hPDI, and uPDI scores, Table S2: Baseline characteristics c participants by healthful plant-based diet index (hPDI) quintiles, Pathways Study, Table S3: Baselir characteristics of participants by unhealthful plant-based diet index (uPDI) quintiles, Pathways Stud

Author Contributions: Conceptualization, I.C.A., I.J.E., and E.M.C.F.; methodology, I.C.A., I.J.E., an E.M.C.F.; formal analysis, I.C.A.; investigation, I.C.A., I.J.E., and E.M.C.F.; resources, E.M.C.F. an L.H.K.; data curation, I.C.A. and I.J.E.; writing—original draft preparation, I.C.A.; writing—revie and editing, I.J.E., M.L.K., J.M.R., C.B.A., L.H.K., and E.M.C.F.; visualization, I.C.A.; supervisio E.M.C.F.; project administration, J.M.R.; funding acquisition, M.L.K., C.B.A., L.H.K., and E.M.C.F. A authors have read and agreed to the published version of the manuscript.

Funding: This research was funded by the American Institute for Cancer Research, Grant Numbe 632996; and the National Cancer Institute, Grant Numbers U01 CA195565, R01 CA105274, K0 CA226155, U24 CA171524, R01 CA251589.

Institutional Review Board Statement: The study was approved by the Institutional Review Boar of Kaiser Permanente Northern California (Protocol Code 1263171 and date of approval 12/11/2020

Informed Consent Statement: Informed consent was obtained from all subjects involved in the stud

Data Availability Statement: The data that support the findings of this study are available on reque from the corresponding author (E.M.C.F). The data for this study is not publicly available due to containing protected health information.

Acknowledgments: The authors thank and acknowledge all of the Pathways participants for thei contributions to this study.

Conflicts of Interest: The authors declare no conflict of interest in analysis or interpretation of dat

References

1. Miller, K.D.; Nogueira, L.; Mariotto, A.B.; Rowland, J.H.; Yabroff, K.R.; Alfano, C.M.; Jemal, A.; Kramer, J.L.; Siegel, R.L. Cance treatment and survivorship statistics. *CA A Cancer J. Clin.* **2019**, *69*, 363–385. [CrossRef] [PubMed]
2. World Cancer Research Fund International/American Institute for Cancer Research. *Diet, Nutrition, Physical Activity, and Breas Cancer Survivors*; World Cancer Research Fund International/American Institute for Cancer Research: London, UK, 2014.
3. Kushi, L.H.; Doyle, C.; McCullough, M.; Rock, C.L.; Demark-Wahnefried, W.; Bandera, E.V.; Gapstur, S.; Patel, A.V.; Andrews, K Gansler, T. American Cancer Society guidelines on nutrition and physical activity for cancer prevention. *CA A Cancer J. Clir* **2012**, *62*, 30–67. [CrossRef] [PubMed]
4. Cespedes, E.M.; Hu, F.B. Dietary patterns: From nutritional epidemiologic analysis to national guidelines. *Am. J. Clin. Nutr.* **201** *101*, 899–900. [CrossRef] [PubMed]
5. Martínez-González, M.A.; Sánchez-Tainta, A.; Corella, D.; Salas-Salvadó, J.; Ros, E.; Arós, F.; Gómez-Gracia, E.; Fiol, M.; Lamuela Raventós, R.M.; Schröder, H.; et al. A provegetarian food pattern and reduction in total mortality in the Prevención con Diet Mediterránea (PREDIMED) study. *Am. J. Clin. Nutr.* **2014**, *100*, 320S–328S. [CrossRef]
6. Satija, A.; Bhupathiraju, S.N.; Rimm, E.B.; Spiegelman, D.; Chiuve, S.E.; Borgi, L.; Willett, W.C.; Manson, J.E.; Sun, Q.; Hu, F.E Plant-Based Dietary Patterns and Incidence of Type 2 Diabetes in US Men and Women: Results from Three Prospective Cohor Studies. *PLoS Med.* **2016**, *13*, e1002039. [CrossRef] [PubMed]
7. Satija, A.; Bhupathiraju, S.N.; Spiegelman, D.; Chiuve, S.E.; Manson, J.E.; Willett, W.; Rexrode, K.M.; Rimm, E.B.; Hu, F.B Healthful and Unhealthful Plant-Based Diets and the Risk of Coronary Heart Disease in U.S. Adults. *J. Am. Coll. Cardiol.* **2017**, *7C* 411–422. [CrossRef] [PubMed]
8. Romanos-Nanclares, A.; Toledo, E.; Sánchez-Bayona, R.; Sánchez-Quesada, C.; Martínez-González, M.Á.; Gea, A. Healthful anc unhealthful provegetarian food patterns and the incidence of breast cancer: Results from a Mediterranean cohort. *Nutrition* **2020** 79–80. [CrossRef] [PubMed]

Rigi, S.; Mousavi, S.M.; Benisi-Kohansal, S.; Azadbakht, L.; Esmaillzadeh, A. The association between plant-based dietary patterns and risk of breast cancer: A case-control study. *Sci. Rep.* **2021**, *9*, 3391. [CrossRef] [PubMed]

10. Sasanfar, B.; Toorang, F.; Booyani, Z.; Vassalami, F.; Mohebbi, E.; Azadbakht, L.; Zendehdel, K. Adherence to plant-based dietary pattern and risk of breast cancer among Iranian women. *Eur. J. Clin. Nutr.* **2021**, 1–10. [CrossRef]
11. Kwan, M.L.; Ambrosone, C.B.; Lee, M.M.; Barlow, J.; Krathwohl, S.E.; Ergas, I.J.; Ashley, C.H.; Bittner, J.R.; Darbinian, J.; Stronach, K.; et al. The Pathways Study: A prospective study of breast cancer survivorship within Kaiser Permanente Northern California. *Cancer Causes Control* **2008**, *19*, 1065–1076. [CrossRef] [PubMed]
12. Ross, T.R.; Ng, D.; Brown, J.S.; Pardee, R.; Hornbrook, M.C.; Hart, G.; Steiner, J.F. The HMO Research Network Virtual Data Warehouse: A Public Data Model to Support Collaboration. EGEMS. *EGEMS* **2014**, *2*, 1049. [CrossRef] [PubMed]
13. Chubak, J.; Ziebell, R.; Greenlee, R.T.; Honda, S.; Hornbrook, M.C.; Epstein, M.; Nekhlyudov, L.; Pawloski, P.A.; Ritzwoller, D.P.; Ghai, N.R.; et al. The Cancer Research Network: A platform for epidemiologic and health services research on cancer prevention, care, and outcomes in large, stable populations. *Cancer Causes Control* **2016**, *27*, 1315–1323. [CrossRef] [PubMed]
14. Team, R.C. *R: A Language and Environment for Statistical Computing*; R Foundation for Statistical Computing: Vienna, Austria, 2020.
15. Therneau, T.M. *A Package for Survival Analysis in R*; R Package Version 3.2-7. Available online: https://CRAN.R-project.org/package=survival (accessed on 21 January 2021).
16. Sjoberg, D.D.; Curry, M.; Hannum, M.; Whiting, K.; Zabor, E.C. *gtsummary: Presentation-Ready Data Summary and Analytic Result Tables*; R Package Version 1.3.5. Available online: https://CRAN.R-project.org/package=gtsummary (accessed on 21 January 2021).
17. Gohel, D. *flextable: Functions for Tabular Reporting*; R Package Version 0.6.1. Available online: https://CRAN.R-project.org/package=flextable (accessed on 21 January 2021).
18. Wickham, H. *ggplot2: Elegant Graphics for Data Analysis*; Springer: New York, NY, USA, 2016.
19. Auguie, B. *gridExtra: Miscellaneous Functions for "Grid" Graphics*; R Package Version 2.3. Available online: https://CRAN.R-project.org/package=gridExtra (accessed on 21 January 2021).
20. Pedersen, T.L. *patchwork: The Composer of Plots*; R Package Version 1.1.1. Available online: https://CRAN.R-project.org/package=patchwork (accessed on 21 January 2021).
21. Terranova, C.O.; Protani, M.M.; Reeves, M.M. Overall Dietary Intake and Prognosis after Breast Cancer: A Systematic Review. *Nutr Cancer* **2018**, *70*, 153–163. [CrossRef] [PubMed]
22. Ergas, I.J.; Cespedes Feliciano, E.M.; Bradshaw, P.T.; Roh, J.M.; Kwan, M.L.; Cadenhead, J.; Santiago-Torres, M.; Troeschel, A.N.; Laraia, B.; Madsen, K.; et al. Diet Quality and Breast Cancer Recurrence and Survival: The Pathways Study. *JNCI Cancer Spectr.* **2021**, *5*, pkab019. [CrossRef] [PubMed]
23. Kwan, M.L.; Weltzien, E.; Kushi, L.H.; Castillo, A.; Slattery, M.L.; Caan, B.J. Dietary patterns and breast cancer recurrence and survival among women with early-stage breast cancer. *J. Clin. Oncol.* **2009**, *27*, 919–926. [CrossRef] [PubMed]
24. Izano, M.A.; Fung, T.T.; Chiuve, S.S.; Hu, F.B.; Holmes, M.D. Are diet quality scores after breast cancer diagnosis associated with improved breast cancer survival? *Nutr. Cancer* **2013**, *65*, 820–826. [CrossRef] [PubMed]
25. Vrieling, A.; Buck, K.; Seibold, P.; Heinz, J.; Obi, N.; Flesch-Janys, D.; Chang-Claude, J. Dietary patterns and survival in German postmenopausal breast cancer survivors. *Br. J. Cancer* **2013**, *108*, 188–192. [CrossRef] [PubMed]
26. Kim, H.; Caulfield, L.E.; Rebholz, C.M. Healthy Plant-Based Diets Are Associated with Lower Risk of All-Cause Mortality in US Adults. *J. Nutr.* **2018**, *148*, 624–631. [CrossRef] [PubMed]
27. Bradshaw, P.T.; Stevens, J.; Khankari, N.; Teitelbaum, S.L.; Neugut, A.I.; Gammon, M.D. Cardiovascular Disease Mortality Among Breast Cancer Survivors. *Epidemiology* **2016**, *27*, 6–13. [CrossRef] [PubMed]
28. Pan, H.; Gray, R.; Braybrooke, J.; Davies, C.; Taylor, C.; McGale, P.; Peto, R.; Pritchard, K.I.; Bergh, J.; Dowsett, M.; et al. 20-Year Risks of Breast-Cancer Recurrence after Stopping Endocrine Therapy at 5 Years. *N. Engl. J. Med.* **2017**, *377*, 1836–1846. [CrossRef] [PubMed]

MDPI
St. Alban-Anlage 66
4052 Basel
Switzerland
Tel. +41 61 683 77 34
Fax +41 61 302 89 18
www.mdpi.com

Nutrients Editorial Office
E-mail: nutrients@mdpi.com
www.mdpi.com/journal/nutrients

www.ingramcontent.com/pod-product-compliance
Lightning Source LLC
LaVergne TN
LVHW070154170726
843515LV00007B/1915

* 9 7 8 3 0 3 6 5 5 2 1 9 4 *